DOMESTIC ANIMAL PHYSIOLOGY

Colin G. Scanes
University of Wisconsin – Milwaukee

Dawn A. Koltes
Iowa State University

WAVELAND

PRESS, INC.
Long Grove, Illinois

For information about this book, contact:
Waveland Press, Inc.
4180 IL Route 83, Suite 101
Long Grove, IL 60047-9580
(847) 634-0081
info@waveland.com
www.waveland.com

Copyright © 2024 by Waveland Press, Inc.

10-digit ISBN 1-4786-5116-4
13-digit ISBN 978-1-4786-5116-1

All rights reserved. No part of this book may be reproduced, stored in a retrieval system, or transmitted in any form or by any means without permission in writing from the publisher.

Printed in the United States of America

7 6 5 4 3 2 1

Dedication

To our teachers who inspired us, to our colleagues who encourage and challenge us, to the students from whom we learn, and, particularly, to our families who support us.

Table of Contents

Preface   *vii*
About The Authors   *ix*

CHAPTER 1   Introduction   1

THEME ONE   Homeostasis via the Blood

CHAPTER 2   Circulatory System (Cardiovascular System)   23

CHAPTER 3   Respiration   48

CHAPTER 4   Endocrine System   65

THEME TWO   Growth and Development

CHAPTER 5   The Nervous and Sensory Systems   95

CHAPTER 6   Muscle   125

CHAPTER 7   Metabolism (in the Liver, Skeletal Muscle, and Adipose Tissue)   150

CHAPTER 8   Skeleton   162

THEME THREE   Interaction With the External Environment

CHAPTER 9   Immune Functioning   179

CHAPTER 10   Skin (Integument)   208

CHAPTER 11   Digestion and the Gastrointestinal Tract   227

CHAPTER 12   Urinary System   254

THEME FOUR   Reproduction (Propagation of the Species)

CHAPTER 13   Male Reproduction (Including Development of Gonads, Associated Ducts, and Accessory Organs)   275

CHAPTER 14   Female Reproduction   297

CHAPTER 15   Lactation   322

Index   *337*

Preface

"Develop a passion for learning. If you do, you will never cease to grow."
Anthony J. D'Angelo (author, born in USA in 1955)

"Tell me and I forget, teach me and I may remember, involve me and I learn."
Chinese proverb

Teaching Philosophy of the Authors

Both Dawn Koltes and Colin Scanes are dedicated to student success coupled with high expectations. Effective student learning is achieved by the creation of a stimulating environment. Both authors have endeavored to make the book as exciting as any textbook is.

This book was developed based on their combined 60 years of experience teaching college-age students in the animal sciences and instruction of domestic animal physiology and human physiology courses. The book resulted from a strong partnership with a team approach improving the clarity of the book as a learning tool for a wide variety of student backgrounds, student interests, and course dynamics. It is hoped that students and other readers will enjoy the book and master the subject.

The book can be read as hard copy or as a PDF on mobile devices when time permits such as in the library, the dorm and even waiting for a bus.

Acknowledgements

The excellent editing by Dakota West at Waveland Press is gratefully acknowledged. The professional support, vision, and encouragement by Don Rosso at Waveland Press are greatly appreciated.

About the Authors

Colin G. Scanes has been a faculty member of multiple universities including the University of Leeds, Rutgers, The State University of New Jersey, Iowa State University, Mississippi State University, and the University of Wisconsin–Milwaukee, teaching classes from 6 to 1,100 students. The large classes focused on anatomy and physiology. He was educated in the United Kingdom with a BSc and DSc (University of Hull) and PhD (University of Wales). He developed innovative approaches to student learning irrespective of whether they are visual or auditory learners.

Dawn A. Koltes joined Iowa State University in 2017 as a faculty member. She obtained her degrees from the University of Arkansas (BS) and Iowa State University (MS, PhD). This followed her role as a research scientist (University of Arkansas) where she also instructed a graduate-level *"Endocrine Physiology of Domestic Animals."* She presently instructs a sophomore-level—*"Domestic Animal Physiology."* This introductory physiology course is a large lecture-based undergraduate course taught both fall and spring where she incorporates active learning techniques to increase engagement of students.

Introduction   1

Learning Objectives

1. Understand what the term *domestication* means (see Section 1.1 and Textbox 1.1).
2. Understand the uses of domestic animals (see Section 1.1 and see Table 1.3).
3. Understand what species are considered domestic animals (see Table 1.1 and Section 1.1.1).
4. Understand the place of major livestock, poultry, and companion animals in biology (i.e., their place in animal classification; see Section 1.2 and Table 1.1 and 1.2).
5. Understand evolutionary history of livestock, poultry, and companion animals, including how, where, and when livestock, poultry, and companion animals were domesticated (see Section 1.3 and Figure 1.1).
6. Understand why livestock, poultry, and companion animals were domesticated and the uses of domesticated animals (see Section 1.4, Figures 1.2 and 1.3, and Tables 1.1 and 1.2).
7. Understand the basic structure of organs and tissues (see Section 1.5 and Figure 1.4).
8. Understand the basic structure of the cell and the role of the major cellular organelles (see Section 1.6).
9. Understand the structure of amino acids and proteins (see Section 1.7).
10. Understand the structure of DNA and RNA (see Section 1.8).
11. Understand the central dogma of molecular biology (see Section 1.9).
12. Know the metric system and its uses (see Section 1.10 and Textbox 1.6).

Chapter 1 Introduction

Table of Contents

- 1.1 Introduction
  - 1.1.1 Species Covered in Book
- 1.2 The Place of Major Livestock, Poultry, and Companion Animals in Biology (Classification)
- 1.3 Evolutionary History of Livestock, Poultry, and Companion Animals
- 1.4 Domestication of Livestock, Poultry, and Companion Animals
- 1.5 Basic Structure of Organs and Tissues
  - 1.5.1 Organ Systems
  - 1.5.2 Organs
  - 1.5.3 Tissues
  - 1.5.4 Epithelial Tissue
  - 1.5.5 Muscle Tissue
  - 1.5.6 Connective Tissues
  - 1.5.7 Nervous Tissue
- 1.6 Intracellular Structure
  - 1.6.1 Introduction
  - 1.6.2 Nucleus
  - 1.6.3 Mitochondria
  - 1.6.4 Cytoskeleton
  - 1.6.5 Other Organelles
  - 1.6.6 Tight Junctions
- 1.7 Proteins Structure
- 1.8 Structure of Deoxyribonucleic Acid (DNA) and Ribonucleic Acid (RNA)
- 1.9 Central Dogma of Molecular Biology
- 1.10 Metric Units

1.1 Introduction

1.1.1 Species Covered in Book

This book will focus on general themes of the physiology of domestic animals, with examples from an individual species of the following species: cattle, sheep, goats, pigs, horses, dogs, cats, chickens, ducks, geese, and turkeys (see Table 1.1 and Figure 1.1).

What does it take for an animal to be **domesticated**?

1. The species must be present in region. If it is not there, there is no chance of domestication.
2. The species must be tamable and/or handleable.

Textbox 1.1 shows glossary related to domesticated animals and other topics in the introduction.

Other livestock, poultry, and companion animal species outside of the previously listed are present but are of limited use in the Americas, Europe, East Asia, and

---

Textbox 1.1 Glossary Related to Domestication, Classification of Animals and Tissues

- **Classification** is grouping living organisms based on their similarities.
- **Companion animals** are domesticated animals that are kept as pets and/or as working animals.
- **Domestication** is a multigeneration process involving taming a wild animal and then selecting which animals it will breed with.
- **Epithelium** is a layer of cells.
- **Feral animals** are domesticated animals living in the wild.
- **Livestock** are domesticated mammals that are used for agricultural purposes.
- **Metric system** is a decimal (factor of ten) system.
- **Organs** are composed of tissues, normally at least two tissues.
- **Organ systems** are composed of two or more organs together conducting one function.
- **Poultry** are domesticated birds that are used for agricultural purposes.
- **Tissues** are groups of cells that have a similar structure and work to achieve a common outcome. They are categorized as epithelium, muscle, connective tissue, and nervous tissue.

Australia. They will not be considered in this book. These include the following:

- Mammalian livestock
    - Alpacas (*Lama pacos*)
    - Camels
        - Bactrian or two-humped camel (*Camelus bactrianus*)
        - Dromedary or one-humped camel (*Camelus dromedaries*)
    - Donkeys (*Equus africanus asinus*).
    - Llamas (*Lama glama*).
    - Water buffalos (*Bubalus bubalis*) (see Figure 1.1.F).
    - Yaks (*Bos grunniens*).
- Mammalian companion animals
    - Donkeys (*Equus africanus asinus*) (see Figure 1.1.D)
    - Guinea pigs (*Cavia porcellus*) (see Figure 1.1.B)
    - Rats (*Rattus norvegicus*)
    - Mice (*Mus musculus domesticus*)
    - Hamsters (*Mesocricetus auratus*)
- Poultry (Avian domestic animals)
    - Geese (*Anser anser domesticus* or *Anser cygnoides domesticus*)
    - Guinea fowl (*Numida meleagris*)
    - Ostrichs (*Struthio camelus*)
- Avian companion animals
    - Pigeons (*Columba livia domestica*)
    - Canaries (*Serinus canaria domestica*)
    - Parrots and Parakeets, e.g., budgerigars (see Figure 1.1H and 1.1.I)
    - Peacock (*Pavo cristatus*)

This book will also not include discussion of reptilian (snakes and lizards), amphibian (frogs, toads, newts), and fish companion animals. Reptiles that are used for food and leather include alligators and snakes. These will also not be covered. Supplements to Chapters 1 to 14 are available covering the physiology of those bony fish that are used in aquaculture.

1.2 The Place of Major Livestock, Poultry, and Companion Animals in Biology (Classification)

Carl Linnaeus (Sweden 1707–1778) developed a **classification** system based on similarities in characteristics between species. This system has been modified but still forms the basis for relationships between species. Examples of classification include the following:

- Animals are in the **kingdom Animalia**.
- Mammals, birds, and fish are in the **phylum Chordata** and **sub-phylum Vertebrata**.
- Mammals (**class Mammalia**) are warm-blooded quadrupeds that produce milk.
- Birds (**class Aves**) are warm-blooded quadrupeds with feathers that typically fly.
- Boney fish (**class Actinopterygii**) are cold-blooded fish with skeletons composed of bone, not cartilage.
- There are multiple orders in a **class**, multiple **families** in an **order**, and multiple genera (plural of **genus**) in a family.

The classification of the major livestock, poultry, and companion animals are shown in Tables 1.1 and 1.2.

Mammals are **homothermic** (self-regulating body temperature) quadrupeds (having four legs) that produce milk for their offspring. They are usually terrestrial

---

Textbox 1.2 A Deeper Dive Into Classification

Types of Organisms (Domains)
Mammals and birds—along with all animals, plants, and fungi—are in the **domain Eukaryota**, as their cells have nuclei. Other domains are Bacteria and Archaea.

Animals are in the **kingdom Animalia**. Within the kingdom, mammals, and birds are in the **sub-kingdom Eumetazoa** (multicellular animals), the **super-phylum Deuterostomia**, the **phylum Chordata**, the **sub-phylum Vertebrata**, **super-class Tetrapoda**, and **class Mammalia** or **Aves**.

Groups of Mammals
There are three major groups of mammals:

- **Monotremes** (sub-class Prototheria, order Monotremata)
- **Marsupials** (sub-class Theria, infraclass Marsupialia)
- **Placental mammals** (sub-class Theria, infra-class Eutheria or Placentalia). These evolved about 120 million years ago.

Figure 1.1 Examples of domestic animals.

A. Examples of (left) a dog (*Canis lupus*) and (right) cat (*Felis catus*). (Shutterstock/New Africa)
B. Guinea pig (*Cavia porcellus*). (Shutterstock/Rita_Kochmarjova)
C. Horse (*Equus ferus*) with rider. (Shutterstock/yoannes surya)
D. Donkey (*Equus africanus*) as a pack animal. (Shutterstock/In Green)
E. Examples of (left) sheep (*Ovis aries*) and (right) cattle (*Bos taurus*). (Shutterstock/rtbilder)
F. Water buffalo (*Bubalus bubalis*). (Shutterstock/David Havel)
G. Chicken (*Gallus gallus*). (Shutterstock/Charlotte Bleijenberg)
H. Blue and yellow macaw (*Ara ararauna*). (Shutterstock/khlungcenter)
I. Blue and green budgerigar (*Melopsittacus undulatus*). Parakeet is a general term for small parrots. In US English, the words parakeet and budgerigar are interchangeable. (Shutterstock/Uliya Krakos)

CHAPTER 1 Introduction

but there are aquatic mammals such as whales, dolphins, seals, and walruses. There are three major groups of mammals:

- **Monotremes** (e.g., platypus) that produce and lay large yolky eggs.
- **Marsupials** that produce an underdeveloped neonate that latch on to the nipple or teat during the rest of development.
- **Placental mammals** that have a placenta allowing them to produce a well-developed neonate(s).

Birds are **homothermic quadrupeds** (animals with four appendages—two wings and two legs). They have **feathers**, their forelimbs are modified as **wings** for flight, and they produce and lay **large yolky eggs**. Birds are usually terrestrial but there are aquatic birds, namely penguins. Existing birds lack teeth and have beaks.

1.3 Evolutionary History of Livestock, Poultry, and Companion Animals

Birds and mammals both **evolved** from reptiles. The **last common ancestor** or extant species for mammals and birds was around about **220 million years ago**, with birds sometimes referred to as feathered flying dinosaurs (Figure 1.2). In contrast, the livestock and companion animals are separated by only about 50 million years (Figure 1.2). **Anseriform** birds (ducks and geese) separated from **Galliform birds** (chickens and turkeys) about 70 million years ago (Figure 1.2).

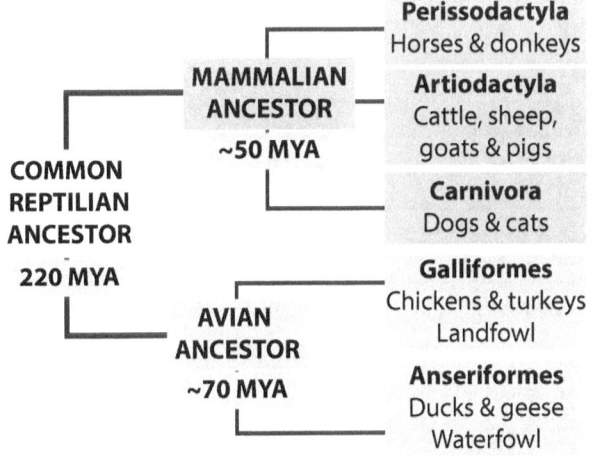

Figure 1.2 Evolutionary history of livestock, poultry, and companion animals.

1.4 Domestication of Livestock, Poultry, and Companion Animals

Humans have had a close interaction with livestock, poultry, and companion animals for thousands of years. We have used these animals for many different reasons, which can be summarized as the following (see Table 1.3):

- Meat, milk, and eggs.
- Protection from predators.
- Killing vermin consuming grain.
- Providing locomotive power.

Table 1.1 summarizes where and when domestic animals were domesticated. The first species to be domesticated is thought to be the **dog**.

Dogs were domesticated from wolves by hunter gatherer peoples in Eurasia. Dogs were the only animal domesticated by **hunter gatherers**. It is theorized that wolves were tamed while consuming meat left over from hunters. Another possibility is that people reared wolf cubs. In turn, the domesticated dogs provided protection for bands of hunter gatherers and aided in their hunting.

Livestock and poultry began to be domesticated after the beginning of the **Neolithic revolution** or the beginning of farming independently in the **Fertile Triangle** and **East Asia** (see Figure 1.3 and Table 1.1). There was also domestication of cattle in **South Asia** later (see Figure 1.3 and Table 1.1). Domestication of livestock and poultry followed domestication of grain as meat from hunting was becoming in short supply. Domesticated livestock provided a stable and increase in food supply. Livestock and poultry had several advantages (see Table 1.3):

- Provided a convenient source of protein.
- Consumed surplus grain and pulses.
- Consumed plant materials that were not digestible by people.
- Easily transported.

Horses were domesticated in the **Eurasian Steppe**. They were invaluable in providing the following (see Figure 1.2 and Table 1.3):

- Animals to ride or draw chariots for hunting and warfare.
- A source of meat and milk.
- Decoy animals to trap wild horses.

6 Chapter 1 Introduction

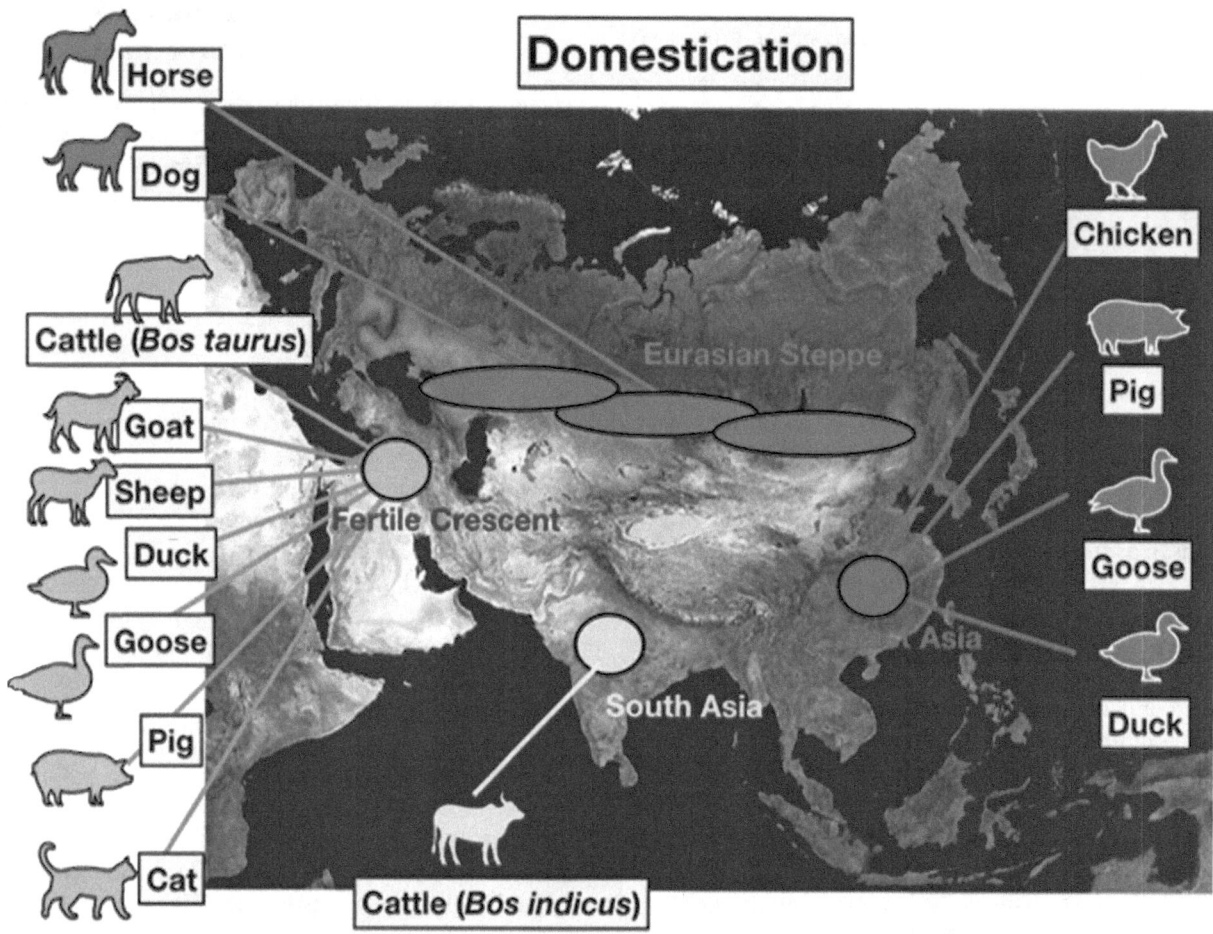

Figure 1.3 Location of domestication of animals.

Table 1.1 Classification and domestication of livestock, poultry, and companion animals.

Domesticated Mammal	Scientific Name for Species	Ancestral species common and scientific name	Time (years before present, or BP) and location of domestication
Dog	*Canis familiaris* or *Canis lupis familiaris*	Grey wolf (*Canis lupis*)	>15,000 BP and possibly over 30,000 BP in Eurasia
Cat	*Felis catus* or *Felis silvestris catus*	African wildcat (*Felis lybica*)	~9500 BP in Near East/Fertile Crescent
Horse	*Equus ferus caballus* or *Equus caballus*	Wild horse (*Equus ferus*)	~5000 to 6000 BP in Eurasia
Cattle	*Bos taurus* *Bos indicus*	Wild cattle (*Bos primigenius*)	Two domestications: 1. ~10,500 BP in Fertile Crescent 2. ~9000 BP in South Asia
Sheep	*Ovis aries*	Wild mouflon (*Ovis gmelini*)	~7000 BP in Fertile Crescent

Table 1.1 Classification and domestication of livestock, poultry, and companion animals (*continued*)

Domesticated Mammal	Scientific Name for Species	Ancestral species common and scientific name	Time (years before present, or BP) and location of domestication
Goat	*Capra hircus*	Wild goats (*Capra aegagrus*)	~10,000 BP in Fertile Crescent
Pigs	*Sus domesticus*	Wild boar (*Sus scrofa*)	At least two domestications: 1. ~10,500 BP in Fertile Crescent 2. >10,000 BP in East Asia
Domesticated Bird			
Chicken	*Gallus domesticus* or *Gallus gallus domesticus*	Red jungle fowl (*Gallus gallus*)	~8000 BP in East Asia and/or SE Asia
Turkey	*Meleagris gallopavo*	Wild turkey (*Meleagris gallopavo*)	~2000 BP in North America
Duck	*Anas platyrhynchos*	Mallard duck (*Anas platyrhynchos*)	Two domestications: 1. >4000 BP in Fertile Crescent 2. >4000 BP in East Asia
Muscovy duck	*Cairina moschata* or *Cairina moschata domestica*	*Cairina moschata*	>500 BP in South and Central America
Goose	Two domestications: 1. *Anser anser domesticus* 2. *Anser cygnoides domesticus*	Two domestications: 1. Greylag goose (*Anser anser*) 2. Swan goose (*Anser cygnoides*)	Two domestications: 1. ~4000 BP in Fertile Crescent 2. ~4000 BP in East Asia

Table 1.2 Classification of humans, livestock, poultry, and companion animals.

	Class	Order
Humans	Mammalia (mammals)	Primates
Cats and dogs	Mammalia (mammals)	Carnivora
Cattle, sheep, goats, and pigs	Mammalia (mammals)	Artiodactyla (even-toed ungulates)
Guinea pigs plus rats, mice, and hamsters	Mammalia (mammals)	Rodentia (rodents)
Horses	Mammalia (mammals)	Perissodactyla (odd-toed ungulates)
Chickens and turkeys	Aves (birds)	Galliformes
Ducks and geese	Aves (birds)	Anseriformes

CHAPTER 1 Introduction

Table 1.3 Examples of products from domestic animals.

Product	Species	Major Constituents	Major Proteins
Meat	Cattle, sheep and goats, chickens, ducks, and geese	Protein, fat, and water	Myosin & actin
Milk	Cattle, sheep, and goats	Protein, fat, carbohydrate, and water	Casein
Eggs	Chickens, ducks, and geese	Protein, fat, and water	Vitellin & phosvitin
Wool	Sheep	Protein	Keratin
Leather	Cattle, pigs, snakes, and alligators	Protein and water	Collagen
Insulation • Feathers • Hair	Ducks and geese Horses	Protein Protein	Keratin Keratin

Table 1.4 Other common and historical roles of domestic animals.

Animal	Function or Role
Cats	Rodent control
Cattle and pigs	Leather for clothing, foot ware, and bags
Chickens	Cockfighting* (illegal in USA)
Chickens	Feathers as absorbent in diapers
Dogs	Herding
Dogs	Hunting
Dogs	Protection
Dogs	Therapy animals
Dogs and miniature horses	Service animals
Horses	Racing as entertainment
Horses	Warfare—light and heavy cavalry*
Horses and cattle (oxen)	Draft animals pulling plows
Horses, donkeys, cattle, and camels	Transportation
Pigs and chickens	Biomedicine, e.g., heart valves (from pigs) & chick embryos (producing vaccines)
Sheep, ducks, and geese	Feathers and hair for clothing and insulation

*Denotes traditionally historical roles

1.5 Basic Structure of Organs and Tissues

1.5.1 Organ Systems

Organ systems are composed of two or more organs with these together conducting one function.

1.5.2 Organs

Organs are composed of **tissues**, normally at least two tissues.

1.5.3 Tissues

Tissues are structurally similar groups of cells that work together to complete a common function. They are categorized as being the following:

1. **Epithelial tissue.**
2. **Muscle tissue.**
3. **Connective tissue.**
4. **Nervous tissue.**

1.5.4 Epithelial Tissue

Epithelial tissue forms glands and membranes and can also serve as a protective barrier. The epithelium can be either of the following:

- **Simple** (consisting of a single layer of cells) (see Figure 1.4).
- **Stratified** (consisting of multiple layers of cells) (see Figure 1.4).

There are the following types of epithelia (see Figure 1.4):

- **Simple squamous** (pavement) **epithelium**—a single layer of thin, flattened, or scale-like cells. Examples of simple squamous (pavement) epithelium include capillaries and alveoli. Their thin nature allows for gases to be quickly exchanged.
- **Simple cuboidal epithelium**—a single layer of cells that are cube shaped or are square when viewed in 2D. Examples of simple cuboidal epithelium are in the nephrons of the kidneys, pancreas, and salivary glands.
- **Simple columnar epithelium**—a single layer of tall (or long) cells. These cells are usually polar in nature. Examples of simple columnar epithelium are the epithelium lining the small intestine and the female reproductive tract.
- **Simple ciliated (pseudostratified) columnar epithelium**—a single layer of cells that are as tall as they are wide and that have cilia, but with the nuclei at different heights giving the appearance of stratified epithelia. The cells have cilia on one surface (the apical or luminal surface). Examples of simple ciliated (pseudostratified) columnar epithelium are in the nose, bronchi, oviduct, and uterus.
- **Simple glandular columnar epithelium**—a single layer of cells that are as tall as they are wide and that have cilia. They can secrete chemicals. Examples of simple glandular columnar epithelium include in the stomach and small intestines.
- **Stratified non-keratinized squamous epithelium**—multiple layers of flattened cells. Examples include vagina, mouth, and esophagus.
- **Stratified keratinized epithelium**—multiple layers with the outermost composed of keratinized dead cells. This type is seen in the epidermis of the skin.
- **Stratified transitional epithelium**—multiple layers of cells and stretchable. This type of epithelium is found in the bladder, urethra, and ureter.

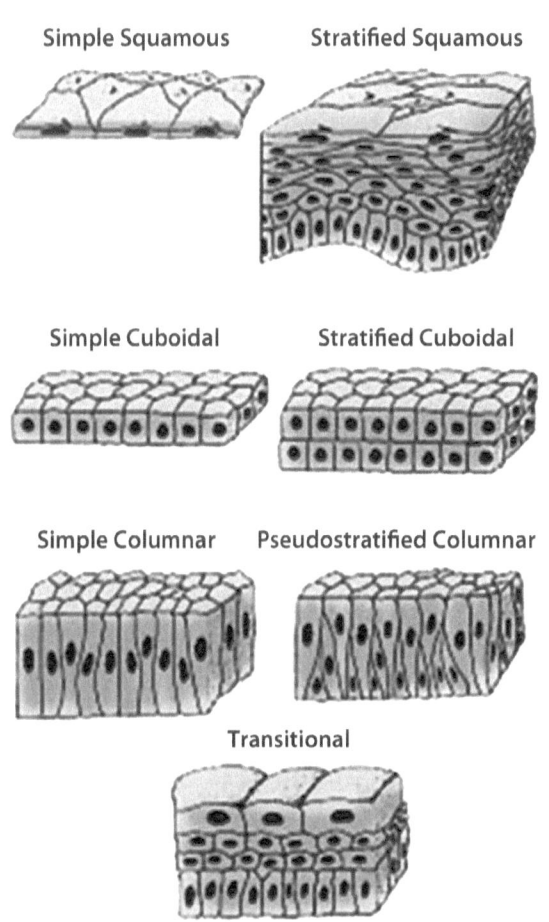

Figure 1.4 Types of epithelia. (Courtesy of the US government)

10 CHAPTER 1 Introduction

1.5.5 Muscle Tissue

Muscle tissue is a group of cells that function to contract leading to movement. The functions and physiological process of how movement occurs will be covered in Chapter 6. Muscle tissue can be one of the following:

- Skeletal
- Cardiac
- Smooth

1.5.6 Connective Tissues

Connective tissues span wide spaces in an animal's body. These tissues function to form a matrix and connects groups of cells together. Connective tissues consist of the following:

- Loose connective tissue
- Adipose tissue
- Blood
- Bone
- Cartilage

1.5.7 Nervous Tissue

Cells of the nervous tissue work together to support and send communications throughout the body. The functions and physiological process of the nervous system will be covered in Chapter 5. Nervous tissue consists of either of the following:

- Neurons
- Glial or neuroglial cells

1.6 Intracellular Structure

1.6.1 Introduction

The overall structure of the cell is shown in Figure 1.5. **Key organelles** and structures described in this book are the following:

- Nucleus
- Mitochondria
- Lysosome
- Rough endoplasmic reticulum
- Golgi apparatus
- Cytoskeleton
- Tight junctions between cells

1.6.2 Nucleus

The **nucleus** contains the genetic material known as DNA (deoxyribonucleic acid). This genetic material is organized into **chromatin** and associated proteins

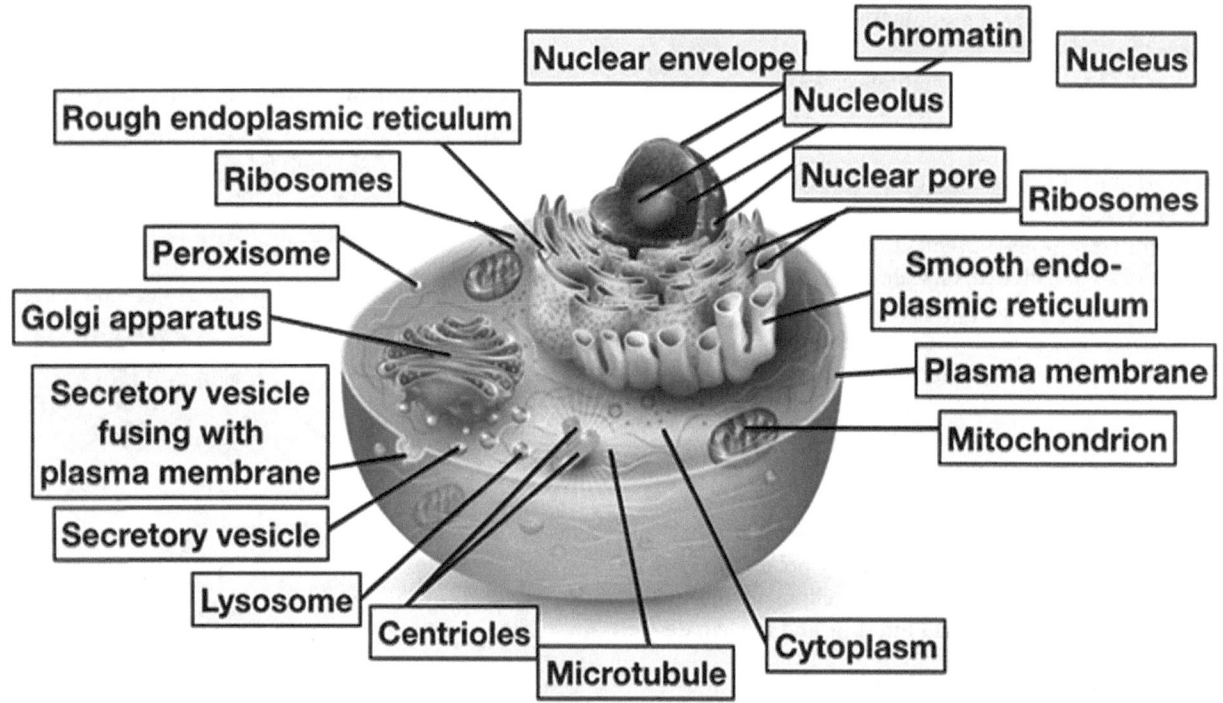

Figure 1.5 The structure of a cell. (Adapted from Shutterstock/TimeLineArtist)

(DNA wrapped around with proteins called histones). In turn, the chromatin is organized into chromosomes. There are multiple **genes** on each chromosome. Additionally, DNA is found in mitochondria which are passed to the offspring through the oocyte but not spermatozoa. All genes are composed of DNA. The DNA can be chemically modified to influence its activity.

The structure of DNA is covered in Section 1.8. During cell division, new copies of the DNA are produced by DNA **replication**, with one copy going to one daughter cell and the other going to the second daughter cell (see Figure 1.6). The sequence of bases contained in DNA in a specific gene can be transcribed into **RNA** (ribonucleic acid) **transcripts** or functional RNA. Transcripts are translated into specific proteins (covered in detail in Section 1.8).

The sequence of the **genomes** of domestic animals has been determined (see Textbox 1.3).

1.6.3 Mitochondria

The **mitochondria** [singular mitochondrion, plural mitochondria] generate energy in the form of **ATP** (adenosine triphosphate) for the cell. The process is called **cellular respiration** or **oxidative phosphorylation**.

A key component of oxidative phosphorylation is **nicotinamide adenine dinucleotide** (**NAD⁺**). This is reduced (hydrogenated) reversibly to the electron donor (coenzyme), **NADH**.

$$NAD^+ + H^+ + 2 \text{ electrons} \leftrightarrows NADH$$

NADH is generated by **glycolysis** and the **citric acid cycle** in the **mitochondria** (discussed in detail in

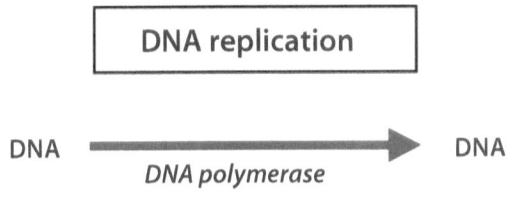

Figure 1.6 Replication of DNA during mitosis or meiosis.

Textbox 1.3 A Deeper Dive Into Genomes of Domestic Animals

Species	Number of base pairs (billion bases or gigabases or Gb)*
Mammals	
Cat	2,521,863,845 (2.5 Gb)
Cattle	2,715,853,762 (2.7 Gb)
Dog	2,343,303,000 (2.3 Gb)
Donkey	2,321,044,345 (2.3 Gb)
Goat	2,922,813,246 (2.9 Gb)
Horse	2,506,966,135 (2.5 Gb)
Pig	2,501,912,388 (2.5 Gb)
Birds	
Chicken	1,048,336,660 (1.0Gb)
Turkey	1,115,474,681 (1.0 Gb)
Fish	
Channel Catfish	783,274,721 (0.8 Gb)
Nile Tilapia	1,005,681,550 (1.0 Gb)

*From Ensembl Genome database on September 4th, 2022.

12 Chapter 1 Introduction

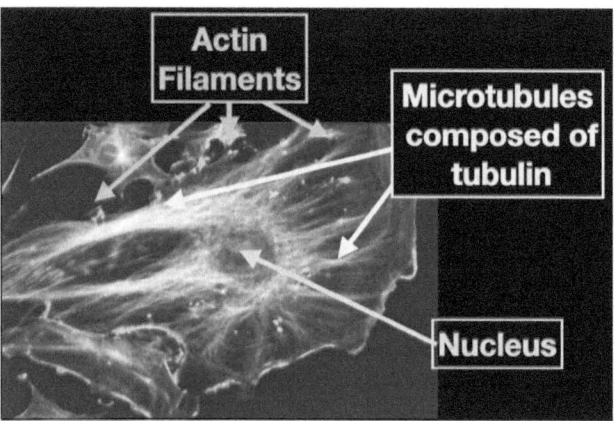
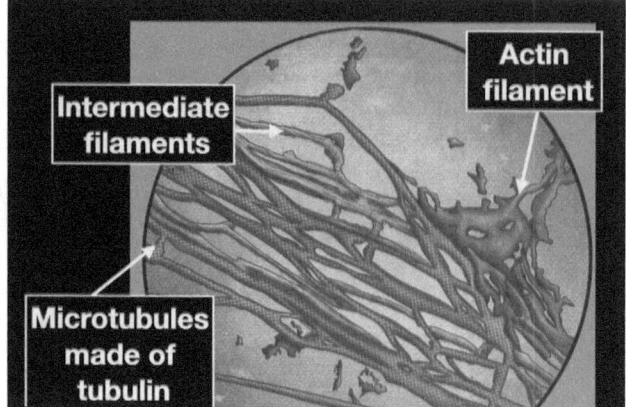

Figure 1.7 Cytoskeleton. (Left) Fluorescence image of cell showing microtubules and actin filaments. (Adapted from the National Institutes of Health) (Right) Schematic of cytoskeleton. (Adapted from Ilse Yohn)

Chapter 7). The last step in oxidative phosphorylation is the formation of adenosine triphosphate (ATP) from adenosine diphosphate (**ADP**).

1.6.4 Cytoskeleton

A **cytoskeleton** is present in the cytoplasm of cells (see Figure 1.7). There are three components of the cytoskeleton, namely the following:

- **Actin filaments** (aka microfilaments) composed of actin.
- **Intermediate filaments**.
- **Microtubules** composed of tubulin.

Among the roles of the cytoskeleton are the following:

- To maintain the positions of the organelles.
- To facilitate transport within the cytoplasm.
- To resist deformation such that the cell maintains its structure.

1.6.5 Other Organelles

The role of the **ribosomes** associated with the **rough endoplasmic reticulum** is to perform protein synthesis (discussed below). **Lysosomes** are responsible for proteolytic

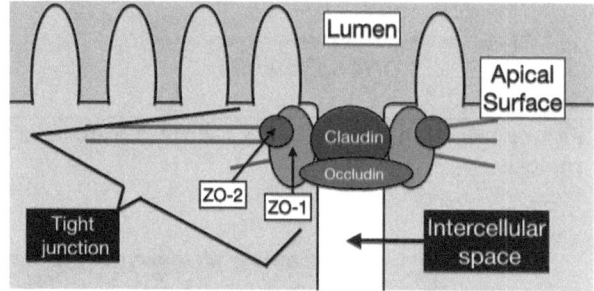

Figure 1.8 Schematic of a tight junction.

degradation of bacteria and other cellular debris. The **secretory granules** are formed in the Golgi apparatus and their contents are released when the secretory granule fuses with the cell membrane.

The **cell membrane** is a bi-layer of phospholipids with receptors, ion channels, and other channels perforating the membrane (see Figure 1.8).

1.6.6 Tight Junctions

There can be **tight junctions** between epithelial cells creating a **barrier** (for a deeper dive into tight junctions see Textbox 1.4).

Textbox 1.4   A Deeper Dive Into Tight Junctions

Tight junctions between epithelial cells create a barrier. The tight junctions consist of multiple proteins including the following (see Figure 1.9):

- Zonula occludens (ZO) proteins
- Claudin proteins
- Occludin

Tight junctions allow some ions to pass through them.

(continued)

Textbox 1.4 (continued)

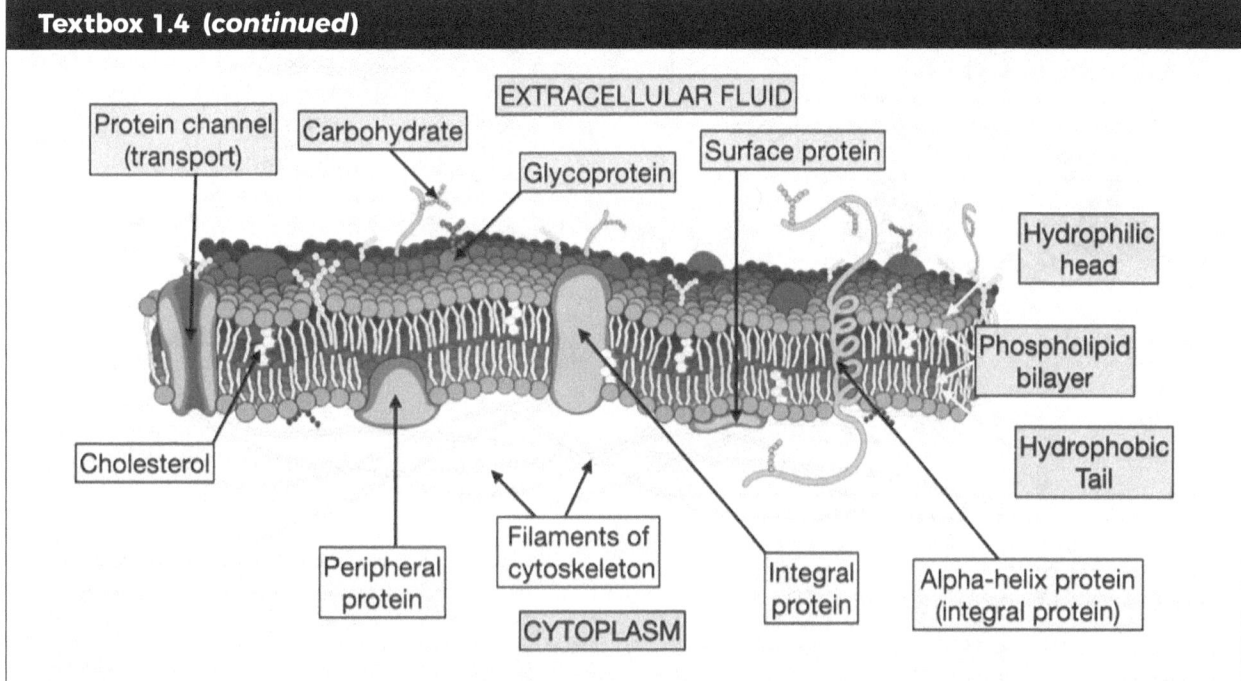

Figure 1.9 Schematic diagram of cell membrane (plasma membrane or plasmalemma). Missing are transmembrane receptors to hormones, cytokines, and neurotransmitters. These are discussed in Chapters 4 (hormones), 5 (neurotransmitters), and 9 (cytokines). (Adapted from LadyofHats/Mariana Ruiz, Wikimedia Commons)

1.7 Proteins Structure

Proteins are composed of amino acids. Figure 1.10 summarizes a generalized structure of an amino acid with both amine and carboxyl groups, while Figure 1.11 shows the structure of all the amino acids in peptides and proteins. Amino acids can be **hydrophobic, basic,** or **acidic**. Amino acids are linked by **peptide bonds** to form polypeptides or proteins as illustrated in Figure 1.12. Textbox 1.5 provides a series of definitions about protein structures.

Amino acids have other roles:

- **Arginine** is a precursor of **nitric oxide** and **polyamines**.
- **Cysteine** is a precursor of **glutathione** which is part of the **antioxidant** system (see Section 7.9 in Chapter 7).
- **Glutamine** is used as an **energy source** by intestinal tissues.
- **Tyrosine** and **phenylalanine** are converted to **dopamine, norepinephrine,** and **epinephrine**.
- **Tyrosine** is the precursor of the hormones, **thyroxine,** and **triiodothyronine**.
- **Tryptophan** is the precursor of **serotonin** and **melatonin**.

Textbox 1.5 Protein Structure Definitions

A protein is not like a random ball of string or Christmas lights. Instead, it is ordered.

- The **primary structure** of a protein is the sequence of amino acid residues. When in a polypeptide or protein, amino acids are called amino acid residues.
- The **secondary structure** of a protein includes α-helix and the β-sheet. These are formed spontaneously.
- The **tertiary structure** of a protein is the three-dimensional structure of a protein held together by hydrogen bonds and disulfide bridges. The disulfide bridges link together cystine residues (see Figure 1.10).
- The **quaternary structure** of a protein is when several proteins (monomers or sub-units) are linked together.

14   Chapter 1   Introduction

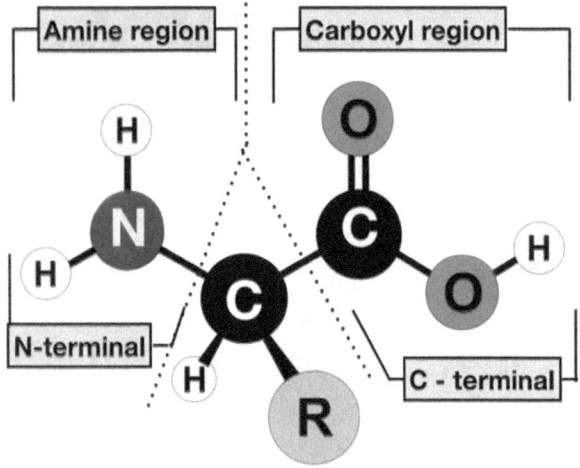

Figure 1.10 Generalized structure of an amino acid [H, hydrogen; C, carbon; N, nitrogen; O, oxygen; and R, residue unique to specific amino acid]. (Adapted from Techguy 78, CC BY 4.0)

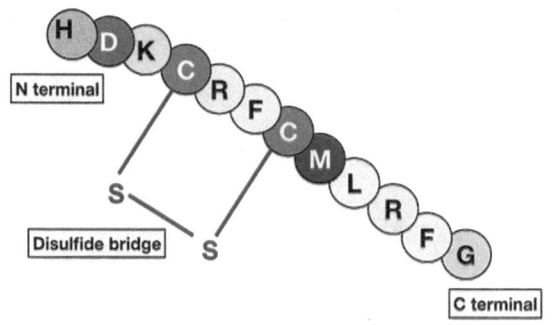

Figure 1.12 Simplified schematic of protein.

The protein may require modification. This is called **post-translational modification** and includes the following:

1. **Proteolytic cleaving** at one or several sites by specific enzymes. This is important in the activation of **zymogens** (see Chapter 11).
2. **Disulfide bridge** (-S-S-) formation.
3. Addition of phosphate groups in **phosphorylation**.
4. Addition of amine groups in **amidation**.

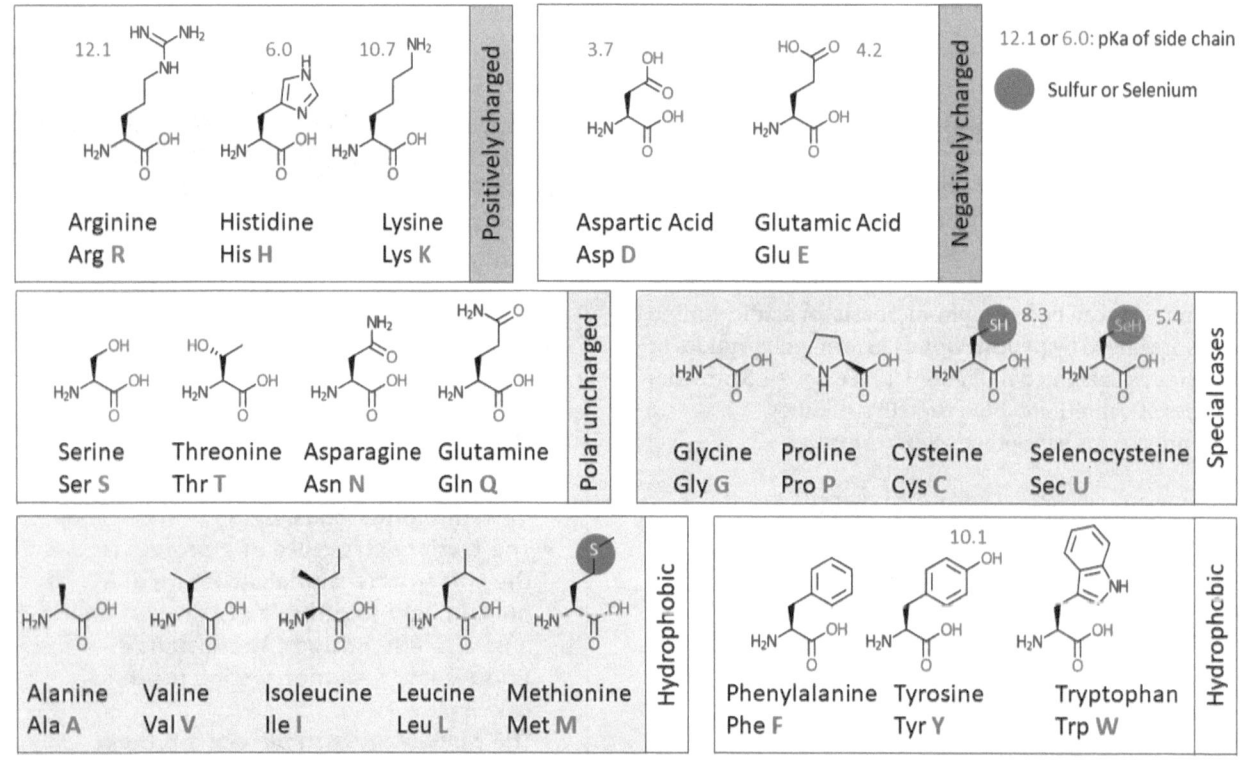

Figure 1.11 Structure of amino acids. (Thomas.rychmans68, CC BY 4.0)

5. Addition of an acyl (fatty acids) group in **acylation**.
6. Addition of small or large carbohydrates in **glycosylation**.

> **Textbox 1.6 Definitions Related to the Central Dogma of Biology**
>
> - **Deoxyribonucleic acid (DNA)**: an organic macromolecule consisting of nucleotides encoding the genetic code.
> - **Down regulation**: decreased expression of a gene.
> - **Expression**: transcription of a specific gene.
> - **Post-translational changes**: chemical modification of a protein such as adding phosphate or carbohydrates.
> - **Up-regulation**: increased expression of a gene.
> - **Ribonucleic acid (RNA)**: an organic macromolecule encoding a copy of a gene complimentary to the DNA sequence.
> - **Transcription**: production of an RNA strand complementary to the sequence within DNA.
> - **Translation**: production of a protein with a specific sequence of amino acids from an RNA template.

1.8 Structure of Deoxyribonucleic Acid (DNA) and Ribonucleic Acid (RNA)

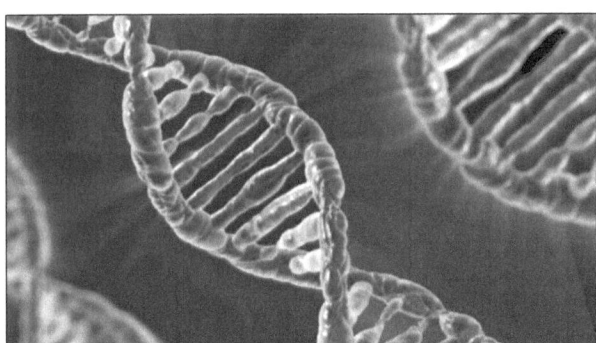

Figure 1.13 DNA has the structure of a double helix. (Courtesy Pixabay, Wikimedia Commons)

DNA is a **double helix** composed of nucleotides (see Figure 1.13). Each nucleotide consists of the following (see Figure 1.13):

- A **deoxyribose sugar** (see Figure 1.14)
- A **base**
- A **phosphate**

There are four different bases with a set complementary bases in DNA and RNA (see Figure 1.12).

Base	Complimentary base	
	In DNA	In RNA
A (Adenine)	T (Thymine)	U (Uracil)
T (Thymine)	A (Adenine)	A (Adenine)
C (Cytosine)	G (Guanine)	G (Guanine)
G (Guanine)	C (Cytosine)	C (Cytosine)

In both DNA and RNA, three nucleotides (a triplet) form a **codon**. Each codon codes for a single amino acid.

DNA can be chemically modified (e.g., methylation) to influence its activity. This can be non-heritable or heritable. In the case of the latter, it is called **epigenetic** modification.

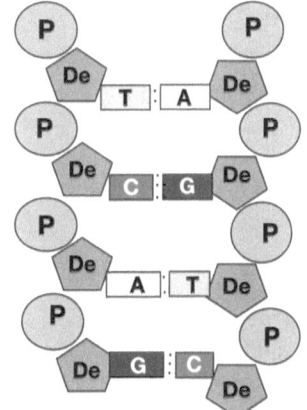 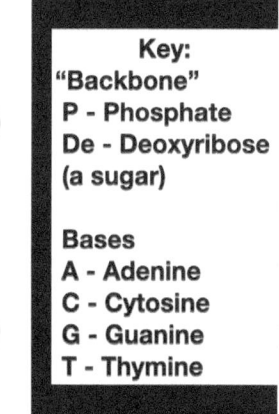

Figure 1.14 Structure of DNA showing backbone of deoxyribose and phosphate and bases with adenine lined up with its complementary base, thymine, and cytosine with its complementary base, guanine.

RNA is composed of a **single chain** of nucleotides. Each nucleotide consists of the following:

- A **ribose sugar** (see Figure 1.15)
- A **base**
- A **phosphate**

There is a chain of ribose sugars alternating with phosphate groups. Each ribose is linked to a one base—**uracil**, **adenine**, **guanine**, or **cytosine**; uracil taking the place of thymine in DNA. The sequence of bases is complimentary to that in DNA. Three nucleotides (a triplet) code for a single amino acid.

There are multiple forms of transcribed RNA:

- **Messenger RNA** (mRNA) which is translated to proteins.
- **Transfer RNA** (tRNA) which bind amino acids during translation.
- **MicroRNA** (miRNA) which functions in the control of transcription, translation, and other aspects of cell physiology.

1.9 Central Dogma of Molecular Biology

The **Central Dogma Theory** is a generalized theory that outlines the flow of information. **Replication** is the first part of this process. During this process, the DNA will replicate itself. **Transcription** is the second part of this process. Copying the DNA to produce RNA is called **transcription** and is catalyzed by **RNA polymerase** (Figure 1.16). For each nucleotide in DNA, there is a **complementary nucleotide** in the RNA transcript (also

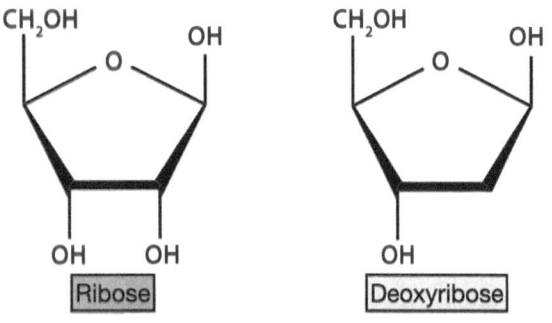

Figure 1.15 Structure of deoxyribose and ribose. (Adapted from NEUROtiker, public domain)

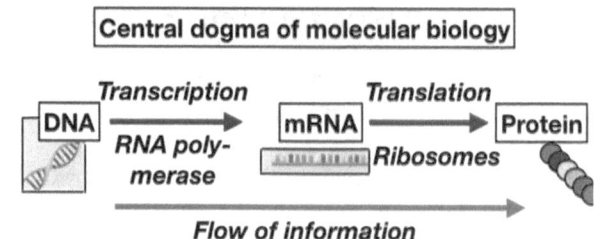

Figure 1.16 Central dogma of molecular biology. RNA is synthesized based on the structure of DNA (the sequence of bases). Proteins are synthesized based on the sequence of bases in RNA.

Textbox 1.7 Interesting History of DNA

Discovery of Deoxyribonucleic Acid
Deoxyribonucleic acid was purified from pus (containing leukocytes and macrophage) in 1869 by a Swiss scientist, Friedrich Miescher. It was initially called nuclein.

Double Helix Structure
The concept of the double helix was developed in 1953 by James Watson (American) and Francis Crick (English). This was based on X-ray crystallography conducted by Rosalind Franklin (English) working in Maurice Wilkins' laboratory in London. The first publication included Rosalind Franklin's work without mentioning her. James Watson, Francis Crick, and Maurice Wilkins received the Nobel Prize in Physiology or Medicine in 1962. However, Rosalind Franklin did not and died in 1952. Many consider that the inclusion of her work, and not attributing it, was not appropriate and that she should have also received the Nobel Prize in Physiology or Medicine posthumously.

DNA as the Genetic Material
Alfred Hershey (American) and Martha Chase (American) determined in 1952 that DNA is the genetic material. George Beadle (American) and Edward Tatum (American) determined that transcripts code for enzymes in 1941.

Central Dogma
Francis Crick postulated the central dogma of molecular biology (DNA → RNA → protein) in the 1950s.

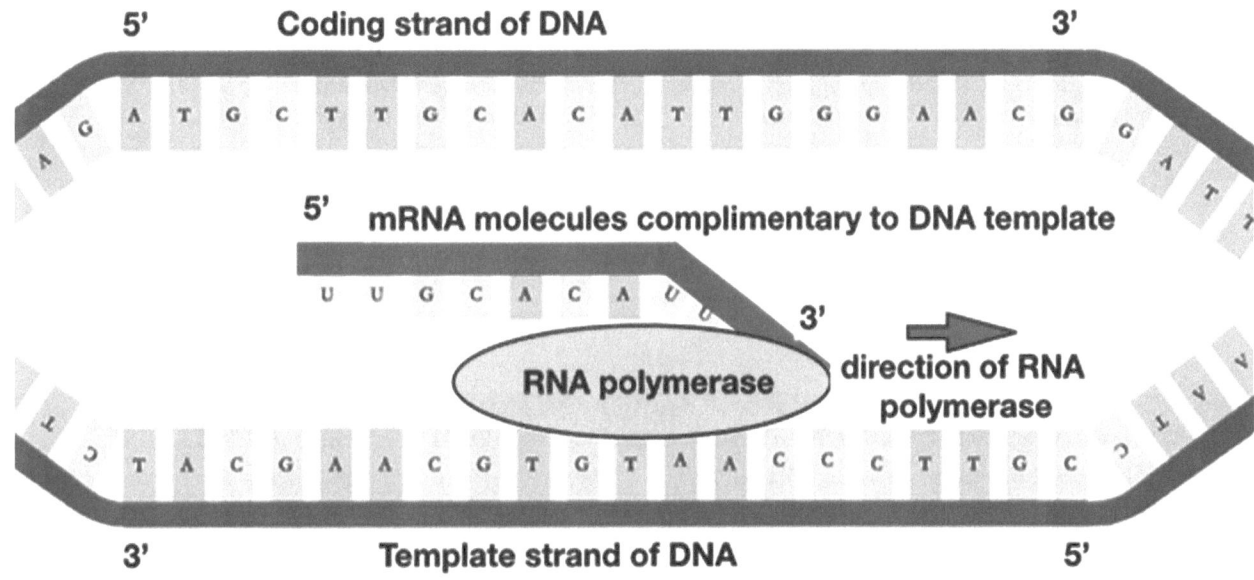

Figure 1.17 Generation of RNA transcript. Nucleotides in DNA: A (adenine), C (cytosine), G (guanine), and T (thymine). Nucleotides in RNA: A (adenine), C (cytosine), G (guanine), U (uracil). (Adapted from Kep17, CC BY 4.0)

see Figure 1.14). In this process, the DNA's thymine is changed to uracil in RNA. The transcribed RNA in eukaryotes is converted to messenger RNA after splicing of exons in the nucleus. The final step in this process is **translation**. Messenger RNA is moved out of the nucleus to a ribosome of the **rough endoplasmic reticulum**. The messenger RNA acts as a template for the formation of specific protein. This is **translation**.

There is a flow of information from DNA to RNA and then to a specific protein (Figures 1.16 and 1.17). Proteins are synthesized with a specific sequence of amino acid residues. Textbox 1.8 provides a deeper

---

Textbox 1.8 A Deeper Dive Into Protein Synthesis

In eukaryotes, the nuclear RNA transcript produced during transcription consists of a series of exons interspaced by introns together with a cap and a tail (Figure 1.18). The introns are cut out and the exons are spliced together to form mature messenger RNA (mRNA) in the nucleus. The mRNA is then exported from the nucleus. Proteins are then synthesized using the mRNA as a template, in association with the ribosomes in the cytoplasm. Translation begins at the translation start and finishes at the translational stop point.

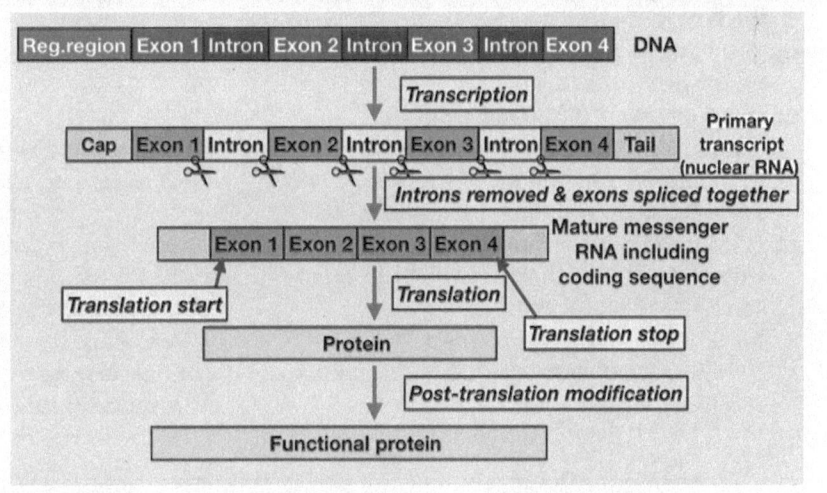

Figure 1.18 A more detailed schematic of protein synthesis. [Key: Reg. region = regulatory region]

dive into protein synthesis. In some viruses, there is a **reverse transcriptase** synthesizing DNA based on the sequence of RNA. This is not found in eukaryotes. No RNA is synthesized based on the sequence in proteins.

1.10 Metric Units

Metric units are used throughout the world except in the United States of America and, to some extent, in the United Kingdom. Metric units are used exclusively in science. It is, therefore, important to know the metric units. The advantage of the metric system is that everything is based on the number ten and decimals. In addition, there is an easy relationship between volume and mass with one **milliliter** (mL) of water having a mass of one **gram** (g). The metric system is about the same as **International Units** (SI) units. Metric units are shown in Textbox 1.9 and Textbox 1.10, while the history of metric units is shown in Textbox 1.11.

Textbox 1.9   Metric Units of Length, Volume, Mass, and Concentrations

What are commonly metric units of length?
The meter (**m**) is basic unit of length.
Millimeter (**mm**) is 10^{-3} m (one thousandth of a meter).
Centimeter (**cm**) is 10^{-2} m (one hundredth of a meter).
Kilometer (**km**) is 10^{3} m (one thousand meters).

What is the relationship between metric units of length and imperial units?
A cm is approximately 4/10 of an inch.
A meter is approximately 39.4 inches.
A kilometer is about 5/8 of a mile.

What are metric units of volume?
One liter (L) = 1 cubic decimeter (10 cm × 10 cm × 10 cm).
dL (deciliter) = 10^{-1} liters or 1 dL = 1 liter/10 (a tenth of a liter).
mL (milliliter) = 10^{-3} liters or 1 mL = 1 liter/1000 (a thousandth of a liter).
µL (microliter) = 10^{-6} liters or 1 µL = 1 liter/1,000,000 (a millionth of a liter).
nL (nanoliters) = 10^{-9} liters or 1 nL = 1 liter/1,000,000,000 (a billionth of a liter).
pL (picoliters) = 10^{-12} liters or 1 pL = 1 liter/1,000,000,000,000 (a trillionth of a liter).
fL (femtoliter) = 10^{-15} liters or 1 fL = 1 liter/1,000,000,000,000,000 (a quadrillionth of a liter)

What is the relationship between metric units of volume and imperial units?
One liter (L) = 2.1 pints (US).

What are units of mass?
1. kg (kilogram) = 10^{3} gram or thousand grams. g (gram).
2. mg (milligram) = 10^{-3} gram or 1 mg = 1 gram/1000 (a thousandth of a gram).
3. µL (microgram) = 10^{-6} gram or 1 µg = 1 gram /1,000,000 (a millionth of a gram).
4. nL (nanogram) = 10^{-9} gram or 1 ng = 1 gram /1,000,000,000 (a billionth of a gram).
5. pL (picogram) = 10^{-12} gram or 1 pg = 1 gram/1,000,000,000,000 (a trillionth of a gram).
6. fg (femtogram) = 10^{-15} gram or 1 fg = 1 gram/1,000,000,000,000,000 (a quadrillionth of a gram).
7. Metric ton (tonne or megagram or Mg) = 1000 (10^{3}) kilogram or thousand kilograms.

What is the relationship between metric units of mass and imperial units?
One kilogram (kg) = 2.2 lb.
One metric ton = 2205 lb.

Moles and molar
A mole is the mass of a compound in grams times the molecular weight.
A molar solution is the mass of a substance times the molecular weight per liter.

Textbox 1.10 Metric Units for Temperature

What is the metric system for temperature?
The metric system for temperature is known as degrees Celsius (°C):
The temperature at which water freezes is 0°C.
The temperature at which water boils is 100oC.
The history of the units is covered in Textbox 1.8.

What is the conversion between degrees Celsius and degrees Fahrenheit (F)?
0°C is 32°F.
20°C is 68°F.
37°C is 98°F.
100°C is 212°F.
The system of **absolute temperatures** is named after William **Kelvin**. This is an SI unit.
0° Kelvin is the lowest temperature that can be achieved.
273° Kelvin = 0°C.
373° Kelvin = 100°C.

Textbox 1.11 Interesting Point— History of Temperature Degrees

In 1724, Daniel Gabriel **Fahrenheit** proposed the temperature scale based on zero degrees being the lowest temperature he could get with an ice, salt, and water mixture and body temperature being one hundred degrees. This was, therefore, a centigrade system.

In 1742, Anders **Celsius** proposed the temperature scale based on zero degrees being the freezing point for water and one hundred degrees being the boiling point for water. This was, therefore, a centigrade system and, superficially, SI.

Today, the Fahrenheit system was tweaked to allow ready conversion between degrees F and degrees C. Today, 180 degrees Fahrenheit is equivalent to 100 degrees Celsius and body temperature of people is 37°C or 98.6°F.

Theme 1

Homeostasis via the Blood

Circulatory System 2
(Cardiovascular System)

Learning Objectives

1. To identify the main functions of the cardiovascular system (Section 2.2).
2. Understand the roles of arteries, veins, and capillaries (Section 2.3).
3. Understand heart structures and their functioning (Section 2.3 and Textbox 3.2).
4. Understand heart functioning (Section 2.4.4).
5. Understand the electric (depolarization) changes in the heart (Section 2.4.3).
6. Understand pressure changes in the heart (Section 2.4).
7. Understand what an EKG signifies (Section 2.5).
8. Understand cardiac output (Section 2.5 and Textbox 2.3).
9. Understand the relationships between cardiac output, heart rate, and stroke volume (Section 2.5 and Textbox 2.3).
10. Understand the relationships between cardiac output, mean arterial pressure (MAP), and total peripheral resistance (Section 2.5 and Textbox 2.3).
11. Understand the role of myoglobin (Section 2.6).
12. Understand types of capillaries (Section 2.8).
13. Understand how blood flow through the capillaries is regulated (Section 2.7.2).
14. Understand how lymph is moved in lymph vessels (Section 2.8).
15. Understand the difference between lymph and plasma (Section 2.8).
16. Understand the difference between serum and plasma (Section 2.9.2).
17. Understand what a hematocrit is (Section 2.9.2).
18. Understand the functioning of erythrocytes (Section 2.11.2).
19. Understand the longevity of erythrocytes (Section 2.11.3).
20. Understand the production of erythrocytes (Section 2.11.4).
21. Understand the control of erythrocytes production (Section 2.11.5).
22. Understand the breakdown of hemoglobin (Section 2.11.6).

Theme 1 Homeostasis via the Blood

23. Know the types of leukocytes (Section 2.12.1).
24. Understand the longevity of leukocytes (Section 2.12.3).
25. Production of white blood cells (Section 2.12.4).
26. Understand the function of leukocytes (Section 2.12.5).
27. Understand the structure of platelets (Section 2.13.1).
28. Understand the role of platelets (Section 2.13.3).
29. Understand the longevity of platelets (Section 2.13.4).
30. Understand the production of platelets (thrombocytes) (Section 2.13.5).
31. Understand the role of plasma proteins (Section 2.14.3).
32. Understand what transport proteins do (Section 2.14.3).
33. Understand why and how blood clots (Section 2.15).
34. Understand examples of diseases of the cardiovascular system in domestic animals (Section 2.16 and Textbox 2.6).
35. Understand examples of diseases of blood in domestic animals (Section 2.16 and Textbox 2.7).

Table of Contents

- 2.1 Introduction
- 2.2 Role of the Circulatory System
  - 2.2.1 Functions of the Cardiovascular System
- 2.3 Components of the Circulatory or Cardiovascular System
  - 2.3.1 Introduction to Components of the Circulatory System
  - 2.3.2 Arteries
  - 2.3.3 Veins
  - 2.3.4 Capillaries
  - 2.3.5 Lymph Vessels
  - 2.3.6 Double Circulation (Pulmonary Circuit and Systemic Circuit)
    - 2.3.6.1 Introduction to Double Circulation
    - 2.3.6.2 Systemic Capillaries
    - 2.3.6.3 Pulmonary Capillaries
  - 2.3.7 Coronary Blood Vessels
  - 2.3.8 Hepatic Portal Veins
- 2.4 Functioning of the Heart
  - 2.4.1 Introduction to Functioning of the Heart
  - 2.4.2 What Are the Mechanisms That Keep Blood Flowing in a Single Direction?
  - 2.4.3 What Causes the Heart to Pump?
  - 2.4.4 Intrinsic Conduction System
  - 2.4.5 What Are the Pressure Changes in the Heart?
  - 2.4.6 Science in Action: The Electrocardiogram
- 2.5 Cardiac Output
- 2.6 Importance of Oxygen to Heart Functioning
  - 2.6.1 Introduction to the Oxygen Needs of the Heart
  - 2.6.2 Role of Myoglobin
- 2.7 Capillaries
  - 2.7.1 Introduction to Capillaries
  - 2.7.2 How Is Blood Flow Through the Capillaries Regulated?
  - 2.7.3 Types of Capillaries
- 2.8 Lymph and Lymph Vessels
- 2.9 Blood
  - 2.9.1 Introduction to Blood
  - 2.9.2 What Is the Difference Between Plasma and Serum?
  - 2.9.3 What Are Hematocrit and Packed Cell Volume and How Are They Related?
- 2.10 Blood Cells/Formed Elements

2.11 Red Blood Cells (Erythrocytes)
    2.11.1 Introduction to Erythrocytes
    2.11.2 Functioning of Erythrocytes
    2.11.3 Lifespan of Erythrocytes
    2.11.4 Production of Red Blood Cells (Erythropoiesis)
    2.11.5 How Is Production of Red Blood Cells Controlled?
    2.11.6 Breakdown of Red Blood Cells
2.12 White Blood Cells (Leukocytes)
    2.12.1 Introduction to Leukocytes
    2.12.2 Number and Proportion of Leukocytes
    2.12.3 Lifespan on Leukocytes
    2.12.4 Production of White Blood Cells
    2.12.5 Function of Leukocytes
2.13 Platelets (Thrombocytes)
    2.13.1 Structure of Platelets
    2.13.2 Thrombocytes in Poultry
    2.13.3 Role of Platelets
    2.13.4 Lifespan of Platelets
    2.13.5 Production of Platelets
2.14 Plasma
    2.14.1 Introduction to Plasma
    2.14.2 What Are the Functions of the Proteins in Plasma?
    2.14.3 Plasma Proteins as Transporters
        2.14.3.1 Metal-Binding Proteins
        2.14.3.2 Vitamin-Binding Proteins
        2.14.3.3 Hormone-Binding Proteins
    2.14.4 Yolk Proteins
2.15 Blood Clotting
    2.15.1 Hemostasis
    2.15.2 Vitamin K and Blood Clotting
2.16 Diseases of Cardiovascular System in Domestic Animals

2.1. Introduction

The circulatory or **cardiovascular system** is essential to domestic animals as it is an intricate and integral system for maintaining homeostasis. It provides tissues with **oxygen** from the lungs and nutrients from the gastrointestinal tract. It takes **carbon dioxide** and other wastes away from organs. It transports hormones from endocrine cells to target tissues. It also circulates **immune cells** that monitor for foreign particles and cell abnormalities together with **antibodies**.

2.2 Role of the Circulatory System

2.2.1 Functions of the Cardiovascular System

The functions of the cardiovascular system are transport, homeostasis, and protection.

Textbox 2.1 Key Vocabulary

- **Aorta**: blood vessel from the left ventricle.
- **Arteries**: blood vessels leading from the heart.
- **Atrium**: one of the four chambers of the heart. There are two atria, right and left.
- **Capillaries**: thin blood vessels where gaseous exchange occurs.
- **Cardiac**: pertaining to the heart.
- **Circulation**: transit of blood around the body.
- **Erythrocyte**: red blood cell.
- **Heart**: pumping organ.
- **Homeostasis**: striving to achieve the optimal internal environment.
- **Leukocyte**: white blood cell.
- **Lymph**: interstitial fluid that is not picked up in the capillaries.
- **Plasma**: the fluid component of blood.
- **Pulmonary**: pertaining to the lung.
- **Systemic**: pertaining to the entire organism including all the organs except the lungs.
- **Thrombocyte**: platelets.
- **Vein**: blood vessels leading to the heart.
- **Ventricle**: one of the four chambers of the heart. There are two ventricles, right and left.

1. Transportation:
   a. Red blood cells carry oxygen in the blood to the organs and tissues of the body.
   b. Blood carries metabolic waste in the form of carbon dioxide (CO_2) from the tissues back to the lung.
   c. **Nitrogenous waste** (urea in mammals, uric acid in birds) is transported to the kidney where it is also excreted in the form of urine.
   d. **Nutrients** (glucose, amino acids, fatty acids, etc., together with micronutrients) are transported after digestion in the gastrointestinal tract or from storage in the liver and adipose tissue.
   e. Blood transports the communication signals called hormones from **endocrine glands** to various tissues.
   f. Unfortunately, the cardiovascular system also transports **pathogens** (viruses, bacteria, protozoa, and helminths), **cancer cells**, and **toxicants**.
2. Homeostatic functions of the cardiovascular system:
   a. One important regulatory role of the blood is to maintain the **blood** or **physiological pH** to within the range from 7.35 to 7.45. This is discussed under the respiratory system in Chapter 3.
   b. Blood also plays a role in the **regulation of ions**, such as sodium, potassium, and calcium.
   c. Blood plays a role in the **regulation of temperature**. While the hypothalamus is the control center for whole body temperature regulation, blood regulates localized body temperature by distributing heat from areas of heat production (e.g., digestion in the gastrointestinal system; exercise of the muscle).
   d. Blood also plays a role in the regulation of **blood volume**.
3. Provides protection through the following:
   a. Blood transports antibodies from where they are produced to a site of infection.
   b. Blood transports white blood cells (leukocytes) to a site of infection.
   c. Blood transports **platelets (thrombocytes)** to a site of damaged tissues.
   d. Moreover, blood plays a role in the **inflammation response** protecting the animal after infection.
   e. **Blood clots** to prevent blood loss.

2.3 Components of the Circulatory or Cardiovascular System

2.3.1 Introduction to Components of the Circulatory System

The circulatory or cardiovascular system consists of the heart (pumping blood) together with **blood vessels** through which **blood flows**. In addition to the heart (discussed below), there are the following blood vessels: arteries, veins, and capillaries.

2.3.2 Arteries

Arteries are blood vessels through which blood flows under pressure away from the heart. **Smooth muscle** surround arteries. When these smooth muscle contract, the blood pressure increases. When these smooth muscles relax, the blood pressure decreases.

2.3.3 Veins

Veins are blood vessels moving blood to the heart. Blood flow in veins is due to the following:

- **Residual pressure** from the heart.
- **Skeletal muscles** squeezing the veins.
- Valves preventing **back-flow** of blood.

2.3.4 Capillaries

Capillaries are very small blood vessels with thin walls. Blood passes through the capillaries and while there, gas exchange occurs, nutrients pass to tissues, hormones move to target cells or move from endocrine cells into the blood stream, and tissues loose waste products.

2.3.5 Lymph Vessels

Lymph vessels transport lymph back into the veins. Lymph is **interstitial fluid** that is not picked up in the capillaries.

2.3.6 Double Circulation (Pulmonary Circuit and Systemic Circuit)

2.3.6.1 Introduction to Double Circulation

In both mammals and birds, there is double circulation (see Figure 2.1). consisting of the following:

- **Pulmonary circulation** or **circuit** is **deoxygenated blood** from the heart passing through the lungs (where it is oxygenated) and then back to the heart.

Chapter 2 Circulatory System

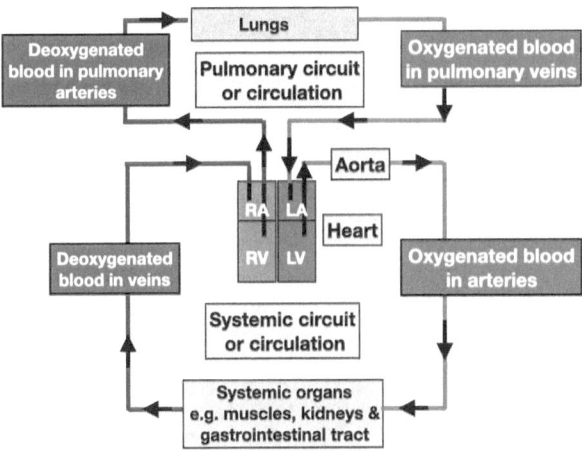

Figure 2.1 Double circulation consisting of the separate systemic circuit (or systemic circulation) and the pulmonary circuit (or pulmonary circulation).
[Key: RA = right atrium, RV = right ventricle, LA = left atrium, LV = left ventricle, → = direction of flow of blood]

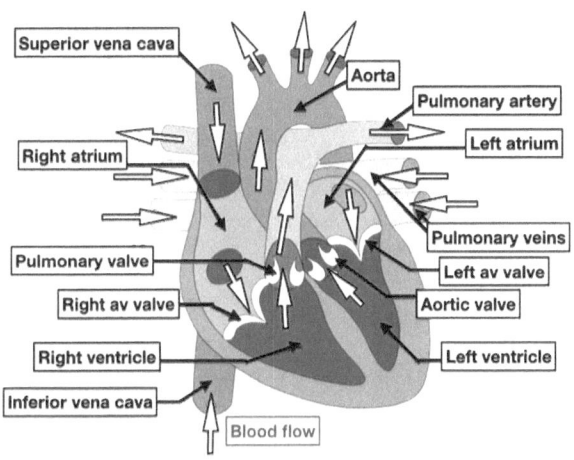

Figure 2.2 Structure of mammalian heart.
[Key: av = atrioventricular valve]
(Courtesy of Christinel Miller, CC BY-SA 4.0)

- **Systemic circulation** or **circuit** is **oxygenated blood** from the heart passing through the organs and tissues of the body (where it loses some of its oxygen) and then back to the heart.

2.3.6.2 Systemic Capillaries

Oxygen, **nutrients**, hormones, etc., pass through the thin-walled capillaries to the cells of the systemic organs. Carbon dioxide (CO_2) and other waste pass from the cells of the systemic organs through the walls **of the capillaries into** the blood.

2.3.6.3 Pulmonary Capillaries

Oxygen passes through the thin-walled capillaries from the **alveoli** of the lungs to the blood. Carbon dioxide passes from the blood through the walls of the capillaries into the alveoli of the lungs. Nutrients, hormones, etc., pass from the blood through the walls of the capillaries to the cells of the lungs.

2.3.7 Coronary Blood Vessels

While the heart chambers are intermittently filled with blood, the cardiac muscle does not get its oxygen from blood in the heart chambers. Instead, the regions of the heart get oxygen together with nutrients from capillaries that are fed by the **coronary arteries**. Blood then passes via the coronary veins to the major veins returning blood to the heart.

2.3.8 Hepatic Portal Veins

The **hepatic portal veins** transport blood draining from the **gastrointestinal tract** to the liver. This is a specialized system as it is composed of two capillary networks back-to-back. Other portal systems include the **hypophyseal portal system** and the **renal portal system** (in birds).

2.4 Functioning of the Heart

2.4.1 Introduction to Functioning of the Heart

The heart, in livestock, companion animals, and poultry species, is a set of two **pumps** that work together to **move blood**. The right side and the left side, but these two pumps essentially will work together to allow for a uniformed response. If the right side pumped faster than the left side it would cause blood to back up in the system. This pump is described as the right side or the **pulmonary circuit** (see Figure 2.1). This circuit is where blood that is going to come into the right side of the heart. Blood will enter the right atrium, flow into the **right ventricle**, and then be pushed out of the heart and go towards the lung where it is going to pick up oxygen. The second pump, **systemic circuit**, is going to be the left side of the heart,

> **Textbox 2.2 Definitions Related to the Heart**
>
> **Major Structures**
>
> - **Atrium (plural atria)**: upper chamber of the heart that is thin walled.
> - **Ventricle**: lower chamber of the heart that is thick walled.
>
> **Electrical Stimulation**
>
> - **Autorhythmic cells**: pacemaker cells that regulate heart rate.
> - **Cardiac contractile cells**: cells in the heart that contract to cause the pumping action of the heart.
> - **Action potential**: an electrical signal that contains a resting membrane potential, depolarized, repolarized, and hyperpolarized state.
> - **Intrinsic conduction system**: the electrical signal of the heart that regulates the cardiac contractile cells and is composed of autorhythmic cells.
>
> **Structures Related to Coordination of the Heart**
>
> - **Sinoatrial node (SA node** or **pacemaker)**: functions to initiate contraction of the atria of the heart by electrical depolarization of the atrial cardiac cells.
> - **Atrioventricular node (AV node)**: functions to stimulate contraction of the ventricles of the heart.
> - **Bundle of His**: part of the system that causes the ventricles to contact with a time delay compared to the atria. Electrical depolarization passes from the atrioventricular node, along the bundle of His and the left and right branches to the Purkinje fibers.
> - **Purkinje fibers**: part of the electrical system of the heart. Electrical depolarization passes along the Purkinje fibers to their terminals at cardiac muscle cells of the ventricles, inducing the cardiac cells to contract in a synchronized manner.
>
> **Heart Functioning**
>
> - **Cardiac output (CO)**: the amount of blood ejected by each ventricle per minute. Cardiac output is equal to stroke volume times heart rate.
> - **Heart rate**: the number of beats of the heart per minute.
> - **Stroke volume**: the amount of blood ejected from the ventricle with each beat.
> - **Isovolumetric contraction**: contraction with no change in the volume.

and this is where the oxygen-rich blood from the lung has entered the left atrium, then the **left ventricle**. Then blood is pumped out of the heart through the systemic circuit (see Figure 2.1).

The heart is a **four-chambered pump** consisting of two atria (right and left) and two ventricles (right and left) with **valves** that are critical to the functioning of the heart (see Figure 2.2). Figure 2.2 shows blood flowing from the major **systemic veins** (**superior vena cava** and **inferior vena cava**) into the right atrium and then to the right ventricle through the **tricuspid** valve. The right ventricle then pumps the deoxygenated blood to the lungs via the **pulmonary arteries**. Oxygenated blood returns to the left atrium via the **pulmonary veins**. It is pumped by the left atrium to the left ventricle via the **mitral valve**. The left ventricle pumps the blood to the aorta via the aortic valve.

2.4.2 What are the Mechanisms that Keep Blood Flowing in a Single Direction?

The four valves are critically important to the functioning of the heart:

- **Right atrioventricular valve** or **tricuspid valve**—between the right atrium and the right ventricle.
- **Left atrioventricular** valve or **mitral valve**—between the left atrium and the left ventricle.
- **Pulmonary semilunar valve**—between the right ventricle and pulmonary circulation.
- **Aortic semilunar valve**—between the left ventricle and systemic circulation.

Atrioventricular valves (AV valves), more specifically the tricuspid valve and **bicuspid valve**, act to separate the

Table 2.1 Examples of blood pressure in horses.

Blood pressure (BP) parameter	Blood pressure (mm Hg)*
Pulmonary BP	40
Systolic BP	135
Diastolic BP	100
Mean arterial pressure (MAP)	117

*A range is usually used but for simplification, an average is shown.

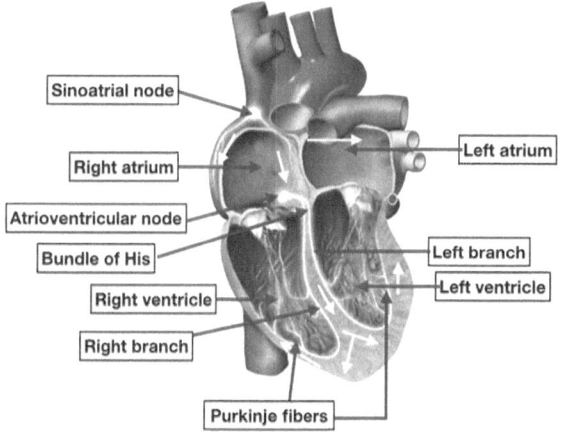

Figure 2.3 Electrical stimulation of the heart chambers to contract.
[Key: → = pathway of depolarization]
(Adapted from Bruce Blaus, CC BY-SA 4.0)

atria from the ventricle and prevent backflow of blood from the ventricle to either the left or right atrium. These values do this by forming a physical barrier during the contraction of the ventricles. During a contraction of the ventricles there is increased pressure (see Figure 2.4) and the atrioventricular valves are forced up into the atrium, forming a plug. There are specific attachment points called **chords of tendineae** (see Figure 2.3) which prevent the valves from inverting on themselves into the atrium and, consequently, allowing blood to flow back into the atrium.

Semilunar valves are the other part of this mechanism that prevents blood from flowing back into the ventricles from the **pulmonary artery** and the aorta (see Figure 2.4). These valves act in a way similar to turnstiles, in that blood is forced through the semilunar valves and they close after, preventing blood from flowing back into the ventricle. The semilunar valves also ensure **blood pressure** does not drop to zero when the ventricles relax, and instead the blood pressure drops to the diastolic blood pressure (see Figure 2.4). (For an example of blood pressures in the horse see Table 2.1).

A **wave of depolarization** passes from the sinoatrial node throughout the atria, causing them to contract. A wave of depolarization passes from the atrioventricular node along the two branches of the bundle of His to the Purkinje fibers to the cardiac muscle cells of the ventricle, causing the ventricles to contract.

2.4.3 What Causes the Heart to Pump?

Pumping of the heart is controlled by two different types of **cardiomyocytes**: the **contractile cardiac cells** and the autorhythmic cells. The contractile cardiac cells represent 99% of cardiomyocytes. These contract as a single unit which lead to the pumping action in the heart. Unlike other myocytes, the **cardiac contractile cells** have an

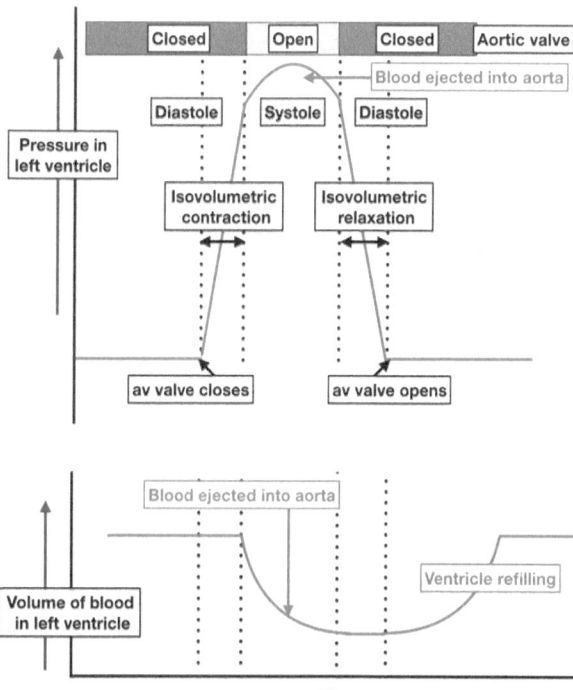

Figure 2.4 Changes in pressure and volume of blood in the left ventricle through the cardiac cycle.

extended absolute **refractory** period (the period of time where cells cannot respond to another stimulus). In contractile cardiac cells, this extended absolute refractory period is due to an **influx of calcium** at the start of **repolarization** (see Figure 2.3). This allows the contractile cardiac cells to enter a period of **relaxation**. The second type of cardiomyocytes are the autorhythmic cells (auto meaning *self*, rhythmic meaning a *rhythm* or *pace*). These represent 1% of cardiomyocytes. These are the pacemaker cells and form part of the intrinsic conduction system that leads the coordinated stimulation of the contractile cardiac cells.

2.4.4 Intrinsic Conduction System

The sinoatrial node (or pacemaker) is located in the wall of the right atrium (see Figure 2.3). The sinoatrial node initiates the process causing a wave of depolarization and, consequently, contraction of the atria (see Figure 2.4). The sinoatrial node only sends a signal to the atrium as there is **insulation** between the atria and ventricles. Contraction of the right atrium forces blood into the right ventricle while contraction of the left atrium forces blood into the left ventricle. Blood cannot flow back because of AV valves. The wave of depolarization in the atria stimulates the atrioventricular node (AV node) (see Figure 2.4). Subsequently, a wave of depolarization passes along the bundle of His, the right and left branches, and Purkinje fibers (see Figure 2.4). This then causes the ventricles to contact. **Ventricular contraction** starts at the apex or bottom of the heart allowing blood to be pushed or ejected through either the pulmonary artery (right ventricle) and aorta (left ventricle) (see Figure 2.3). The time from stimulation of the AV node and passage of the wave of depolarization along the bundle of His, the right and left branches, and Purkinje fibers creates a delay between contractions of the atria and that of the ventricles. The entire **cardiac cycle** takes about 0.22 seconds.

2.4.5 What are the Pressure Changes in the Heart?

The cardiac cycle for the left ventricle is summarized in Figure 2.4. The stages of the cycle for the left ventricle are the following:

1. Blood passes from the left atrium to the left ventricle through the left atrioventricular or mitral valve. The aortic (semilunar) valve is closed.
2. The ventricle begins to contract in an **isovolumetric manner** (the volume remains the same). The left atrioventricular or mitral valve closes. The aortic valve (semilunar) remains closed.
3. Pressure in the ventricle builds up until a critical pressure is reached such that the **aortic valve** (semilunar valve) opens (this is **systole**).
4. Blood squirts from the ventricle into the aorta causing the pressure to drop.
5. When a critical pressure is reached, the **aortic valve** (semilunar) closes (This is **diastole**).
6. The ventricle relaxes in an isovolumetric manner.

An example of the blood pressures for horses during systole and diastole are shown in Table 2.1.

The cardiac cycle for the **right ventricle** consists of the following:

1. Blood passes from the right atrium to the right ventricle through the right atrioventricular valve. The pulmonary valve is closed.
2. The ventricle begins to contract in an isovolumetric manner (the volume remains the same). The right atrioventricular valve closes. The pulmonary valve remains closed.
3. Pressure in the ventricle builds up until a critical pressure is reached such that the pulmonary valve opens. **Note**: The critical pressure is lower than for the left ventricle.
4. Blood squirts from the ventricle into the pulmonary arteries causing the pressure to drop.
5. When a critical pressure is reached, the pulmonary valve closes.
6. The right ventricle relaxes in an isovolumetric manner.

An example of the pulmonary blood pressure is shown in Table 2.1. **Note**: the pulmonary blood pressure is much lower than in the systemic circuit during either systole or diastole.

2.4.6 Science in Action: The Electrocardiogram

An **electrocardiograph** (ECG or EKG with the K being the German form of *cardio*) depicts the electrical stimulation of the heart or the intrinsic conduction (see Figure 2.5). The EKG starts with the **P wave**. This reflects **atrial depolarization** after stimulation from sinoatrial node (and consequently contraction). Next comes the **QRS complex**. This sharp signal reflects the electronic signal stimulating the ventricle to contract together with the repolarization of the atria. The **S and T waves** reflect repolarization of the ventricles.

EKGs are commonly used by veterinarians, particularly with horses. It has been used in research, for instance,

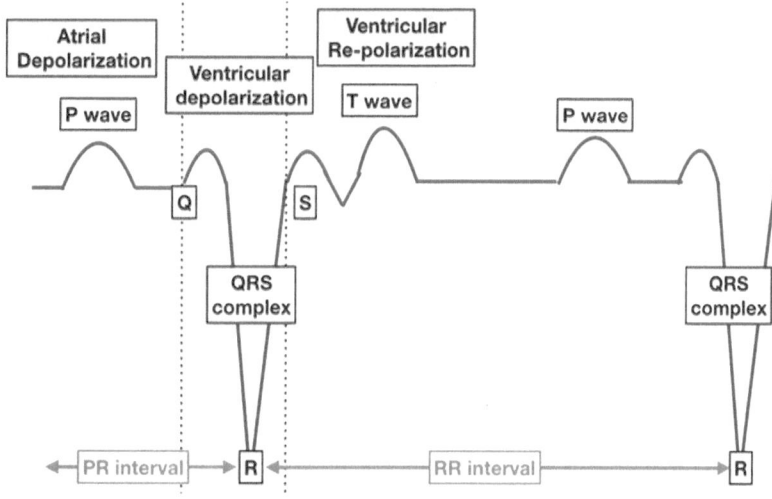

Figure 2.5 Electrocardiogram (EKG) of horse. Note that the QRS wave looks different from that in people. This is due to the positioning of the electrodes. (Adapted from Arioneo)

to identify pigs with the halothane tolerance gene. These pigs have increased risks of low-quality meat. This is **pale, soft, and exudative (PSE)** meat which is not well received by consumers.

2.5 Cardiac Output

The heart pumps blood either around the body (in the systemic circuit) or through the lungs (in the pulmonary circuit). Cardiac output is the quantity of blood pumped per minute by a ventricle.

Cardiac output (CO) = Heart rate × Stroke volume (in liters)

Also see Textbox 2.3 for consequences of cardiac output.

The units of cardiac output are liters min.^{-1} (or liters per minute). Heart rate is the number of beats (contractions of left ventricle) per minute. The stroke volume is the volume pumped by the left ventricle in a single contraction. Examples of cardiac output in domestic animal are shown in Table 2.2. below.

The quantity of blood by the left ventricle in a single contract is the same of that pumped by the right ventricle. In cattle, the heart is pumping about 33 liters (70 pints or 8.75 gallons) of blood around the body per minute.

Cardiac output is also related to mean arterial pressure and peripheral resistance (see Textbox 2.3). Cardiac output is increased due to increases in both the heart rate and

Table 2.2 Cardiac output in selected domestic animals.

Animal	Cardiac output liters per minute
Cats	0.2
Dogs	3.4
Cattle	33
Horses	36
Broiler chicken	0.4
Laying hen	0.25

stroke volume. A consequence of the increase in cardiac output seen in Textbox 2.4.

Heart rate can be altered by hormones, ionic changes, age, exercise, sex, and body temperature. For instance, heart rate is increased during physical activity by the release of **norepinephrine** from **sympathetic terminals** near the sinoatrial node (pacemaker). Heart rate is increased also by the fight or flight hormone, **epinephrine**. Conversely, heart rate is decreased by **acetylcholine** released from parasympathetic terminals near the sinoatrial node. Changes in intra- and extracellular ion concentrations can alter heart rate and they can affect extended absolute refractory period or pacemaking potential in cardiomyocytes.

Textbox 2.3 Cardiac Output

Cardiac output is also equal to mean arterial pressure (MAP) divided by total peripheral resistance (TPR) [also known as systemic vascular resistance].

$$\text{Cardiac output} = \frac{\text{Mean arterial pressure (MAP)}}{\text{Total peripheral resistance (TPR)}}$$

Definitions

- **Systolic blood pressure (BP)** is arterial blood pressure when the left ventricle is contracting and the aortic valve is *open*.
- **Diastolic blood pressure** is arterial blood pressure when the left ventricle is either contracting or relaxing but the aortic (semilunar) valve is *shut*.
- **Mean arterial pressure (MAP)** is the average blood pressure in the systemic circuit over the cardiac cycle. MAP = diastolic BP + 1/3(systolic BP − diastolic BP).
- **Pulse pressure** is equal to systolic minus diastolic blood pressure

Cardiac output can also be calculated by Fick's principle by the amount of oxygen used by an animal (VO_2) multiplied by the difference in oxygen in atrial and venous systemic blood (a − v O_2 difference).

$$CO = \frac{VO_2}{(a - v\ O_2\ \text{difference})}$$

During exercise, cardiac output is increased. While there is a small increase in MAP, the biggest change is a decrease in total peripheral resistance.

Textbox 2.4 Interesting Factoid

Cardiac Parameters in Racehorses
There are marked effects of exercise on cardiovascular parameters in horses (see Table 2.3).

Figure 2.6 A racehorse galloping. (Courtesy of Jeff Kubina, CC BY-SA 2.0)

Table 2.3 Cardiac parameters in racehorses.

	Basal	During exercise (galloping)
Heart rate (beats per minute)	30–40	Up 6.3 fold
Cardiac output (liters per minute)	30	Up 10 fold
Blood pressure (BP)		
Systolic mm mercury	115	Up 80%
Diastolic mm mercury	85	Up 40%

Stroke volume is influenced by 3 factors: preload, contractility, and afterload. Preload is the amount of blood that the ventricle can hold when cardiomyocytes are not stretched. **Contractility** is the force with which the ventricle can contract and eject blood in the stroke volume at the prevailing **afterload** (**systolic arterial BP**) and **preload** (end of diastolic volume). Afterload is the pressure required to eject blood out of the ventricle.

2.6 Importance of Oxygen to Heart Functioning

2.6.1 Introduction to the Oxygen Needs of the Heart

Oxygen is critically important to cardiac functioning as the myocardial cells only function **aerobically**.

There must be high concentrations of oxygen close to the mitochondria of the myocardial cells.

Cardiac oxygen demand has to be met by **cardiac oxygen supply**. As cardiac output increases, such as with exercise, there is increased oxygen demand, which has to be met by increased oxygen supply from the coronary arteries. There are three factors that influence cardiac oxygen demand:

- Heart rate.
- Contractility.
- **Ventricular-wall tension.**

The primary source of oxygen is oxygenated hemoglobin in cardiac capillaries (the role of hemoglobin in transporting oxygen is discussed Section 2.11.2). In addition, myoglobin acts as a temporary store of oxygen (see below).

2.6.2 Role of Myoglobin

Myoglobin acts as a temporary storage of oxygen. Myoglobin is also found in striated muscle, including cardiac muscle. It is not normally found in blood but can be following a muscle injury. Myoglobin is a protein with a single **ferrous iron** containing **heme unit**. Myoglobin binds oxygen reversibly.

$$\text{Myoglobin} + O_2 \leftrightarrows \text{Myoglobin} \sim O_2$$

2.7 Capillaries

2.7.1 Introduction to capillaries

Blood passes from the arteriole through the capillary to a venule (Figure 2.7). Oxygen, carbon dioxide, nutrients, etc., pass to the cells through the interstitial fluid surrounding them (Figure 2.8).

As the blood passes from the arteriole through the capillary to the venule, the following occurs:

- There is a drop in the **hydrostatic pressure** (the pressure derived from the contraction of the heart).
- There is **pressure** of the interstitial fluid pushing fluid back into the capillary.
- Some plasma (without the proteins) passes into the interstitial fluid in the tissues.
- **Colloidal osmotic pressure** bringing fluid from the interstitial fluid into the plasma.

2.7.2 How Is Blood Flow Through the Capillaries Regulated?

Movement of blood into and out of the capillary bed is regulated by valves or sphincters. When these sphincters

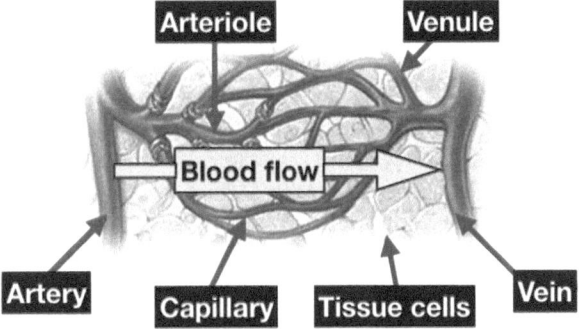

Figure 2.7 Schematic of capillary showing blood flowing from the arteriole, through the capillary to a venule. (Courtesy National Cancer Center)

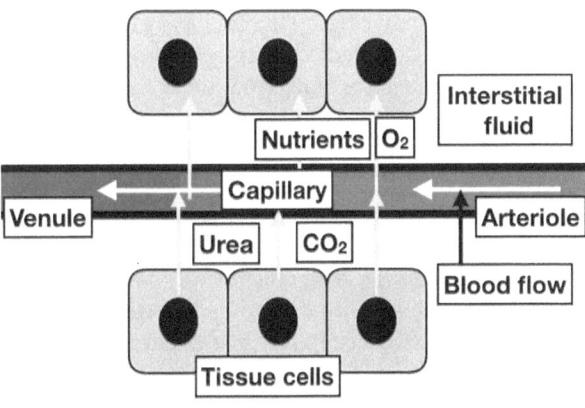

Figure 2.8 Schematic of capillary showing transport of oxygen and nutrients from the capillary through the interstitial fluid to the cells and carbon dioxide and other waste in the opposite direction.

are open, blood will flow into this capillary bed and allow for this exchange of respiratory gases, nutrients pass to the cells, nitrogenous waste passes from the cells, etc. If gases, hormones, nutrients, and metabolic waste reach an equilibrium and we do not need more exchange, the sphincters will be closed and blood to bypass these areas.

2.7.3 Types of Capillaries

There are three different types of capillaries:

- **Continuous capillaries** (found in the skin, muscles, lungs, and the central nervous system). These are the *least permeable*, with nutrients passing through small openings between capillary cells.
- **Fenestrated capillaries** (found in **areas of absorption** including kidney tubules and small intestine).

These have fenestrations or opening that allow ready movement of small molecules such as nutrients into or out of the capillaries at a rapid rate.
- **Sinusoid capillaries** (found in the bone marrow, liver, spleen, and adrenal medulla). These are the *most permeable* allowing for very rapid movement of ions and constituents across these blood vessels. Moreover, they allow for movement of larger items, such as red blood cells or white blood cells, from areas where they are produced into the circulation where they evoke their actions.

2.8 Lymph and Lymph Vessels

In capillaries, the volume of the arterial blood entering is greater than that of the venous blood leaving. About 90% of the volume entering the capillary exits in the venous drainage. But what of the remaining 10%? There is inadequate drainage of fluids from the interstitial spaces into the veins. This excess fluid close to capillaries is effectively scavenged and placed into the lymph vessels as lymph.

Lymph is identical to interstitial fluid and also similar in composition to blood plasma but with a lower concentration of proteins (i.e., water together with ions and some proteins). In the absence of lymphatic system, or if it is blocked, fluid will accumulate in the interstitial spaces causing edema or swelling.

The lymphatic system can be viewed as **unidirectional** starting at either the capillaries or small intestine. The lymph flows in a series of **lymphatic vessels**. The initial lymph vessels have thin non-muscular walls. Lymph is transported in **lymphatic ducts** then **lymphatic trunks**. Ultimately lymph drains into the **right and left subclavian veins**, returning the constituents of lymph into the blood stream. The subclavian veins, in turn, flow into the vena cava. The lacteals drain the small intestine. Absorbed lipids, particularly **triacylglycerides**, are converted to **chylomicrons** in the enterocytes in the small intestine. The chylomicrons transit into the lacteals and are transported to the liver.

Lymph is moved in the lymph vessels due to the following:

- **Intrinsic contractions of smooth muscle** in the collecting lymphatics in a regular "cardiac-like" contraction.
- Contraction of skeletal muscles squeezing the lymph vessels.
- Some lymph passes to or from the blood in the lymph nodes.

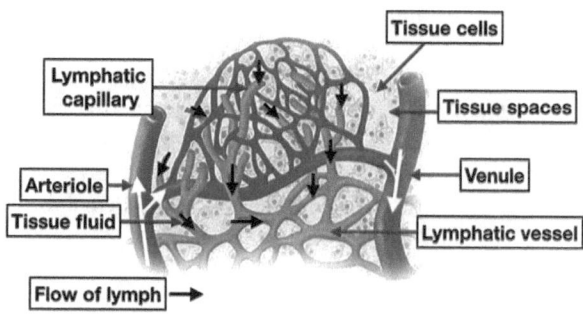

Figure 2.9 Schematic of capillary showing lymph vessels draining excess fluid in a capillary. (Adapted from National Cancer Center)

Lymph vessels have **valves** in livestock, companion animals, and other mammals. These prevent back flow of lymph. Moreover, lymph is filtered by passage through the **lymph nodes**. In contrast, in poultry and other birds, there do not appear to be either valves in the lymph vessels or lymph nodes.

2.9 Blood

2.9.1 Introduction to Blood

Blood is composed of the following:
- Cells or formed elements.
- A liquid called plasma.

This is easily seen when blood is taken and mixed with an anticoagulant and then centrifuged. The red blood cells go to the bottom of the tube, with white blood cells above them and plasma on top (see Figure 2.10).

2.9.2 What is the Difference Between Plasma and Serum?

Plasma is the liquid in blood. It can be obtained by taking a blood sample and mixing it with an **anticoagulant** to prevent blood clotting. **Serum** is the fluid released from blood that has clotted and the clot is retracting. Serum and plasma have similar compositions except that serum lacks the major clotting protein, fibrinogen.

2.9.3 What are Hematocrit and Packed Cell Volume and How are they Related?

When blood is taken into a tube containing heparin or another anticoagulant (see Textbox 2.5), it can be

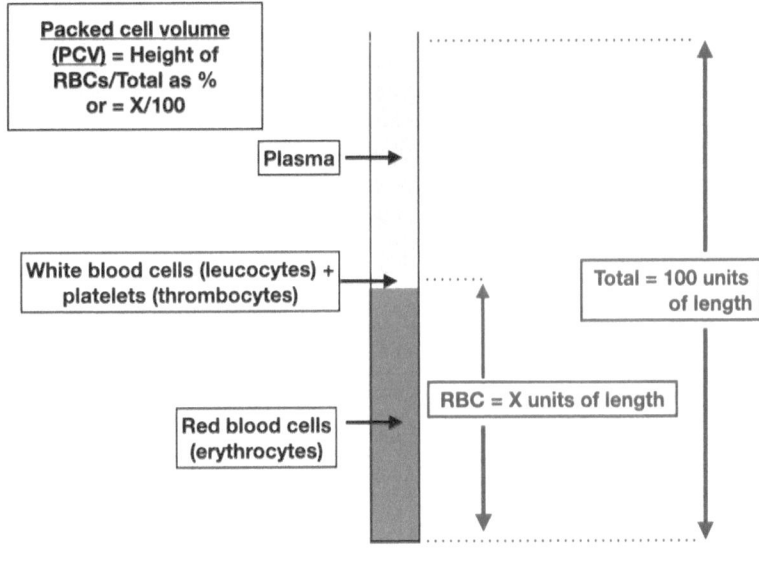

Figure 2.10 Schematic of PCV as a measure of erythrocyte status. Blood is collected with an anticoagulant and then centrifuged so that the cells fall to the bottom of the hematocrit tube.

centrifuged in a **hematocrit tube** or **microhematocrit tube** (Figure 2.10). The percentage of red blood cells is the **packed cell volume (PCV)** (see Figure 2.10). Another indicator of the volume of erythrocytes is the **hematocrit (Hct)**. This index is calculated from the **number of erythrocytes per unit volume** (RBC) and the average volume of erythrocytes or **mean corpuscular volume (MCV)**.

$$Hct = MCV \times RBC/10$$

Table 2.4 includes the average PCV and MCV in domestic animals and adult humans for comparison. The average PCV for domesticated mammals is about 38%. The MCV of domesticated mammals is about 50 fL, while that of chickens is more than twice as large. For completeness, Table 2.4 also includes average numbers of the blood concentrations of hemoglobin and mean corpuscular hemoglobin (MCH). Textbox 2.6 provide definitions of some blood parameters. These are also discussed in Section 2.11.

Hematocrits or PCVs are useful for humans and domestic animals (Figure 2.11 and Textbox 2.5). Low PCVs are indicative of inadequate amounts of iron in the diet, diseases, or blood loss from injury or parasites.

Textbox 2.5 Definitions Related to Erythrocytes and Hemoglobin

- **Blood concentrations of hemoglobin** is determined chemically from a blood sample taken into a tube containing anticoagulant.
- **Mean corpuscular hemoglobin (MCH)** is the average amount (mass) of hemoglobin in each erythrocyte.
- **Mean corpuscular volume (MCV)** is the average volume of each erythrocyte (also see Section 2.9.3).
- **Packed cell volume (PCV)** is determined directly by centrifuging blood containing anticoagulant in a hematocrit tube or microhematocrit tube (also see Section 2.9.3).
- **Hematocrit (Hct)** is the product of MCV and number of erythrocytes per unit volume (RBC) divided by 10 (also see Section 2.9.3).

$$Hct = MCV \times RBC$$

Table 2.4 Packed cell volume, hemoglobin, mean corpuscular volume (MCV), and mean corpuscular hemoglobin (MCH) in domestic animals, with ranges for men and women shown for comparison.

Species	PCV %	Hemoglobin g per dL (g dL-1)	MCV fL	MCH pg
Cat	46	12.6	47	15
Cattle	37	11.5	50	14
Dog	37	15.4	71	24
Horse	35	11.5	43	16
Pig	35	13.1	59	19
Sheep	40	13.0	34	10
Chicken	44	10.1	147	32
Women	35–45	12–15	80–94	27–33
Men	38–49	14–17	80–94	27–33

Key: PCV = packed cell volume; MCV = mean corpuscular volume; MCH = mean corpuscular hemoglobin

Textbox 2.6 Blood Tubes

Blood tubes have different tube color tops to indicate what is in the tube. If the blood tube has a red tube, it means the tube does not contain any chemicals and will allow the blood to clot and separate into blood cells and serum. A tube with a purple or lavender top tube contains the anticoagulant potassium EDTA (ethylenediaminetetraacetic acid) (see Figure 2.11). A tube with a green top contains another anticoagulant, **sodium heparin** (see Figure 2.11). When centrifuged blood with an anticoagulant, separates into a red blood cell, white blood cell, and plasma fraction.

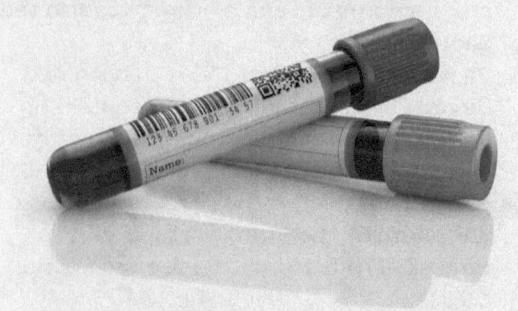

Figure 2.11 Blood tubes with different color tops indicating what is in the tube (anticoagulant or not). (Shutterstock/Zakharevych Vladyslav)

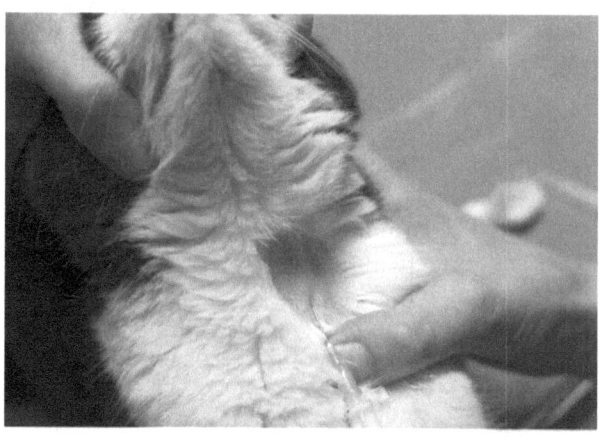

Figure 2.12 Blood sample being taken from a cat. (Shutterstock/Henk Vrieselaar)

2.10 Blood Cells/Formed Elements

There are three types of blood cells/formed elements:

1. Erythrocytes or red blood cells (see Figure 2.13).
2. Leukocytes or white blood cells (see Figure 2.14).
3. Thrombocytes or platelets (see Figure 2.15).

The structures and functions of the erythrocytes, leukocytes, and thrombocytes are shown in, respectively, Tables 2.5 and 2.6.

2.11 Red Blood Cells (Erythrocytes)

2.11.1 Introduction to Erythrocytes

Erythrocytes play a critical in the transport of respiratory gases. This is also covered under respiration in Chapter 3. Textbox 2.6 shows definitions related to erythrocytes.

Information on the following characteristics of erythrocytes is provided in tables: structure in Table 2.5, functioning in Table 2.6, and quantitative parameters in domestic animals in Tables 2.4 and 2.7.

2.11.2 Functioning of Erythrocytes

Oxygen is transported bound reversibly to the protein, hemoglobin, in the red blood cells. Textbox 2.6 shows definitions related to hemoglobin.

Figure 2.13 Mammalian red blood cells showing structure as bi-concave disks filled with hemoglobin. In birds, erythrocytes are spheroids containing nuclei and other intracellular organelles and filled with hemoglobin. (Shutterstock/Kateryna Kon)

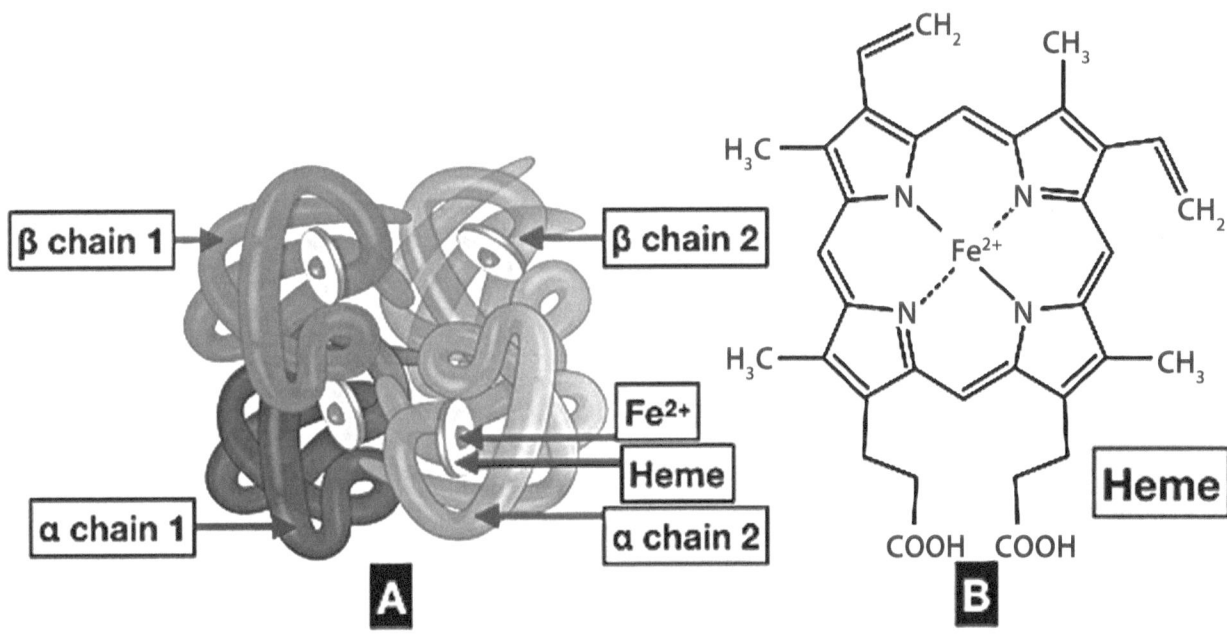

Figure 2.14 (A) Structure of hemoglobin and (B) the heme group (containing ferrous iron Fe^{2+}). (Adapted from OpenStax College, Wikimedia Commons)

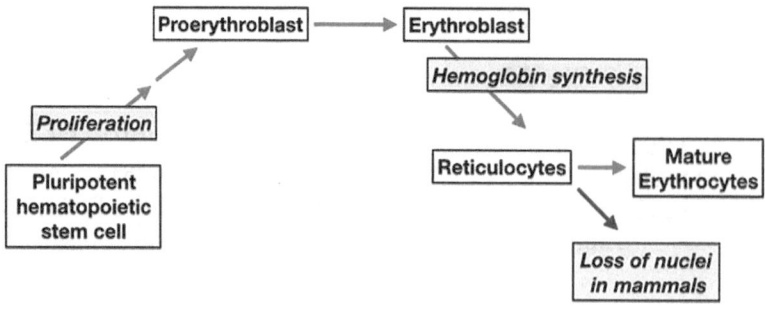

Figure 2.15 Erythropoiesis (production of red blood cells).

Table 2.5 Structures of cells and cell fragments in blood (formed elements).

	Domestic Mammals (Cats, Cattle, Dogs, Horses, Pigs, Sheep Etc.)	Poultry (Chickens, Ducks, Geese, Turkeys Etc.)
Erythrocytes (Red Blood Cells)	Biconcave, hemoglobin filled, no-nucleus or other organelles (mitochondria, endoplasmic reticulum) present	Ellipsoid, hemoglobin filled, nucleus or other organelles (mitochondria, endoplasmic reticulum) present
Leukocytes (White Blood Cells)		
Neutrophils/Heterophils (A Phagocytic Type of Granulocytes)	Neutrophils	Heterophils
	Multi-lobed nucleus with granules in cytoplasm	
Lymphocytes (A Type of Agranulocytes)	Compact ovoid nucleus, little cytoplasm.	
Basophils (A Type of Granulocytes)	Takes up/stained by the blue/purple dye—hematoxylin. Multi-lobed nucleus with granules in considerable cytoplasm.	
Eosinophils (A Type of Granulocytes)	Takes up/stained by the red dye—eosin. U-shaped nucleus with granules in considerable cytoplasm.	
Macrophage (A Phagocytic Type of Agranulocytes)	Ovoid nucleus. Marked cytoplasm contains vacuoles, mitochondria, and some granules.	
Monocytes (A Phagocytic Type of Agranulocytes That Matures Into Macrophage)	Single compact kidney-shaped nucleus, little cytoplasm.	
Thrombocytes (Platelets)	Small cell fragments each with cell membrane but no nucleus. Contains mitochondria, microtubules, and granules that can be released.	Cells with nucleus or other organelles (mitochondria, endoplasmic reticulum) present. Considerable cytoplasm contains granules that can be released.

Table 2.6 Functions of cells and cell fragments in blood (formed elements).

Blood cells	Function
Erythrocytes (red blood cells)	Transport of oxygen bound to hemoglobin plus transport of carbon dioxide.
Leukocytes (white blood cells)	
Neutrophils/Heterophils	Phagocytosis—engulfing and destroying bacteria and fungi.
Lymphocytes	Three types: • B lymphocytes—produce antibodies, for example, against viruses and bacteria. • T lymphocytes—killing virus infected cells, regulating immune response, producing cytokines. • Natural killer cells (NK cells)—killing pathogens.
Macrophage	• Phagocytosis—engulfing and destroying foreign organisms such as bacteria. • Initiating immune responses.
Monocytes	Migrate into tissue where they develop into either macrophage that engulf (phagocytosis) and destroy bacteria and dead tissue fragments or dendritic cells that present antigens to lymphocytes.
Eosinophils	• Plays a role in fighting parasitic infections. • Release factors such as cytokines to induce inflammation. • Receptors interact with pathogen-associated molecular patterns (PAMPs).
Basophils	Combating infections by: • Regulating immune functions. • Releasing enzymes including heparin (preventing blood clotting). • Releasing the vasodilator, histamine.
Thrombocytes (platelets)	Initiates clotting of blood.

Table 2.7 Numbers of erythrocytes, leukocytes, and platelets/thrombocytes per unit volume in blood of domestic animals with ranges for men and women shown for comparison.

Species	Erythrocytes $\times 10^6$ per µL	Leukocytes $\times 10^3$ per µL	Thrombocytes/Platelets $\times 10^3$ per µL
Cat	7.5	12.5	550
Cattle	7.5	8.0	450
Dog	6.4	9.5	416
Horse	8.2	8.3	186
Pig	6.5	16.5	350
Sheep	12.0	6.0	950
Chicken	3.2	25.5	34.4
Women	4.2–5.4	4.5–11.0	150–450
Men	4.7–6.1	4.5–11.0	150–450

Hemoglobin consists of the following (Figure 2.14):

- **Four protein sub-units** (2 α and 2 β sub-units).
- **Four heme sub-units** each containing a ferrous iron.

Carbon dioxide is transported as **bicarbonate** in both the plasma and erythrocytes. The enzyme, **carbonic anhydrase**, catalyzing carbon dioxide combining with water to form carbonic acid close to the tissues (and in the opposite direction in pulmonary capillaries). The carbonic acid rapidly dissociates to bicarbonate plus a proton (hydrogen ion).

$$CO_2 + H_2O \leftrightarrows H_2CO_3 \leftrightarrows HCO_3 + H^+$$

Carbonic anhydrase

2.11.3 Lifespan of Erythrocytes

Erythrocytes have a finite lifespan in the circulation. Lifespans of erythrocytes vary by species as can be seen from the following:

- Cats: 70 days.
- Cattle: 130–160 days.
- Dogs: 110–120 days.
- Horses: 140–150 days.
- Pigs: 86 days.
- Sheep: 120 days.
- Chicken: 35 days.
- Humans: 120.

Surprisingly and despite having a nucleus and other intracellular organelles, avian erythrocytes have a shorter lifespan than mammalian erythrocytes.

2.11.4 Production of Red Blood Cells (Erythropoiesis)

In mammals and birds, red blood cells are made in the bone marrow in a process called **erythropoiesis**. This is summarized in Figure 2.15.

2.11.5 How is Production of Red Blood Cells Controlled?

When blood is lost (e.g., by injury) or if the animal is moved to high altitudes, the amount of oxygen in the blood goes down. To address this, the animal produces more red blood cells, but how? Kidney cells sense low oxygen in the blood and release the hormone **erythropoietin** (see Figure 2.16). This travels in the blood to the bone marrow where it stimulates the production of new red blood cells (see Figure 2.16).

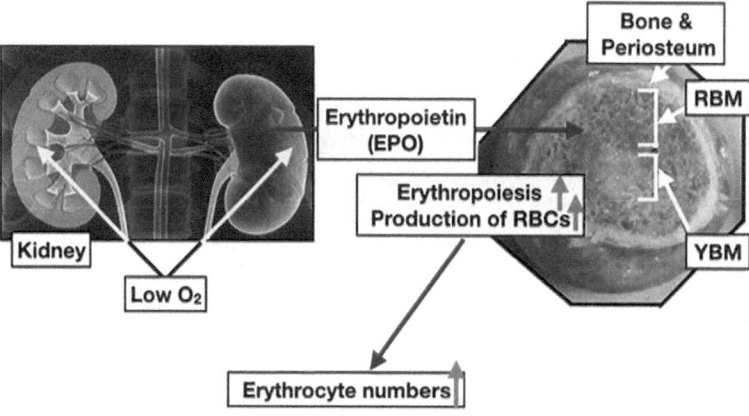

Figure 2.16 Control of the production of red blood cells by the hormone erythropoietin (EPO) in response to low oxygen tension in the blood.
[Key: RBM = red bone barrow, YBM = yellow bone marrow]
(Image of kidney courtesy of the Centers for Disease Control)

2.11.6 Breakdown of Red Blood Cells

Old erythrocytes are removed from the circulation by filtering in the spleen and **phagocytosis** by **macrophage**. The old cells are identified by exposure of proteins on the cell membrane and/or the erythrocytes becoming deformed. The old red blood cells are engulfed by macrophage by phagocytosis (Figure 2.17; Phagocytosis is described in Chapter 9, Section 9.6).

The erythrocytes are broken down in the **phagolysosome**. Hemoglobin is broken down, generating amino acids and iron (ferrous ions) that release into the blood and are then available to be re-used. The heme units are degraded to a green bile pigment, **biliverdin**. Biliverdin is released into the blood stream or converted to a red/orange bile pigment, **bilirubin**. Both biliverdin and bilirubin travel from the spleen to the liver in the blood stream and are secreted into the **bile** by the liver (Figure 2.17). Bile is released into the small intestine and the bile pigments give the feces its color.

2.12 White Blood Cells (Leukocytes)

2.12.1 Introduction to Leukocytes

Leukocytes contain nuclei and other cellular organelles. Some leukocytes contain granules that can be released from the cell in the process of secretion. These are called **granulocytes**. Other leukocytes do not contain granules and are called **agranulocytes**.

There are six types of white blood cells:

- **Neutrophils** (called heterophils in poultry): These granulocytes have a polymorphic nucleus and multiple secretory granules. These are phagocytic cells.
- **Lymphocytes**: These have no granules, little cytoplasm, and a spheroid nucleus.
- **Macrophage**: These have ovoid nuclei and substantial cytoplasm.
- **Monocytes**: These have no granules. These are phagocytic cells that phagocytose pathogens. Moreover, they can be transformed into macrophage.
- **Eosinophils**: These granulocytes have a polymorphic nucleus and multiple secretory granules. They are stained by the dye eosin and appear red.
- **Basophils**: These granulocytes have a polymorphic nucleus and multiple secretory granules. They are stained by basic dyes and appear blue.

The structure of five types of leukocytes is shown in Figure 2.18 and Table 2.5.

Phagocytic cells are cells that phagocytose pathogens and tissue debris. The process of phagocytosis is covered in Section 9.6 in Chapter 9.

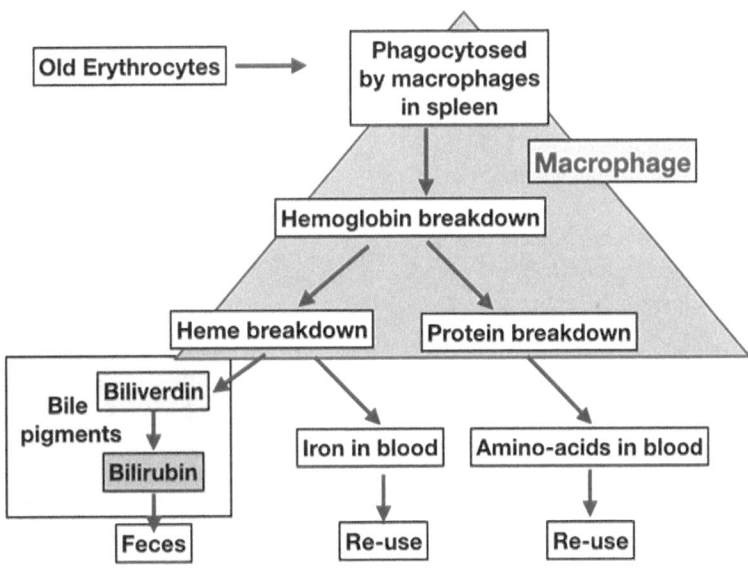

Figure 2.17 The processing of old erythrocytes by macrophage.

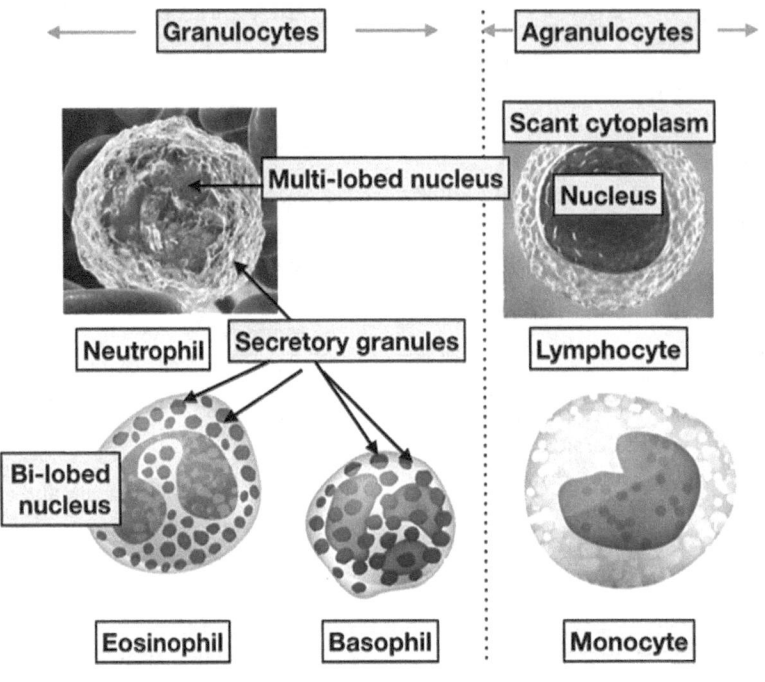

Figure 2.18 Types of white blood cells (leukocytes). (Shutterstock/Anna Holmart; Shutterstock/Kateryna Kon)

2.12.2 Number and Proportion of Leukocytes

The number of leukocytes per unit volume of blood in domestic animals is shown in Table 2.7. The most abundant leukocytes are heterophils and lymphocytes. The least abundant leukocytes are the basophils. Table 2.8 shows the percentages of neutrophils, lymphocytes, monocytes, eosinophils, and basophils in domestic animals.

The number of white blood cells can change with disease status. The number of eosinophils in blood is increased with responses to allergen or parasites.

2.12.3 Lifespan on Leukocytes

The lifespan of leukocytes varies considerably:

- Neutrophils: 5 days.
- Lymphocytes: 1–8 weeks.
- Monocytes: 1 day.
- Eosinophils: 2–5 days.
- Basophils: 1–2 days.

To match the short lifespan of leukocytes, there is a consequent need to replace them (see below).

2.12.4 Production of White Blood Cells

Leukocytes are produced in the bone marrow (see Figures 2.18 and 2.19).

2.12.5 Function of Leukocytes

Leukocytes are essential to the ability of an animal to resist infection. Among the actions of leukocytes responding to infections include the following:

- Phagocytosing bacteria and other pathogens (neutrophils, monocytes, and the macrophage).
- Producing antibodies against the pathogen (lymphocytes).
- Inducing **inflammatory** response (e.g., in the respiratory tract).
- Releasing **histamine**, enzymes, and cytokines (to ramp up the immune response, e.g., basophils and eosinophils) together with growth factors.

The functions of the five types of leukocytes are summarized in Table 2.6. The actions of leukocytes are also covered under the immune system in Chapter 9.

Table 2.8 Percentages of neutrophils, lymphocytes, monocytes, eosinophils, and basophils, with ranges for men and women shown for comparison.

Species	Neutrophils %	Lymphocyte %	Monocytes %	Eosinophil %	Basophils %
Cat	54	31	2.5	2	0.5
Cattle	24	60	4	10	1.0
Dog	71	14.5	6	4.5	0.5
Horse	61	31.5	3	3.5	1.0
Pig	37	50.5	6	5.7	1.0
Sheep	30	60	3	5	1.5
Chicken	26[a]	58	6	1.7	2.4
Adult people	50–70	25–35	4–6[b]	1–3	0.4–1.0

[a] Where they are called heterophils.
[b] Includes macrophage.

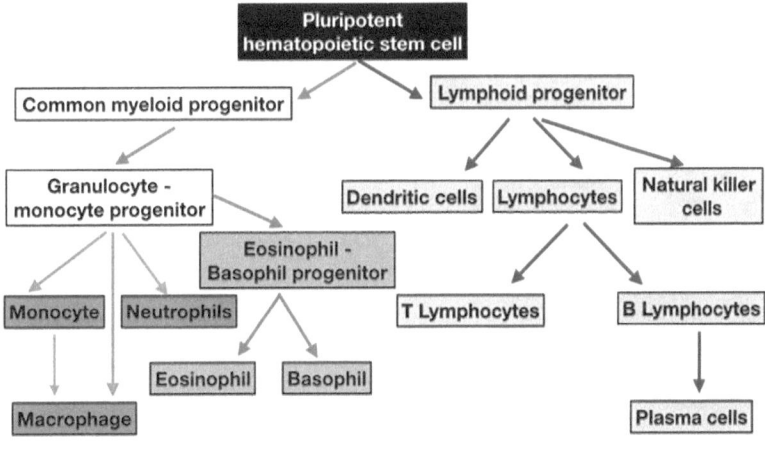

Figure 2.19 Development of blood cells (NK cells are natural killer cells).

2.13 Platelets (Thrombocytes)

2.13.1 Structure of Platelets

Mammalian platelets are small. Their volume is 7 fL (femtoliters) or 7 trillionths of a liter (for discussion of metric units see Textbox 1.7). The mammalian platelets have a volume that is about 14% of the volume of an erythrocyte (see Table 2.4). They lack nuclei. Platelets might be thought of as merely fragments of cytoplasm. **This would be wrong!** They are very important and finely controlled. Platelets have the following structures:

- A **cell membrane** with receptors.
- **Mitochondria** producing ATP.
- **Microtubules** allowing the formation of **pseudopodia** when the platelet is activated (see Figure 2.20).
- **Secretory granules** that are released when the platelet is activated. There are two types of granules:
  - α **granules** containing pro-inflammatory chemicals, proteins that promote platelets to stick together (adhering proteins) and growth factors to promote tissue repair.
  - **Dense granules** containing ATP, calcium, serotonin, and histamine.

2.13.2 Thrombocytes in Poultry

In contrast to platelets (thrombocytes) in mammals, the thrombocytes of birds are not called platelets. Moreover, thrombocytes in birds have nuclei and an endoplasmic reticulum and consequently can make new proteins.

2.13.3 Role of Platelets

Platelets are activated at sites of injury to blood vessels and the exposure of collagen in the endothelium. Once activated, they change shape, putting out pseudopodia (see Figure 7.21). Firstly, they stick together (**adhesion**) and **aggregate** at the site of injury (aggregation), forming a temporary plug. They play the critical role in **initiating blood clotting** (discussed under blood clotting). Platelets also release antibacterial peptides.

2.13.4 Lifespan of Platelets

Platelets in mammals have a lifespan of about 7–10 days. This compares with five days for neutrophils, one day for monocytes and one to two days for basophils (see Section 2.12.3), and about 100 days for erythrocytes (see Section 2.11.3). The result of the short lifespan of platelets is the need for the continual production of platelets.

2.13.5 Production of Platelets

Platelets are made in the bone marrow and, also, the lungs. The precursor cells are the very large megakaryocytes (see Figure 2.21). Each megakaryocyte can produce a thousand platelets.

2.14 Plasma

2.14.1 Introduction to Plasma

Plasma consists of the following:

- Ions such as **sodium**, **chloride**, bicarbonate, phosphate, magnesium, and potassium.
- Nutrients including **glucose**, **amino acids**, and other organic molecules.
- Proteins as follows:
  - **Albumen**.
  - α-**globulin**.
  - β-**globulin**.
  - γ-**globulin**.
  - **Fibrinogen**.
- Minerals, such as iron, bound to **binding proteins**.

2.14.2 What Are the Functions of the Proteins in Plasma?

The proteins in blood plasma have important functions:

- **Transporters**.
- **Enzymes**.
- **Immunoglobulins** or antibodies (γ-globulin).
- Hormones (see Chapter 4).

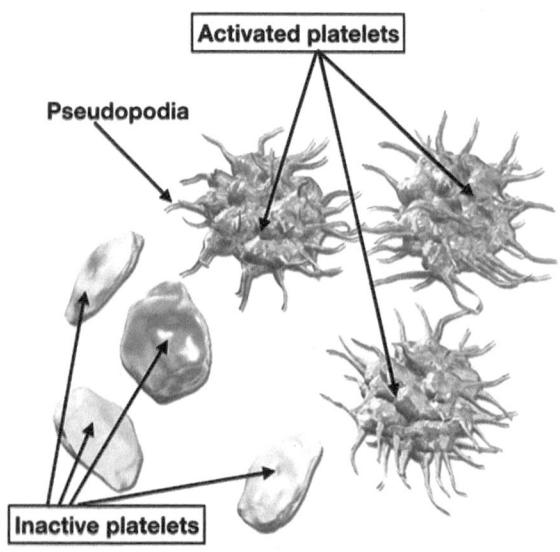

Figure 2.20 Structure of inactive and activated mammalian platelets. (Vigilius, CC BY 4.0)

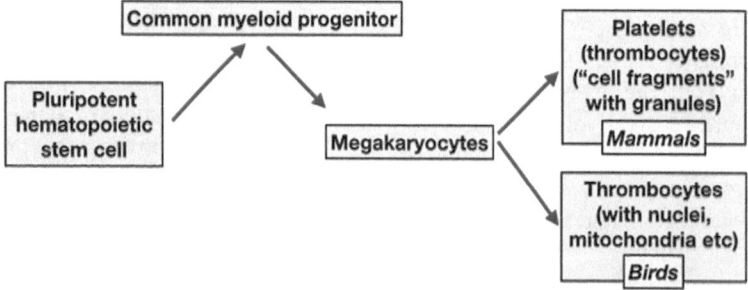

Figure 2.21 Production of platelets in the bone marrow.

2.14.3 Plasma Proteins as Transporters

2.14.3.1 Metal-Binding Proteins

- Iron (ferrous ions) are transported bound to the protein, transferrin (a β-globulin).
- Copper ions are transported bound to the protein ceruloplasmin (an α-globulin).

2.14.3.2 Vitamin-Binding Proteins

- Vitamin A is transported bound to retinol-binding protein (RBP) which can complex with prealbumin.

2.14.3.3 Hormone-Binding Proteins

- Cortisol in livestock and companion animals together with corticosterone in poultry bind to albumen and **corticosteroid binding globulin** (CBG) or transcortin.
- Thyroid hormones bind to both albumin (mammals and birds), thyroxine-binding globulin (mammals but not birds), and transthyretin (mammals and birds).
- **Testosterone** and **estradiol** bind to **sex steroid binding globulin** in mammals but this protein is not present in birds.

2.14.4 Yolk Proteins

In female chickens (hens), the yolk proteins are made in the liver and are transported to the ovaries in the blood. **Yolk precursors** in blood plasma consist of the following:

- **Vitellogenin** (broken down to lipovitellin and phosvitin in the egg).
- **Very low-density lipoprotein** (VLDL).

As mammals do not have yolk in their eggs, they do not produce yolk precursors.

2.15 Blood Clotting

2.15.1 Hemostasis

Hemostasis is stopping bleeding by blood coagulating or going from a liquid state to a gel. Blood needs to clot rapidly at or near the site of injury otherwise too much blood will be lost. However, blood should not clot at other times. The first part of clotting is the **activation of platelets** and the formation of a **platelet plug** (see platelet functioning above).

Clotting is the formation of **fibrin fibers** that trap red blood cells (see Figure 2.22). The fibrin fibers are the

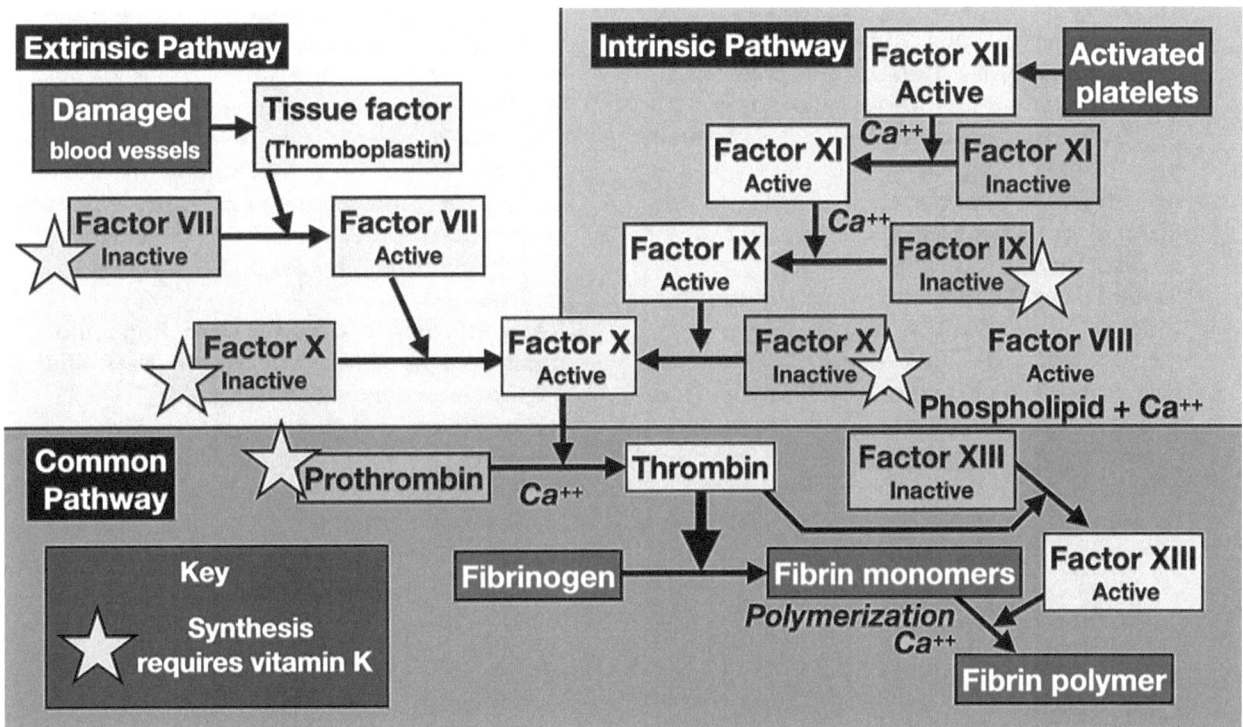

Figure 2.22 Blood clotting cascades.

result of **fibrin polymerization**. This is stimulated by **clotting factor VIIIa**; "a" standing for activated. Fibrin is generated from its precursor, **fibrinogen** due to the action of the proteolytic enzyme **thrombin**. Thrombin is one of a series of clotting factors that are normally inactive zymogens in the plasma but when activated are **proteases**. The activation is by another clotting factor; itself a protease when active. There is a cascade of activating clotting factors with a multiplier effect at each stage. One of the final stages is the activation of the protease thrombin. This involves conversion of prothrombin (inactive) to thrombin. This is catalyzed by activated clotting factor Xa. The process of clotting is initiated either by the exposure of a clotting factor in the sub-endothelium following injury to the blood vessel or to actions of **activated platelets**.

2.15.2 Vitamin K and Blood Clotting

Vitamin K is essential to the synthesis of clotting factors [prothrombin (II), VII, IX, and X]. A vitamin K deficiency leads to problems with blood clotting.

It was observed that cattle and sheep fed sweet clover died after either dehorning or castration due to their blood not clotting. This is due to a fungal chemical that prevented vitamin K from acting. A related chemical is **warfarin**, and this is used to either kill rodents (rats and mice) or decrease blood clotting in people.

2.16 Diseases of Cardiovascular System in Domestic Animals

Textboxes 2.7 and 2.8, respectively, summarize diseases of the cardiovascular system and of blood in domestic animals.

Textbox 2.7   Diseases of the Cardiovascular System in Domestic Animals

- **Aneurism** in horses is where there is weakening of an artery. This can lead to ballooning of the artery wall and, ultimately, hemorrhage and even death.
- **Ascites** is an accumulation of fluid in the abdomen. This is seen in broiler chicken (mortality 1–2%) particularly at high altitudes (mortality ~ 30%). One cause of ascites is right ventricular insufficiency and pulmonary hypertension. Another cause of ascites in poultry is toxins produced by fungi in the feed (myotoxins).
- **Arrhythmia** is an abnormality of the synchronization of contractions of the atria and ventricles. Arrhythmia is seen in horses.
- **Bloodworms** (*Strongylus vulgaris*) are nematodes that are parasitic in horses. Migrating bloodworm larva induce arteritis (inflammation of the arteries) and this can reduce blood flow, for instance, in the coronary arteries (supplying the heart with oxygen) or cerebral arteries (supplying the brain with oxygen). Dewormers such as Ivermectin are effective in treating bloodworms.
- **Cardiac troponins** are released into blood from cardiac muscle undergoing atresia (break down). Plasma concentrations of cardiac troponins are a diagnostic indicator of **heart atresia**.
- **Edema** is localized swelling due to accumulation of lymphatic fluid. Edema can be due to injury, inflammation, liver, lung or kidney disease, circulatory problems such as venous insufficiency or congestive heart failure and bacterial toxins. In horses, malignant edema is caused by toxins from clostridial myositis. Another example of edema in horses is bone edema due to injury. This is found in racehorses or jumping horses due to repeated trauma.
- **Heart murmurs** are sounds indicating abnormalities of blood flow through the heart and can be seen in dogs and horses.
- **Hypertension** is abnormally high blood pressure.
- **Tachycardia** in dogs is a rapid heart rate outside the normal range. The heart rate can go up to 400 in dogs. This is associated with arrythmia.

Textbox 2.8 Diseases of Blood in Domestic Animals

- **Anemia in horses** is a deficiency of erythrocytes, hemoglobin, and the ability of the blood to carry oxygen. It can be due to inadequate production of erythrocytes due to either abnormalities in the bone marrow or low blood concentrations of erythropoietin (the hormone that stimulates red blood cell production). Another reason for anemia is blood cell parasites.
- **Bloodworms** (*Strongylus vulgaris*) in horses cause an increase in the number of eosinophils in the blood.
- A **thrombus** or **blood clot** is found in attached to blood vessels, particularly veins, of domestic animals such as horses. Blood clots restrict blood flow. An example is a thrombus blocking the pulmonary artery (pulmonary arterial thrombosis) and hence reduces blood flow through one lung. A blood clot can break off and travel to other sites in the body in the blood stream. It is called an **embolus**. An embolus can block a blood vessel where there is narrowing and thereby reduce oxygen supply.

Respiration

3

Learning Objectives

1. Understand the functioning of the respiratory system (Section 3.1.1).
2. Know the composition of air (Section 3.2.2).
3. Understand what the partial pressure of a gas is (Section 3.2.3).
4. Understand the gas laws (Section 3.2.1).
5. Understand the components of the respiratory system (Section 3.3).
6. Understand how air is inspired and expired (Section 3.4).
7. Understand external respiration in birds and how it is different from livestock and companion animals (Section 3.5).
8. Understand gas exchange (Section 3.8).
9. Understand how oxygen is transported in blood bound to hemoglobin (Section 3.9).
10. Understand transport of carbon dioxide bound to hemoglobin (Section 3.9).
11. Understand how carbon dioxide is transported in blood as bicarbonate ions (Section 3.10).
12. Understand the importance of the enzyme carbonic anhydrase in the transport of carbon dioxide in blood (Section 3.10).
13. Understand how respiration is controlled (Section 3.11).
14. Understand diseases of the respiratory system (Section 3.12).

Chapter 3 Respiration

Table of Contents

- 3.1 Introduction
  - 3.1.1 What Is the Purpose of Respiration?
- 3.2 Gas Laws
  - 3.2.1 Overview of Gas Laws
  - 3.2.2 What Is the Partial Pressure of a Gas?
  - 3.2.3 Composition of Air
- 3.3 Functional Anatomy of the Respiratory Organs in Livestock and Companion Animals
  - 3.3.1 Nostrils
  - 3.3.2 Guttural Pouch
  - 3.3.3 Epiglottis
  - 3.3.4 Pharynx (Throat)
  - 3.3.5 Larynx (Voice Box)
  - 3.3.6 Trachea (Wind Pipe)
  - 3.3.7 Bronchus (Plural, Bronchi)
  - 3.3.8 Bronchioles
  - 3.3.9 Alveolus (Plural, Alveoli)
- 3.4 Mechanism of Respiration (Pulmonary Ventilation)
  - 3.4.1 Inspiration
  - 3.4.2 Expiration
  - 3.4.3 How Is Air Moved Into and Out of the Alveoli?
  - 3.4.4 What Is the Consequence of the Volume of the Thoracic Cavity Being Increased?
  - 3.4.5 What Is the Consequence of Increased Volume?
  - 3.4.6 How Is Air Moved Out of the Alveoli?
  - 3.4.7 What Is the Consequence of Decreased Volume?
- 3.5 Pulmonary Ventilation in Chickens
  - 3.5.1 Introduction to Pulmonary Ventilation in Chickens
  - 3.5.2 What Causes Air to Flow in Birds?
  - 3.5.3 Differences Between Pulmonary Ventilation in Chickens and Domesticated Mammals
- 3.6 Central Control of Breathing
- 3.7 Other Roles of the Respiratory System
- 3.8 Gas Exchange
  - 3.8.1 Introduction to Gas Exchange
  - 3.8.2 Solubility of Gases in Water
- 3.9 Transport of Oxygen
  - 3.9.1 Overview of Transport of Oxygen
  - 3.9.2 How Is This Bohr Effect Employed?
- 3.10 Transport of Carbon Dioxide
  - 3.10.1 How Is Carbon Dioxide Transported in the Blood?
  - 3.10.2 Carbonic Anhydrase and Carbon Dioxide Transport as Bicarbonate
  - 3.10.3 Carbon Dioxide Binding to Hemoglobin
- 3.11 Integrative Control of Respiration
  - 3.11.1 Overview of Integrative Control of Respiration
  - 3.11.2 Why Does the Body Use pH as a Proxy for Carbon Dioxide (and Oxygen)?
  - 3.11.3 What Is the Effect of Panting?
- 3.12 Diseases Related to Respiration

3.1 Introduction

3.1.1 What Is the Purpose of Respiration?

Respiration brings oxygen (O_2) into the body and removes carbon dioxide (CO_2). The respiratory cycle is composed of the following:

- **Pulmonary ventilation** is the physiological act of breathing and associated processes. It brings oxygen into the body and rids it of carbon dioxide.
- **External respiration** is the process of gas exchange across the lung.
- **Transport of gases** through the blood following the gas laws.
- **Internal respiration** is the process of gas exchange between blood and the tissues, bringing oxygen to the tissues and ridding them of carbon dioxide.

In addition, there is **cellular respiration** in tissues using oxygen and nutrients to generate a chemical form of energy, **adenosine triphosphate** (**ATP**). This is not covered in this chapter (see Chapter 7 for details). Textbox 3.1 provides a glossary of terms used in respiration.

Textbox 3.1   Key Vocabulary

- **Alveolus**: site of movement of oxygen into the blood from the alveolar lumen and loss of carbon dioxide from the blood into the alveolar lumen.
- **Atmospheric pressure**: the pressure exerted by the external environment.
- **Bronchiole**: small tubes conducting air to and from the alveoli.
- **Bronchus**: two tubes (one per lung) connecting trachea to the bronchioles. Respiratory gases pass through the bronchi to the bronchioles during inspiration and from the bronchioles to the trachea during expiration.
- **Bicarbonate**: ion transporting carbon dioxide and buffering the pH in blood.
- **Carbon dioxide**: a gas produced by the oxidation of glucose, amino acids, and fatty acids.
- **Diaphragm**: dome-shaped muscle that contracts during inspiration, bringing air into the lungs.
- **Expiration**: breathing out.
- **Gas**: a compressible fluid that follows a series of laws.
- **Hemoglobin**: protein containing heme units and ferrous ions. Oxygen, and some carbon dioxide, are transported in the blood bound reversibly to hemoglobin in the erythrocytes.
- **Inspiration**: breathing in.
- **Lung**: organ for gaseous exchange; the lungs contain alveoli and bronchioles.
- **Oxygen**: the gas needed for cellular metabolism.
- **Pulmonary**: pertaining to the lung.

Textbox 3.2   Gas Lawsv

1. **Gases are compressible** (unlike liquids).
2. **Gases flow from areas of high pressure** to areas of lower pressure.
3. **Boyle's Law**
   At a constant temperature, the pressure of a gas (P) is inversely proportional to the volume (V).

   $$P = \frac{\text{constant } K}{V}$$

   Thus, if the volume is increased the pressure drops, or if the volume is decreased the pressure increases.
4. **Charles's Law**
   The volume of a gas (V) is proportional to the absolute temperature (T).

   $$V \propto T$$

5. **Dalton's Law**
   The pressure of a mixture of gases is equal to the sum of the partial pressures of the individual components of the mixture.
6. **Henry's Law**
   At a liquid gas junction and at a constant temperature, the amount of the gas dissolving in the water depends on the partial pressure of the individual gas and its solubility in the liquid.

Before addressing respiration and the respiratory organs, it is important to understand how gases behave. The gas laws are scientific descriptions of how gases behave. (Scientific laws have been established to describe physical phenomena and are often described by a formula.)

3.2 Gas Laws

3.2.1 Overview of Gas Laws

The gas laws are summarized in Textbox 3.2.

Let's consider what happens when a gas is in a closed container and the volume changes:

- Based on Boyle's Law (see Textbox 3.2), if the volume of the container is decreased, the pressure goes up. If the volume of the container is increased, the pressure goes down. This is illustrated in Figure 3.1.

Let's consider what happens when air is in a container with an opening and the volume changes:

- If the volume of the container is decreased, the pressure goes up and air flows out. If the volume of the container is increased, the pressure goes down and air flows in. This is illustrated in Figure 3.2.

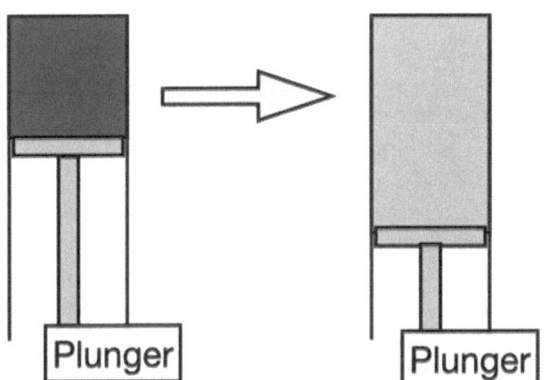

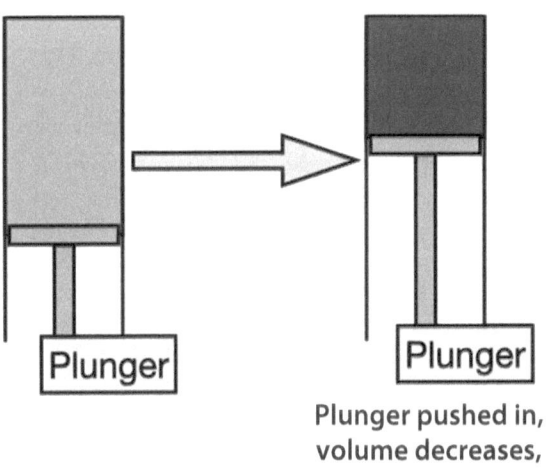

Figure 3.1 Relationship between volume and pressure in a closed system.

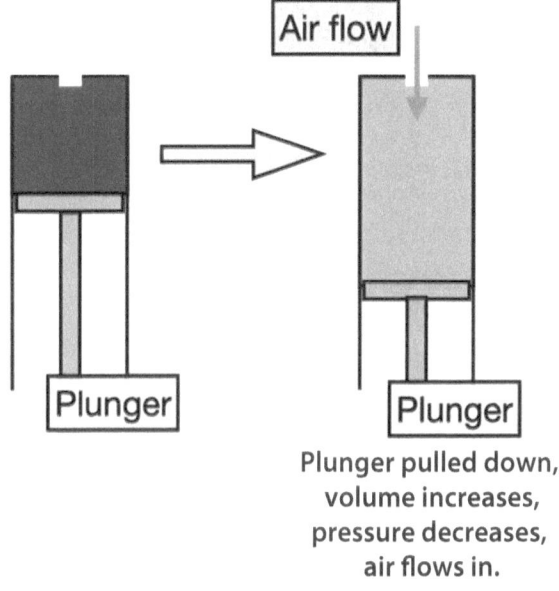

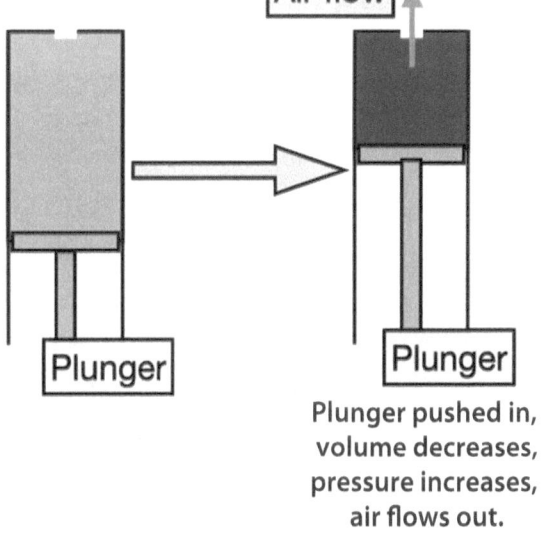

Figure 3.2 Relationship between volume and pressure in an open system.

3.2.2 What Is the Partial Pressure of a Gas?

Based on Dalton's Law (see Textbox 3.2), the partial pressure of a gas is the pressure exerted by a single gas within the whole composition of the gas. Partial pressure of an individual gas is the atmospheric pressure multiplied by percentage of the gas divided by one hundred. Example:

- Atmospheric pressure is 760 mm mercury (Hg) at sea level. The partial pressure of oxygen at sea level = 760 × 21/100 mm mercury = 160 mm mercury.

Other examples are shown in Textbox 3.3 and 3.4.

Textbox 3.3 The Impact of High Altitudes on the Partial Pressure of Oxygen

Planes are pressurized to the equivalent of being at an altitude of 8,000 feet (1.5 miles high). Atmospheric pressure is 544 mm mercury at 8,000 feet. The partial pressure of oxygen at 8,000 feet in a pressurized plane = 544 × 21/100 mm mercury = 114 mm mercury (29% lower than at sea level). This may impact horses during transportation and service and other dogs on planes.

Figure 3.3 Horses and dogs flying.

Are Dogs Allowed on Planes?
The answer is yes, but dogs and cats are only allowed on passenger planes if they are small enough to fit under the seat. In addition, service dogs are allowed in the cabin of planes provided they are fully trained. According to the United States Department of Transportation and the Air Carrier Access Act, airlines must accept service animals to accompany people with disabilities. Service dogs are used for visual impairments, auditory impairments, seizures, mobility impairments, and post-traumatic stress disorder.

Emotional therapy dogs are allowed in the cabin if the owner has a certificate from a medical professional stating that the owner of the dog has a mental health disability and needs to be accompanied by their therapy dog. Dogs can also be transported in carriers in the hold.

Horses are transported in special planes. There are thousands of horses transported in this way.

Textbox 3.4 The Impact of High Altitudes on the Partial Pressure of Oxygen: Living at High Latitudes

Figure 3.4 The Tibetan plateau. (Photo courtesy of StateStreet, CC BY-SA 3.0)

The Tibetan plateau is at 13,100 feet above sea level. The atmospheric pressure is 464 mm mercury. The partial pressure of oxygen is 97 mm mercury (or 39% lower than at sea level). This creates a problem.

Tibetan chickens are native chickens that are adapted to live under high altitude conditions with hypoxia (low oxygen) with markedly improved embryonic survival and high blood carbon dioxide. There are similar adaptations to high altitude in Tibetan native sheep and pigs.

3.2.3 Composition of Air

Nitrogen	78%
Oxygen	21%
Argon	0.9%
Carbon dioxide	0.04%
All gases	100%

3.3 Functional Anatomy of the Respiratory Organs in Livestock and Companion Animals

The structure of the respiratory organs of the horse is shown in Figure 3.5.

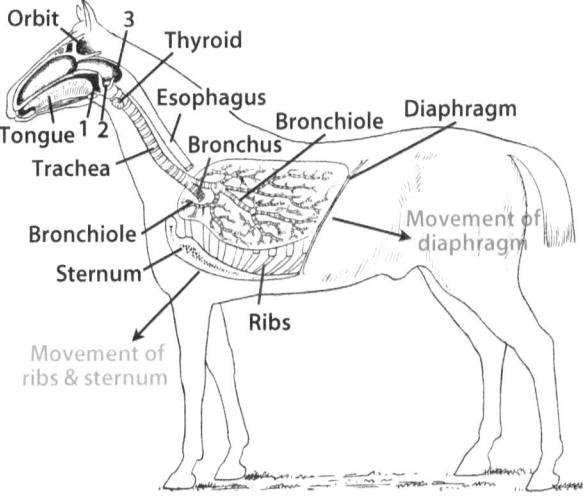

Figure 3.5 Respiratory system of the horse showing lobar bronchi and bronchioles in the lungs; the ducts from the exterior beginning at the nasal cavity, followed by the pharyngeal cavity, the larynx, trachea, and bronchi; and the diaphragm, ribs and sternum.
[Key: 1 = epiglottis; 2 = larynx, 3 = guttural pouch]
(Adapted from Henry Thompson)

The respiratory organs are either of the following (see Figure 3.6):

- **Conducting zone**: organs or tissues (e.g., trachea, bronchus [plural bronchi], bronchioles, and terminal bronchioles) through which air flows to and from the respiratory zone.
- **Respiratory zone**: organs or tissues (alveoli and respiratory bronchioles) where gas exchange occurs (oxygen coming into the blood and carbon dioxide leaving the blood).

The anatomical features of the respiratory organs follow in order of air flow into the lungs (Sections 3.3.1 through 3.3.9).

3.3.1 Nostrils

Nostrils contain conchae. These increase the surface area to allow warming of air during inspiration. In addition, the mucus-covered conchae trap large dust particles. This mucus is moved by the cilia and then swallowed. Within this region is the olfactory epithelium, which sends nerve stimulation to the brain to provide the animal with smells (see Chapter 5).

3.3.2 Guttural Pouch

Guttural pouch is an extension of the Eustachian tube. It is found in horses and donkeys but not in other domestic animals. The guttural pouch allows equalization of pressure between the throat and the middle ear.

3.3.3 Epiglottis

The **epiglottis** closes over the trachea when food or water are being swallowed. This prevents food or water passing down the trachea.

3.3.4 Pharynx (Throat)

The **pharynx** (**throat**) functions to allow movement of air.

3.3.5 Larynx (Voice Box)

The **larynx** (**voice box**) functions to allow transit of air and vocalization.

3.3.6 Trachea (Wind Pipe)

The **trachea** (**wind pipe**) is characterized by the presence of multiple C-shaped rings of cartilage that keep the trachea open. The trachea functions to allow transit of air to the lungs. In addition, the trachea traps pathogens, microorganisms, and dust into a mucus which is produced by goblet cells. Mucus is moved by **cilia** towards the esophagus, where it is swallowed.

3.3.7 Bronchus (Plural, Bronchi)

There is one main or primary bronchus for each lung (see Figure 3.6). Again, there are C-shaped rings of cartilage to keep the bronchi open. There are multiple lobar and segmental bronchi (see Figure 3.6). These have a progressively smaller diameter than in the trachea. The bronchi allow transit of air to and from the lungs.

3.3.8 Bronchioles

There are pulmonary bronchioles (one per lobule of the lung). From these, there are multiple terminal bronchioles and smaller respiratory bronchioles terminating in the alveoli (see Figure 3.6 and 3.7). Bronchioles function to allow transit of air. Bronchioles have ciliary epithelium to move mucus.

3.3.9 Alveolus (Plural, Alveoli)

Alveoli are one cell thick with pulmonary capillaries in close proximity (Figures 3.6–3.8). There are large numbers

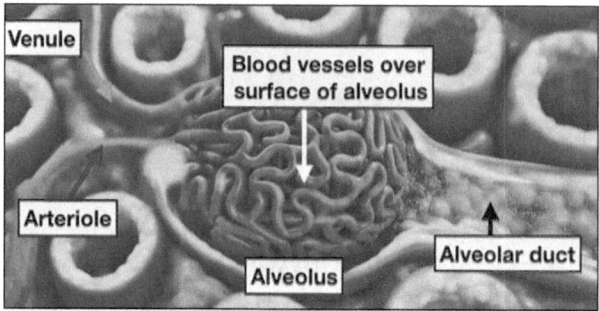

Figure 3.8 Alveolus showing blood supply. There are multiple blood vessels over the surface of the alveolus. (Adapted from Shutterstock)

of alveoli in the lungs of mammals (e.g., one million alveoli in horses). The advantage of a large number of small alveoli is that this provides a large respiratory surface for gaseous exchange. Alveoli allow oxygen to pass into the pulmonary blood and carbon dioxide to pass from the pulmonary blood into the alveoli. In addition, there are leukocytes phagocytizing pathogens in the alveoli.

3.4 Mechanism of Respiration (Pulmonary Ventilation)

3.4.1 Inspiration

During inspiration, air is drawn through the nostrils (or mouth) passing through the pharynx and larynx through the trachea, then through the bronchi, and then through the bronchioles, and finally into the alveoli.

3.4.2 Expiration

During expiration, air passes from the alveoli, to the bronchioles, to the bronchi, to the trachea, to the pharynx and larynx, and out of the body through the nostrils (or mouth).

3.4.3 How Is Air Moved Into and Out of the Alveoli?

In livestock and companion animals, the volume of the thoracic cavity is increased when the following occur:

1. The external intercostal muscles between the ribs contract. This moves the sternum and ribs forward, enlarging the thoracic cavity (see Figures 3.5 and 3.9).
2. The curved diaphragm (a muscular sheet) contracts (see Figures 3.5 and 3.9). This causes the volume of the thoracic cavity to get markedly greater.

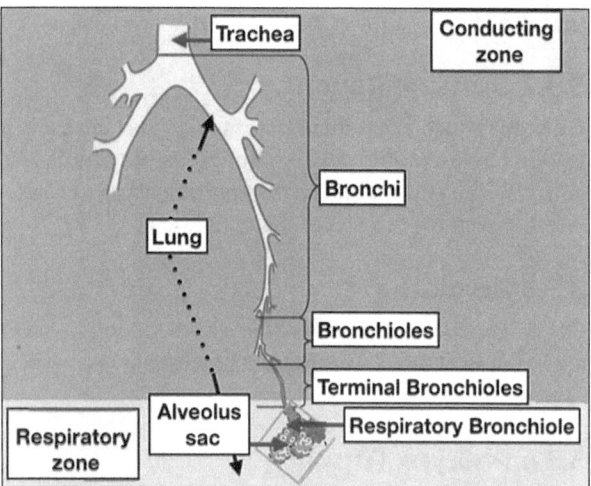

Figure 3.6 Aspects of the respiratory tree showing relative positions of trachea, bronchi, bronchiole, and terminal bronchioles in the conducting region together with respiratory bronchioles and alveolus (plural, alveoli) in the respiratory region. (Shutterstock/ScientificStock)

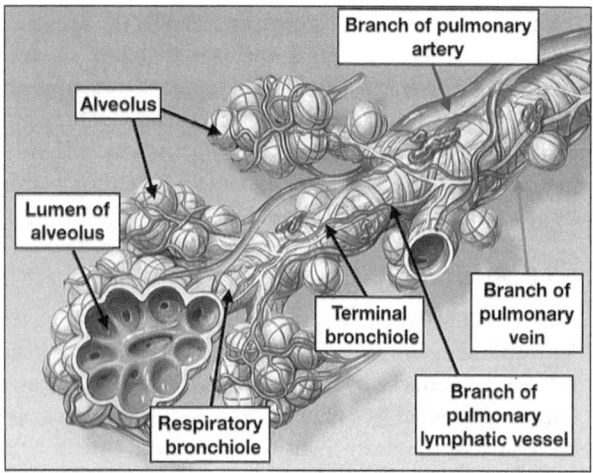

Figure 3.7 Structure of alveoli (the plural of alveolus) terminal ducts and respiratory bronchioles.
(Adapted from Patrick J. Lynch, CC BY 2.5)

3.4.4 What Is the Consequence of the Volume of the Thoracic Cavity Being Increased?

The lungs are surrounded by **pleural membranes**. The outer pleural membrane is attached to the thoracic wall. So, when the volume of thoracic cavity becomes greater, the volume of the lungs becomes greater.

3.4.5 What Is the Consequence of Increased Volume?

Remembering Boyle's law, when the volume of a gas (in the lungs) is increased, the pressure is reduced. When the pressure in the lungs is decreased, air flows in (Figure 3.9).

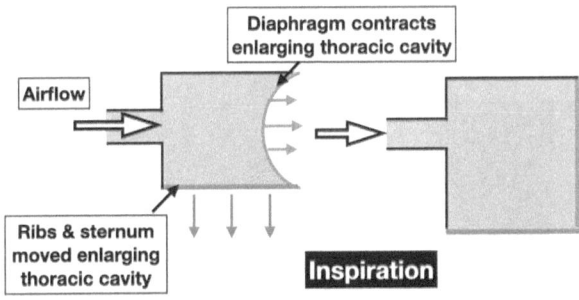

Figure 3.9 Schematic of inspiration in companion animals and livestock.

3.4.6 How Is Air Moved Out of the Alveoli?

In livestock and companion animals, the volume of the thoracic cavity is decreased when the following occur:

1. The **internal intercostal muscles** contract.
2. The **external intercostal muscles** between the ribs relax. This moves the sternum and ribs back, reducing the volume of the thoracic cavity (see Figures 3.5 and 3.10).
3. The diaphragm (a muscular sheet) relaxes back to a curved state (see Figure 3.5 and 3.10). This causes the volume of the thoracic cavity to decrease.
4. There is **elastic recoil** of the alveoli returning it to its original volume. This is due to the stretched **elastic fibers** returning to their unstretched lengths.

3.4.7 What Is the Consequence of Decreased Volume?

Remembering Boyle's law, when the volume of a gas (in the lungs) is decreased, the pressure is increased. When the pressure in the lungs is increased, air flows out (Figure 3.10).

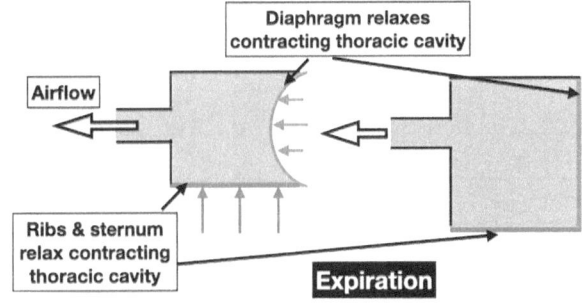

Figure 3.10 Schematic of expiration in companion animals and livestock.

Textbox 3.5 Summary of Pulmonary Ventilation

Inspiration
1. Contractions of the intercostal muscles and the diaphragm increase the volume of the thoracic cavity.
2. The lungs (and the elastic fibers within it) are stretched, increasing their volumes.
3. Based on Boyle's Law, the pressures in the lungs, specifically the alveoli, decrease.
4. Air flows from the outside (high pressure) into the lungs (low pressure).

Expiration
1. Relaxation of the intercostal muscles and the diaphragm decreases the volume of the thoracic cavity.
2. The lungs are relaxed, decreasing their volumes.
3. The elastic fibers in the lungs relax in what is called elastic recoil.
4. Based on Boyle's Law, the pressures in the lungs, specifically the alveoli, increase.
5. Air flows from the lungs (high pressure) into the outside (low pressure).

56 Theme 1 Homeostasis via the Blood

Table 3.1 Basal respiration rates in domestic animals together with humans for comparison.

Species	Basal Respiration Rate (Rate of Breathing or Breaths per Minute)
Cat	28
Cattle	38
Dog	26
Horse*	12
Sheep	25
Chicken	36
Human adults	12–20

*In horses, the rate of breathing can be increased to 120 breaths per minute in vigorous exercise.

Definitions related to respiration are covered in Textbox 3.4 and illustrated in Figure 3.11. Table 3.1 shows the respiration rate in domestic animals.

During inspiration, incoming air is mixed with air in the residual volume of the lung or dead space. What is the implication of this? The **concentration of oxygen** (pO_2) in the alveoli of mammals is markedly lower than in atmospheric air (Table 3.2). Moreover, the **concentration of carbon dioxide** (pCO_2) is much higher in the alveoli than in the atmosphere. Blood picks up oxygen in the alveoli while carbon dioxide is lost from the blood. The reverse process occurs in the capillaries of tissues. Table 3.2 also summarizes the concentrations oxygen and carbon dioxide in systemic arteries and veins.

Respiratory definitions are summarized in Textbox 3.6. and Figure 3.11. An example of parameters related to respiration in dogs and horses is shown in Textbox 3.7.

Textbox 3.6 Respiration in Dogs and Horses

Respiration in Horses
Capacity of lungs—55 liters
Tidal volume:
Basal at rest—8.5 liters
During vigorous exercise—16 liters
Minute volume:
Basal at rest—80 liters
During vigorous exercise—1,800 liters per minute

Elite Dogs and Horses
Maximal oxygen uptake (VO₂max):
Dogs—15 mL kg^{-1} min.$^{-1}$
Elite dogs (Alaskan huskies)—240 mL kg^{-1} min.$^{-1}$
Horses—122 mL kg^{-1} min.$^{-1}$
Elite horses—220 mL kg^{-1} min.$^{-1}$

Figure 3.11 Alaska huskies. (Frank Kovalchek, CC BY 2.0)

Table 3.2 Concentrations of oxygen and carbon dioxide in air (at sea level), alveoli, blood vessels, and tissues.

	Partial pressure of oxygen and carbon dioxide [mm mercury (Hg) or torr]	
	Partial pressure of oxygen (pO_2)	Partial pressure of carbon dioxide (pCO_2)
Atmosphere	160	3
Alveoli	100	40
Pulmonary venous blood/ systemic arteries	100	40
Tissue	20–70*	>45
Systemic veins	20–70	>45

*Brain concentrations are lower (35–40 mm Hg)

3.5 Pulmonary Ventilation in Chickens

3.5.1 Introduction to Pulmonary Ventilation in Chickens

In chickens and other birds, air flows through air capillaries, or **parabronchi**, in the lungs due to pumping actions of the air sacs (powered by the inspiratory muscles). Oxygen and carbon dioxide pass through the thin walls into pulmonary blood vessels with blood and air passing in opposite directions.

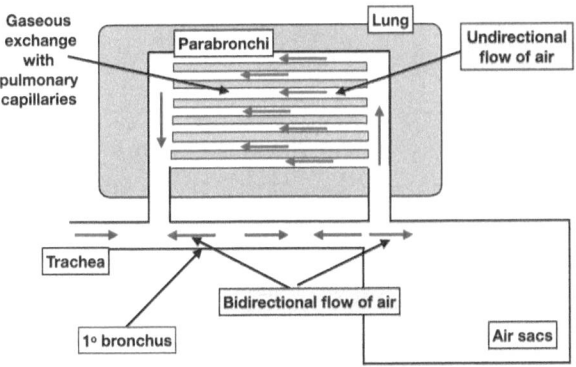

Figure 3.12 Schematic of the respiratory system of chickens.

There are nine air sacs in birds: clavicular (2), cervical (1), cranial thoracic (2), caudal thoracic (2), and abdominal (2).

During pulmonary ventilation, the airs sacs expand and contract drawing air in and out of the air sacs and, consequently, the lungs. The air sacs function in a "bellows like" manner, pumping air through the parabronchi (see Figure 3.12). The residual volume is shifted from the lungs in mammals (see Figure 3.13) into the air sacs in birds (see Figure 3.12). This increases the pO_2 in the air as it passes through the parabronchi in a unidirectional manner (pO_2 is defined in Section 3.2.2). **Gaseous exchange** is further enhanced by the air and blood flowing in opposite directions.

Such is the low residual volume (see Figure 3.13 for mammalian equivalent) in the caudal or abdominal air sacs that even when inspired air mixes with the residual air in the air sacs, the oxygen concentration is greater than 125 mm Hg. This is markedly greater than in the mammalian alveolus.

3.5.2 What Causes Air to Flow in Birds?

The mechanism is the following. The volume of lungs does not change during either inspiration or expiration. During inspiration in chickens, the vertebral and sternal ribs (plus the sternum) are moved up (due to muscle contractions), increasing the volume of the single cavity, the thoracic-abdominal cavity; there is no diaphragm in birds. The pressures in the air sacs decline and air flows into them. During expiration, air flows from the air sacs through the parabronchi (tertiary bronchi) in the lungs.

3.5.3 Differences Between Pulmonary Ventilation in Chickens and Domesticated Mammals

These differences are summarized in Textbox 3.7.

Textbox 3.7   Differences Between Pulmonary Ventilation in Chickens and Domesticated Mammals

1. Presence of air sacs in chickens (not present in mammals).
2. Absence of diaphragm (present in mammals).
3. Lungs do not expand in chickens (in contrast, the lungs expand during inspirations and contract during expiration).

Textbox 3.8   Definitions

Figure 3.13 summarizes respiratory definitions.

- **Expiratory reserve volume** is the volume of the lung that can be increased from the tidal volume during breathing out.
- **Inspiratory reserve volume** is the volume of the lung that can be increased from the tidal volume during breathing in.
- **Maximal oxygen uptake (VO₂max)** is the maximal oxygen uptake during exercise.
- **Minute ventilation** is tidal volume times respiration rate.

(*continued*)

Textbox 3.8 (continued)

- **Residue volume** is the volume of air left in the lung after exhaling as much as possible.
- **Respiration cycle** is one breath in and one breath out.
- **Respiration rate** is the number of breaths per minute.
- **Tidal volume** is the volume of air moved into and out of the lungs in a single respiratory cycle.
- **Total lung capacity** is the maximum amount of air in the lungs. This is equal to the vital capacity plus the residual volume.
- **Vital capacity** is the maximum volume of air moved into and out of the lungs in a respiratory cycle.

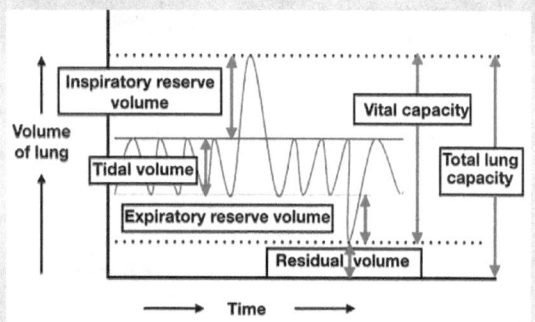

Figure 3.13 Respiration in livestock and companion animals showing terminology. (Adapted from Stephanie Greenwood, public domain).

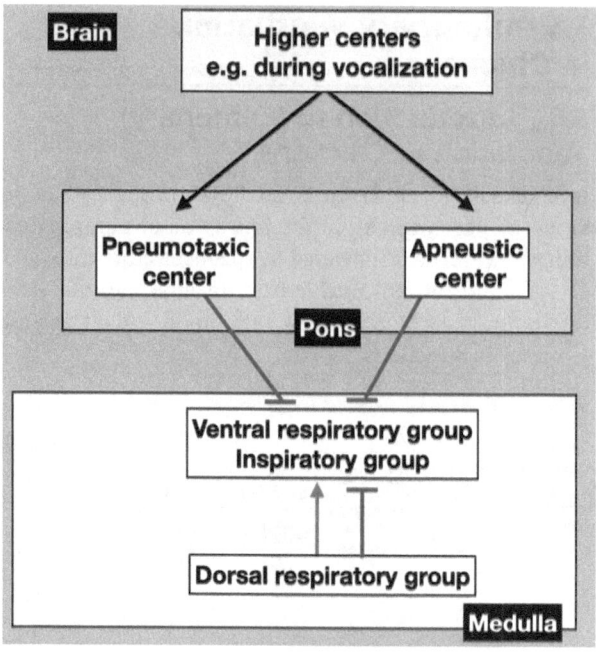

Figure 3.14 Components of the respiratory center showing interactions.
[Key: ↑ arrow is positive, ⊤ is negative]

3.6 Central Control of Breathing

Pulmonary ventilation is controlled by the neurons in the brain, specifically in the respiratory center(s). The respiratory center is in the brain, specifically in the pons and medulla. It consists of the following (Figure 3.14):

- Pontine respiratory group (PRG) in pons.
  - Pneumotaxic center.
  - Apneustic center.

Textbox 3.9 A Deeper Dive Into the Control of Pulmonary Ventilation

The pneumotaxic center (part of the pontine respiratory group) modifies the output of the ventral respiratory group. Specifically, when there is high output from the pneumotaxic center, the stimulatory actions of the inspiratory neurons and, hence, inspiration is curtailed. This can be viewed as stimulating the inspiratory "off switch". Conversely, when there is low output from the pontine respiratory group, the stimulatory actions of the inspiratory neurons are slowed and, hence, inspiration is prolonged.

The apneustic center initiates the inspiratory output from the inspiratory neurons of the ventral respiratory group.

The inspiratory neurons stimulate the inspiratory muscles. The expiratory neurons stimulate the expiratory muscles (see Figure 3.15). There is a neural circuit such that when the inspiratory neurons are firing, the expiratory neurons are not firing. Similarly, when the expiratory neurons are firing, the inspiratory neurons are not firing.

(continued)

Textbox 3.9 (*continued*)

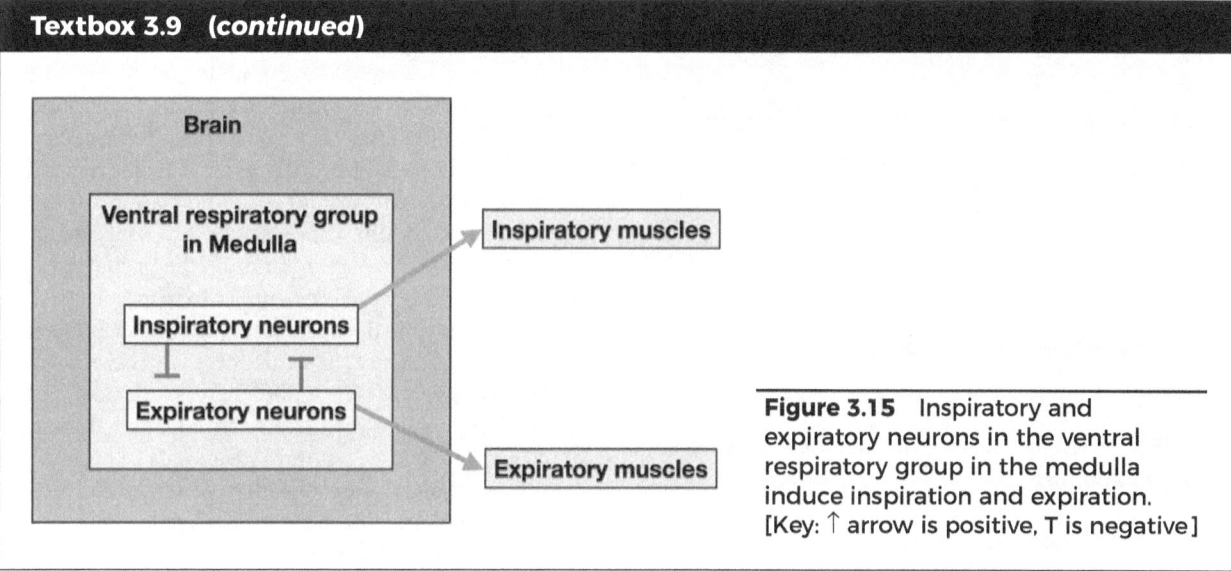

Figure 3.15 Inspiratory and expiratory neurons in the ventral respiratory group in the medulla induce inspiration and expiration. [Key: ↑ arrow is positive, ⊤ is negative]

- **Ventral respiratory group** in medulla.
  - **Inspiratory neurons**.
  - **Expiratory neurons**.
- **Dorsal respiratory group** in medulla.

3.7 Other Roles of the Respiratory System

1. **Angiotensin converting enzyme (ACE)** in the lungs converts **angiotensin I** to **angiotensin II**.
2. **Olfaction**: the act of inspiration brings air across the olfactory epithelium. In addition, air passes by the vomeronasal organ in horses and cattle. This is located at the base of the nasal septum. This detects pheromones (discussed in detail in Chapter 5).
3. **Vocalization**: air passing across the larynx is used for vocalization and, consequently, communication. This is found in all domestic animals.
4. **Acid-base homeostasis**: without expelling carbon dioxide, there is a build-up of acidity of hydrogen ions (H⁺) (protons).

$$CO_2 + H_2O \rightarrow H_2CO_3 \rightarrow H^+ + HCO_3^{-1}$$

Carbon dioxide + Water → Carbonic acid → Proton + Bicarbonate ion

5. **Aiding movement** of venous blood and lymph, due to reduced pressure in the thoracic cavity during inspiration.
6. **Defense against pathogens**, with leukocytes in the alveoli and a mucus sheet trapping pathogens in the trachea.

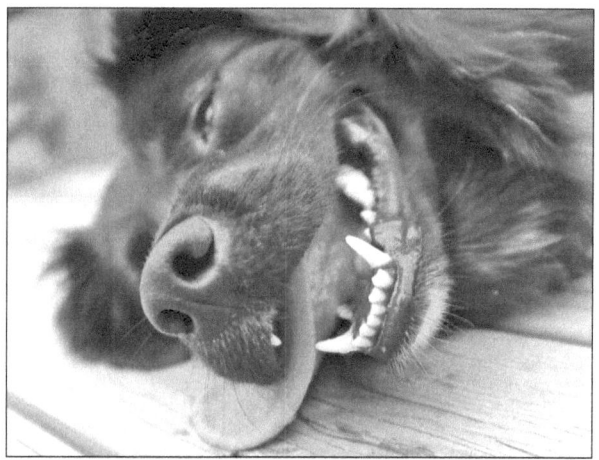

Figure 3.16 Dog panting to remove excess heat. (Courtesy of xlibber, CC BY 2.0)

7. **Reduced weight**: the presence of air sacs in bones reduces their weight in chickens and other birds.
8. **Dissipate heat**: panting is a method of removing excess heat (see Figure 3.16). A consequence of panting is excessive loss of carbon dioxide and, therefore, the risk that blood pH will rise (alkalosis).

3.8 Gas Exchange

3.8.1 Introduction to Gas Exchange

It is critically important for animals to get enough oxygen to their tissues and lose carbon dioxide generated by the tissues. Simply relying on dissolved oxygen and carbon dioxide in the blood is grossly insufficient to meet an animal's needs.

3.8.2 Solubility of Gases in Water

As the body temperature in an animal rises, there is a progressive decline in the **solubility** of both oxygen and carbon dioxide. For instance, the solubility of carbon dioxide is less than a third at 37°C than at 0°C. In contrast, as the temperature increases, an animal needs more oxygen and produces more carbon dioxide. Domestic animals have high body temperatures. Therefore, there must be mechanisms to transport oxygen and carbon dioxide.

3.9 Transport of Oxygen

3.9.1 Overview of Transport of Oxygen

Oxygen is primarily transported bound to hemoglobin. Hemoglobin consists of the following (see Figure 3.17):

- **Globins** (4 protein sub-units).
- **Heme units** (4 heme units).
- **Ferrous ions** (4 Fe^{2+}).

Each hemoglobin molecule binds up to four oxygen molecules in a reversible manner (see Figure 3.17). It is an association, not a chemical bond. The binding of one molecule of oxygen affects the ability to bind the second molecule of oxygen. In turn, the binding of two molecules of oxygen affects the ability to bind a third molecule of oxygen, and the binding of three molecules of oxygen influences the binding of a fourth molecule of oxygen. Hemoglobin picks up oxygen at the capillaries surrounding the alveoli in the lungs (Figure 3.18) and releases oxygen in the capillaries. Additionally, the converse (removal of oxygen) is true. This produces a dissociation curve (see Figure 3.19), not a linear or straight-line relationship.

The binding of oxygen depends on the **partial pressure** of oxygen. As the partial pressure of oxygen increases, the percentage of hemoglobin saturated with oxygen increases (Figure 3.19). This oxygen saturation or dissociation curve is not a linear relationship, rather there is a sigmoidal curve (Figure 3.19). In the alveolus, there are high partial pressures of oxygen and, consequently, oxygen binds to hemoglobin and the percentage saturated with oxygen reaches 100%.

As the blood passes through the capillaries in tissues, oxygen is released from the hemoglobin. This is possible because the partial pressure of oxygen in the tissues is low and, therefore, much but not all of the oxygen dissociates from hemoglobin (Figure 3.19). Note: there is still some oxygen bound to hemoglobin in the systemic veins after the blood passes through the tissues.

There is an additional factor. The pH (acidity or alkalinity) of the blood affects the oxygen saturation curve (Figure

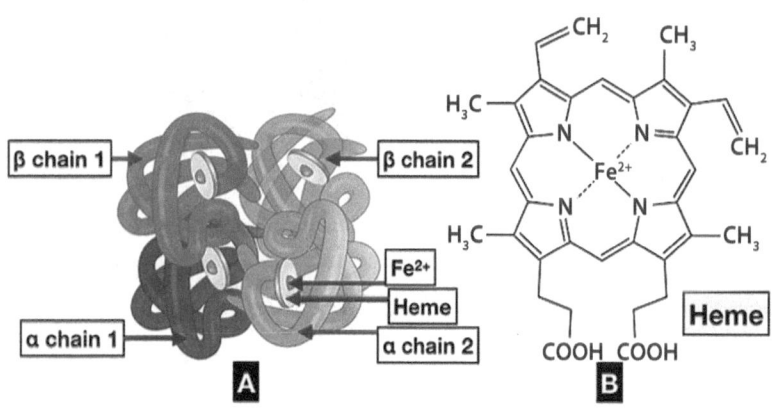

Figure 3.17 Structure of hemoglobin (A) and the heme group (containing ferrous iron Fe^{2+}) (B). (Adapted from OpenStax College, CC BY 4.0)

At lungs

$$Hb + O_2 \rightarrow Hb \sim O_2 + O_2 \rightarrow Hb \sim [O_2]_2 + O_2 \rightarrow$$
Hemoglobin
$$Hb \sim [O_2]_3 + O_2 \rightarrow Hb \sim [O_2]_4$$

At tissues

$$Hb + O_2 \leftarrow Hb \sim O_2 + O_2 \leftarrow Hb \sim [O_2]_2 + O_2$$
Hemoglobin
$$\leftarrow Hb \sim [O_2]_3 + O_2 \leftarrow Hb \sim [O_2]_4$$

Figure 3.18
Oxygen binds to hemoglobin in the lungs and dissociates from the hemoglobin at the tissues. (Note the reversed direction of the reaction!)

3.19). At lower pH, there is lower amounts of oxygen bound to hemoglobin (Figure 3.19). This is called the **Bohr effect**.

3.9.2 How Is This Bohr Effect Employed?

As blood passes to capillaries in tissues, the pH drops because of the release of carbon dioxide and increased concentrations of acids such as lactic acid and pyruvic acid. This enhances the amount of oxygen released from hemoglobin (see Figure 3.20).

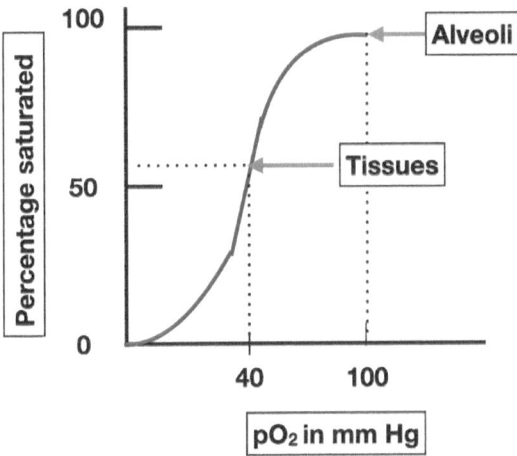

Figure 3.19 Saturation or dissociation curve for oxygen. At high partial pressure of oxygen, hemoglobin is maximally (100 %) saturated with oxygen. As the partial pressure of oxygen drops, there is a lower percentage of oxygen bound to hemoglobin. At the tissues, there is a low partial pressure of oxygen and, therefore, less oxygen is bound to hemoglobin.

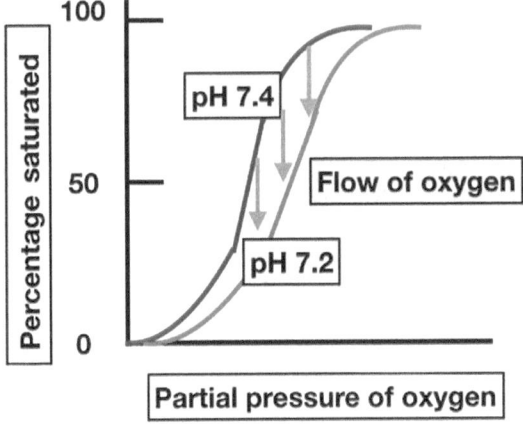

Figure 3.20 Effect of reducing blood pH on the percentage saturation of hemoglobin (the Bohr effect).

3.10 Transport of Carbon Dioxide

3.10.1 How Is Carbon Dioxide Transported in the Blood?

Carbon dioxide is transported in the blood in three forms in order of importance:

- As bicarbonate ions (~70%).
- Bound to hemoglobin as carbaminohemoglobin (20%).
- Simple solution (10%).

3.10.2 Carbonic Anhydrase and Carbon Dioxide Transport as Bicarbonate

A key enzyme in the transport of carbon dioxide (CO_2) is **carbonic anhydrase**. It catalyzes the formation of carbonic acid (H_2CO_3) from carbon dioxide and water. This is found in the erythrocytes.

$$H_2O + CO_2 \underset{}{\overset{\text{Carbonic anhydrase}}{\rightleftarrows}} H_2CO_3$$

At the tissues, the cells produce large quantities of carbon dioxide. This moves into the erythrocytes in capillaries passing down a concentration gradient (see Figure 8.20). In the erythrocytes in tissue capillaries, the reaction is as follows:

$$H_2O + CO_2 \overset{\text{Carbonic anhydrase}}{\rightarrow} H_2CO_3$$

The carbonic acid rapidly and spontaneously dissociated to a hydrogen ion (H^+ or proton) and a bicarbonate ion (HCO_3^-) (see Figure 8.20.) in a manner as follows:

$$H_2CO_3 \rightarrow H^+ + HCO_3^-$$

The hydrogen ions are buffered by the protein in the erythrocyte including hemoglobin. The bicarbonate ions leave the erythrocyte via a bicarbonate chloride antiport (a transporter for chloride and bicarbonate) (see Figure 3.21). To maintain the electric change of the erythrocyte, chloride moves into the erythrocyte also via the bicarbonate chloride antiport (see Figures 3.21 and 3.22). This is referred to as the **chloride shift** (see Figure 3.21).

In the lungs, as blood passes through the pulmonary capillaries (Figure 3.22), carbon dioxide moves into the alveoli. How does that occur? Firstly, carbon dioxide moves down as concentration gradient. Carbon dioxide is generated from carbonic acid catalyzed by carbonic anhydrase in the erythrocytes.

$$\text{H}_2\text{CO}_3 \xrightarrow{\text{Carbonic anhydrase}} \text{H}_2\text{O} + \text{CO}_2$$

The carbonic acid is formed rapidly from bicarbonate ions as follows (see Figure 3.21.).

$$\text{H}^+ + \text{HCO}_3^- \rightarrow \text{H}_2\text{CO}_3$$

Bicarbonate ions pass into the erythrocytes, passing down a concentration gradient and due to the bicarbonate chloride antiport (see Figure 3.21.). To maintain electrical balance, chloride ions move out of the erythrocytes due to the bicarbonate chloride antiport (see Figure 3.22). This is the chloride shift.

3.10.3 Carbon Dioxide Binding to Hemoglobin

Carbon dioxide also binds reversibly to hemoglobin forming **carbaminohemoglobin** as illustrated in the equations below.

$$\begin{array}{cc} \text{At tissues} & \text{At alveolus} \\ \text{Hb} + \text{CO}_2 \rightarrow \text{Hb} \sim \text{CO}_2 & \rightarrow \text{Hb} + \text{CO}_2 \end{array}$$

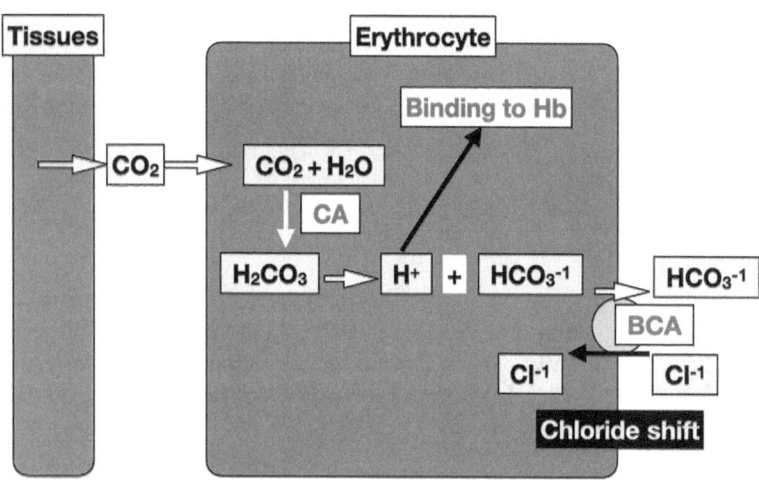

Figure 3.21 Flow of carbon dioxide (CO_2) in the capillaries adjacent to the tissues showing the roles of carbonic anhydrase (CA) and bicarbonate ions together with the chloride shift.
[Key: BCA = bicarbonate chloride anion exchanger; CA = carbonic anhydrase]

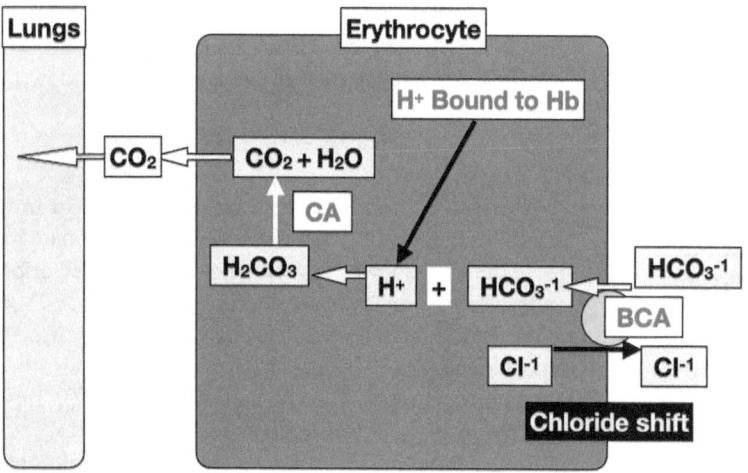

Figure 3.22 Flow of carbon dioxide (CO_2) in the capillaries adjacent to the alveolus in the lungs showing the roles of carbonic anhydrase (CA) and bicarbonate ions together with the chloride shift.
[Key: BCA = bicarbonate chloride anion exchanger; CA = carbonic anhydrase]

3.11 Integrative Control of Respiration

3.11.1 Overview of Integrative Control of Respiration

The amount of oxygen brought into an animal can be increased by two processes (also see Textbox 3.5):

1. Increasing the **rate of respiration** within certain limits.
2. Increasing the **depth of breathing** or the tidal volume (also see Textbox 3.5 and Figure 3.11).

This is an attempt by the animal to meet its needs for oxygen, for instance, during exercise. There are both central and peripheral chemoreceptors that ensure respiration is meeting the needs of the animal.

Peripheral chemoreceptors in the **carotid** and **aortic bodies** respond to low oxygen and low pH (as a proxy for carbon dioxide); this leads to an increased rate of respiration (see Figure 3.23). **Central chemoreceptors** respond to low oxygen and low pH; this leads to an increased rate of respiration.

3.11.2 Why Does the Body Use pH as a Proxy for Carbon Dioxide (and Oxygen)?

As the carbon dioxide concentrations rise, there is an increase in hydrogen ions (protons) and, consequently, a reduction in the pH (see Figure 3.24).

$$CO_2 + H_2O \rightarrow H_2CO_3 \rightarrow HCO_3^- + H^+$$

Figure 3.24 Carbon dioxide is converted to carbonic acid in the brain, then rapidly dissociates, generating bicarbonate and hydrogen ions. With the increased concentrations in hydrogen ions, the pH declines.

There are **mechanoreceptors** in the airways (trachea, lung, and pulmonary vessels). These detect stretching and thereby prevent overinflation of the respiratory system. Output from **stretch receptors** in the lung prevents the lung from being overinflated by inhibiting the inspiratory neurons (see Figure 3.23). There are also responses to **irritants** in the airway that lead to coughing and sneezing.

3.11.3 What Is the Effect of Panting?

When an animal is panting to reduce their temperature, there is increased voiding of carbon dioxide. What does this do? The loss of carbon dioxide reduces both the plasma concentrations of bicarbonate and the concentrations of hydrogen ions (protons). With the decrease in concentrations of hydrogen ions, there is an increase in the pH in the plasma. This can lead to alkalosis.

3.12 Diseases Related to Respiration

Textbox 3.10. summarizes diseases related to respiration.

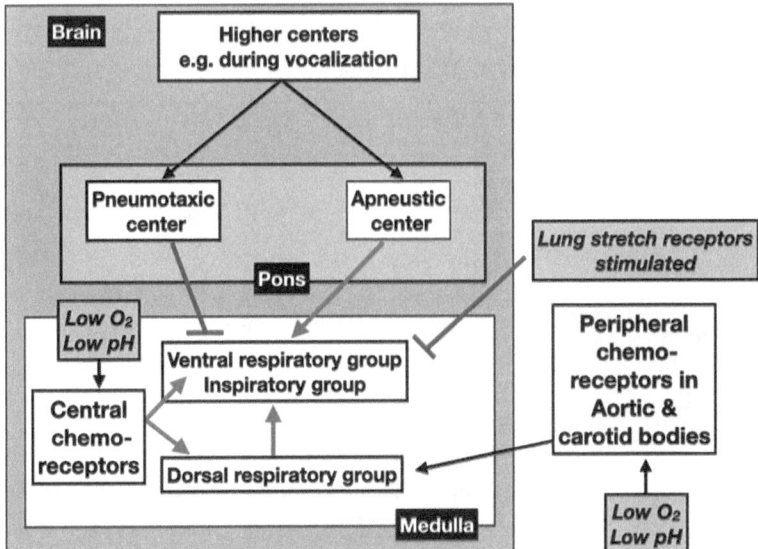

Figure 3.23 Influence of chemoreceptors and stretch receptors on the respiratory centers. [Key: ↑ is positive, ⊤ is negative]

Textbox 3.10 Diseases Related to Respiration

- **Hypoxia**: insufficient oxygen in the arterial blood to supply the needs of the tissues.

Respiratory Diseases in Horses
- **Equine asthma**: difficulty in breathing due to narrowing of airways and production of excessive mucus.
- **Heaves**: recurrent airway obstruction and inflammatory airway disease.
- **Pneumonia**: infection of the lungs by bacteria, viruses, and fungi. This is most commonly found in foals.
- **Pneumothorax**: collapsed lung(s).
- **Pulmonary edema**: an accumulation of fluid in the lungs.
- **Upper respiratory obstruction**: blockage in the trachea.

Respiratory Diseases in Dogs
- **Respiratory Alkalosis/Acidosis**: the normal pH of the blood is 7.36–7.44. In contrast, there can be acidemia (low blood pH < 7.36) or alkalemia (high blood pH > 7.44). Acidemia and alkalemia can be due to either respiratory or metabolic issues.
- **Respiratory acidosis**: there is an increase in blood pCO_2 due to respiratory insufficiency with inadequate movement of carbon dioxide from blood into the alveolus and out the body.
- **Respiratory alkalosis**: there is a decrease in blood pCO_2. This can be due to hyperventilation, for instance in a heat stressed dog.
- **Heat stroke**: dogs suffering heat stroke have core body temperatures of greater than 41°C. Dogs pant to lose heat but there is a limit to this. Heat stroke may cause the dog to collapse or have seizures.
- **Brachycephalic Obstructive Airway Syndrome (BOAS)**: Dogs with brachycephalic obstructive airway syndrome (BOAS) have respiratory problems including (1) shortness of breath, particularly when exercising; (2) overheating; and (3) respiratory noises such as snorting or snoring. Dogs with brachycephaly (short muzzle)—such as the bulldog, pug, (see Figure 3.24), and French bulldog—are much more likely to have BOAS.

Short Muzzle

Long Muzzle

Figure 3.25 Examples of dogs with short muzzles (pug) and long muzzles (bull terrier). (Top: Nancy Wong, CC BY-SA 4.0; Bottom: Goldmull, CC BY-SA 4.0)

Examples of dog breeds with short muzzles (brachycephalic breeds) includes Boston terriers, boxers, bull mastiffs, English bulldogs, French bulldogs, Pekingese, pugs, and shih tzus. Breeds of dogs that do not have this condition include collies, Dalmatians, German shepherds, golden retrievers, great Danes, and Labrador retrievers.

Endocrine System 4

Learning Objectives

1. Understand what a hormone is (see Section 4.1.1).
2. Know the differences between a hormone, a paracrine factor, and an autocrine factor (see Section 4.1.1).
3. Know the major endocrine glands and their hormones (see Section 4.1.2).
4. Understand the following for each hormone: (1) their chemistry; (2) what they do; and (3) how their release is controlled.
5. Understand how protein and peptide hormones are made (see Section 4.1.3.2).
6. Understand how modified amino acid hormones are made (see Section 4.1.3.3).
7. Understand how protein, peptide, and modified amino acid hormones are released (see Section 4.1.3.4).
8. Understand how protein, peptide, and modified amino acid hormones act (see Section 4.1.3.5).
9. Understand how steroid hormones are made (see Section 4.1.3.2).
10. Understand how steroid and related hormones act (see Section 4.1.3.4).
11. Understand how steroid hormones are released (see Section 4.1.3.4).
12. Understand how steroid hormones act (see Section 4.1.3.4).
13. Understand negative and positive feedback (see Section 4.1.4).
14. Understand the differences between the anterior pituitary gland and the posterior pituitary gland.
15. Understand the differences between protein/peptide and steroid hormones.
16. Understand the consequences of lack of a hormone or excess amounts of a hormone in domestic animals.

Table of Contents

4.1 Introduction
    4.1.1 Definition of a Hormone
    4.1.2 Introduction to the Endocrine System
    4.1.3 Chemistry, Synthesis, and Mode of Action of Hormones
        4.1.3.1 Overview of Proteins, Peptide, and Modified Amino Acid Hormones
        4.1.3.2 Synthesis of Proteins and Peptide Hormones
        4.1.3.3 Synthesis of Modified Amino Acid Hormones
        4.1.3.4 Release of Proteins, Peptide, and Modified Amino Acid Hormones
        4.1.3.5 Mechanism of Action of Proteins, Peptide, and Modified Amino Acid Hormones
        4.1.3.6 Overview of Steroids and Related Hormones
        4.1.3.7 Synthesis and Release of Steroid Hormones
        4.1.3.8 Mechanism of Action of Steroid Hormones
    4.1.4 Negative and Positive Feedback
4.2 Pituitary Gland
    4.2.1 Anterior Pituitary Gland
        4.2.1.1 Overview of Anterior Pituitary Gland
        4.2.1.2 Adrenocorticotropic Hormone (ACTH)
            4.2.1.2.1 Synthesis of Adrenocorticotropic Hormone (ACTH)
            4.2.1.2.2 Action of ACTH
            4.2.1.2.3 Control of ACTH Release
        4.2.1.3 Follicle-Stimulating Hormone (FSH)
            4.2.1.3.1 Chemistry of FSH
            4.2.1.3.2 Actions of FSH
            4.2.1.3.3 Control of FSH Release
        4.2.1.4 Luteinizing Hormone (LH)
            4.2.1.4.1 Chemistry of LH
            4.2.1.4.2 Actions of LH
            4.2.1.4.3 Control of LH release
        4.2.1.5 Growth Hormone (GH)/Somatotropin (ST)
            4.2.1.5.1 Chemistry of Growth Hormone
            4.2.1.5.2 Actions of Growth Hormone
            4.2.1.5.3 Control of Growth Hormone Release
        4.2.1.6 Prolactin (PRL)
            4.2.1.6.1 Chemistry of Prolactin
            4.2.1.6.2 Actions of Prolactin
            4.2.1.6.3 Control of Prolactin Release
        4.2.1.7 Thyroid-Stimulating Hormone (TSH)
            4.2.1.7.1 Chemistry of TSH

- 4.2.1.7.2 Actions of TSH
- 4.2.1.7.3 Control of TSH Release
- 4.2.2 Posterior Pituitary Gland
  - 4.2.2.1 Introduction to the Posterior Pituitary Gland
  - 4.2.2.2 Pars Intermedia
  - 4.2.2.3 Pars Nervosa
  - 4.2.2.4 Chemistry of Pars Nervosa Hormones
  - 4.2.2.5 Oxytocin (Mesotocin in Chickens)
    - 4.2.2.5.1 Actions of Oxytocin
    - 4.2.2.5.2 When Is Oxytocin Released?
  - 4.2.2.6 Arginine Vasopressin (AVP)/Antidiuretic Hormone (ADH)
    - 4.2.2.6.1 What Does AVP Do?
    - 4.2.2.6.2 When Is AVP Released?
- 4.3 Thyroid Gland
  - 4.3.1 Thyroid Gland
  - 4.3.2 Thyroid Hormones
  - 4.3.3 Synthesis of Thyroxine
  - 4.3.4 Actions of Thyroid Hormones
  - 4.3.5 Transport of T_4 and T_3 in the Plasma
  - 4.3.6 Control of the Release of T_4 and T_3
- 4.4 Parathyroid Gland
  - 4.4.1 Parathyroid Hormone
    - 4.4.1.1 Chemistry of Parathyroid Hormone
    - 4.4.1.2 Action of Parathyroid Hormone
    - 4.4.1.3 Control to the Release of Parathyroid Hormone
- 4.5 Calcitonin
- 4.6 Adrenal Gland
  - 4.6.1 Introduction to the Adrenal Glands
  - 4.6.2 Where Are Adrenal Hormones Made?
  - 4.6.3 Epinephrine and Norepinephrine
    - 4.6.3.1 Where Are Epinephrine and Norepinephrine Synthesized?
    - 4.6.3.2 What is the Structure of Epinephrine and Norepinephrine?
    - 4.6.3.3 What Does Epinephrine and Norepinephrine Do in Animals?
    - 4.6.3.4 Neuronal Control of Epinephrine and Norepinephrine Release
    - 4.6.3.5 When Do Animals Release Epinephrine and Norepinephrine?
  - 4.6.4 Cortisol (Corticosterone in Poultry)
    - 4.6.4.1 What Does Cortisol (or Corticosterone in Poultry) Do?
    - 4.6.4.2 How Is Release of Cortisol (or Corticosterone in Poultry) Controlled?
  - 4.6.5 Aldosterone
    - 4.6.5.1 What Does Aldosterone Do?
    - 4.6.5.2 How Is Aldosterone Synthesis Controlled?
- 4.7. Islets of Langerhans (Pancreatic Islets or Endocrine Pancreas)
  - 4.7.1 Introduction to the Islets of Langerhans
  - 4.7.2 Glucagon
    - 4.7.2.1 Chemistry of Glucagon

68 THEME 1 Homeostasis via the Blood

- 4.7.2.2 What Does Glucagon Do?
- 4.7.2.3 When Is Glucagon Released?
- 4.7.3 Insulin
  - 4.7.3.1 Structure and Synthesis
  - 4.7.3.2 What Does Insulin Do?
  - 4.7.3.3 When Is Insulin Released?
  - 4.7.3.4 How Do Intestinal Hormones Influence Insulin Release?
- 4.8 Gastrointestinal Hormones
  - 4.8.1 Introduction to Gastrointestinal Hormones
  - 4.8.2 What Does Gastrin Do?
  - 4.8.3 How Is Gastrin Release Controlled?
- 4.9 Other Endocrine Glands
  - 4.9.1 Liver
  - 4.9.2 Pineal
  - 4.9.3 Kidneys
- 4.10 Endocrine Differences Between Livestock/Companion Animals and Poultry
- 4.11 Examples of Endocrine Diseases in Domestic Animals

4.1 Introduction

The endocrine system communicates signals or hormones to regulate physiological functions such as metabolism and reproduction.

4.1.1 Definition of a Hormone

A **hormone** is a chemical that is released from an endocrine cell directly into the blood stream. The hormone then passes to a target tissue, where it binds to a receptor and exerts a **specific effect**. An **autocrine factor** is a chemical that is released from an autocrine cell and acts on the same cell. A **paracrine factor** is a chemical that is released from a paracrine cell and acts on an adjacent cell.

4.1.2 Introduction to the Endocrine System

Textbox 4.1 provides a glossary of terms related to the endocrine system while common endocrine abbreviations are listed in Textbox 4.2.

The major endocrine glands are the following:

1. **Pituitary gland.**
2. **Thyroid and parathyroid glands.**
3. **Adrenal gland.**
4. **Endocrine pancreas.**
5. **Endocrine gonads** (also discussed in Chapters 13 and 14).

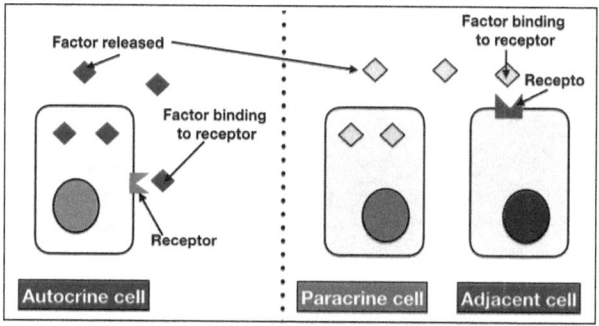
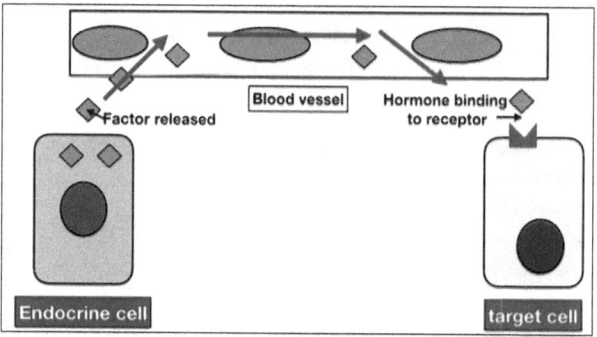

Figure 4.1 Schematic representation of the differences between a paracrine factor or an autocrine factor (above) and a hormone (below).

Textbox 4.1 Glossary of Terms Related to the Endocrine System

- **Endocrine cells**: Cells that produce hormones, releasing them into the bloodstream.
- **Endocrine gland**: A gland that makes and releases hormones.
- **Endocrine system**: Aggregate of endocrine glands.
- **Feedback**: This is when hormone A stimulates the release of hormone B and hormone B affects the release of hormone A.
- **Hormone**: A specific chemical released from an endocrine cell, passing via the bloodstream to target tissues exerting a specific effect.
- **Negative Feedback**: This is when hormone A stimulates the release of hormone B and hormone B decreases the release of hormone A. This allows for hormone A to maintain a setpoint.
- **Positive Feedback**: This is when hormone A stimulates the release of hormone B and hormone B increases the release of hormone A. This allows for amplification until a physiological outcome is complete.
- **Peptide**: A hormone made up of amino acids in a short chain.
- **Protein hormone**: A hormone made up of amino acids in a long chain.
- **Secretion of hormones**: Release of a hormone.
- **Steroid hormone**: A hormone based on cholesterol.
- **Synthesis of hormones**: Production of hormones.
- **Target tissue**: Hormones act on specific or target tissue.

Textbox 4.2 Endocrine Abbreviations

ACTH: Adrenocorticotropic hormone.
AVP: Arginine vasopressin.
cAMP: Cyclic adenosine monophosphate.
FSH: Follicle-stimulating hormone.
GH: Growth hormone.
LH: Luteinizing hormone.
PTH: Parathyroid hormone.
ST: Somatotropin.
T_3: Triiodothyronine.
T_4: Thyroxine.
TSH: Thyroid-stimulating hormone.

In addition, there are organs that have endocrine functions, but these are not their primary functions:

1. **Gastrointestinal tract**.
2. **Pineal gland**.
3. **Liver**.
4. **Kidneys**.
5. **Adipose tissue**.

There are three mechanisms of hormonal release, namely the following:

1. **Humoral stimuli**: ions (e.g., calcium (Ca^{2+}) or sodium (Na^+) or other chemicals) stimulating or inhibiting synthesis and/or release of a second hormone.
2. **Hormonal stimuli**: a hormone stimulates or inhibits synthesis and/or release of a second hormone.
3. **Neural stimuli**: a neurotransmitter or neuropeptide stimulates or inhibits synthesis and/or release of a hormone.

Additionally, hormones interact with each other by three known mechanisms:

- **Permissive**: one hormone is required to exert its action before a second hormone can exert a response.
- **Synergistic**: two hormones acting together to provide a greater response than the aggregate of each alone.
- **Oppositional**: two hormones working to exert opposite effects than the other.

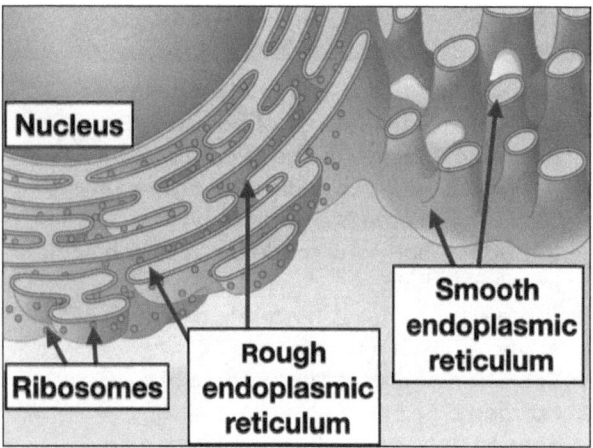

Figure 4.2 Protein hormones are synthesized on ribosomes in the rough endoplasmic reticulum (adjacent to the nucleus) while steroid hormones are synthesized in the smooth endoplasmic reticulum. (Adapted from OpenStax Anatomy and Physiology, CC BY 4.0)

4.1.3 Chemistry, Synthesis, and Mode of Action of Hormones

Hormones can be classified into two groups based on their synthesis, release, structure, and how they act:

- Proteins, peptide, and modified amino acid hormones.
- Steroids, related hormones, and thyroid hormones.

4.1.3.1 Overview of Proteins, Peptide, and Modified Amino Acid Hormones

Proteins and peptide hormones are derived through translation of mRNA. This category of hormones also includes peptides cleaved from a protein precursor, glycoproteins (containing carbohydrates), and acylated peptides (containing a fatty acid) or sulfated peptides (containing a sulfate). In addition, this category of hormones includes modified amino acids such as epinephrine and melatonin.

4.1.3.2 Synthesis of Proteins and Peptide Hormones

Proteins and peptide hormones are synthesized in the **rough endoplasmic reticulum** (see Figure 4.2). They are packaged into **secretory vesicles** or secretory granules in the Golgi apparatus in the cytoplasm of the endocrine cell (see Figure 4.3). The hormone in the vesicles is stored until the time it is needed.

The secretory vesicles function to do the following (See Figure 4.3):

- Store the protein/peptide hormone.
- Transport the protein/peptide hormone in the cytoplasm.
- Release the contents of the secretory vesicles.

Prior to, or after, this packaging, there is proteolytic cleavage and/or other chemical modification to generate the mature hormone.

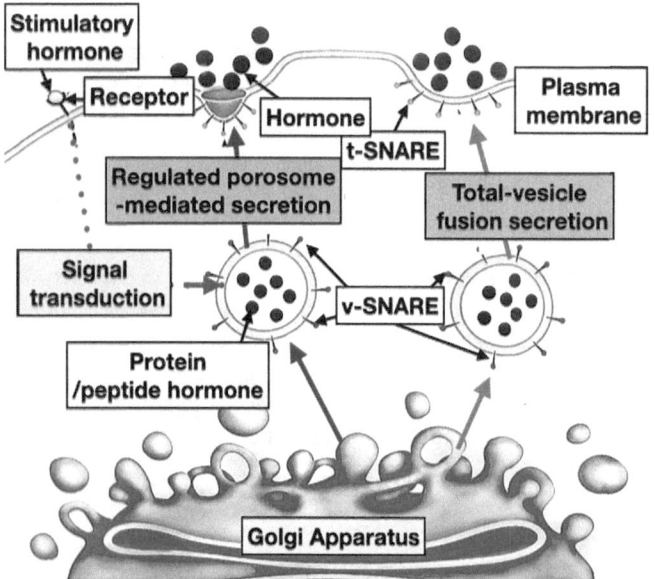

Figure 4.3 Protein and peptide hormones are packaged into secretory vesicles in the Golgi apparatus. When the endocrine cell is stimulated by a hormone (or another stimulus), the v-SNARE and t-SNARE bind such that the contents of the secretory vesicles are released.
[Key: v-SNARE = vesicle SNARE; t-SNARE = target SNARE, a SNAP receptor with SNAP being soluble N-ethylmaleimide-sensitive factor attachment proteins]
(Adapted from Wonjinee, CC BY-SA 4.0)

4.1.3.3 Synthesis of Modified Amino Acid Hormones

Modified amino acid hormones are synthesized through a series of enzymatic conversions both in the **cytoplasm** and in secretory vesicles. The modified amino acid hormones are taken up by secretory vesicles. They are stored until needed.

4.1.3.4 Release of Proteins, Peptide, and Modified Amino Acid Hormones

Protein, peptide hormone, and modified amino acids are released by the process of **exocytosis**. When either a stimulating hormone binds to receptors on the cell membrane or other stimulatory agent activates the **signal transduction system** of the cell. Examples of intracellular signal transduction are **cyclic adenosine monophosphate** (**cAMP**) or **calcium ions** (see Figure 4.3).

> **Textbox 4.3 A Deeper Dive Into the Release of Protein, Peptide, and Modified Amino Acid Hormones**
>
> **SNARE protein** molecules are important for the release of protein, peptide, and modified amino acid hormones; SNARE referring to SNAP Receptor, and SNAP for soluble N-ethylmaleimide-sensitive factor attachment proteins. When the signal transducer system is activated (such as cAMP), the vesicles fuse with the cell membrane with **v-SNARE** (vesicle SNAREs) binding to t-SNARE (target SNAREs on the cell membrane) (Figure 4.3). The t-SNAREs are either in the plasma membrane or in organelles in the plasma membrane called **porosomes** (Figure 4.3).

4.1.3.5 Mechanism of Action of Proteins, Peptide, and Modified Amino Acid Hormones

Proteins, peptide, and modified amino acid hormones act by binding to **extracellular receptors** on the surface of target cells. The hormone receptor complex initiates a series of changes inside the cells or signal transduction. It frequently involves an intracellular messenger such as cAMP and **phosphorylation** (activation) of intracellular proteins (see Figure 4.4).

4.1.3.6 Overview of Steroids and Related Hormones

Steroids and related hormones are synthesized from cholesterol. Unlike protein and peptide hormones, they have a low solubility in water. Steroids and related hormones include the following hormones:

1. Steroid hormones that influence **sodium homeostasis** (**mineralocorticoids** such as aldosterone).

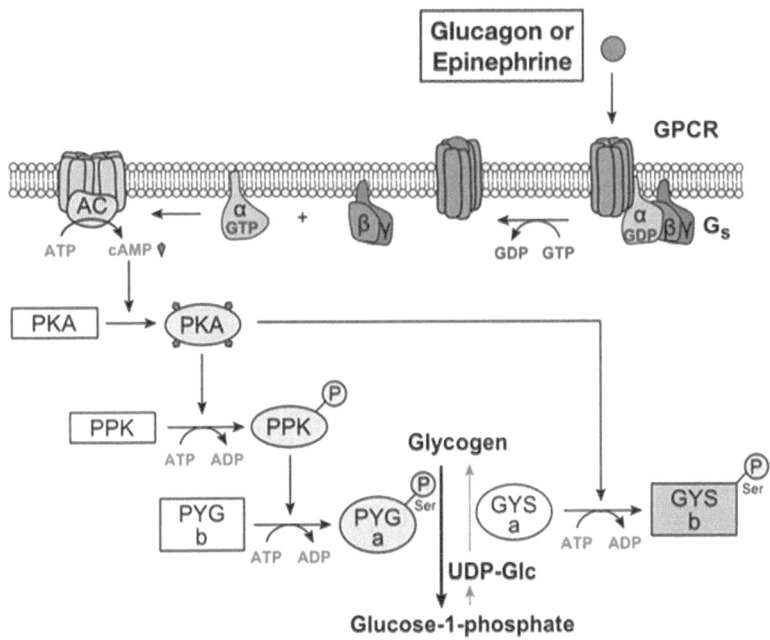

Figure 4.4 Example of the mechanism of action of protein/peptide hormone glucagon.
[Key:
AC = adenylate cyclase;
ATP = adenosine triphosphate;
cAMP = cyclic adenosine monophosphate;
GDP = guanosine-diphosphate;
GTP = guanosine-triphosphate;
GYS = glycogen synthase;
PKA = protein kinase A.]
(Adapted from FrozenMan, CC BY-SA 4.0)

2. Steroid hormones influencing glucose metabolism (**glucocorticoids** such as cortisol).
3. Steroid hormones inducing female characteristics (**estrogens** such as estradiol).
4. Steroid hormones inducing male characteristics (**androgens** such as testosterone).
5. Steroid hormones maintaining pregnancy (**progestogens** such as progesterone).
6. The steroid-related and vitamin D hormone **1,25 dihydroxy vitamin D** (1,25-dihydroxycholecalciferol or calcitriol).

4.1.3.7 Synthesis and Release of Steroid Hormones

Steroid hormones are synthesized from cholesterol in the smooth endoplasmic reticulum (see Figures 4.2 and 4.5). Once produced, the hormones are released. There is very little storage of steroid hormones in the cells that produce them. (Synthesis of thyroid hormones is considered in Section 4.12).

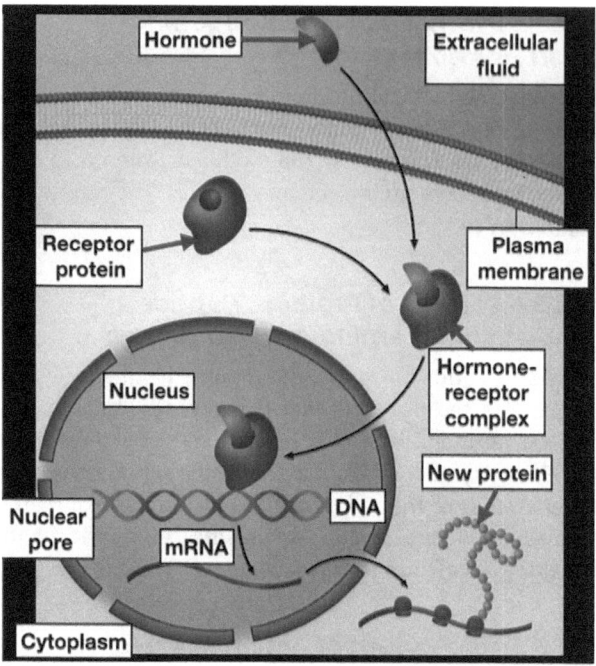

Figure 4.6 Mechanism of action of steroid hormones. (Adapted from Ali Zehan, CC BY-SA 4.0)

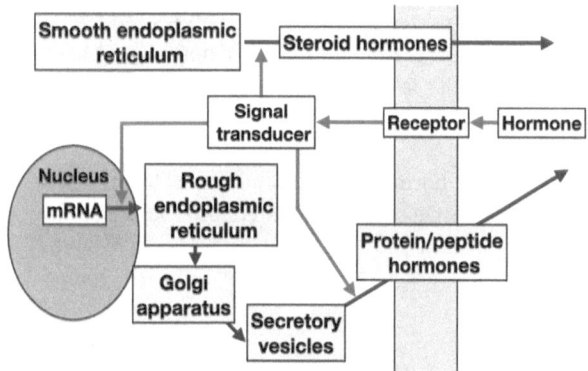

Figure 4.5 Comparison of synthesis of protein/peptide with that of steroid hormones.

4.1.3.8 Mechanism of Action of Steroid Hormones

These act by binding to intracellular receptors associated with specific genes in the nucleus of target cells. Binding of the hormone results in increased transcription of the gene (see Figure 4.6). In addition, steroid hormones can bind to receptors and exert effects in the cytoplasm.

Steroid hormones and thyroid hormones (thyroxine and triiodothyronine) are frequently transported in the blood bound to specific binding proteins. Thyroid hormones act in a manner similar to steroid hormones.

4.1.4 Negative and Positive Feedback

The concept of negative feedback is important in physiology and is the primary feedback mechanism. This can be summarized as follows:

A stimulates B which inhibits A

More commonly, it is of the following pattern.

A stimulates B which stimulates C which inhibits A

Examples of negative feedback are with reproductive hormones. The hypothalamus stimulates the anterior pituitary gland to release luteinizing hormone (LH) (see Figure 4.7). In turn, LH stimulates the testes to produce testosterone which has a negative feedback effect on the hypothalamus and anterior pituitary gland (see Figure 4.7). Similarly, in females, LH stimulates the corpus luteum to produce progesterone (see Figure 4.7). This progesterone then has a negative feedback effect. Another example of negative feedback is discussed under adrenal cortex (below).

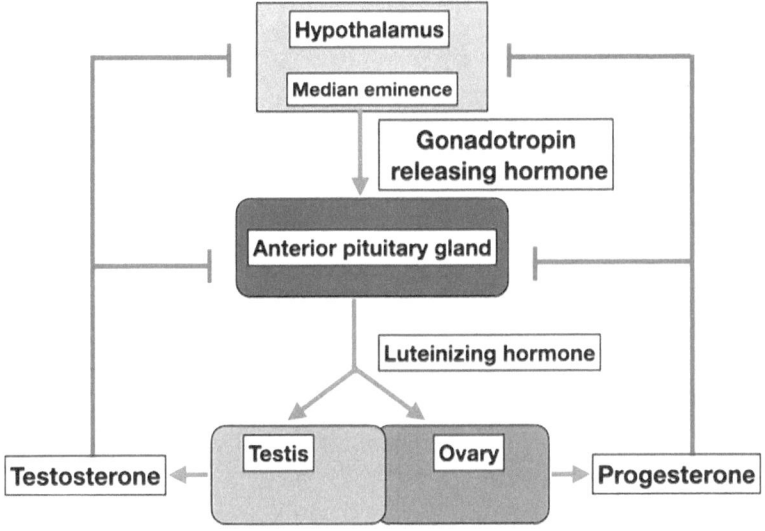

Figure 4.7 Negative feedback of gonadal hormones. Testosterone synthesized in the interstitial (endocrine) cells in males and progesterone from the corpus luteum (in the ovary) of female mammals feed back on hypothalamic release of gonadotropin-releasing hormone (GnRH) and pituitary release of luteinizing hormone (LH). [Key: → is positive, T is negative]

Positive feedback is when a hormone stimulates the release of a second hormone and this stimulates release of the first hormone. This can be summarized as follows:

A stimulates B which stimulates A

An example of positive feedback is with estradiol (see Figure 4.8). Gonadotropin-releasing hormone (GnRH) stimulates release of LH (see Figure 4.8). The LH stimulates release of estradiol from the follicle (see Figure 4.8). The estradiol stimulates the release of both gonadotropin-releasing hormone and LH (see Figure 4.8).

4.2 Pituitary Gland

There are three major parts of the pituitary gland:

- **Anterior lobe** (or pars distalis).
- **Intermediate** component (pars intermedia).
- **Pars nervosa**.

The pars intermedia and pars nervosa make up the posterior lobes of the pituitary gland (see Figure 6.4).

4.2.1 Anterior Pituitary Gland
4.2.1.1 Overview of Anterior Pituitary Gland

The anterior pituitary gland is located at the base of the brain (see Figure 4.9). It receives blood in blood

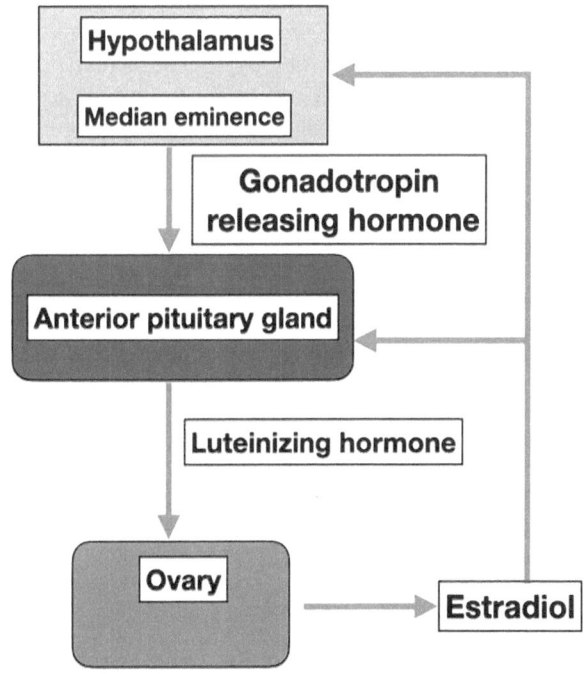

Figure 4.8 Positive feedback of estradiol from the mature follicle in the ovary of female mammals. The feedback stimulates the hypothalamic release of gonadotropin-releasing hormone (GnRH) and pituitary release of luteinizing hormone (LH). [Key: ↑ is positive]

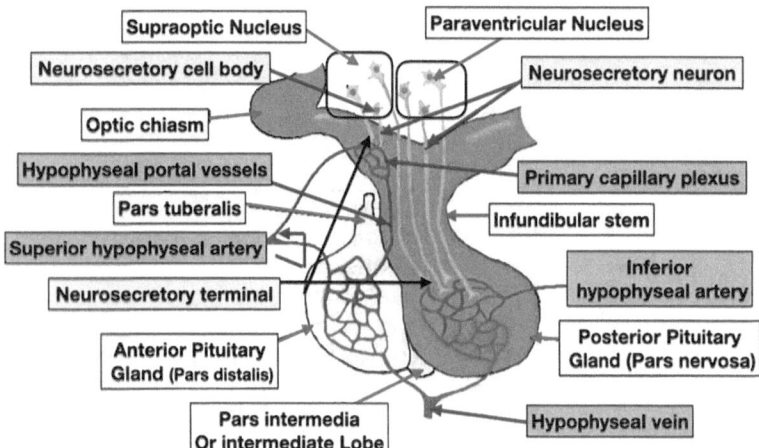

Figure 4.9 Schematic diagram of the hypothalamus and pituitary gland. (Takuma-sa, CC BY 3.0)

vessels that bathe the **median eminence** (part of the hypothalamus) (see Figure 4.9). Releasing factors are secreted from neurosecretory terminals from the median eminence into the **hypophyseal portal blood** and these control the release of anterior pituitary hormones. Note: A **portal blood vessel** is defined as a vein that passes from one tissue or organ to another tissue or organ. A **neurosecretory cell** releasing hormones is shown in Figure 4.10.

There are six hormones produced by the anterior pituitary gland (summarized in Table 4.1):

- **Adrenocorticotropic hormone (ACTH).**
- **Follicle-stimulating hormone (FSH).**
- **Luteinizing hormone (LH).**
- **Growth hormone (GH)/somatotropin (ST).**
- **Prolactin (PRL).**
- **Thyroid-stimulating hormone (TSH).**

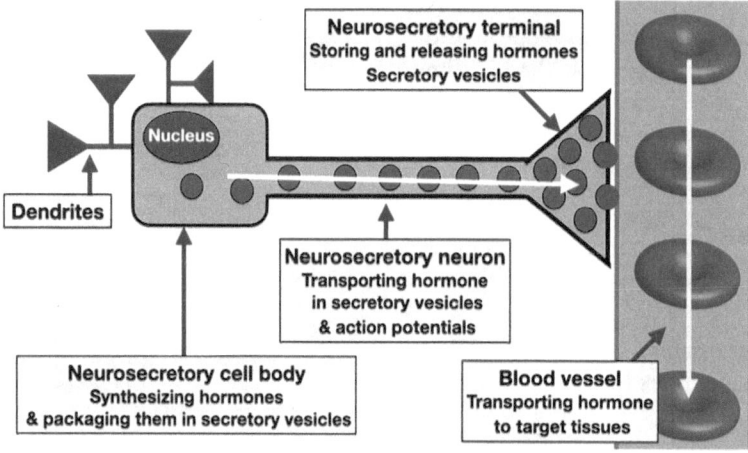

Figure 4.10 Conceptual diagram of a neurosecretory cell. The hormone is synthesized in the rough endoplasmic reticulum of the cell bodies. It is packaged into secretory vesicles in the Golgi apparatus. The secretory vesicles containing the hormone pass down the axon of the neurosecretory neuron and are stored in the terminal of the neurosecretory cell. When stimulated by action potentials passing down the neuron, the secretory vesicles fuse with the plasma membrane of the terminal and pass into the blood and then to the target tissues. (Images of erythrocytes from Togo picture gallery, maintained by Database Center for Life Science.)

Table 4.1 Hormones of the pituitary gland.

Name	Function in livestock and major mammalian companion animals
Anterior pituitary gland (pars distalis)	
Adrenocorticotropic hormone (ACTH) aka corticotropin	Synthesis and release of glucocorticoids, such as cortisol, together with mineralocorticoids (aldosterone) from adrenal cortical cells increased (↑).
Follicle stimulating hormone (FSH)	Female (♀) • Follicle growth ↑ and oocyte development Male (♂) • Spermatogenesis (production of sperm) ↑.
Growth hormone (GH) aka somatotropin	• Growth indirectly ↑ via increase in insulin-like growth factor-1 (IGF-1) from the liver. • Lipolysis. • Gluconeogenesis. • Milk production in cattle ↑.
Luteinizing hormone (LH)	♀ • Ovulation • Causes corpus luteum to form from remnants of follicle after ovulation. • Progesterone synthesis ↑. ♂ • Production of testosterone by testes ↑.
Prolactin	Initiation and maintenance of milk production.
Thyroid stimulating hormone (TSH) aka thyrotropin	Thyroxine (T_4) and triiodothyronine (T_3) release ↑.
Posterior pituitary gland	
Pars intermedia (not present in birds)	
α-melanocyte stimulating hormone (α-MSH)	Darkening of skin or hair
Pars nervosa	
Oxytocin	• Contraction of uterus during parturition. • Milk let-down by contraction of myoepithelial cells surrounding alveoli in mammary glands.
Arginine vasopressin aka antidiuretic hormone (ADH)	Reabsorption of water in the collecting ducts of the kidneys.

Anterior pituitary hormones are produced by separate cells in the gland. Their release is controlled by a series of releasing factors as summarized in Table 4.2.

4.2.1.2 Adrenocorticotropic Hormone (ACTH)

4.2.1.2.1 Synthesis of Adrenocorticotropic Hormone (ACTH)

Adrenocortical hormone (ACTH) is also known as corticotropin. ACTH is synthesized as a large polypeptide, **proopiomelanocortin** (**POMC**) (see Figure 4.11). This is cleaved by proteases to generate ACTH and other peptides.

4.2.1.2.2 Action of ACTH

As part of the response to long-term stress, ACTH acts to increase synthesis and release of cortisol (or corticosterone in poultry) from cortical cells in the adrenal gland (in cortex of adrenal in mammals). In addition, ACTH

Table 4.2 Hormones of the hypothalamus controlling the release of anterior pituitary hormones.

Name	Function
Stimulatory releasing hormones	
Corticotropin releasing hormone (CRH)	Release of ACTH increased (↑)
Gonadotropin releasing hormone (GnRH)	Release of LH ↑ and FSH ↑
Growth hormone releasing hormone (GHRH)	Release of GH ↑
Thyrotropin releasing hormone (TRH)	Release of TSH ↑
Inhibitory releasing hormones	
Prolactin releasing inhibitory factor (PIF) = Dopamine	Release of prolactin decreased (↓)
Somatostatin (GHIH)	Release of GH ↓
Possible releasing hormones	
Prolactin releasing peptide (PrRP)	Release of prolactin ↑
Ghrelin	Release of GH ↑

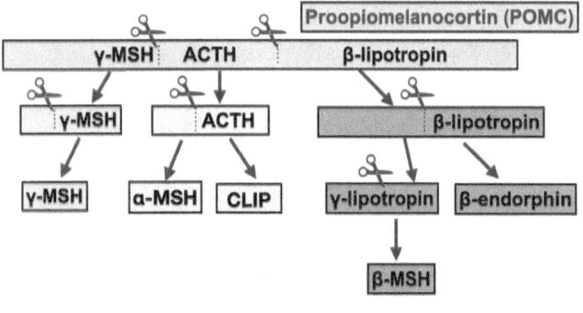

Figure 4.11 Proopiomelanocortin (POMC) is the translation peptide produced when the POMC gene is expressed. POMC is expressed and processed in the anterior pituitary gland to adrenocorticotropic hormone (ACTH) and γ-lipotropin. In the intermediate lobe of the pituitary (pars intermedia), hypothalamus, and skin, ACTH is processed to β-melanocyte-stimulating hormone (β-MSH). γ-lipotropin is processed to β-endorphin. In addition, β-MSH is generated.

stimulates the production of **aldosterone** from cortical cells and thereby increases sodium reabsorption (see Section 4.6.4 for more details about cortisol and aldosterone).

4.2.1.2.3 Control of ACTH Release

ACTH release is stimulated by two releasing factors:

- **Corticotropin-releasing hormone (CRH).**
- **Arginine vasopressin (AVP)** in livestock and companion animals (instead there is lysine vasopressin in pigs and arginine vasotocin in poultry).

In addition, cortisol inhibits ACTH release in a negative feedback manner.

4.2.1.3 Follicle-Stimulating Hormone (FSH)

4.2.1.3.1 Chemistry of FSH

Follicle-stimulating hormone (FSH) is a **glycoprotein hormone** composed of two subunits (α- and β-). It is one of the two gonadotropins along with luteinizing hormone (LH).

4.2.1.3.2 Actions of FSH

FSH stimulates the following:

- Follicular growth in females.
- Estrogen production by theca and granulosa cells of the follicle.
- Spermatogenesis in males by supporting the Sertoli cells and production of **androgen-binding protein**.

4.2.1.3.3 Control of FSH Release

FSH release is stimulated by the releasing hormone **gonadotropin-releasing hormone (GnRH)**. FSH release is inhibited by the gonadal hormone **inhibin** in a negative feedback manner. Moreover, release of FSH is stimulated by estrogens and inhibited by progesterone.

4.2.1.4 Luteinizing Hormone (LH)

4.2.1.4.1 Chemistry of LH

Luteinizing hormone (LH) is a glycoprotein hormone composed of two subunits (an α- and a β- sub-unit).

4.2.1.4.2 Actions of LH

LH stimulates the following in females:

- Ovulation (mammals and birds).
- Formation of the **corpus luteum** (mammals only).
- Synthesis of progesterone by the corpus luteum (mammals only).

LH also stimulates the interstitial endocrine cells (a.k.a. Leydig cells) of the testes to produce testosterone.

4.2.1.4.3 Control of LH release

LH release is stimulated by the releasing hormone gonadotropin-releasing hormone (GnRH). LH release is inhibited by testosterone and inhibin in males. In females, release of LH is stimulated by estrogens and inhibited by progesterone and inhibin.

4.2.1.5 Growth Hormone (GH)/ Somatotropin (ST)

4.2.1.5.1 Chemistry of Growth Hormone

Growth hormone (GH)/somatotropin (ST) is a protein hormone.

4.2.1.5.2 Actions of Growth Hormone

Growth hormone (GH)/somatotropin (ST) is essential to growth, particularly that of the skeleton. It acts in an indirect manner. GH stimulates the liver and other organs to produce insulin like **growth factor 1 (IGF-1)**. In turn, IGF-1 stimulates growth, particularly of bone. GH also reduces triacylglyceride in adipose tissue. GH acts by binding to the GH receptor. In some cases, animals lack a functional GH receptor resulting in proportionate dwarfism. This is seen in both cattle and chickens.

In ruminants, milk production (**galactopoiesis**) is increased by growth hormone (GH)/somatotropin (ST). In addition, GH stimulates **lipolysis** (break down of triglyceride) in adipose tissue (lipolysis is discussed in Section 7.7.1 in Chapter 7).

4.2.1.5.3 Control of Growth Hormone Release

Release of growth hormone is controlled by the following releasing hormones:

- **GH-releasing hormone** (GHRH) (stimulatory).
- **Ghrelin** (stimulatory).
- **Somatostatin** (inhibitory).

In addition, IGF-1 inhibits GH release in a negative feedback manner.

4.2.1.6 Prolactin (PRL)

4.2.1.6.1 Chemistry of Prolactin

Prolactin is a protein hormone.

4.2.1.6.2 Actions of Prolactin

Prolactin has a number of effects related to lactation:

- It is one of the hormones that stimulates the development of the mammary glands inducing lobular alveolar growth (mammogenesis).
- It is essential to the initiation of lactation (lactogenesis).
- It maintains milk production (galactopoiesis).

Prolactin has other effects. For instance, prolactin inhibits reproduction. In addition, in turkeys, prolactin induces incubation behavior, or **broodiness** (sitting on eggs). Prolactin is also related to growth of both wool and horns. Prolactin acts via binding to **prolactin receptors**.

4.2.1.6.3 Control of Prolactin Release

Prolactin release is controlled by at least two releasing factors:

- Prolactin-releasing factor(s) (PRF) stimulates prolactin release. Prolactin release is stimulated by **vasoactive intestinal peptide** and **thyrotropin-releasing hormone**.
- **Dopamine** inhibits prolactin release. (Dopamine is a neurotransmitter—see Chapter 5).

4.2.1.7 Thyroid-Stimulating Hormone (TSH)

4.2.1.7.1 Chemistry of TSH

Thyroid-stimulating hormone (TSH) is also known as thyrotropin. TSH is a glycoprotein hormone made up of two subunits (an α- and a β-).

4.2.1.7.2 Actions of TSH

TSH has two effects:

- Increasing secretion of thyroid hormones
- Increasing growth of the thyroid.

4.2.1.7.3 Control of TSH Release

TSH release is controlled by one stimulatory releasing factor, TRH. In addition, thyroid hormones inhibit TSH release in a negative feedback manner.

4.2.2 Posterior Pituitary Gland

4.2.2.1 Introduction to the Posterior Pituitary Gland

The posterior pituitary gland consists of the following:

- **Pars intermedia.**
- **Pars nervosa.**

4.2.2.2 Pars Intermedia

In mammals, there is an intermediate lobe (pars intermedia) which is part of the posterior pituitary gland (see Figure 4.9). The cells of the pars intermedia produce alpha MSH (α-MSH) (see Figure 4.11). Camargue horses, an ancient breed of horses from France, lose their dark coat progressively with age, becoming white. During this time, plasma concentrations of α-MSH also decline. There is also a link between α-MSH and seasonal breeding in sheep. Pituitary pars intermedia dysfunction (PPID) is seen in horses and there are very high plasma concentrations of α-MSH. α-MSH is also released from keratocytes (discussed in Chapter 10) but also functions as a neuropeptide in the brain influencing reproduction and feed intake.

4.2.2.3 Pars Nervosa

The pars nervosa is a neural or neurosecretory gland that is essentially a down-cropping of the hypothalamus. The hormones of the pars nervosa are the following:

- **Oxytocin.**
- **Arginine vasopressin** (or equivalent in other species) or antidiuretic hormone (ADH).

4.2.2.4 Chemistry of Pars Nervosa Hormones

These hormones are peptides with nine amino acids (**nonapeptide**) and are synthesized in the rough endoplasmic reticulum of neurosecretory cell bodies in the hypothalamus (Figure 4.10). They are packaged into secretory vesicles in close association with **neurophysins**. These pass down the axons of neurosecretory neurons and are stored in neurosecretory terminals (Figure 4.10). When the **neurosecretory nerve** is stimulated, the secretory vesicles fuse with the plasma membrane. The peptide hormones are dissociated from the neurophysins and are released into the bloodstream (Figure 4.10).

4.2.2.5 Oxytocin (Mesotocin in Chickens)

The equivalent of oxytocin in birds is mesotocin. As poultry do not produce milk or give birth, mesotocin does not have a role related to either of these. Instead, mesotocin acts in the process of egg laying.

4.2.2.5.1 Actions of Oxytocin

Oxytocin has the following actions:

- Stimulates milk let down by inducing contraction of myoepithelial cells in the mammary gland.
- Causes contractions of the myometrium in the uterus during labor.

There is also evidence that oxytocin has behavioral effect; for instance, aiding bonding between people and dogs and increasing docility and friendliness in horses. In addition, there is release of oxytocin during ejaculation. Oxytocin induces contractions of the vas deferens and increases transit of spermatozoa.

4.2.2.5.2 When Is Oxytocin Released?

Oxytocin is released in **neuroendocrine reflexes** in response to either stimulation of the teat or of the cervix. Release of oxytocin is inhibited by stress in dairy cattle. Whereas fetal stress is suspected to induce the release of oxytocin during parturition. Oxytocin is also released during ejaculation.

4.2.2.6 Arginine Vasopressin (AVP)/ Antidiuretic Hormone (ADH)

Antidiuretic hormone (ADH) is arginine vasopressin (AVP) in cattle, sheep, horses, dogs, and cats. The equivalent in pigs is lysine vasopressin and, in chickens, is arginine vasotocin. AVP/ADH is produced by expression of the AVP gene in the hypothalamus.

4.2.2.6.1 What Does AVP Do?

AVP acts as an antidiuretic hormone by increased water resorption in the collecting ducts and distal convoluted tubules (discussed in Chapter 12). AVP also has a pressor effect acting on vascular smooth muscle. In chickens, arginine vasotocin stimulates the laying of eggs.

4.2.2.6.2 When is AVP Released?

AVP is released when **plasma osmolarity** increases above a set point; the osmolarity being detected by the sensory neurosecretory neurons in hypothalamus. The increased resorption of water corrects the osmolarity. There are other factors influencing AVP release. These include angiotensin-II and hypoglycemia (low blood glucose).

4.3 Thyroid Gland

4.3.1 Thyroid Gland

The thyroid gland is located adjacent to the trachea. There is a single thyroid gland in mammals but two in birds. The thyroid gland is made up of multiple follicles. The follicles are composed of **follicular cells** surrounding a **lumen** filled with colloid (see Figure 4.12). In mammals, there are also some cells between the follicles, called **parafollicular cells** or C cells that produce the hormone **calcitonin**. Also in mammals, there are a total of four nodules on the surface of the thyroid. These are the parathyroid glands.

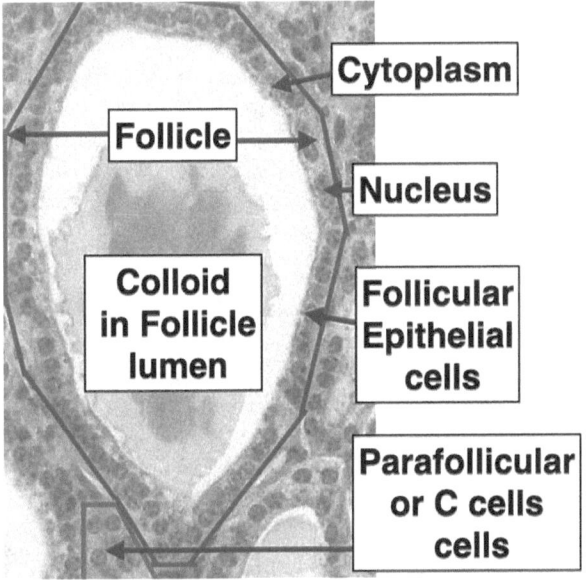

Figure 4.12 Histology of horse thyroid. (Adapted from Uwe Gille, CC BY-SA 3.0)

4.3.2 Thyroid Hormones

There are two thyroid hormones:

- Thyroxine (T_4).
- Triiodothyronine (T_3).

The structure of T_4 and T_3 is shown in Figure 4.13. Both T_4 and T_3 are made in the thyroid gland. In addition, T_4 is converted to the more active thyroid hormone, T_3, in the liver. Like many other hormones, both T_3 and T_4 are **amino acid-based hormones**. They are unusual as they contain **iodine atoms** (three in T_3 and four in T_4). However, T_3 acts as if it were a steroid hormone binding to a **nuclear receptor** and, thereby, increases expression of specific genes.

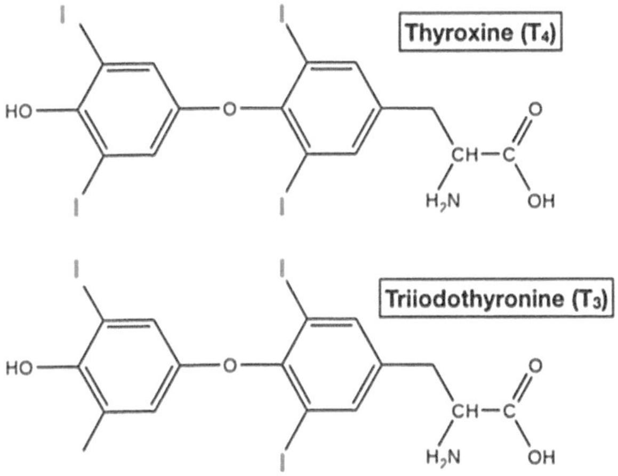

Figure 4.13 Structure of thyroxine (T_4) and triiodothyronine (T_3). The I indicates iodine atoms. (Adapted from Wesalius, public domain)

4.3.3 Synthesis of Thyroxine

Figure 4.14 summarizes how T_3 and T_4 are made. The backbone for the synthesis of T_3 and T_4 is the protein **thyroglobulin**. This is made in the rough endoplasmic reticulum of the follicular cells in the thyroid. Thyroglobulin is secreted into the colloid in the **lumen of the follicles** (Figure 4.12). **Iodide** is transported into the follicular cells by a sodium iodide symporter and then to the lumen by the transporter pendrin. In the lumen, the iodide is oxidized and iodinates tyrosine residues in thyroglobulin. The tyrosine residues are linked together. Thyroglobulin containing T_3 and T_4 is taken up the follicular cells. This is broken down by **proteinases** generating T_3 and T_4 which transit into interstitial space and then into the blood stream. Synthesis and release of T_3 and T_4 is stimulated by the pituitary hormone, thyroid-stimulating hormone (TSH).

4.3.4 Actions of Thyroid Hormones

Thyroid hormones maintain metabolism at the optimal level increasing **metabolism** when the animal is **cold** and decreasing metabolism when the animal is either hot or not getting enough to eat. Thyroid hormones are important for allowing growth, reproduction, and lactation.

4.3.5 Transport of T_4 and T_3 in the Plasma

T_4 and T_3 have limited solubility in water and are transported in the blood bound reversibly to carrier proteins: **thyroxine-binding globulin** (TGB), **transthyretin**, and serum albumin.

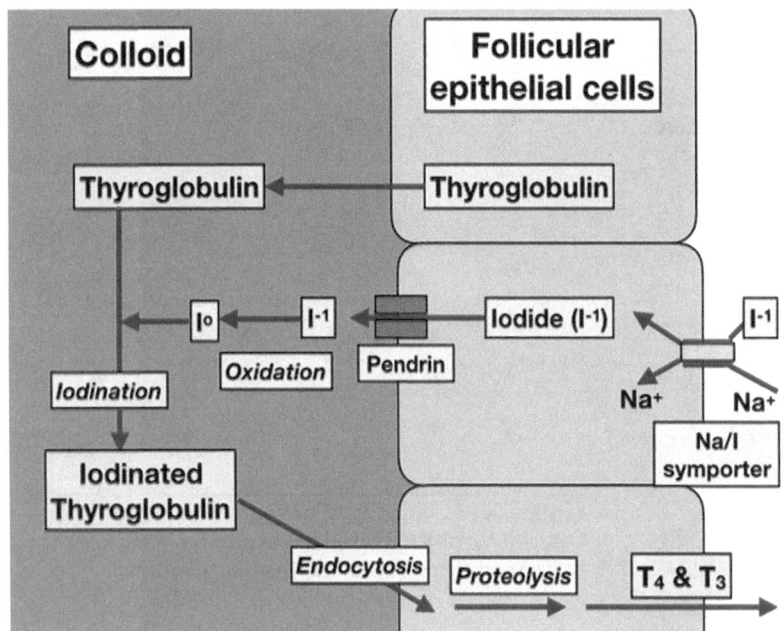

Figure 4.14 Synthesis of thyroxine (T_4) and triiodothyronine (T_3).

Free   *Bound*

TGB or transthyretin or albumin + T_4/ T_3 ⇆ T_4/ T_3 bound to TGB or transthyretin or albumin

4.3.6 Control of the Release of T_4 and T_3

Release of T_3 and T_4 is stimulated by the pituitary hormone thyroid-stimulating hormone (TSH).

4.4 Parathyroid Gland

The parathyroid glands are on the surface of the thyroid glands of mammals but are distinct endocrine organs in birds. The parathyroid glands produce the hormone **parathyroid hormone** (PTH).

4.4.1 Parathyroid Hormone

4.4.1.1 Chemistry of Parathyroid Hormone

Parathyroid hormone (PTH) is a protein with 84 amino acid residues that plays a major role in **calcium homeostasis**. PTH is encoded by the PTH gene. There are also PTH related peptide genes. PTH related peptide also affects calcium homeostasis.

4.4.1.2 Action of Parathyroid Hormone

PTH is the *most important* hormone controlling plasma concentrations of calcium. PTH release is stimulated by low plasma concentrations of calcium (see Figure 4.15). In turn, PTH has multiple actions to bring plasma calcium concentrations back to the required level (see Figure 4.15). Without this action of PTH in bringing plasma concentrations of calcium back to the normal level, many bodily functions cease to act. PTH increases plasma concentrations of calcium by mobilizing calcium in bones.

PTH acts in the following ways (see Figure 4.15):

- Stimulating the **osteoclasts** to increase **bone resorption**. Thereby, PTH increases calcium release from bones.
- Stimulating the kidney (the proximal convoluted tubules and collective ducts) to increase reabsorption of calcium.
- Stimulating the kidney to produce 1,25 dihydroxy vitamin D (calcitriol). This travels to the small intestine via the blood stream and increases absorption of calcium.

PTH acts by binding to the PTH receptor.

4.4.1.3 Control to the Release of Parathyroid Hormone

Release of PTH is increased when there are low plasma concentrations of calcium (see Figure 4.15).

4.5 Calcitonin

Calcitonin is a hormone that is produced by **C cells** that are either distributed between the follicles of the thyroids (mammals) or from the paired **ultimobranchial**

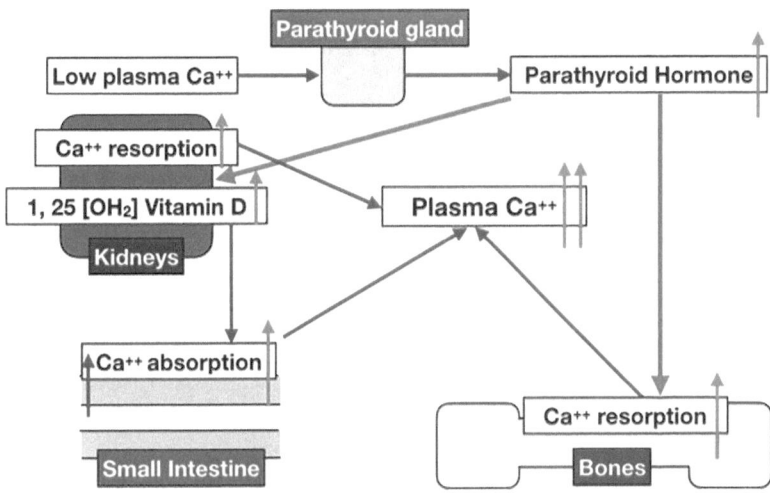

Figure 4.15 Actions of parathyroid hormone controlling blood concentrations of calcium (Ca^{++}). This is achieved partly by an increase of 1,25 dihydroxy vitamin D (1,25 $(OH)_2$ vitamin D).

glands (birds). Calcitonin is a 32-amino-acid-containing peptide.

Calcitonin is released when blood calcium concentrations are too high. Calcitonin only plays a minor role in maintaining plasma concentrations of calcium. Calcitonin act to protect the animal against hypercalcemia. Calcitonin acts via a specific receptor in the cell membrane, the calcitonin receptor.

There is also a peptide called the **Calcitonin Gene Related Peptide** (CGRP). This is a neuropeptide made by alternative splicing of the expressed calcitonin RNA. CGRP acts via a receptor similar to the calcitonin receptor.

4.6 Adrenal Gland

4.6.1 Introduction to the Adrenal Glands

There are two paired adrenal glands. These small organs are located on the anterior surface of the kidneys. There are two distinct components of adrenal glands (see Figure 4.16):

- **Cortical cells** in the adrenal cortex surrounding the medulla.
- **Chromaffin cells** in the adrenal medulla in the center of the adrenal glands.

4.6.2 Where Are Adrenal Hormones Made?

Chromaffin cells in the adrenal medulla produce the following hormones:

- **Epinephrine** (a modified amino acid).
- **Norepinephrine** (a modified amino acid).

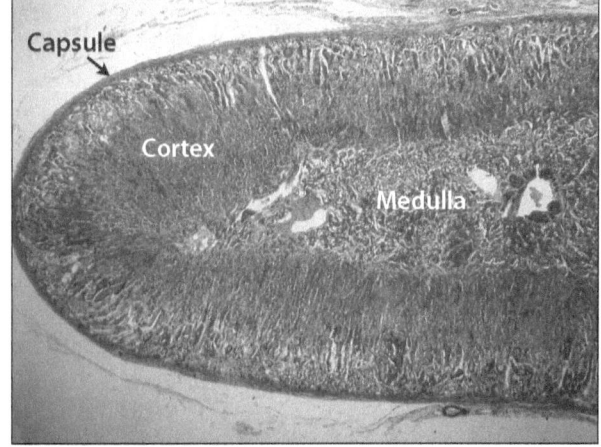

Figure 4.16 Section through the mammalian adrenal showing central medulla composed of chromaffin cells and the cortex composed of cortical cells.

Cortical cells in the adrenal cortex in domesticated mammals produce the following hormones:

- Cortisol (a steroid).
- Aldosterone (a steroid).
- Adrenal androgens (a steroid) in both males and females (see Textbox 4.4 for more details).

4.6.3 Epinephrine and Norepinephrine

4.6.3.1 Where Are Epinephrine and Norepinephrine Synthesized?

Epinephrine and norepinephrine are both modified amino acids. They are synthesized from **tyrosine** in the chromaffin cells.

Textbox 4.4 A Deeper Dive into Adrenal Androgens

The adrenal cortex also produces a series of weak androgens such as the following:

- Androstenedione (A_4), a chemical that has been abused by human and animal athletes.
- Androstenediol (A_5).
- Dehydroepiandrosterone (DHEA).
- Dehydroepiandrosterone sulfate (DHEAS).
- 11-ketotestosterone (11-KT).

These are weak androgens but can be converted in multiple tissues to the male hormone testosterone, and/or to the female hormones estradiol and other estrogens. These in turn have effects.

Textbox 4.5 Interesting History: Discovery of the Fight or Flight Response

The American physiologist Walter Cannon discovered and developed the concept of "fight or flight" based on studies with dogs and cats. When animals are exposed to a physical emergency, there is a physiological response to danger (or acute response to stress).

Figure 4.18 A frightened cat. (Shutterstock/shymar27)

Among the fight or flight responses are the following:

- Increased heart rate.
- Increased respiration rate.
- Shifts in blood flow to organs and muscles and away for the skin and gastrointestinal tract.
- Increased blood concentrations of glucose.

Walter Cannon also developed the concept of homeostasis.

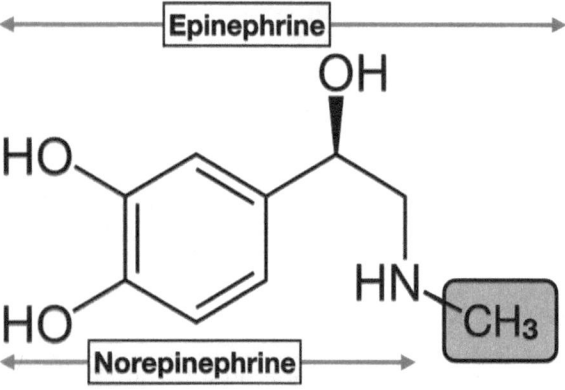

Figure 4.17 Structure of epinephrine and norepinephrine. Note: Norepinephrine lacks a methyl (CH_3) group. The methyl group is circled and highlighted.
(Adapted from NEUROtiker, public domain)

4.6.3.2 What is the Structure of Epinephrine and Norepinephrine?

The structure of epinephrine is shown in Figure 4.17. The structure of norepinephrine is the same as epinephrine except that the methyl group (CH_3) is missing.

The two hormones of the adrenal medulla are the following:

- Epinephrine (adrenaline in UK English).
- Norepinephrine (noradrenaline in UK English).

4.6.3.3 What Does Epinephrine and Norepinephrine Do in Animals?

Epinephrine and norepinephrine have multiple effects that equip animals to maximize their chances of survival in an emergency situation. These effects include the following:

- Decrease in blood flow to skin and intestine.
- Increase in blood flow to skeletal muscle.
- Increase in heart rate.

- Increase in respiration rate.
- Mobilizing glucose from glycogen stores in the liver.
- Mobilizing triacylglycerol to free fatty acids and glycerol.
- Dilated pupils that allow more light into the eyes (improving peripheral vision).

Effects of epinephrine and norepinephrine are mediated by them binding to adrenergic receptors. There are multiple types include α1, α2, β1, β2 and β3.

4.6.3.4 Neuronal Control of Epinephrine and Norepinephrine Release

Acetyl choline is released from nerve terminals adjacent to the chromaffin cells and stimulates the release of epinephrine and norepinephrine.

4.6.3.5 When Do Animals Release Epinephrine and Norepinephrine?

The **fight or flight response** is the response of animals to a dangerous situation. This concept was developed using cats and dogs (see Textbox 4.4). Today, we cannot ethically expose animals to physical dangers to see how they respond. Instead, we can measure plasma concentrations of epinephrine and norepinephrine in domestic animals undergoing husbandry procedures to determine their effects on animal welfare. For instance, plasma concentrations of epinephrine and norepinephrine are greatly increased during transportation of pigs. In addition, plasma concentrations of epinephrine and norepinephrine are influenced by physical activity being, for instance, increased during exercise in horses. Plasma concentrations of epinephrine are increased in dogs that are standing and alert compared to dogs laying down.

4.6.4 Cortisol (Corticosterone in Poultry)

Cortisol and corticosterone are a type of steroid hormones that are called **glucocorticoids**. These hormones are called **glucocorticoids** due to their ability to increase the blood concentration of *gluc*ose, are produced in the *cort*ex and is a ster*oid*.

Glucocorticoid hormones are transported in the blood plasma bound to **corticosteroid binding globulin** and serum albumin. The structure of cortisol is shown in Figure 4.19.

4.6.4.1 What Does Cortisol (or Corticosterone in Poultry) Do?

The major effects of cortisol are the following:

- Increasing gluconeogenesis (synthesis of glucose from many amino acids, glycerol, and lactate) in the liver. **Gluconeogenesis** is converting many but not all amino acids into glucose.
- Increasing plasma concentrations of glucose. This synthesis of glucose from amino acids results in increased plasma concentrations of glucose and greater glycogen in the liver.
- Increased **proteolysis** (breakdown of muscle protein) to generate amino acids that are then used to make glucose.
- Increased lipolysis (breakdown of triacylglycerol in adipose tissue) to generate free fatty acids and glycerol. The fatty acids are used to produce energy, substituting for glucose, while the glycerol is converted to glucose.
- Decreased utilization of glucose.

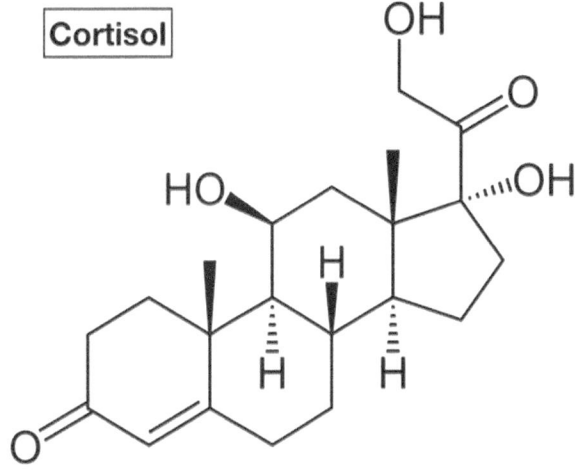

Figure 4.19 Structure of cortisol. (Courtesy of NEUROtiker)

4.6.4.2 How Is Release of Cortisol (or Corticosterone in Poultry) Controlled?

Cortisol (or corticosterone in poultry) is released during acute phase stress. Its release is stimulated by the anterior pituitary hormone, adrenocorticotropic hormone (ACTH), which is released in response to two releasing hormones, corticotropin-releasing hormone (CRH) and arginine vasopressin (AVP) (or arginine vasotocin [VT] in poultry) (see Figure 4.20). These are released from, respectively, the hypothalamus and the posterior pituitary gland.

4.6.5 Aldosterone

Aldosterone is a steroid hormone in the type called mineralocorticoids.

84   THEME 1   Homeostasis via the Blood

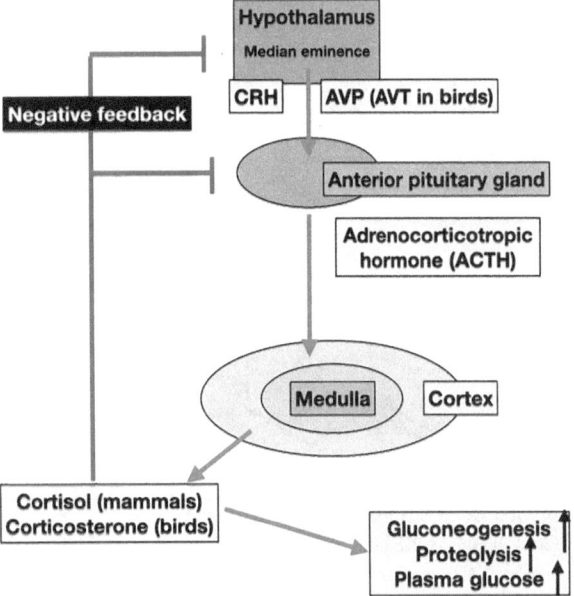

Figure 4.20 Control and actions of cortisol (corticosterone in poultry) production including negative feedback at both hypothalamic and pituitary levels. [Key: AVP = arginine vasopressin (lysine vasopressin in pigs); AVT = arginine vasotocin; and CRH = corticotropin releasing hormone, ↑ is positive, T is negative]

4.6.5.1 What Does Aldosterone Do?

Aldosterone increases reabsorption of sodium in the kidneys by acting on the **proximal convoluted tubules** and **collecting ducts** to increase reabsorption of sodium ions. Aldosterone increases the expression of sodium channels and the sodium/potassium pump (sodium/potassium ATPase). Chloride ions and some water pass passively with the sodium ions. The increase in sodium, chloride, and water in the plasma will then result in an increase in the blood pressure.

4.6.5.2 How Is Aldosterone Synthesis Controlled?

Aldosterone production is increased by **angiotensin II**. The juxtaglomerular cells in the kidneys release the protease, **renin**, in response to **sympathetic nervous stimuli** and low blood pressure. This enzyme converts **angiotensinogen** to **angiotensin I**. In turn, angiotensin I is converted to angiotensin II by **angiotensin converting enzyme (ACE)** during passage through the lungs. In addition, ACTH stimulates aldosterone production.

4.7. Islets of Langerhans (Pancreatic Islets or Endocrine Pancreas)

4.7.1 Introduction to the Islets of Langerhans

Not only does the pancreas have an important role in digestion (see Chapter 11) but also it is one of the important endocrine glands. Distributed through the pancreas are endocrine cells arranged in islets of Langerhans (pancreatic islets). The two main cell types in these islets are the following (see Figure 4.21):

- **Alpha (α) cells.** These produce the hormone **glucagon**. This hormone increases plasma concentrations of glucose.
- **Beta (β) cells.** These produce the hormone **insulin**. This hormone decreases plasma concentrations of glucose.

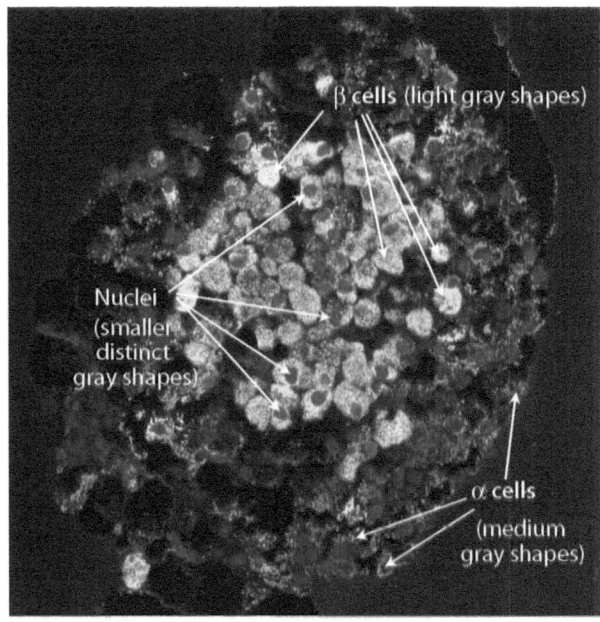

Figure 4.21 Islets of Langerhans in the pancreas. Nuclei are stained blue. Beta (β) cells that produce insulin are stained green. Alpha (α) cells that produce glucagon are stained red. (Note: Colors only visible in eBook). (Adapted from Masur, CC BY 2.5)

4.7.2 Glucagon

4.7.2.1 Chemistry of Glucagon

Glucagon is a 29-amino-acid-containing polypeptide. It is produced by cleavage of the product of the glucagon gene in the islet α cells (see Figure 4.22).

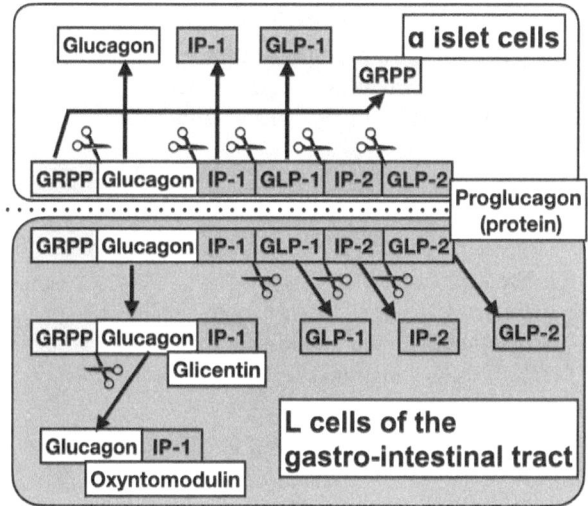

Figure 4.22 Comparison of the processing of the protein, pro-glucagon, in α islet cells with that in L cells of the intestine. [Key: GRRP = glicentin-related polypeptide; GLP-1 = glucagon-like peptide 1; GLP-2 = glucagon-like peptide 2; IP = intervening peptide]

4.7.2.2 What Does Glucagon Do?

Glucagon acts to increase plasma concentrations of glucose by the following:

- Stimulating **glycogenolysis** (breakdown of glycogen releasing glucose-1-phosphate which in turn is converted to glucose) in the liver.
- Stimulating gluconeogenesis (synthesizing glucose from amino acids, glycerol, and lactate) in the liver.
- Inducing a shift from utilizing glucose to using fatty acids as the energy source.
- Increasing lipolysis (breakdown of triacylglycerol) in adipose tissue in some species (e.g., chickens).
- Inhibiting **lipogenesis** (synthesis of fatty acids).

Glucagon acts via binding to the **glucagon receptor** (GSGR) (see Figure 4.2). In turn, **adenylate** cycle is activated via **G proteins** (see Figure 4.2). Adenylate cycle catalyzes the conversion of adenosine triphosphate (ATP) to cyclic adenosine monophosphate (cyclic AMP or cAMP) (see Figure 6.2). The cAMP activates **protein kinase A** (see Figure 4.2). Glycogen synthase is inactivated via a series of protein kinases (see Figure 4.2).

4.7.2.3 When Is Glucagon Released?

Glucagon is released when plasma concentrations of glucose are low. This makes release of glucose under humoral or regulatory control. In contrast, secretion of glucagon is stopped when plasma concentrations of glucose are high. In addition, release of glucagon is stimulated by epinephrine but suppressed in the presence of somatostatin or insulin. Parenthetically, insulin is produced by beta cells in the islets of Langerhans (see Figure 4.21), and somatostatin is produced by delta cells in the islets of Langerhans and acts in a paracrine manner.

4.7.3 Insulin

The discovery of insulin is discussed in Textbox 4.6.

Textbox 4.6 Interesting History: Dogs and the Discovery of Insulin

Four researchers were involved in the discovery of insulin:

- Frederick Banting (physician).
- Charles Best (Banting's assistant).
- John MacLeod (physician, professor, and laboratory director at the University of Toronto).
- James Collip (an associate professor in biochemistry from the University of Alberta, he was brought into the team by Professor MacLeod).

Banting and Best used two sets of dogs in their goal to establish the causative hormone for diabetes:

1. Pancreatectomized dogs (pancreases surgically removed) developed diabetes with high blood concentrations of glucose (Group A).
2. Dogs with their pancreatic ducts tied off. This led to the atrophy of the exocrine tissues but left the islets intact (Group B).

They found that a pancreatic extract from the Group B dogs was effective in lowering the blood glucose concentrations in Group A dogs. This was the first demonstration of insulin.

Crude extracts of the pancreas were not effective with diabetic patients because of the proteases from the exocrine pancreas. Dr. Collip played a critical role in developing methods to purify insulin and this was used

(continued)

Textbox 4.6 (continued)

to successfully treat diabetics. The source for insulin from the 1920s was cattle and pig pancreases because of their availability in packing plants. This insulin was called beef/pork insulin.

A Sour Note
The Nobel Prize was awarded to only Frederick Banting and John MacLeod for the discovery of insulin. Dr. Banting did share half his prize with Best, but poor Dr. Collip was left out completely.

4.7.3.1 Structure and Synthesis

Insulin is a protein hormone consisting of two polypeptide chains:

- **A chain** with 21 amino acid residues.
- **B chain** with 30 amino acid residues.

These are linked together by disulfide bridges (see Figure 4.23).

Insulin is synthesized as **pre-pro-insulin**. The B and A chains become aligned (see Figure 4.23) as disulfide bridges are formed. The C chain and signal peptide are then clipped off by specific proteases and insulin is formed.

4.7.3.2 What Does Insulin Do?

Insulin reduces plasma concentration of glucose, pushing energy into storage. This is summarized in Figure 4.25. Among the effect of insulin are the following:

- Increasing synthesis of glycogen (the storage form of glucose) in the liver and skeletal muscles.
- Decreasing glycogenolysis (breakdown of glycogen) in the liver and skeletal muscles.
- Increasing uptake of glucose by myocytes and adipocytes by moving glucose transporter 4 (GLUT 4) into the cell membranes.

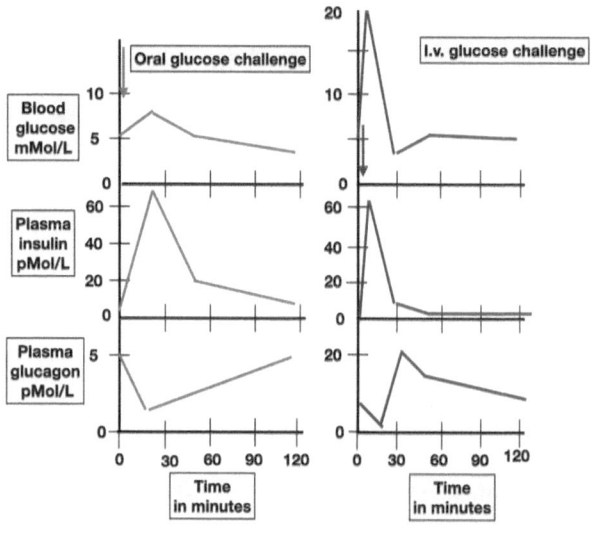

Figure 4.24 Changes in plasma concentrations of glucose, insulin, and glucagon in fasted pigs receiving either an oral or intravenous challenge with glucose.

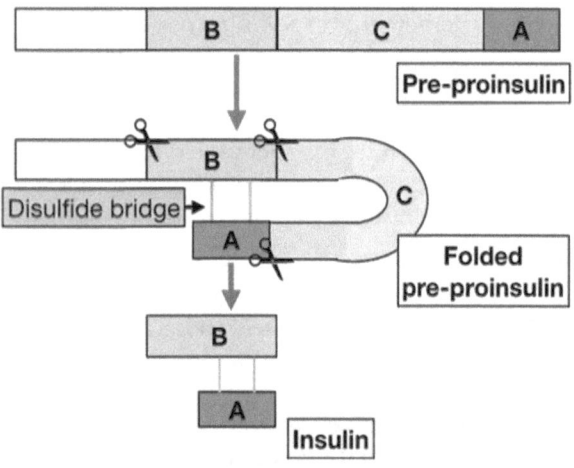

Figure 4.23 Synthesis of insulin as two peptide chains linked together by disulfide bridges. (A, B, and C are A chains; B = B chains; and C = the linking C chain.)

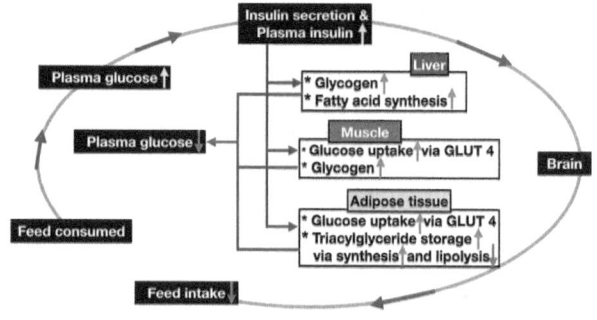

Figure 4.25 Schematic representation of the effects of insulin on carbohydrate and lipid metabolism.

- Increasing storage of **triacylglycerol** in adipose tissue by increasing synthesis and decreasing lipolysis (breakdown of triacylglycerol).

In addition, insulin has effects in the brain, such as facilitating development and growth. Moreover, insulin depresses feed intake.

4.7.3.3 When Is Insulin Released?

Insulin is released after a meal. Specifically, high concentrations of glucose stimulate the release of insulin from the β islet cells.

4.7.3.4 How Do Intestinal Hormones Influence Insulin Release?

The magnitude of the increase in plasma concentrations of insulin in response to a glucose challenge is greater if the glucose is given by mouth compared to intravenous administration (see Figure 4.19). This is because gastrointestinal hormones such as glicentin, gastric inhibitory peptide, and glucagon-like peptides are released when there is food in the intestinal tract. These stimulate insulin release. This is the **incretin effect**.

4.8 Gastrointestinal Hormones

4.8.1 Introduction to Gastrointestinal Hormones

The first protein/peptide hormone to be discovered was a gastrointestinal hormone (see Textbox 4.7).

> **Textbox 4.7 Interesting History: Secretin—The First Hormone Discovered**
>
> In 1901, two English physiologists, William Bayliss and Ernest Starling, discovered secretin. This was the first hormone described and was based on experimentation with dogs. They called the hormone *secretin*.

There are multiple types of hormones producing enteroendocrine cells (see Table 4.3). These produce a series of hormones that are important to gastrointestinal functioning (see Table 4.3). These are located in the epithelium

Table 4.3 Summary of gastrointestinal hormones.

Hormone	Cell type producing	Where produced	Action
Somatostatin	D or delta cells	Stomach and duodenum (part of small intestine)	Multiple inhibitory effects
Gastrin	G cell	Stomach	Increasing release of stomach acid
Cholecystokinin	I cell	Duodenum and jejunum (parts of small intestine)	Stimulating contraction of the gall bladder and relaxation of the hepato-pancreatic sphincter causing bile to pass from gall bladder into the small intestine
Gastric inhibitory peptide (GIP)	K cells	Duodenum and jejunum (parts of small intestine)	Increasing insulin release and depresses gastric muscle contractions
Glicentin, oxyntomodulin, GLP-1 & GLP-2	L cells	Colon and ileum (part of small intestine)	Increasing insulin release
Ghrelin	A and D cells	Stomach, duodenum, jejunum (small intestine) and possibly the crop in poultry	Appetite

(continued)

Table 4.3 Summary of gastrointestinal hormones (*continued*)

Hormone	Cell type producing	Where produced	Action
Histamine	Enterochromaffin-like cells (ELC cells)	Stomach	Multiple including stimulating release of gastrin
Motilin	M or Mo cells	Duodenum and jejunum	Controls gastric muscle contractions
Neurotensin	N cell	Small intestine	Smooth muscle contraction
Leptin	P cells	Stomach	Metabolism
Secretin	S cell	Duodenum	Exocrine pancreatic secretions (bicarbonate) released; stomach secretions inhibited
Serotonin	EC cells (Enterochromaffin cells)	Stomach	Multiple

and represent only about 1% of epithelial cells. Collectively enteroendocrine cells are the largest endocrine organ in animals. An important gastrointestinal hormone covered below is gastrin.

4.8.2 What Does Gastrin Do?

Gastrin has a number of roles controlling the functioning of the stomach. Gastrin stimulates the following:

- Release of hydrochloric acid from the **parietal cells** in the stomach.
- Release of **histamine** from ELC cells and this also stimulates secretion of hydrochloric acid.
- **Growth of gastric mucosa** by increasing proliferation and inhibiting cell death (apoptosis).
- Contractions of the stomach.
- Gastrin acts via the cholecystokinin B receptor.

4.8.3 How Is Gastrin Release Controlled?

Gastrin release is stimulated by **distension** (stretching) of the stomach, peptides and amino acids in the stomach, low acidity in the stomach (high pH), **vagal nerves**, and gastrin-releasing peptide (a neuropeptide from nerves). In contrast, gastrin release is inhibited by somatostatin and high luminal acidity (low pH).

4.9 Other Endocrine Glands

Hormones of the ovaries, testes, and placenta are discussed under male and female reproduction (see, respectively, Sections 13.6 in Chapter 13 and Section 14.5 in Chapter 14). Other endocrine glands include the liver, pineal, and kidneys.

4.9.1 Liver

The liver releases insulin-like growth factor-1 (IGF-1) in response to growth hormone (GH). The IGF-1 stimulates muscle and bone growth.

4.9.2 Pineal

The pineal gland releases **melatonin** during the night (scotophase or period of darkness). Melatonin plays a role in seasonal breeding.

4.9.3 Kidneys

The kidneys have multiple endocrine roles:

- Converting 25-hydroxy vitamin D_3 to the active form, 1,25-dihydroxy vitamin D [1,25 $(OH)_2$ vit D_3] at times of low plasma concentrations of glucose. The 1,25 $(OH)_2$ vit D_3 increases calcium absorption in the small intestine.
- Releasing **erythropoietin** at times of low oxygen tension in the blood. In turn, the erythropoietin increases formation of red blood cells.
- Releasing **renin** at times of low blood pressure.

4.10 Endocrine Differences Between Livestock/Companion Animals and Poultry

Differences in the endocrine system between mammalian domestic animals and poultry are summarized in Tables 4.4 and 4.5.

Table 4.4 Examples of difference in hormones between livestock and poultry

Hormones produced		Hormone actions/roles	
Livestock, cats & dogs	Poultry & other birds	Livestock, cats & dogs	Poultry & other birds
Anterior pituitary gland (pars distalis)			
Prolactin	Prolactin	Initiates milk production	Initiates milk production by the crop in pigeons but not in other birds. Incubation & brooding behavior ↑
Posterior pituitary gland			
Pars nervosa			
Oxytocin	Mesotocin	Induces contraction of myometrium in uterus	Induces contraction of myometrium in oviduct
Arginine vasopressin (Lysine vasopressin in pigs)	Arginine vasotocin	Increases water retention; increases blood pressure	Increases water retention: Induces contraction of myometrium in oviduct causing oviposition (egg laying)
Pars intermedia			
Alpha MSH (α-MSH)	Absent	Coloration?	Not applicable
Adrenal cortex			
Cortisol*	Corticosterone	Increases blood glucose ↑, Liver glycogen ↑, Gluconeogenesis ↑, Adipose tissue ↑	Increases blood glucose ↑, Liver glycogen ↑, Gluconeogenesis ↑, Adipose tissue ↑
Ovary			
Estradiol (E_2)	Estradiol (E_2)	Growth of uterus (myometrium) and mammary glands ↑. LH (GnRH) release ↑ Follicular development	Yolk precursor synthesis ↑. Oviduct produces egg white proteins ↑. With testosterone, calcium deposited into medullary bone ↑. With progesterone (P4). growth of oviduct ↑.
Progesterone (P_4)	Progesterone (P_4)	Growth of uterus (endometrium) and mammary glands ↑. LH (GnRH) release ↓	LH (GnRH) release ↑ Production of one egg white protein, avidin ↑
Very low, if present	Testosterone (T)	Little or none	With E_2 deposition of medulla bone

(continued)

Table 4.4 Examples of difference in hormones between livestock and poultry (*continued*)

Hormones produced		Hormone actions/roles	
Livestock, cats & dogs	Poultry & other birds	Livestock, cats & dogs	Poultry & other birds
Relaxin	Not present	Relaxation of cervix and pubic ligaments	Not applicable
Progesterone	Not present	Growth of uterus (endometrium) and mammary glands ↑. LH (GnRH) release ☐	Not applicable
Relaxin	Not present	Relaxation of cervix and pubic ligaments	Not applicable
Pancreas islets of Langerhans			
Insulin	Insulin	Uptake of glucose in muscle and adipose tissue ↑ via GLUT-4	No GLUT-4
Insulin/glucagon	Insulin/glucagon	Set point for plasma concentrations of glucose	Set point for plasma concentrations of glucose but much higher

*Corticosterone in rats, mice, and hamsters

Table 4.4 Examples of difference in hormones between livestock and poultry

Hormones produced		Hormone actions/roles	
Livestock, cats & dogs	Poultry & other birds	Livestock, cats & dogs	Poultry & other birds
Anterior pituitary gland (pars distalis)			
Prolactin	Prolactin	Initiates milk production	Initiates milk production by the crop in pigeons but not in other birds. Incubation & brooding behavior ↑
Posterior pituitary gland			
Pars nervosa			
Oxytocin	Mesotocin	Induces contraction of myometrium in uterus	Induces contraction of myometrium in oviduct
Arginine vasopressin (Lysine vasopressin in pigs)	Arginine vasotocin	Increases water retention; increases blood pressure	Increases water retention; Induces contraction of myometrium in oviduct causing oviposition (egg laying)
Pars intermedia			
Alpha MSH (α-MSH)	Absent	Coloration?	Not applicable
Adrenal cortex			
Cortisol*	Corticosterone	Increases blood glucose ↑, Liver glycogen ↑, Gluconeogenesis ↑, Adipose tissue ↑	Increases blood glucose ↑, Liver glycogen ↑, Gluconeogenesis ↑, Adipose tissue ↑
Ovary			
Estradiol (E_2)	Estradiol (E_2)	Growth of uterus (myometrium) and mammary glands ↑, LH (GnRH) release ↑ Follicular development	Yolk precursor synthesis ↑, Oviduct produces egg white proteins ↑. With testosterone, calcium deposited into medullary bone ↑. With progesterone (P_4), growth of oviduct ↑.
Progesterone (P_4)	Progesterone (P_4)	Growth of uterus (endometrium) and mammary glands ↑, LH (GnRH) release ↓	LH (GnRH) release ↑ Production of one egg white protein, avidin ↑
Testosterone (T)	Very low, if present	Little or none	With E_2 deposition of medulla bone

(continued)

Table 4.4 Examples of difference in hormones between livestock and poultry (*continued*)

Hormones produced		Hormone actions/roles	
Livestock, cats & dogs	Poultry & other birds	Livestock, cats & dogs	Poultry & other birds
Relaxin	Not present	Relaxation of cervix and pubic ligaments	Not applicable
Placenta			
Progesterone	Not present	Growth of uterus (endometrium) and mammary glands ↑. LH (GnRH) release □	Not applicable
Relaxin	Not present	Relaxation of cervix and pubic ligaments	Not applicable
Pancreas islets of Langerhans			
Insulin	Insulin	Uptake of glucose in muscle and adipose tissue ↑ via GLUT-4	No GLUT-4
Insulin/glucagon	Insulin/glucagon	Set point for plasma concentrations of glucose	Set point for plasma concentrations of glucose but much higher

*Corticosterone in rats, mice, and hamsters

Table 4.5 Differences in endocrine tissues between domestic mammals and poultry.

	Livestock and companion animals	Poultry
Pars intermedia (intermediate part of pituitary gland)	Present	Absence
Thyroid gland	Single thyroid	Paired thyroid glands
Parathyroid glands	On the surface of thyroid gland	Independent of thyroid
Parafollicular or C cells	Distributed in the thyroid	In specific organ—the ultimobranchial glands or ultimobranchial bodies.
Adrenal gland	Chromaffin cells are located in the adrenal medulla (middle):Cortical cells are in layers around the medulla.	Chromaffin cells are located in clumps interspersed by cortical tissue.
Ovary	Corpus luteum but inverted structure in horses	No corpus luteum

4.11 Examples of Endocrine Diseases in Domestic Animals

Textbox 4.8 summarizes examples of endocrine diseases in domestic animals.

Textbox 4.8 Examples of Endocrine Diseases in Domestic Animals

Addison's Disease (or Hypoadrenocorticism)
Addison's disease is insufficient production of the adrenal cortical hormones, cortisol, and/or aldosterone.

Insufficient aldosterone will be followed by loss of sodium and, consequently, water. These can lead to cardiovascular and renal problems. Among the symptoms of this include the following: dehydration, excess urination, excess thirst, irregular heart rate, painful abdomen, and weak pulse. Treatment is administration of mineralocorticoids.

Insufficient cortisol will be followed by disruption of glucose homeostasis and low blood concentrations of glucose (**hypoglycemia**). Among the symptoms of these include weight loss, lethargy, lack of appetite, and depression. Treatment is the administration of glucocorticoids.

Cushing's Disease (or Hyperadrenocorticism)
Cushing's disease is due to excessive production of the adrenal cortical hormone cortisol.

Cushing's Disease in dogs: Among the symptoms are increases in appetite, thirst, urination, hair loss, thin skin, and multiple skin infections. Cushing's disease in dogs can be caused by the following:

- Pituitary dependent; ~80–85% due to a tumor in the anterior pituitary gland and, consequently, high release of ACTH (adrenocorticotropic hormone) and then release of cortisol from the adrenal cortex
- Pituitary independent; ~15–20% due to a tumor in the adrenal cortex.

One drug, Trilostane, is approved to treat Cushing's disease in dogs. It suppresses cortisol production.

(continued)

> **Textbox 4.8** *(continued)*
>
> **Cushing's syndrome in older horses**: Cushing's syndrome is also observed in older horses. In this case, it is caused by a tumor in the pituitary releasing ACTH and causing excessive production of cortisol. Among the symptoms are development of long shaggy coat, loss of muscle, and susceptibility to pathogens. Treatment is the administration of either pergolide mesylate (dopamine agonist) or cyproheptadine (serotonin antagonist).
>
> **Diabetes**
> **Diabetes** is due to either lack of insulin or an inadequate response to insulin.
>
> **Diabetes in older dogs and cats**: Symptoms include excessive urination, decreased appetite, weight loss, high blood glucose (hyperglycemia), and glucose in urine. Causes include inadequate release of insulin, and can be secondary to Cushing's disease. Treatment is the administration of a daily injection of insulin and a special diet.
>
> **Dwarfism**
> Proportionate dwarfism is when animals or people are proportionately smaller in size. Dwarfism is found in cattle, goats, chickens, and other species. It is primarily caused by defects in the GH-IGF-1 axis. Proportionate dwarfism in cattle and chickens is due to a defect in the GH receptor gene and, consequently, a loss or reduced functioning of the GH receptor.
>
> **Hypothyroidism**
> **Hypothyroidism** is inadequate production of thyroid hormones. Hypothyroidism is found in domestic animals.
>
> **Canine hypothyroidism** is caused by inadequate production of thyroxine. Symptoms include loss of hair, flakey skin, weight changes, sluggishness. Treatment is the administration of thyroxine.
>
> **Iodine Deficiency in Cattle, Pigs, and Sheep**
> There is hypothyroidism (low levels of thyroxine in the blood) in animals receiving inadequate amounts of iodine in their diet. This is due to lack of iodine in the soil of either the pastures or the land on which the feed is produced. The lack of iodine is accompanied by abnormal growth of the thyroid glands with the large thyroids called goiters. Treatment is iodized salt blocks.
>
> **Hyperthyroidism**
> **Hyperthyroidism** is excessive production of thyroid hormones. Hyperthyroidism is found in cats. Symptoms include weight loss, increased appetite, pacing, anxiety, and poor coat conditions.

Theme Two

Growth and Development

The Nervous and Sensory Systems

5

Learning Objectives

1. Know the different parts of the nervous system.
2. Understand how the nervous system develops in embryonic development (see Section 5.2).
3. Understand how the brain develops during embryonic development (see Section 5.2).
4. Know the cells in the central nervous system (see Section 5.3).
5. Know the functions of neurons (see Section 5.3.1).
6. Understand what an action potential is and how it works (see Section 5.4).
7. Understand the role of the synapse (see Section 5.5).
8. Understand the functional anatomy of the central nervous system (see Section 5.6).
9. Understand the roles of the meninges (see Section 5.6.2).
10. Understand the roles of the cerebrospinal fluid (see Section 5.7.2).
11. Understand the roles of the blood–brain barrier (see Section 5.7.3).
12. Understand what a circadian rhythm is (see Section 5.7.6.2).
13. Understand what light is (see Section 5.11.3.1).
14. Understand reception of light (see Section 5.11).
15. Understand how the eyes function (see Section 5.11).
16. Understand what sound waves are (see Section 5.14.1).
17. Understand reception of sound waves (see Section 5.14).
18. Understand how the ears function (see Section 5.14.4).
19. Understand reception of chemicals (see Section 5.15).
    a. Olfaction (see Section 5.15.1).
    b. Taste (see Section 5.15.3).
    c. Other chemoreception (see Section 5.15.5).
20. Understand what pheromones are and what they do (see Section 5.15.4).

21. Understand mechanoreception (touch, stretch, and pressure).
22. Understand what vibrissae do (see Section 5.16.1).
23. Understand nociception (pain) (see Section 5.16).
24. Understand the differences between senses in domestic mammals and those of the chicken (see Section 5.17).

Table of Contents

5.1 Introduction
    5.1.1 Overall Introduction
    5.1.2 Introduction to the Nervous System
5.2 Development of the Nervous System
5.3 Cells of the Nervous System
    5.3.1 Neurons
        5.3.1.1 What Is a Neuron?
    5.3.2 Glial Cells
5.4 Nerve Conduction/Action Potential
    5.4.1 Overview of Nerve Conduction/Action Potential
    5.4.2 Recovery After the Action Potential
    5.4.3 Refractory Periods
5.5 Functioning of the Synapse
    5.5.1 Introduction to the Synapse
    5.5.2 Neurotransmitters and Their Functioning
    5.5.3 Functioning of the Synapse
5.6 Central Nervous System
    5.6.1 Introduction to the Central Nervous System
    5.6.2 Meninges
5.7 Brain
    5.7.1 Introduction to the Brain
    5.7.2 Cerebrospinal Fluid
    5.7.3 Blood–Brain Barrier
    5.7.4 Functions of the Brain
    5.7.5 Cerebrum and Cerebral Cortex
    5.7.6 Hypothalamus
        5.7.6.1 Introduction to the Hypothalamus
        5.7.6.2 What is a Circadian Rhythm?
    5.7.7 Pineal Gland
    5.7.8 Thalamus
    5.7.9 Mesencephalon-Derived Brain Structures
    5.7.10 Pons
    5.7.11 Cerebellum
    5.7.12 Medulla Oblongata—A Myelencephalon-Derived Brain Structure
5.8 Spinal Cord
5.9 Autonomic Nervous System
    5.9.1 Introduction to the Autonomic Nervous System
    5.9.2 Sympathetic Nervous System
    5.9.3 Parasympathetic Nervous System
    5.9.4 Enteric Nervous System
5.10 Introduction to the Senses
5.11 Vision/Sight
    5.11.1 Retina
    5.11.2 Maintaining the Structural Integrity and Physiological Functioning of the Eye
    5.11.3 Protecting the Retina and Other Parts of the Eye
    5.11.4 Photoreception
        5.11.4.1 What Is Light?
        5.11.4.2 Where Is Light Detected?
        5.11.4.3 How Is Light Detected?
        5.11.4.4 Brain Photoreceptors
5.12 Eyelids
5.13 Tears
    5.13.1 The Composition of Tears
    5.13.2 Function of tears

5.14 Hearing
    5.14.1 What Is Sound?
    5.14.2 What is Hertz?
    5.14.3 Do Animals Detect the Same Frequencies of Sound?
    5.14.4 Functioning of the Ear
    5.14.5 Do Animals Detect the Level of Sound?
5.15 Chemoreception
    5.15.1 Olfaction (Smell)
    5.15.2 Molecular Olfaction Receptors
    5.15.3 Taste
        5.15.3.1 Is Taste Reception the Same Across Domestic Animals?
    5.15.4 Pheromones
    5.15.5 Other Chemoreception
5.16 Nociceptors or Pain Receptors
    5.16.1 Vibrissae (Whiskers)
5.17 Comparison of Senses in Livestock/Mammalians Companion Animals With Those in Poultry

5.1 Introduction

5.1.1 Overall Introduction

The function of the nervous system is communication within the animal's body. This encompasses communication of the internal and external environments and control of movement. Textbox 5.1. provides a glossary of terms used in discussing the nervous system.

Textbox 5.1 Glossary of Terms Related to the Nervous System

- **Action potential**: a wave of depolarization passing along an axon or electrical signal.
- **Autonomic nervous system**: the involuntary nervous system and is made up of the sympathetic, parasympathetic, and enteric nervous systems.
- **Axon**: long thin nerve fiber.
- **Blood-brain barrier**: a unique feature of the microvascular of the brain that tightly regulates what chemicals can pass from the blood into the central nervous system (CNS). This anatomical and functional barrier protects the CNS by blocking some drugs, toxins, pathogens, and immune cells from entering it.
- **Brain**: a critical part of the central nervous system. The brain controls most functions of an animal's body.
- **Central nervous system (CNS)**: the brain and the spinal cord.
- **Cerebrospinal fluid**: fluid in the ventricles of the brain and the spinal cord.
- **Chemoreception**: detection of specific chemicals by chemoreceptor cells.
- **Dendrite**: part of a neuron that receives input from other neurons.
- **Enteric nervous system**: controls the functioning of the gastrointestinal tract.
- **Glial cell**: cell in the central and peripheral nervous system that are not neurons. These cells support the neurons and their functions.
- **Mechanoreceptor**: receptor cells containing mechanically gated ion channels responding to touch, pressure, stretching, motion, and sound waves.
- **Motor Nervous System**: the parts of the nervous system that allow movement (muscle contractions).
- **Nerve**: collection of axons passing to organs.
- **Neuron**: nerve cell consisting of dendrites, a cell body with a nucleus synthesizing proteins, and an axon.
- **Neural tube**: an embryonic structure that gives rise to the central nervous system.
- **Neurotransmitter**: a chemical that passes across the synaptic gap between neurons.
- **Nociception**: detection of pain reception by nociceptor cells (pain receptor cells).
- **Parasympathetic nervous system**: one of the components of the autonomic nervous system.

(continued)

Textbox 5.1 (continued)

- **Peripheral nervous system**: consists of the autonomic nervous system and the motor nervous system.
- **Pheromone**: chemical released into the environment that elicits a response in other animals of the same species.
- **Photoreception**: detection of light by photoreceptor cells.
- **Sensory nervous system**: the part of the nervous system that communicates stimuli from receptors detecting aspects of the internal and external environments.
- **Somatic nervous system**: made up of spinal and cranial nerves, the autonomic nervous system, and motor neurons between the central nervous system and the internal organs.
- **Sympathetic nervous system**: one of the components of the autonomic nervous system.
- **Synapse**: the junction between neurons. There is a gap and neurotransmitters pass across the synaptic gap.

5.1.2 Introduction to the Nervous System

The nervous system of mammals and birds consists of the following components:

- Central nervous system (CNS).
  - Brain.
  - Spinal cord.
- Peripheral nervous system.
  - Efferent (motor) division (nervous impulses passing from the CNS to tissues in the periphery).
- Autonomic nervous system (involuntary nervous system).
  - Sympathetic division.
  - Parasympathetic division.
  - Enteric nervous system.
- Somatic nervous system (voluntary nervous system).
  - Afferent (sensory) division (nervous impulses passing from receptor cells in the periphery to the CNS).

The nervous system is intimately related to the multiple senses, and these are also covered in this chapter.

5.2 Development of the Nervous System

The central nervous system originates from the neural tube. The neural tube develops from the neural plate along the notochord in the process of neurulation (see Figure 5.1). The lumen of the tube develops into the ventricles of brain and the central canal of the spinal cord. The peripheral nervous system develops from the neural crest.

Neurons develop from **neural progenitor cells**, themselves derived from either **neural crest** or neural tube cells. The neural progenitor cells undergo the following:

- Division.
- Outgrowth of axon and dendrites.
- Establishment and refinement of **synaptic connections** (synaptogenesis).

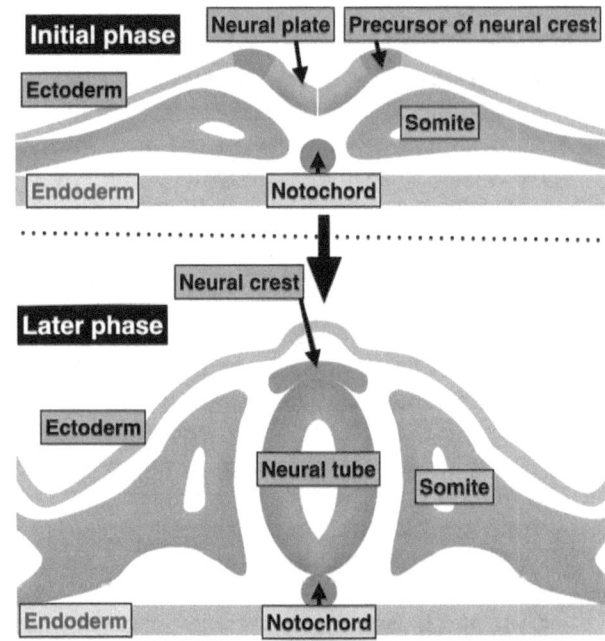

Figure 5.1 Development of the neural tube and neural crest during embryonic development. The neural plate develops into a neural fold and, ultimately, the neural tube. This process is called neurulation.
[Key: BMP = bone morphogenic protein]
(Adapted from Abdulrahman112, CC BY-SA 3.0)

Textbox 5.2 A Deeper Dive Into the Development of the Neural Tube

Within the neural tube, there is expression of signaling proteins. Some cells produce bone morphogenetic protein and this forms a gradient from dorsal to ventral in the neural tube. Other cells produce a factor called sonic hedgehog (see Figure 5.2). There is a gradient of sonic hedgehog proteins but in the opposite direction (see Figure 5.2).

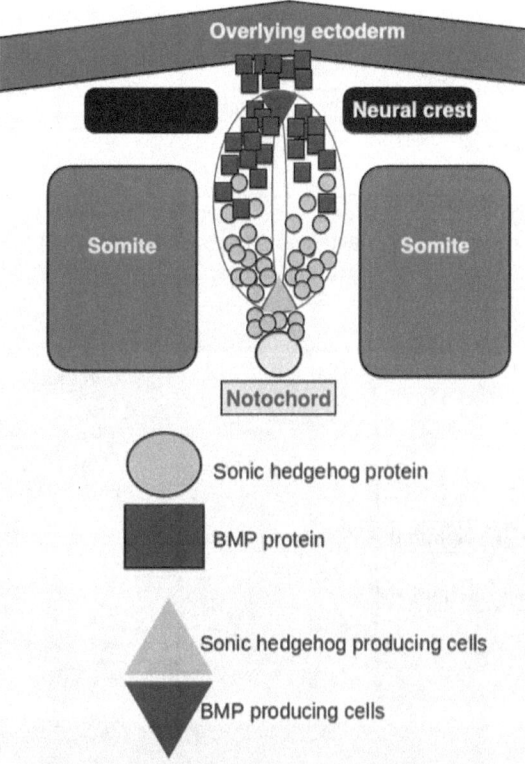

Figure 5.2 Development of the neural tube and neural crest during embryonic development. Doral cells of the neural tube produce and release bone morphogenic protein (BMP) establishing a gradient from top to bottom. Ventral cells of the neural tube (above the notochord) produce and release sonic hedgehog, establishing a gradient from bottom to top.
(Adapted from Catcasillas, CC BY 3.0).

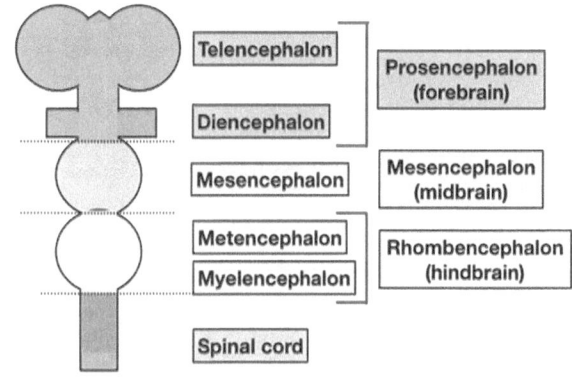

Figure 5.3 Development of the brain.

There are three regions of the brain initially during development (Figure 5.3):

- **Prosencephalon** (forebrain).
- **Mesencephalon** (midbrain).
- **Rhombencephalon** (hindbrain).

There is further development of the regions of the brain:

- The **telencephalon** gives rise to the **cerebrum** and **cerebral cortex**.
- The **diencephalon** differentiates into the **thalamus**, **hypothalamus**, **epithalamus**, and **subthalamus**.
- The mesencephalon becomes the front of the brain stem (**tectum**, **cerebral aqueduct**, **tegmentum**, and **cerebral peduncles**).
- The **metencephalon** gives rise to the **pons** and the **cerebellum**.
- The **myelencephalon** gives rise to the **medulla oblongata**.

Brain development depends on transcription factors, signaling molecules as inductive signals, and controllers of gene expression.

Textbox 5.3 A Deeper Dive Into Signaling Molecules

The signaling molecules include fibroblast growth factor 8 (FGF 8), and the Wnt; the latter being a signaling protein across metazoan (multicellular) animals. Wnt is a portmanteau acronym for *wingless* in fruit flies and its vertebrate homolog (*int*egrated or *int-1*). It binds to a member of the frizzled (Fz) receptor family.

5.3 Cells of the Nervous System

There are two major cell types in the nervous system:

- Neurons (see Section 5.4.1)
- Glial cells (see Section 5.4.2).

5.3.1 Neurons

5.3.1.1 What Is a Neuron?

Neurons are the **functional units** of the nervous system. Neurons are electrically excitable cells whose fundamental role is information transfer. Neurons communicate with other neurons and muscle cells via neurotransmitters released from a specialized structure at the end of the neuron, the pre-synaptic nerve terminal (see Section 5.4.1) (also see Figure 5.4).

The structure of a simple neuron is shown in Figure 5.4. Neurons consist of the following:

- A **cell body** (or **soma**) with a nucleus, endoplasmic reticulum that synthesizes proteins and neurotransmitters, and mitochondria that produce ATP.
- Axons are long thin tubes along which pass electric signals (action potentials) or waves of depolarization. In addition, proteins and ATP synthesized in the cell body pass down the axon along microtubules.
- Dendrites are branched structures that extend from the cell body. The sum of the inputs to the dendrites governs if an action potential is generated.
- Synapses are the space between the synaptic terminals of two neurons. A neurotransmitter is a chemical message that passes across the synaptic cleft to stimulate (or inhibit) the adjacent neuron.
- The **cell membrane** of the axon is the **axolemma** (Figure 5.5).
- Myelinated axons have **myelin sheaths** (see Figure 5.4). For peripheral neurons, the myelin sheath is produced by the Schwann cells. For neurons in the central nervous system, the myelin sheath is produced by a specific type of glial cells, the oligodendrocytes.
- **Axon terminals** are the ends of the axon that store and then release neurotransmitters after an action potential has passed down the axon.
- Axons have **nodes of Ranvier** (between the myelin sheaths) (see Figure 5.4). The myelin sheaths are composed of lipids and proteins.
- **Axon hillock** is the area where the soma and the axon meet. During an action potential, this is the site of activation.

There are multiple types of neurons, such as the following (see Figure 5.6):

- **Anaxonic**.
- **Bipolar**.
- **Unipolar**.
- **Multipolar**.

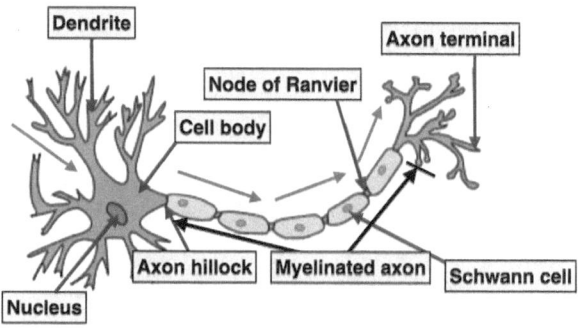

Figure 5.4 Structure of a neuron showing a wave of depolarization (shown by arrows). (Adapted from Quasar Jarosz, CC BY-SA 3.0)

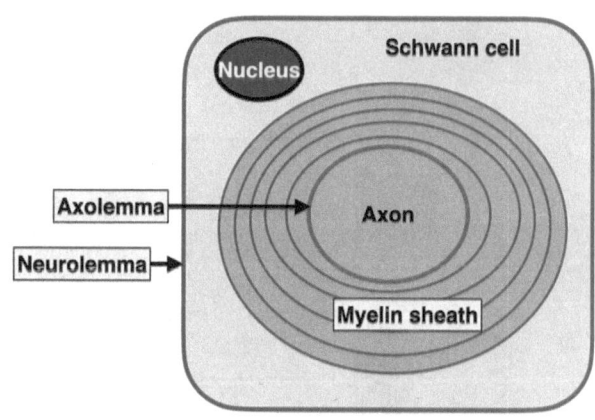

Figure 5.5 Axons are wrapped around by a myelin sheath produced by the Schwann cells. The myelin sheath not only protects the axon but also facilitates transmission of action potentials. The Schwann cells produce the myelin sheath and also protect the axon.

5.3.2 Glial Cells

Glial cells, or neuroglial cells, are non-neuronal cells in the brain and spinal cord together with Schwann cells in the peripheral nervous system and satellite cells in the anterior pituitary gland. The word *glia* means "glue" in Greek. There are four types of glial cells (see Figure 5.7):

CHAPTER 5   The Nervous and Sensory Systems   101

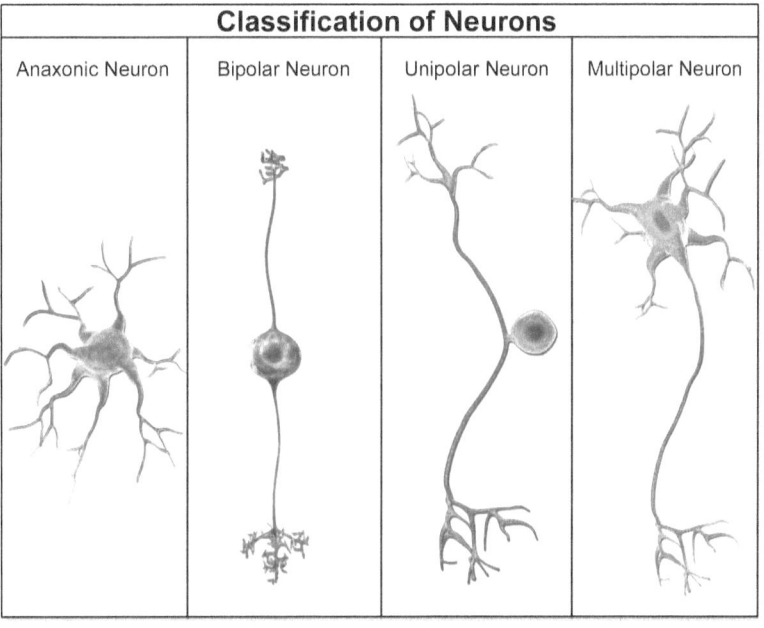

Figure 5.6   Types of neurons. (BruceBlaus, CC BY-SA 4.0)

- **Astrocytes**: star-shaped cells that provide nutrients and support to neurons.
- **Microglial cells**: the phagocytic cells of the central nervous system.
- **Oligodendrocytes**: provide a lipid coat around axons in a manner similar to the Schwann cells.
- **Ependymal cells**: these line the spinal cord and ventricles in the brain.

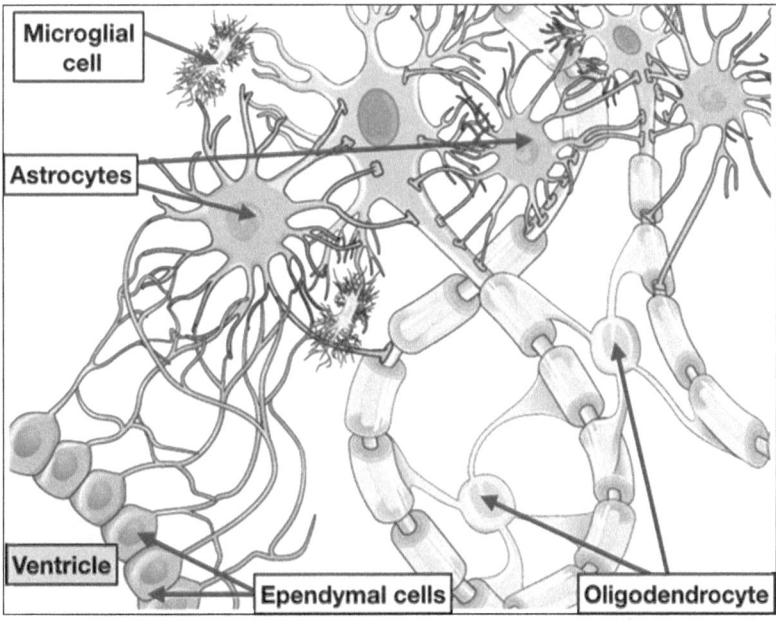

Figure 5.7   Glial cells in the brain.
(Adapted from OpenStax, CC BY 4.0)

Glial cells make up much of the brain (see Figure 5.8). They differ from neurons in that they do not form synapses or have action potentials. However, they do interact with neurons through gap junctions, neurotransmitters, neurotropic factors, and cytokines.

There are **Schwann cells** closely associated with peripheral neurons. These are glial cells associated with peripheral neurons. Schwann cells are derived from neural crest cells (see Figure 5.1) and can be either myelinating or non-myelinating cells.

5.4 Nerve Conduction/Action Potential

5.4.1 Overview of Nerve Conduction/Action Potential

An action potential, or wave of depolarization, moves along the axon due to activation of fast voltage-gated sodium channels allowing sodium to cross the axolemma entering the axon (see Figures 5.9 and 5.10). An action potential is the following cycle (see Figure 5.9):

1. Resting membrane potential.
2. Depolarization.
3. Repolarization.
4. Hyperpolarization.
1. Resting membrane potential.

An action potential moves along the axolemma (see Figure 5.10).

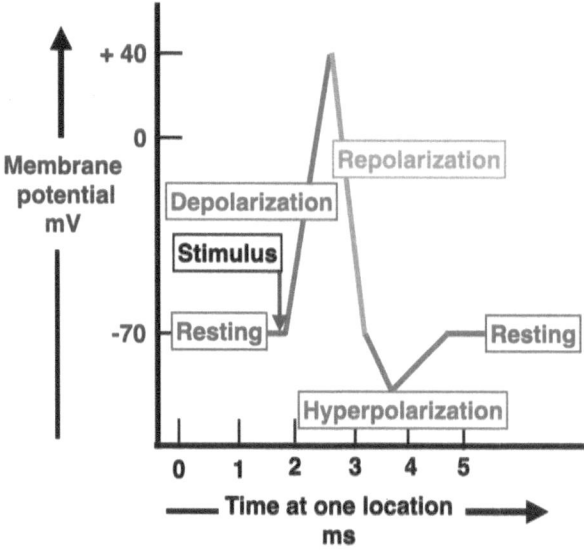

Figure 5.9 Schematic of action potential showing membrane potential in millivolts (mV) across axolemma at a single location over time in milliseconds (ms) during depolarization, repolarization, hyperpolarization, and resting. The action potential follows a stimulus above the threshold voltage.

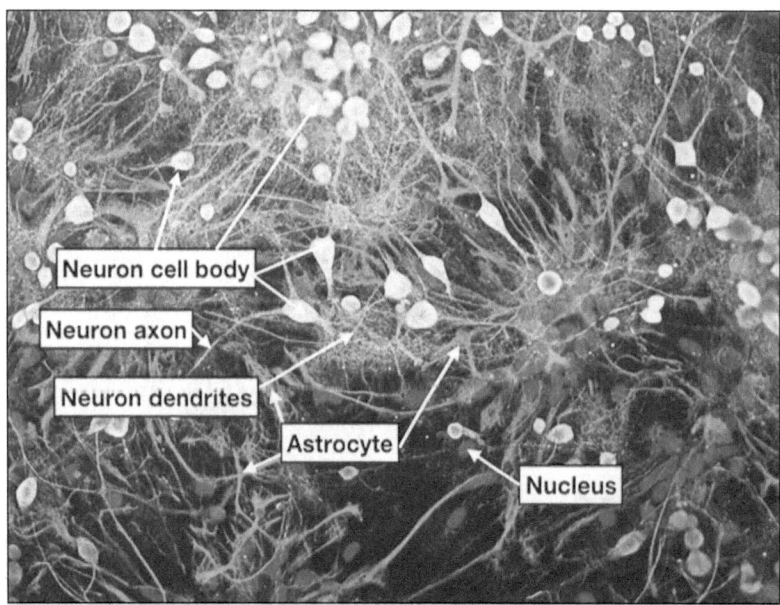

Figure 5.8 Cells of the mammalian brain showing neurons and astrocytes (glial cells). (Adapted from GerryShaw, CC BY-SA 4.0)

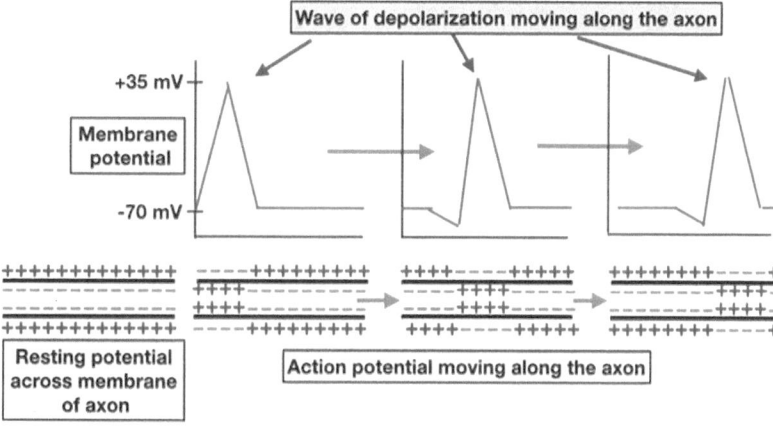

Figure 5.10 An action potential, or wave of depolarization, moves along the axon due to activation of voltage-gated sodium channels, allowing sodium to enter the axon.

1. **Resting membrane potential** is the baseline membrane potential across the axolemma (-70 mV) (see Figures 5.9, 5.10, and 5.11). In the resting stage, there is an unequal distribution of ions across the axolemma. The interior is negatively charged while the exterior is positively charged (see Figures 5.10 and 5.11). This is due to the following:
   - Potassium ions (K^+) diffusing out of the interior of the axon (intracellular fluid) due to leakage potassium channels and down a concentration gradient.
   - Large anions in the interior of the axon that cannot cross the axolemma.
   - Sodium ions (Na^+) in the exterior (extracellular fluid) diffusing into intracellular fluid of the axon due to leakage sodium channels.
   - Both the fast voltage-gated sodium and the slow voltage-gated potassium channels are closed.
   - Residual effect of the sodium pump (Na^+/K^+ ATPase) transporting three Na^+ ions in exchange for two K^+ ions.
2. **Depolarization** is when the membrane potential increases (to +40 mV) (see Figure 5.10 and 5.11). This is due to an influx of sodium ions (Na^+) across the axolemma through fast voltage-gated Na^+ channels. (The Na^+ passes down a concentration gradient and is attracted by the negative charges in the axon.) Maximal membrane potential is reached when all the fast voltage-gated Na^+ channels are open and the maximal amount of Na^+ enters the axon (see Figures 5.10 and 5.11). The slow voltage-gated potassium (K^+) channels are still closed with their opening delayed (see Figures 5.10 and 5.11). Depolarization of a neuron is an "all or nothing" effect initiated by sodium crossing the plasma membrane of an axon.
3. **Repolarization** starts when slow voltage-gated K^+ channels open (see Figures 5.10 and 5.11). (These are activated at the same time as the fast voltage-gated Na^+ channels but open up later, hence they are called slow). This allows K^+ to leave. The fast voltage-gated Na^+ channels close, preventing more Na^+ from entering the axon (see Figures 5.10 and 5.11).

5.4.2 Recovery After the Action Potential

After the action potential passes, the axon has to recover (see Figures 5.10 and 5.11). The sodium channel closes, preventing more sodium from moving in. The **sodium pump** (Na^+/K^+ ATPase) moves sodium ions out of the intracellular fluid and pulls more potassium in a 3:2 ratio (Figure 5.10). The sodium pump uses about 70 % of the energy of the nervous system. Recovery takes about 3 milliseconds.

5.4.3 Refractory Periods

The **refractory period** is when the axon cannot respond or cannot respond fully to a second stimulus. The refractory period is initially an **absolute refractory period** when there cannot be a response. This is followed by a **relative refractory period**. The refractory period ensures that the action potential passes in one direction once depolarization has occurred; the area that the action potential has passed is refractory.

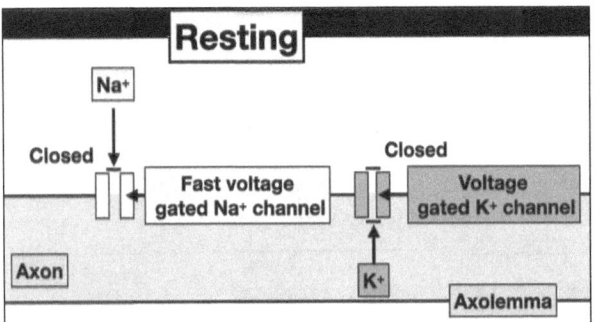

(A) At resting, the fast-voltage Na⁺ channels are closed (not allowing Na⁺ to cross the axolemma into the axon) in the resting phase.

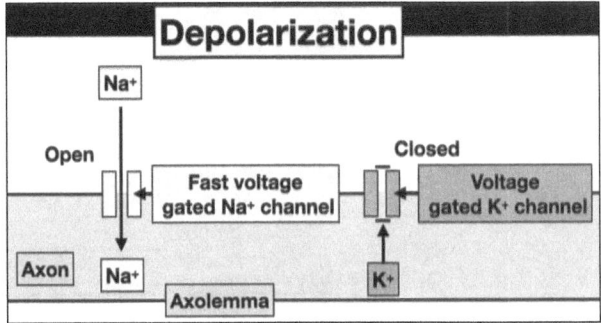

(B) At the beginning of depolarization, the fast-voltage Na⁺ channels open (allowing Na⁺ to cross into the axon).

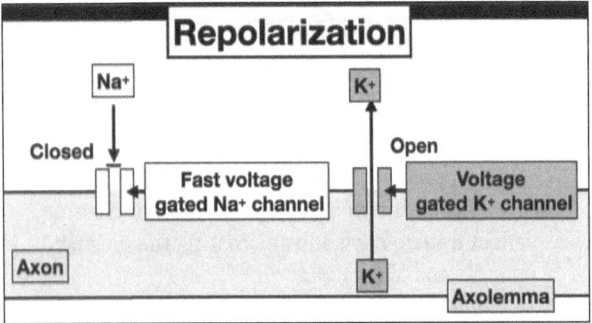

(C) At the end of the beginning of repolarization, the fast-voltage Na⁺ channels close and the slow-voltage K⁺ channels open.

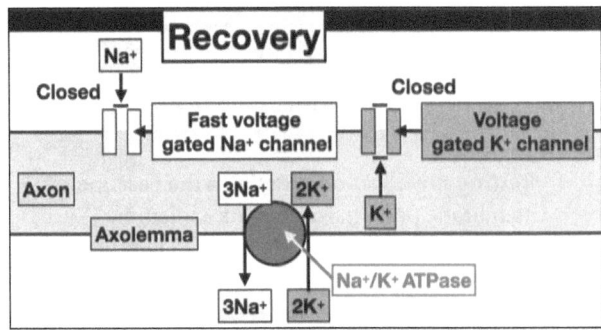

(D) During the recovery, Na⁺ is pumped out and K⁺ is pumped into the axon in a 3:2 ratio by the Na⁺/K⁺ ATPase.

Figure 5.11 Role of fast-voltage gated sodium (Na+) channels, slow-voltage gated potassium (K+) channels, and Na+/K+ ATPase (the sodium pump) in the action potential, specifically resting, depolarization, repolarization, and recovery.

- Absolute refractory period: During depolarization, there is opening of the voltage-gated Na$^+$ channels and this is followed by their inactivation. Consequently, they cannot be activated. The absolute refractory period lasts about 1.5 milliseconds.
- Relative refractory period: During repolarization, there are partial responses. This is the relative refractory period. The responses become progressively greater until the axon is fully repolarized due to the reactivation of the sodium (Na$^+$) channels. The axon becomes capable of a complete response to a stimulus.

In axons with myelin sheaths, the wave of depolarization jumps from one node of Ranvier to the next (Figures 5.4 and 5.12).

The neuron is not just a simple conduit for a signal to pass along from one neuron to another. There are multiple dendrites per neuron (see Figure 5.13). Each receive inputs from other neurons. These inputs pass to

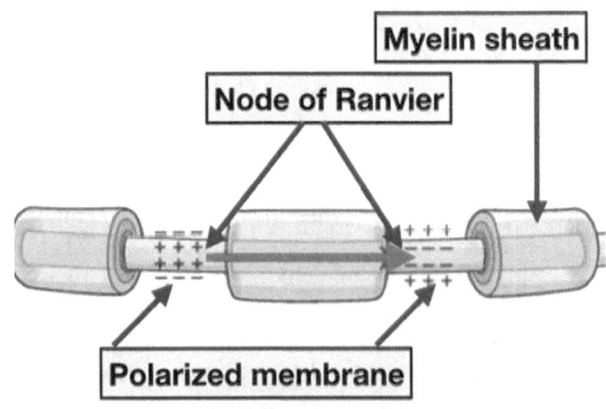

Figure 5.12 Actions potentials pass along the axon in the nodes of Ranvier, jumping to the next node across the myelin sheath.
(Adapted from Openstax, CC BY-SA 4.0)

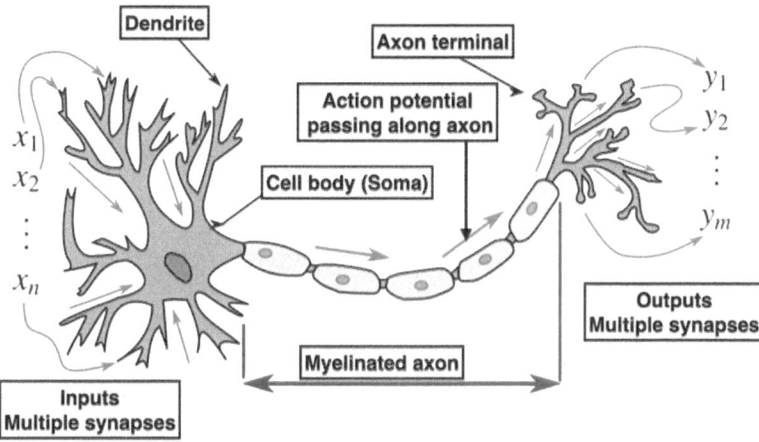

Figure 5.13 Generalized structure of a nerve cell with myelinated axon, cell body, and synapses. (Adapted from Egm4313.s12 [Prof. Loc Vu-Quoc], CC BY-SA 3.0)

the cell body that effectively integrates the inputs (see Figure 5.13). If a threshold is reached, an action potential passes along the myelinated axon to the multiple output synapses.

Excitatory postsynaptic potential (EPSP) is depolarization of a postsynaptic neuron following binding of an excitatory neurotransmitter to receptors. This generates an action potential. **Inhibitory postsynaptic potential (IPSP)** is hyperpolarization of a postsynaptic neuron following binding of an inhibitory neurotransmitter to receptors. This depresses the ability of the postsynaptic neuron to generate action potentials.

5.5 Functioning of the Synapse

5.5.1 Introduction to the Synapse

The synapse consists of three components (Figure 5.14):

- **Pre-synaptic axon terminal** with secretory or synaptic granules containing neurotransmitters.
- **Synaptic cleft**, the gap between the two neurons.
- **Post-synaptic dendrite.**

5.5.2 Neurotransmitters and Their Functioning

Neurotransmitters are specific chemicals that functions as messengers. They are released from the **pre-synaptic terminal** of a neuron in response to nerve impulses. These signals pass across the synaptic cleft by diffusion. They bind to receptors in dendrites of the second neuron, where it promotes a nerve impulse. There are also receptors to the neurotransmitter on the pre-synaptic terminal. Binding of neurotransmitters to these pre-synaptic receptors regulates release of neurotransmitters.

Neurotransmitters can be either of the following:

- **Non-peptidergic**, where the neurotransmitter is not a peptide and is frequently a modified amino acid (see Table 5.1).
- **Peptidergic**, where the neurotransmitter is a peptide (see Textbox 5.5).

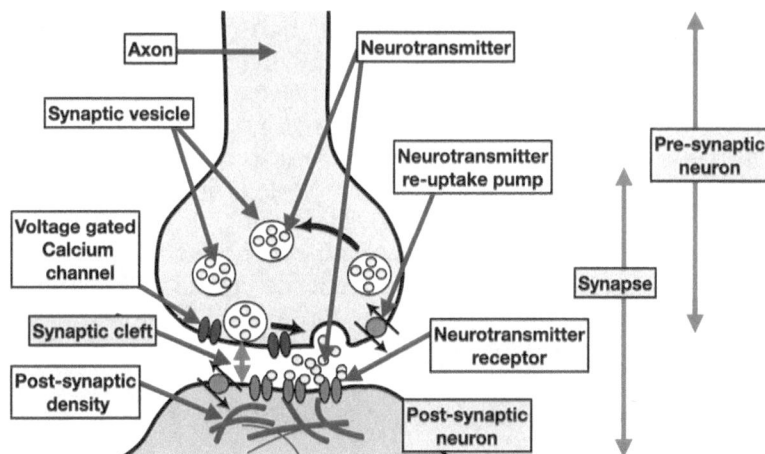

Figure 5.14 A synapse and its functioning. (Adapted from Nrets, CC BY-SA 3.0)

Table 5.1 Examples of important nonpeptidergic neurotransmitters.

Neurotransmitter	Structure	Nutrient Derived From
Acetylcholine	Ester of choline and acetate	Choline
Dopamine	Modified amino acid	Amino acid tyrosine
Epinephrine	Modified amino acid	Amino acid tyrosine
GABA (gamma-Aminobutyric acid or γ-aminobutyric acid)	Amino acid	Amino acid glutamate
Glutamate	Amino acid	Amino acids glutamate and glutamine
Glycine	Amino acid	Amino acid glycine
Histamine	Amino acid	Amino acid histidine
Nitrous oxide	NO	Amino acid arginine
Norepinephrine	Modified amino acid	Amino acid tyrosine
Serotonin	Modified amino acid	Amino acid tryptophan

Neuromodulators are chemicals, such as peptides, that modify or enhance or depress neuronal functioning. They do not act as neurotransmitters.

5.5.3 Functioning of the Synapse

Within the pre-synaptic axon terminal, there are **secretory granules** containing neurotransmitters (see Figure 5.14). These are released when an action potential reaches the end of the axon. This activates **voltage-gated calcium channels** (see Figure 5.14). The increased intracellular concentrations of calcium ions (Ca^{2+}) induce release of the neurotransmitters. The neurotransmitters diffuse rapidly across the **narrow synaptic cleft** (see Figure 5.14). Neurotransmitters bind to receptors on the surface of the **post-synaptic membrane**, stimulating a graded potential in the post-synaptic neuron (see Figure 5.14).

What happens to excess neurotransmitters or their metabolites? There are three routes for excess neurotransmitters or their metabolites:

- Re-uptake by the pre-synaptic axon terminal (see Figure 5.14).

Textbox 5.4   A Deeper Dive Into Synapses

The functioning of an adrenergic and a cholinergic synapse is shown in, respectively, Figures 5.15 and 5.16. The adrenergic synapse is considered first. When an action potential reaches the pre-synaptic end of the axon, the neurotransmitter, norepinephrine (NE), is released from synaptic vesicles due to vesicular monoamine transporter. The NE passes across the narrow synaptic cleft and then binds to the post-synaptic $α_1$-adrenergic receptors, $α_2$-adrenergic receptors, and β-adrenergic receptors, and thereby activates the neuron. There are also pre-synaptic $α_2$-adrenergic receptors. Binding of NE to the pre-synaptic $α_2$-adrenergic receptors decreases release of NE. There is re-uptake of NE by the pre-synaptic norepinephrine transporter.

Figure 5.16 shows a schematic diagram of the functioning of the cholinergic synapse. When an action potential reaches the pre-synaptic axon terminal, there is release of acetylcholine in the pre-synaptic vesicles is released into the synaptic cleft. It rapidly moves across the narrow gap to bind to muscarinic and/or nicotinic cholinergic receptors, activating the post-synaptic terminal, and hence the neuron. There are also pre-synaptic muscarinic cholinergic receptors that suppress release of acetylcholine. Acetylcholine in the synaptic cleft is cleaved by cholinesterase to choline and acetate. The choline is taken up by the pre-synaptic terminal by choline transporter and then acetyl choline is reformed.

(continued)

CHAPTER 5 The Nervous and Sensory Systems 107

Textbox 5.4 (continued)

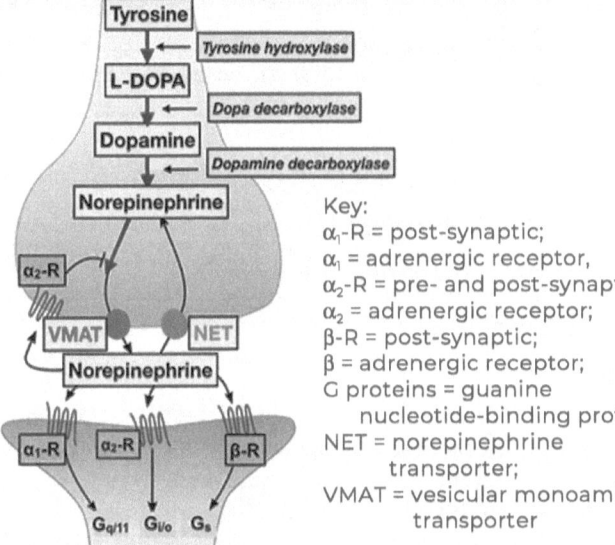

Figure 5.15 Schematic view of an adrenergic synapse. (Adapted from Pancrat, CC BY-SA 3.0)

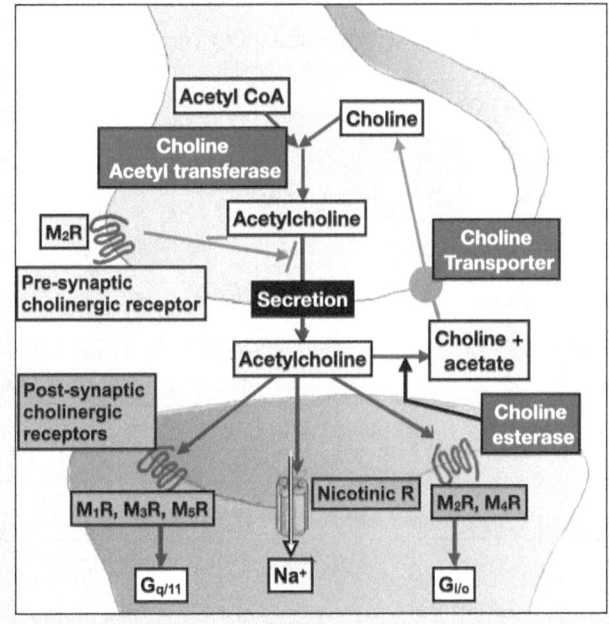

Figure 5.16 Schematic view of a cholinergic synapse. [Key: $G_{i/o}$ and $G_{q/11}$ = G proteins; M = muscarinic receptors; Na+ = sodium ions; R = receptors] (Adapted from Pancrat CC BY-SA 3.0)

Textbox 5.5   A Deeper Dive Into Peptidergic Neurotransmitters

Examples of important neurotransmitters that are peptides (neuropeptides) include the following:

- β-endorphin.
- Cocaine- and amphetamine-regulated transcript (CART).
- Dynorphin.
- Enkephalin.
- Galanin.
- Gastrin-releasing peptide (GRP).
- α-Melanocyte-stimulating hormone (α-MSH)
- Neuropeptide Y (NPY)
- Oxytocin
- Somatostatin
- Substance P

- Neurotransmitters drift away and are taken up by glial cells.
- Broken down by enzymes in the synaptic cleft.

Neurotransmitters are specific chemicals that are produced by the animal. These either transit across the synapse from one neuron to another or directly affect a bodily function. Neurotransmitters act on the post synaptic neuron in the following ways:

- **Stimulation.**
- **Inhibition** (the major inhibitory neurotransmitter is GABA).
- **Modulation.**

There are nine types of adrenergic receptors (see Table 5.2). These respond to epinephrine and/or norepinephrine. In addition, they show selectivity to synthetic ligands (drugs). Binding of epinephrine to β-adrenergic receptors on adipose cells stimulates breakdown of fat (lipolysis). However, the effect is mediated by different receptors in different species:

- β_1-AR in porcine adipose tissue.
- β_2-AR in bovine adipose tissue.
- β_3-AR in human and mouse adipose tissue.

5.6 Central Nervous System

5.6.1 Introduction to the Central Nervous System

The central nervous system consists of the following:

- Brain.
- Spinal cord.

5.6.2 Meninges

There are three layers of meninges surrounding the brain and spinal cord, protecting both. They are composed of connective tissue. The three meninges (see Figure 5.17), together with the spaces between them, are the following:

- **Dura mater** is tough connective tissue surrounding the brain and the spinal cord. In the case of the brain, the dura mater is under the skull.
- **Epidural space** between the dura mater and arachnoid mater.
- **Arachnoid mater**.
- **Subarachnoid space** containing **cerebrospinal fluid (CSF)**.
- **Pia mater.**

Other functions of the membranes include limiting movement of the brain during injury.

Table 5.2 Types of adrenergic receptors.

Receptor Type	Location
α_{1A} adrenergic receptors	Brain, skin, sphincters of gastrointestinal tract, renal artery
α_{1B} adrenergic receptors	Brain and regulate blood flow in the periphery
α_{1D} adrenergic receptors	Brain
α_{2A} adrenergic receptors	Acting centrally in brain and peripheral both pre-and post synaptically
α_{2B} adrenergic receptors	
α_{2C} adrenergic receptors	
β_1 adrenergic receptors	Heart increasing cardiac output (both heart rate and contractility/stroke volume, kidneys increasing release of renin
β_2 adrenergic receptors	Bronchial smooth muscle (inducing dilation), arteries of skeletal muscle, heart, liver, vascular smooth muscle, gastrointestinal smooth muscle, uterine smooth muscle.
β_3 adrenergic receptors	Thermogenesis (brown adipocytes); relaxation of small intestine.

Adrenergic receptors bind both norepinephrine and epinephrine.
β_1 and β_2, but not β_3, adrenergic receptors are blocked by propranolol.

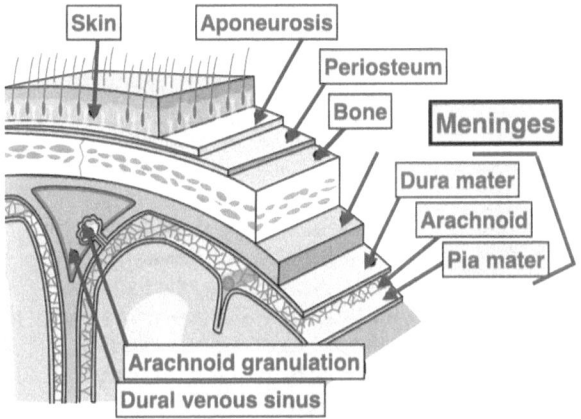

Figure 5.17 The meninges protect the brain.

5.7 Brain

5.7.1 Introduction to the Brain

The brain is composed of neurons and glial cells (see above). The number of neurons in the brain varies by species as follows:

- Human: 86 billion.
- Dog: 2.2 billion.
- Cat: 1.2 billion.
- Chicken: 0.22 billion.
- Mallard: 0.36 billion.

The brain depends on glucose as its energy supply and oxygen for aerobic respiration.

5.7.2 Cerebrospinal Fluid

The hollow spaces of the ventricles of brain and central canal of the spinal cord are filled with cerebrospinal fluid (CSF). This colorless fluid is similar in composition to the following:

1. **Plasma** without the proteins.
2. **Interstitial fluid**.
4. **Lymph**.

However, the concentrations of sodium (Na^+), chloride (Cl^-), and magnesium (Mg^{2+}) ions are higher, while concentrations of potassium (K^+) and calcium (Ca^{2+}) ions are lower.

The CSF functions to protect the brain and spinal cord by acting as a shock absorber. It also provides a buoyancy by reducing the effective weight of the brain. In addition, the CSF aids homeostasis of the interstitial fluid.

> **Textbox 5.6 The Encephalization Quotient—An Index of Relative Brain Sizes in Different Animals**
>
> Based on the encephalization quotient (an index of the relative size of brains to body weight) the brains of domestic animals are much smaller than in humans (Table 5.3).
>
> Table 5.3 The encephalization quotient (EQ) in various domestic animals.
>
> | Species | EQ (Encephalization Quotient)* |
> |---|---|
> | Human | 7.44 |
> | Chicken | 1.39 |
> | Dog | 1.17 |
> | Cat | 1.0 by definition |
> | Horse | 0.86 |
> | Sheep | 0.81 |
>
> *EQ is an index of brain size relative to body weight.

The flows of fluid in the brain are summarized in Figure 5.18. The **chorion plexus** produces CSF by a web of **ependymal cells**. Water transit is facilitated by **aquaporins** (water channels). An additional source of CSF is interstitial fluid. CSF passes out of the hollow spaces in a unidirectional manner by the **arachnoid villi** and into the lymph. There is bidirectional flow between the CSF and the interstitial fluid.

5.7.3 Blood–Brain Barrier

There is a blood–brain barrier between the capillaries in the brain and the interstitial space (see Figure 5.18). The blood–brain barrier protects the CNS from some drugs, toxins, and pathogens, regulating entering chemicals moving from the blood into the CNS (also see Textbox 5.1 for definition). Anatomically, the blood–brain barrier is the **tight junctions** of the capillary epithelium. Fluid, ions, and other small compounds pass from the blood into the interstitial fluid by ion channels, and specific transporters. **Aquaporin 1** (a channel allowing water to enter cells) is expressed in the **vascular endothelium** facilitating water movement outside the brain. However, Aquaporin 1 is *not* found in the vascular epithelium in the brain except for the circumventricular organ, the choroid plexus. There is an additional component of the barrier consisting of aquaporin containing astrocytes.

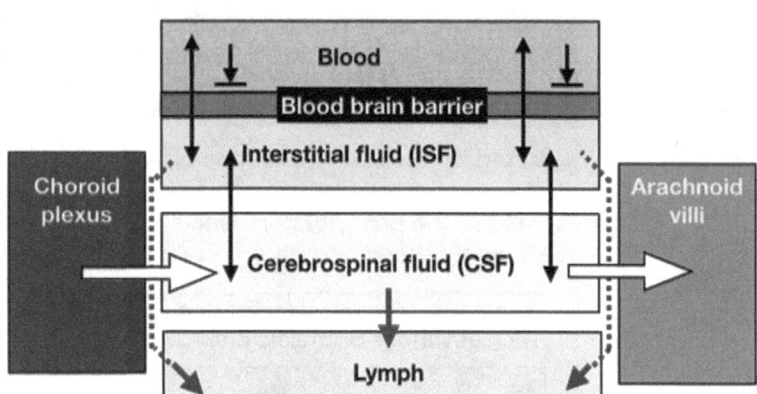

Figure 5.18 Flow of fluid between blood, interstitial spaces, cerebrospinal spaces, and lymph. The blood-brain barrier (consisting of tight junctions between vascular epithelial cells together with astrocytes) prevents many constituents of bloods passing into the interstitial space and cerebrospinal spaces.

5.7.4 Functions of the Brain

Functions of the brain will be discussed under the brain regions (also see Figure 5.19).

5.7.5 Cerebrum and Cerebral Cortex

Cerebrum and cerebral cortex are derived from the embryonic telencephalon. The number of neurons in the cerebrum and cerebral cortex has been estimated in some domestic animals as follows:

- Dogs: 0.5 billion neurons (22% of total neurons in the brain).
- Cats: 0.25 billion neurons (36% of total neurons in the brain).
- Humans: 17 billion neurons (20% of total neurons in the brain).

There are two hemispheres connected by the corpus callosum. Among the areas of importance in the cortex are the following:

- **Visual cortex.**
- **Auditory cortex.**
- **Olfactory cortex.**

The functioning of the cerebrum and cerebral cortex of domestic animals is beginning to be studied using trained dogs (see Textbox 5.7). The surface area of the cortex based on the folding affects cognitive abilities.

5.7.6 Hypothalamus

5.7.6.1 Introduction to the Hypothalamus

The hypothalamus is derived from the embryonic diencephalon. The hypothalamus is the control locus for important functions of the body, including the following:

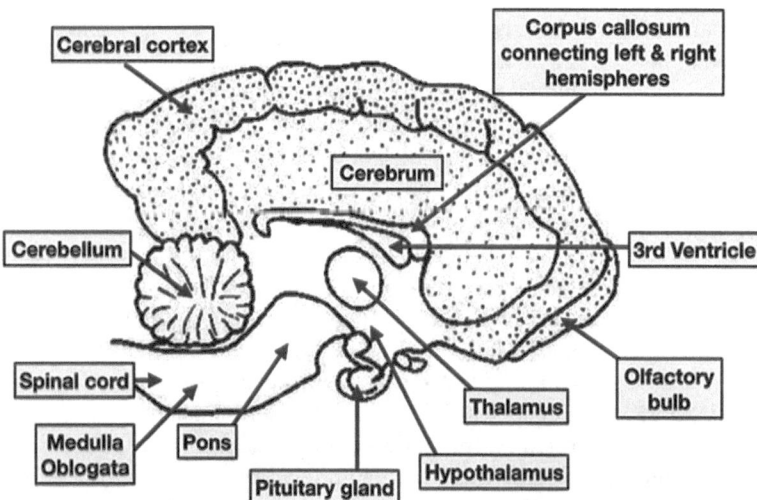

Figure 5.19 Longitudinal section of a dog brain showing major structures. (Adapted from Ruth Lawson, Otago Polytechnic, CC BY 3.0)

Textbox 5.7 Interesting Factoids: Special Features of the Brain in Dogs

Functional Magnetic Resonance Imaging (fMRI) and Dogs

Functional magnetic resonance imaging (fMRI) has been used to determine the regions of the brain of trained dogs that respond to specific stimuli. Examples of stimuli that induce activation of a specific region of the cortex include the following:

- The odor of their owner (the odor obtained by wiping under the arm) is much greater than that of other people or of other dogs ("rear ends").
- Activation in two areas in the cerebrum (the parietotemporal region and primary auditory cortex) in response to key words, such as "good dog."

Fluoxetine and Dogs

Fluoxetine (Prozac in humans) is a selective serotonin reuptake inhibitor. It is widely used in veterinary medicine for dogs in such situations as separation anxiety and other behavioral issues.

Brain Olfactory System

The olfactory system is 2% of brain weight in dogs and 7% of brain weight in pigs (compared to 0.03% in humans).

- Feeding and food consumption.
- Drinking and, consequently, water balance.
- Sleep and wakefulness.
- Body temperature.
- Reproductive behavior.

In addition, the **neuroendocrine cells** in the hypothalamus produce and secrete releasing hormones that control the release of the hormones of the anterior pituitary gland (see Chapter 4). The hypothalamus synthesizes hormones (oxytocin and arginine vasopressin or their homologues) that are released from the posterior pituitary gland (see Chapter 4). The **suprachiasmatic nucleus** in the hypothalamus is the location of circadian oscillators controlling the circadian rhythms of an animal.

5.8.6.2 What is a Circadian Rhythm?

A **circadian rhythm** is an endogenous rhythm (inside the animal). In what seems to be counter intuitive, a circadian rhythm has a periodicity of *about* 24 hours. It is then entrained to *exactly* 24 hours by environmental factors such as the light–dark cycle. In addition, it can be shifted with a change in latitude as interpreted by environmental factors such as the light–dark cycle. This is the basis of jetlag, when the circadian rhythm is out of phase with the local environment.

Figure 5.20 shows circadian rhythms for two hormones, **cortisol** and **melatonin**. The circadian rhythms of these affect the functioning of the brain and the body.

5.7.7 Pineal Gland

The pineal gland (or body) is part of the epithalamus (derived from the embryonic diencephalon). It is composed

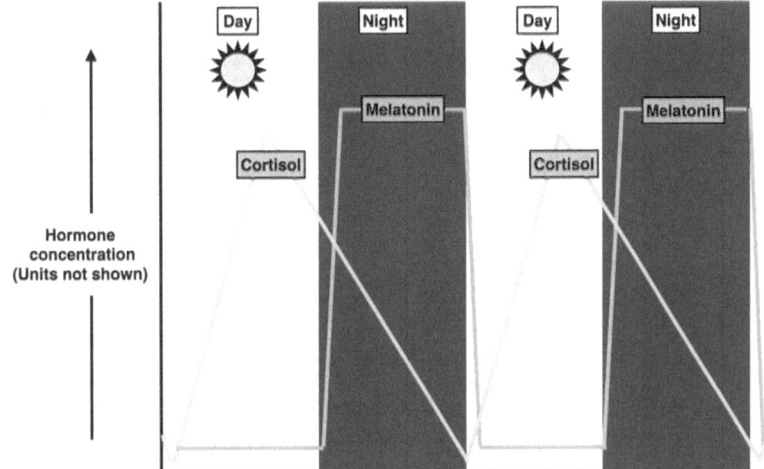

Figure 5.20 Circadian rhythms of plasma concentrations of melatonin and cortisol in livestock and companion animals.

of **pinealocytes** together with glial cells. It is an endocrine gland producing melatonin during the night or period of darkness (scotophase) (see Figure 15.20). Melatonin is synthesized from the neurotransmitter **serotonin**. Melatonin is involved in the control of the sleep wake cycle.

How does the pineal know it is day and night? Light acts on the retina. There is neuronal stimulation of circadian nucleus in the hypothalamus. From this, there are neurons to the pineal gland.

5.7.8 Thalamus

The thalamus relays **sensory inputs** and **motor outputs** between the cerebrum and brain stem. It is also the center for sleep/wakefulness.

5.7.9 Mesencephalon-Derived Brain Structures

This is the front of the brain stem and comprises the tectum, cerebral aqueduct, tegmentum, and cerebral peduncles.

5.7.10 Pons

The pons is derived from embryonic metencephalon. The pons is the brain structure relaying information connecting the medulla oblongata and cerebellum. It also contains the following:

- **Micturition center** (controls urination).
- **Respiratory center** (controls the respiratory rhythm of breathing).

5.7.11 Cerebellum

The cerebellum (the name is from the Latin word *cerebellum*, meaning "little brain") is derived from the embryonic metencephalon. The cerebellum is the major controller of motor function. It is the second biggest part of the brain and contains over 80% of the neurons.

5.7.12 Medulla Oblongata—A Myelencephalon-Derived Brain Structure

The medulla oblongata plays an important role with the senses of taste and touch, together with the movement of tongue. It plays important roles in regulating the following respiration rate, heart rate, and digestive processes.

There is a **neuronal loop** regulating heart rate. **Baroreceptors** in the carotid artery detect blood pressure. It is communicated to the medulla oblongata via **glossophygeal nerve**. The reflex is completed by the vagus nerve from the medulla oblongata to the heart, where **acetylcholine** is released, slowing the heart rate.

5.8 Spinal Cord

The spinal cord is a long tube of **nervous tissue**. It is surrounded by a series of vertebrae: cervical, thoracic, and lumbar, respectively.

Sensory nervous impulses pass along the spinal cord to the brain. While **motor impulses** pass along the spinal cord to the periphery. The spinal cord also conducts sympathetic and parasympathetic impulses from the brain to peripheral organs and back again. Figure 5.21. shows

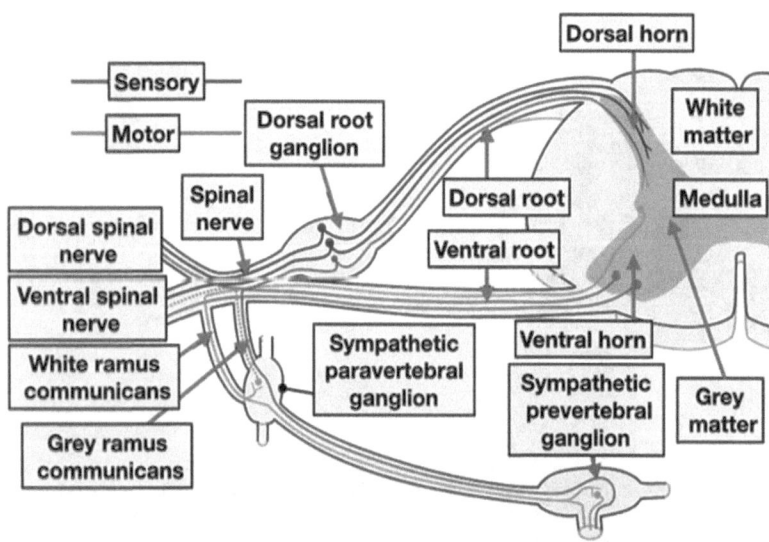

Figure 5.21 Ganglia and spinal cord. (Adapted from Jmarchn, CC BY-SA 3.0)

an illustrative view of motor and sensory nerves in and around the spinal cord and the associated ganglia.

5.9 Autonomic Nervous System

5.9.1 Introduction to the Autonomic Nervous System

The autonomic nervous system consists of the following:

- **Sympathetic nervous system** (with two neurons).
- **Parasympathetic nervous system** (with two neurons).
- **Enteric nervous system** (with three or more neurons).

5.9.2 Sympathetic Nervous System

In the sympathetic nervous system, the **pre-synaptic neurotransmitter** is acetylcholine and the **post synaptic neurotransmitter** is either norepinephrine or acetylcholine. In turn, the norepinephrine acts on adrenergic receptors in the effector tissues. There are multiple types of adrenergic receptors: α_{-1}, α_{-2}, β_{-1}, β_{-2} and β_{-3} adrenergic receptors. In addition to those sympathetic neurons that release norepinephrine, there are also sympathetic neurons that release neuropeptides, such as neuropeptide Y (NPY) or somatostatin.

5.9.3 Parasympathetic Nervous System

In the parasympathetic nervous system, acetylcholine is the neurotransmitter at both pre-synaptic and post-synaptic levels. In turn, acetylcholine acts on either muscarinic or nicotinic cholinergic receptors. In addition to those parasympathetic neurons that release acetylcholine, there are also parasympathetic neurons that release neuropeptides such as vasoactive intestinal peptide (VIP), neuropeptide Y (NPY), or calcitonin gene-related peptide.

5.9.4 Enteric Nervous System

The enteric nervous system largely functions independently of other nerves, but there can be effects of sympathetic and/or parasympathetic neurons. The enteric nervous system consists of the following:

- **Enteric ganglionated plexuses** (between the circular and longitudinal muscle layers of the gastrointestinal tract).
- **Submucosal ganglionated plexuses** (in the submucosa of the gastrointestinal tract).

The neurons of the enteric nervous system employ the following neurotransmitters:

- Acetylcholine.
- Serotonin.
- Nitrous oxide.

In addition, opioid peptides are present in the enteric nervous system. These delay gastric emptying, slow gut transit, suppress bile transport into the duodenum, and decrease gastric secretion via the μ-opioid receptor.

5.10 Introduction to the Senses

Animals respond to both their external and internal environment and to what's going on within them (summarized in Textbox 5.6). Livestock, companion animals, and poultry have the same overall senses, namely the following:

- Sight (visual).
- Hearing (auditory).
- Taste.
- Smell (olfaction).
- Touch.
- Pain.

These senses allow the animal to attempt to live in a safe environment.

Animals have the following sensor cells:

- **Photoreceptors** (or photoreceptor cells) are light receptors (or light receptor cells). These are located in the retina of the eye (to enable vision) and in specific brain areas. In vision, multiple photoreceptor cells work together as part of a system. There are associated anatomical features (the lens of the eye and associated muscles) to focus light.
- **Mechanoreceptors** (mechanoreceptor cells) respond to vibrations in hearing. The associated ears aid the transmission of sound. In addition, there are low-threshold mechanoreceptors in the skin that respond to touch.
- **Chemoreceptors** (chemoreceptor cells) responding to specific chemicals.
  - Taste.
  - Olfaction.
  - Trigeminal (pain receptors activated by irritants such as capsaicin).
  - Respiratory gases.

114 THEME 1 Growth and Development

- **Proprioceptors** responding to the positioning of areas within the body.
- **Baroreceptors** (baroreceptor cells) detecting blood pressure.
- **Temperature-sensitive neurons**.
- **Nociceptors** (nociceptor cells) are pain receptors.

5.11 Vision/Sight

The principal, but not the only, organs responding to light are the two eyes. The overall structure of eyes is shown in Figure 5.22. The components of the eye function to achieve the following:

- The lens and ciliary muscles focusing the light from far, near, or intermediate distances.
- Transmission of light through the following transparent and avascular parts of the eye:
  - Cornea.
  - Aqueous humor.
  - Lens.
  - Vitreous humor to the retina.
- Photoreception in the retina.
- Ensuring the structural integrity and physiological functioning of eye.
- Protection of the retina and other parts of the eye.

5.11.1 Retina

5.11.2 Maintaining the Structural Integrity and Physiological Functioning of the Eye

Examples of the how the structures in the eye maintain physiological functioning of eye include the following:

- The **aqueous humor** providing nutrients to the lens and cornea.
- The aqueous humor maintaining the pressure in the eye.
- The aqueous humor and **vitreous humor** maintaining the shape of the eyes.
- The vitreous humor keeping the retina in place.

Humors are liquids, with the name coming from the Latin for "liquid."

5.11.3 Protecting the Retina and Other Parts of the Eye

Examples of the how the structures in, or associated with, the eye protect the eye include:

- The iris limits the quantity of light hitting the retina.
- The cornea protects the eye from invasion by microorganisms.
- The gelatinous vitreous humor acts to absorb shock.

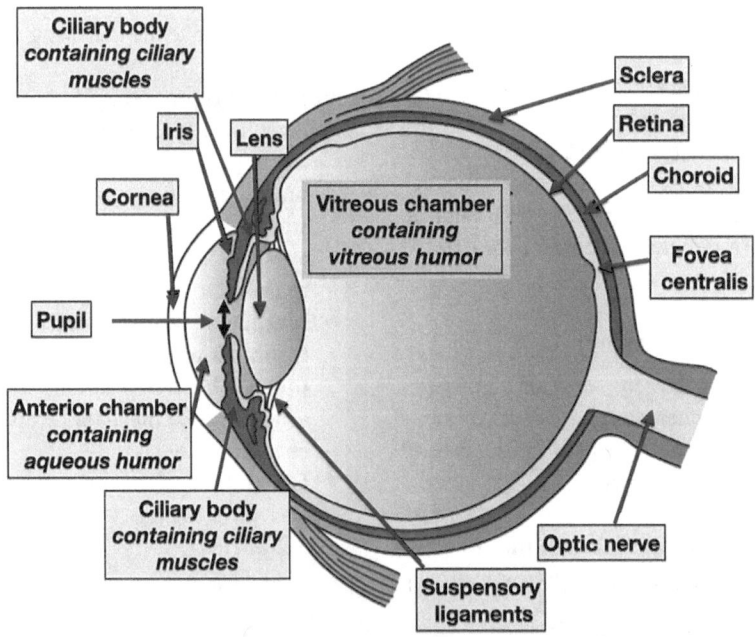

Figure 5.22 Structure of the eye of livestock and companion animals. (Adapted from Holly Fischer, CC BY 3.0)

Textbox 5.8 Color Vision

Humans (and other primates) can detect a spectrum of colors (see below). This is because we have **trichromatic** vision based on three types of cones. This is not the case in livestock, companion animals, or poultry.

Dogs, cats, horses, and livestock have two types of cones and, hence, can see two colors. Their vision is **bichromatic**.

Chickens and other birds have four types of cones and, hence, four colors (red, green, blue, and ultraviolet). In addition, chickens have double cone structures to allow them to track movement better. Their vision is **tetrachromatic**.

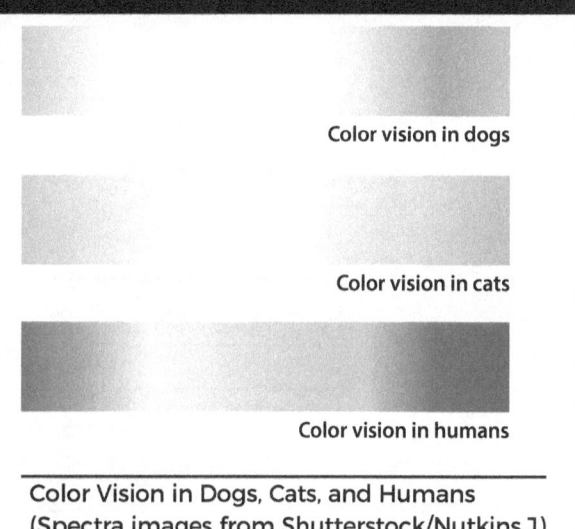

Color Vision in Dogs, Cats, and Humans (Spectra images from Shutterstock/Nutkins.J)

- The eyelids protect the eyes (see Section 5.13).
- Lacrimal fluid (tears) lubricates the eyes, keeping the eyes wet (see Section 5.14).

5.11.4 Photoreception

5.11.4.1 What Is Light?

Visible light is **electromagnetic radiation** in the range of wavelengths between 380 nanometers (nm) to 750 nm (see Figure 5.23A). Electromagnetic radiation is both a wave phenomenon and consists of particles called photons. Vision only uses a very small portion of electromagnetic radiation (see Figure 5.23). Ultraviolet light has wavelengths below 380 nm, while infrared light has wavelengths above 750 nm.

5.11.4.2 Where Is Light Detected?

Light is detected by light sensitive cells in the retina (see Figure 5.24). These cells are either:

- **Cones** having at least one type for each **color**. For example, there are two types of cones in dogs.
- **Rods** are responsible for night vision and shades of black, white, and grey. Rods are concentrated at the periphery of the retina.

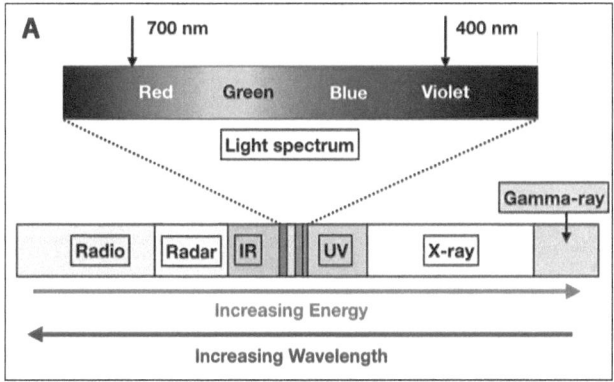
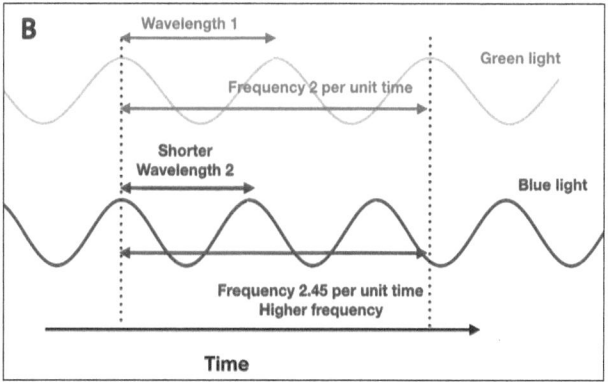

Figure 5.23 Spectra of electromagnetic radiation and visible light.
A. Spectra of electromagnetic radiation with that of visible light ranges from a wavelength of 380 nm to 750 nm [Key: IR infrared, UV ultra violet] (Adapted from Danniel Curze, CC BY-SA 3.0).
B. Wavelength and frequency are inversely related.

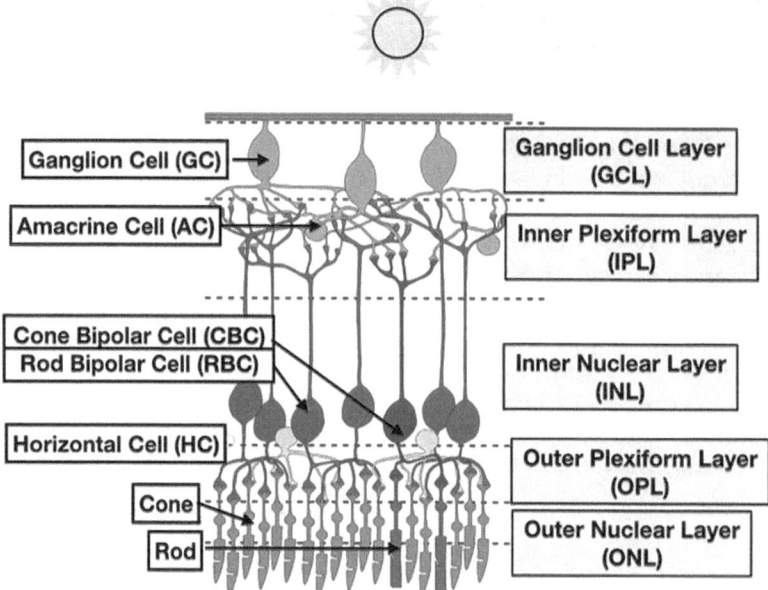

Figure 5.24 Anatomy of the retina showing light responsive rods and cones located on the interior of retina.
(Adapted from Jorg Encke, CC BY-SA 4.0)

5.11.4.3 How Is Light Detected?

The light responsive molecules (photoreceptor molecules) in the photoreceptor cells are the chromophore, **retinal**, and the closely associated apoprotein, **opsin**.

Light photons cause **1-cis-retinal isomer** to be converted to **all-trans-retinal isomer**, activating the opsin (Figure 5.25). In turn, the active opsin activates G protein transducin (a heterotrimer) in the cell membrane (Figure 5.26). Then the activated G protein transducin activates **phosphodiesterase**, the enzyme that converts **cyclic GMP** (cyclic guanylyl monophosphate) to GMP (guanylyl monophosphate) and, thereby, closes the **cGMP-gated, nonselective cation channels** in the cell membrane. This then causes hyperpolarization of the cell and cessation of release of the neurotransmitter **glutamate**.

Textbox 5.9   A Deeper Dive to Photoreception

This has seven transmembrane domains and is located in the cell membrane. The retinal opsin complex acts as a G-protein-coupled receptor. The retinal is derived from carotenoids in the diet. In the basal state, the retinal is 1-cis-retinal. Light photons bring about the conversion of 1-cis-retinal to all-trans-retinal isomer with a marked change in its shape (Figure 5.25). The all-trans-retinal is converted back to 1-cis-retinal.

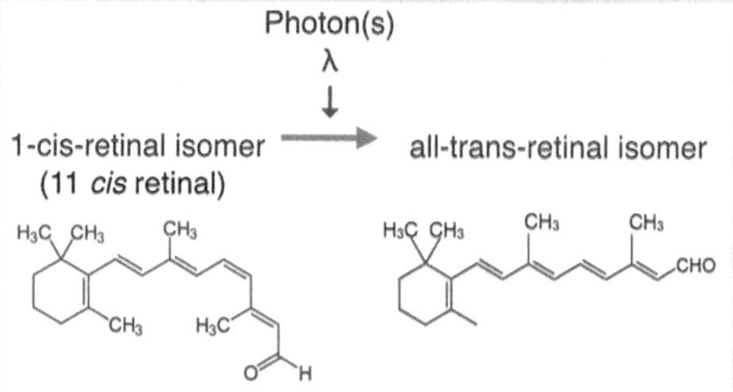

Figure 5.25 Effect of light photons converting the chromophore, 1-cis-retinal isomer, to all-trans-retinal isomer.

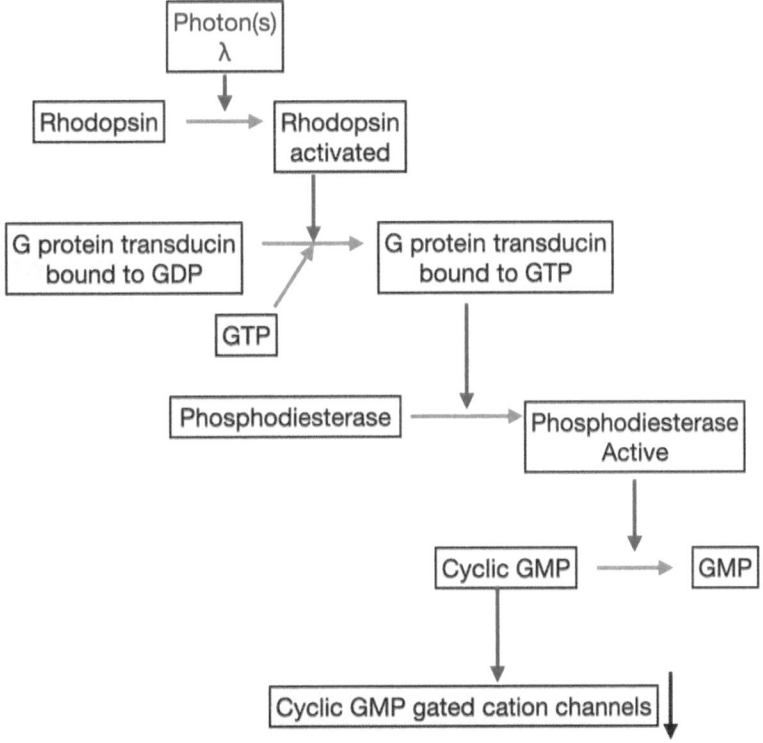

Figure 5.26 Mode of action of light photons on rhodopsin, via G protein transducin, phosphodiesterase, and cyclic GMP (cyclic guanosine monophosphate). Light leads to reduced cyclic GMP, closing of the cyclic GMP-gated cation channels, reduced sodium entry, and consequently hyperpolarization.

There are different opsins in rods and in the different types of cones:

- LW opsin (long-wavelength-sensitive opsin) in cones responding to green.
- SW opsin (short-wavelength-sensitive opsin) in cones responding to blue.
- **Rhodopsin** in rods.

5.11.4.4 Brain Photoreceptors

Photoreceptor molecules (opsins) are also present in the hypothalamus. In poultry and other birds, the specific opsin is VA opsin, or vertebrate ancient opsin. This plays an important role in detecting the length of day (or photoperiod) and the control of seasonal breeding. There are also photoreceptor molecules in other regions of the brain (e.g., pinopsins in the pineal gland).

5.12 Eyelids

The eyelids protect the entire eye, and by blinking it removes debris, dust, or irritants from the surface of the eyes. In some domestic animals, such as dogs and cats, there are three eyelids. Two are similar to those in humans, but the third is the nictitating membrane. The eyelids protect the eyes and ensure that there is a liquid across the cornea. They are highly vascularized such that they are both resistant to bacteria and repair quickly after injury.

5.13 Tears

Lacrimal fluid (tears), aided by blinking of the eyelids, keep the cornea wet and provides lubrication for eye movement. There is a basal rate of production of tears. This is increased in a reflex manner when there are dusts or other irritants in the eye.

The lacrimal or tear glands are **epithelial glands** that produce tears in humans, livestock, companion animals, and poultry. In dogs, the lacrimal gland produces much of the tears, with the remainder coming from the accessory lacrimal gland of the third eyelid (see Section 5.13).

5.13.1 The Composition of Tears

The composition of tears has been determined in dogs and horses. Tears contain the following:

- **Water.**
- **Ions** including sodium, potassium, chloride, bicarbonate, magnesium, and calcium, with higher concentrations of potassium and calcium than in plasma.
- **Proteins** including:
  o Plasma albumin.
  o The bacteriolytic protein, lysozyme.
  o Lactoferrin.
  o Lipocalin.
  o Immunoglobulin A (IgA).
- **Lipids** including cholesterol.
- **Mucin** (glycoproteins).
- **Urea.**

There is some glucose, albeit at much lower concentrations than in the plasma.

5.13.2 Function of tears

The functions of tears are the following:

- Protecting the eyes against dust and grit.
- Protecting the eyes against bacteria by either of the following:
  o Antimicrobial proteins.
  o Flushing them away.
- Creating a tear film across the cornea, thereby aiding vision through a smooth layer.
- Preventing water loss from the eye.
- Providing lubrication for eye lids; this being important during blinking.

5.14 Hearing

Hearing is detecting sound waves in the air. The organ detecting sound is the ear.

5.14.1 What Is Sound?

Sound is vibrations in the air, a liquid, or a solid. It can vary in the following:

- Pitch or wavelength in Hertz (Hz).
- Loudness in decibels (db).
- Direction from which the sound is coming.

5.14.2 What is Hertz?

Hertz is the number of vibrations per second. Hz is the frequency or pitch of a sound.

5.14.3 Do Animals Detect the Same Frequencies of Sound?

The simple answer is no (see Table 5.4).

5.14.4 Functioning of the Ear

Ears function to detect sounds or vibrations in the air in the external environment. The structure of the ear of a domestic animal is shown in Figure 5.27. The **external ears**, or pinna, collect the sound and channel it through the ear canal. The sound causes the **tympanic membrane** to vibrate. In turn, this causes the **ossicles** (the bones **malleus**, **incus**, and **stapes**) to vibrate. The ossicles are essentially amplifying the sound. In the **cochlea**, the vibrations stimulate sound hair cells by acting on the **stereocilia** on the hair cells.

The three **semicircular canals** detect angular acceleration or deceleration of the head in **three directions**. This is critically important to ensuring balance.

The **eustachian tube** is a canal between the middle ear and the throat. Its function is to maintain the pressure in the middle ear and, hence, allow the normal transmission of sound in the middle ear.

The **external ears** of some breeds of dogs (see Figure 5.28), together with those of cats, can be moved to follow the direction of the sound. Chickens and other poultry lack pinnas.

5.14.5 Do Animals Detect the Level of Sound?

Dogs and cats are able to hear sounds that are quieter than can be perceived by people.

Table 5.4 Comparison of the range of wavelength detected by different domestic animals.

Species	Auditory range (Hz)
Human	64–23,000
Dog	65–45,000
Cat	45–64,000
Sheep	100–30,000
Horse	55–33,500
Chicken	125–2,000

Chapter 5 The Nervous and Sensory Systems

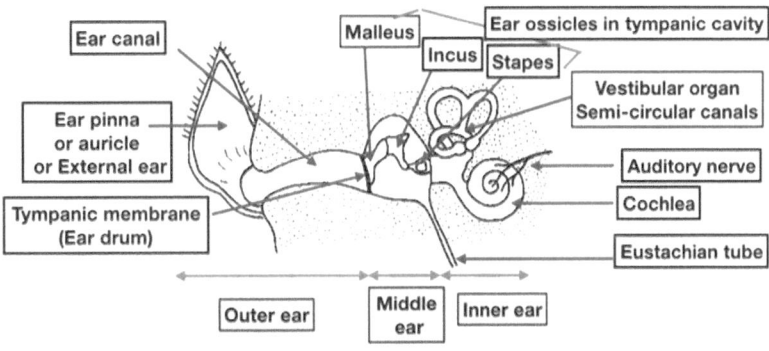

Figure 5.27 Structure of the ear in dogs. (Adapted from Ruth Lawson, CC BY-SA 3.0)

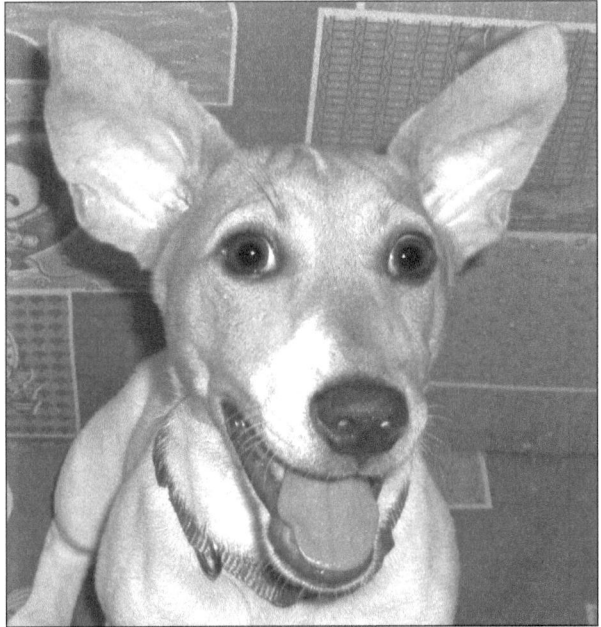

Figure 5.28 Dog showing large external ears (pinnas). These function to focus and amplify sound in a manner similar to a parabolic radio telescope. Dogs can recognize the direction of a sound by movement of the ears and the difference in time for the sound to reach one ear or the other. (Boichico, CC BY-SA 3.0)

5.15 Chemoreception

5.15.1 Olfaction (Smell)

Olfaction is the detection of **volatile chemical substances**. The olfactory system consists of the following:

- Nasal cavity.
- Olfactory epithelium with the bipolar **olfactory receptor neurons**.
- **Vomeronasal organ**.
- **Olfactory bulb**.

This is illustrated in Figures 5.29, 5.30, and 5.31.

5.15.2 Molecular Olfaction Receptors

There is a single type of receptor in each olfactory receptor neuron. When a specific odorant binds to the receptor in the olfactory cilia, adenylyl cyclase is activated through the conversion of ATP to cyclic AMP (see Figure 5.31). In turn, there increases entry of both calcium (via the calcium channel) and sodium ions into the neuron and an action potential is generated. This passes along the neuron to the olfactory bulb.

Dogs are over ten thousand times better than people in detecting odors. This is based on the following (see Table 5.5):

- Greater numbers of olfactory receptor neurons.
- Greater numbers of olfactory cilia per neuron.

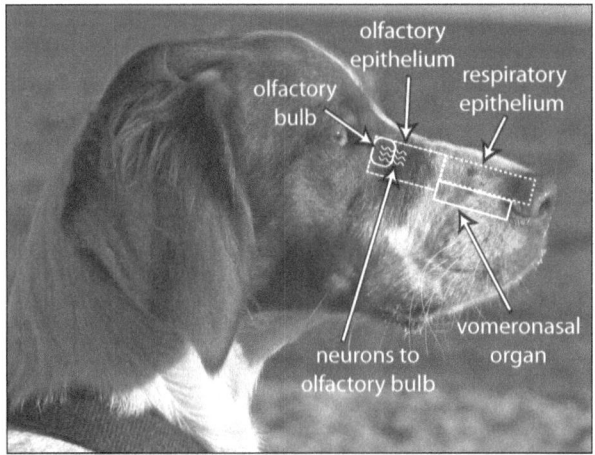

Figure 5.29 Schematic of olfaction in dogs. (Adapted from Uber Phot, CC BY-SA 2.0)

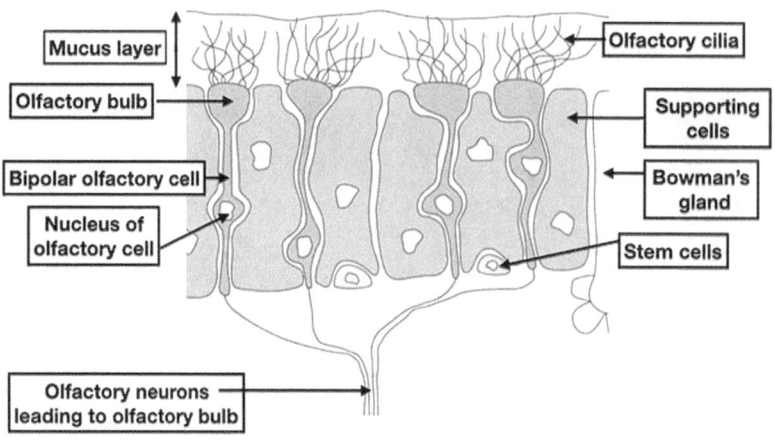

Figure 5.30 Olfactory epithelium showing olfactory receptor cells. (Adapted from MarianSigler, CC BY-SA 3.0)

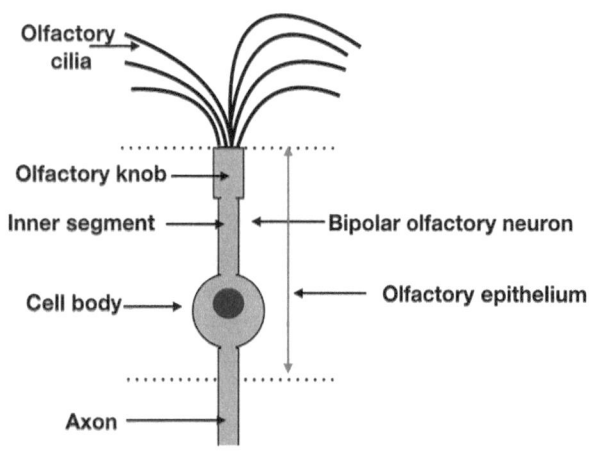

Figure 5.31 Schematic of olfactory receptor cells.

- Greater numbers of olfactory molecular receptor genes.
- Fewer olfactory molecular pseudogenes.

5.15.3 Taste

Taste buds (or gustatory papillae) are small sensory organs located on the tongue (see Figure 5.34). They detect taste in foods. The taste buds contain chemical receptors, the gustatory neurons/taste receptors that respond to the following tastes:

- Sweet (sucrose, fructose, glucose).
- Sour (acid/hydrogen ions, e.g., hydrochloric acid).
- Bitter (e.g., quinine).
- Salty (sodium chloride).
- Umami (e.g., glutamate).

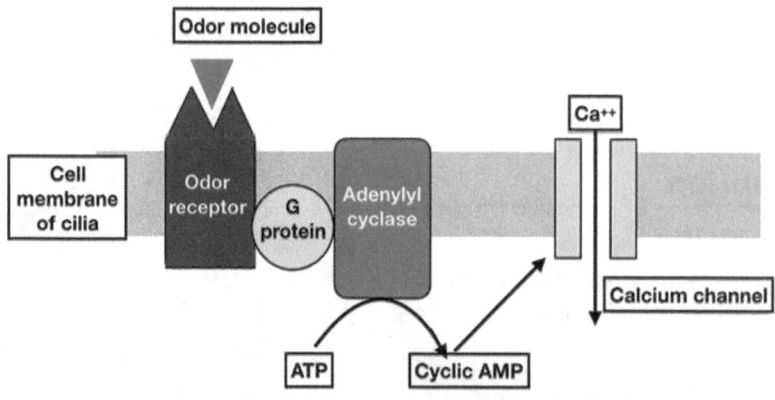

Figure 5.32 Mechanism of how an odor molecule is detected.

Chapter 5 The Nervous and Sensory Systems

Table 5.5 Comparison of olfaction in dogs, humans, and poultry.

Parameter of olfaction	Dogs	Humans	Poultry
Number of olfactory neurons in millions	220[a]	6	5.8 (duck)
Number of cilia per olfactory receptor cell	>150	25	NA
Number of genes in specific odorant receptors gene families	1300	340	80 (chicken)
Number of pseudogenes[b] in specific odorant receptors gene families	120	300	480 (chicken)

NA = data not available.

[a] In contrast, there are 300 million olfactory receptor neurons in bloodhounds.
[b] A pseudogene is similar to a gene but does not code a protein and, therefore, is a non-functional gene.

Textbox 5.10 Olfaction in Dogs

Among the best breeds of dogs for olfactory acuity (sense of smell) are bloodhound, basset hound, beagle, German shepherd, and Labrador retriever. These and other breeds with a high scenting ability are used to detect explosives, drugs, and even cancer, together with search and rescue (see Figure 5.33).

Figure 5.33 Dogs being used to detect drugs, explosives, and illegal foods. (Courtesy of USDA APHIS)

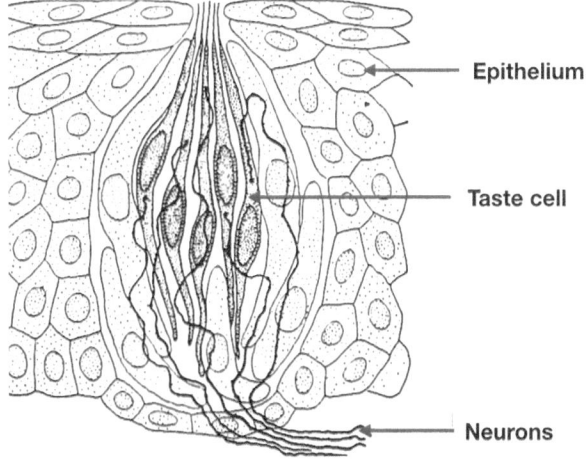

Figure 5.34 Taste bud showing taste cells and neurons. (Douglas Marsland)

There are about 50 to 100 taste receptors per taste bud. There are also taste receptors in the epithelium elsewhere in the mouth, pharynx, larynx, and upper part of the esophagus.

Tastes are detected at the molecular level by the following:

- Sweet is detected by a sweet receptor. The sweet receptor is a dimer of two units, TAS1R2 (taste receptor family or type 1 member 2) and TAS1R3 (taste receptor family 1 member 3).
- Sour is detected by proton- (hydrogen ion) gated cation and chloride channels.
- Bitter taste reception is mediated by over 20 different genes of the type 2 taste receptor families.
- Salt is detected by **epithelial-type sodium channels (ENaC)**.

- The **umami receptor** is a dimer of two units TAS1R1 (taste receptor family 1 member 1) and TAS1R3 (taste receptor family 1 member 3).

Taste is mediated by taste buds or gustatory papillae on the surface of the tongue. Specific molecular receptors are responsible for taste. When a specific stimulus binds to the receptor, there is depolarization of the taste receptor cell and activation of a neuron (in branches of three cranial nerves—VII [facial], IX [glossopharyngeal], and X [vagus]) via the intervening synapse. Taste receptors are **G-protein-coupled receptors** and are also found in the gastrointestinal tract.

5.15.3.1 Is Taste Reception the Same Across Domestic Animals?

In some felid species, including cats, there is no taste reception to sugar and other sweeteners. This is due to the gene TAS1R2 being inactive but present as a **pseudogene** (a pseudogene not being expressed). There are very few taste buds in poultry and very low numbers of taste receptors.

5.15.4 Pheromones

Pheromones are specific chemicals, or cocktail of chemicals, that are produced by one animal to influence another animal of the same species. They are normally airborne and cue physiological changes and/or behaviors. It is well established that some mammals produce pheromones. However, their presence has not been established in any bird.

Examples of pheromones include the following:

- Male goats produce a pheromone/scent that influences GnRH neural networks controlling reproduction in female goats.
- Male pigs produce pheromones (e.g., androstenone and androstanol). This is found in their saliva. The pheromone induces sexual maturation in female pigs and lordosis (standing in the position for mating).
- Dogs and cats produce pheromones that reduce anxiety and indicate reproductive status.
- Male cattle, buffalos, and horses (see Figure 5.31) show a "**flehmen**" response (wrinkling of the nose and moving upper lip to show the teeth) to the chemicals in the urine of females at the time of estrus. The name "flehmen" comes from German, meaning "to show front teeth in the upper jaw" (see Figure 5.31).

Figure 5.35 Flehmen response shown by a stallion. (Courtesy of Wausberg, CC BY-SA 3.0)

In rodents and bulls, pheromones are perceived in the **vomeronasal organ (VNO)**. It is not clear if the VNO has a role in detection of pheromones in other livestock or companion animals. Instead, pheromones are detected by the olfactory system.

5.15.5 Other Chemoreception

Carbon dioxide (via hydrogen ions) can be detected in the blood by the respiratory center of the brain. Blood concentrations of oxygen are detected by specific receptors in the carotid body.

5.16 Nociceptors or Pain Receptors

Animals respond to pain irrespective of whether they are livestock, companion animals, or poultry. Pain signals either damage or potential injury to tissues. **Nociceptors**, or pain receptors, are specialized cells that respond to noxious stimuli, e.g., from damaged tissues. Nociceptor axons (Aδ) have free endings in the skin. Nociceptors respond to three overall types of stimuli:

- Mechanical (high-threshold mechanoreceptors).
- Chemical, responding to chemicals released from damaged tissues or noxious chemicals.
- Thermal, responding to high temperatures that will burn.

Chemical nociceptors respond to chemicals from damaged tissues, e.g., histamine or prostaglandins. In addition, chemical nociceptors respond to noxious chemicals and irritants. Examples of irritants are capsaicin (the heat of spicy foods) together with veratryl acetamide and veratryl amine. Birds exhibit little aversion to capsaicin, eating feed containing it. They do, however, avoid feed containing either veratryl acetamide or veratryl amine.

The body also has neurotransmitters that act to block or reduce the perception of pain. These are the endogenous opioids, peptides/proteins that mimic morphine or heroin. The endogenous opioids are the following:

- **β-endorphins** acting via mu (μ) opioid receptors.
- **Met-enkephalin** acting via delta (δ) opioid receptors.
- **Dynorphins** acting via kappa (κ) opioid receptors.

5.16.1 Vibrissae (Whiskers)

Livestock and companion animals have **vibrissae** (whiskers) (see Figure 5.36). These are not found in poultry and other birds. These are stiff coarse hairs that are attached to **tactile nerves**. They aid vision and navigation.

5.17 Comparison of Senses in Livestock/Mammalians Companion Animals With Those in Poultry

Table 5.6. compares senses in livestock/mammalians companion animals with those in poultry.

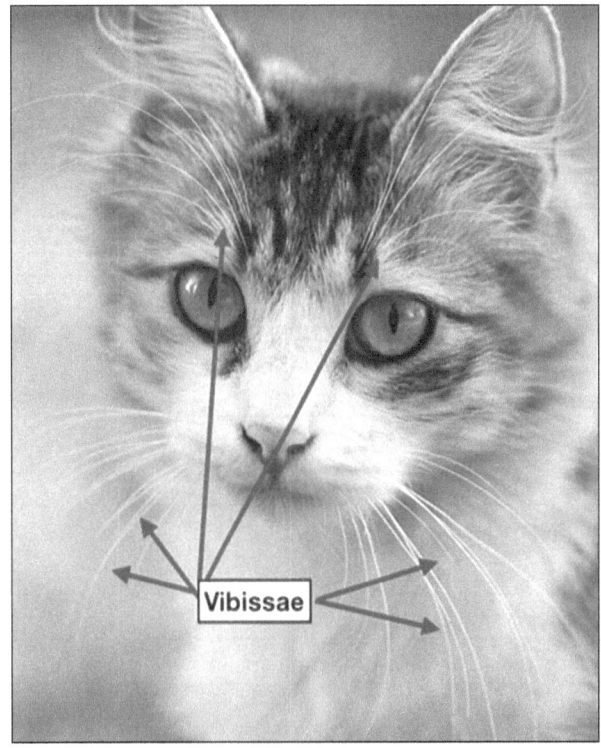

Figure 5.36 Cat showing vibrissae above the eyes and on either side of the mouth/nostrils. (Courtesy of AdinaVoicu)

Table 5.6 Comparison of senses in livestock/mammalians companion animals with those in poultry.

	Livestock & companion mammals	Poultry
Sight		
Light or photoreception (rods & cones)	Yes	Yes
Color reception (number of colors perceived + types of cones)	2 colors	4 colors, including UV
Hearing		
Ears	Yes	Yes
Chemical reception		
Taste (gustation)	Yes	Yes
Taste buds	>8000	<400
Receptors for sweet (T1R2 receptor)	Yes	No
Receptors for bitter, e.g., quinine (a series of T2Rs)	Yes	Yes but limited
Receptors for sour (PCKD channels)	Yes	Yes

(continued)

Theme 1 Growth and Development

	Livestock & companion mammals	Poultry
Receptors for salt (sodium chloride) (ENaC receptors)	Yes	Yes
Receptors for umami (T1R1 and T1R3 receptors)	Yes	Yes
Receptors for fatty acids (G-protein-coupled receptor-120)		
Trigeminal (chemesthesis - responding to irritants like we do with spicy foods)	Yes	Yes, but with no capsaicin response
Olfaction (smell)	Yes	Yes
Reception of pheromones	Yes	No
Oxygen	Yes	Yes
pH	Yes	Yes
Pressure/touch		
Temperature		
Set-point for temperature	Yes	Yes, but higher
Temperature sensitive neurons in hypothalamus	Yes	Yes
Pain		
Nociceptors (pain receptors)	Yes	Yes

Textbox 5.11 Diseases of the Nervous System

- **BSE (bovine spongiform encephalopathy)** is a prion disease affecting the brain of cattle.
- **Brain tumors** are found in dogs, particularly short-snouted breeds such as English bulldogs, boxers, pugs, and Boston terriers, but also in golden retrievers.
- **Chronic wasting disease** is a prion disease affecting the brain of deer and elk. There is loss of neurons with astrocytes replacing the dead neurons creating "holes."
- **Inherited diseases** are found in dogs and include epilepsy.
- **Paralysis** in dogs and horses due to toxins such as botulism, tetanus, or snake venom.
- **Dementia** in dogs (Canine Cognitive Dysfunction or CCD), with symptoms including confusion, irritability, and blankly staring.

Diseases and the Senses

- **Infectious Bovine Keratoconjunctivitis** (IBK, "Pink eye," "New Forest Disease") is caused by the bacterium Moraxella bovis and can be spread by flies. The cattle with this are in pain when in full sunlight. This leads to a reduced growth rate. There can be redness in the eyeball and even blindness.
- **Glaucoma in dogs** is a painful condition in which there is increased pressure in the eye. This can lead to blindness.
- **Cataracts in dogs** are found in old dogs. The eye has a milky appearance as the lens becomes progressively less transparent.

Muscle

6

Learning Objectives

1. To understand the biological roles of muscles (see Section 6.1).
2. To understand the special importance of muscles in domestic animals (see Sections 6.2, 6.12, and 6.13).
3. To understand the difference between smooth, cardiac, and skeletal muscle cells (see Section 6.2).
4. To understand the functional anatomy of skeletal and cardiac muscle (see Section 6.3).
5. To understand the functional anatomy of smooth muscle (see Section 6.8).
6. To understand the role of the sarcoplasmic reticulum (see Section 6.4).
7. To understand pre- and post-natal muscle development (myogenesis) (see Section 6.5).
8. To understand stimulation of contraction of skeletal or cardiac muscle (see Section 6.6).
9. To understand the mechanism of contraction of skeletal or cardiac muscle (see Section 6.7).
10. To understand stimulation of contraction of smooth muscle (see Sections 6.9.2 and 6.9.3).
11. To understand the mechanism of contraction of smooth muscle (see Section 6.9.1).
12. To understand how energy is generated for muscle contraction (see Section 6.10).

Table of Contents

6.1 Introduction
    6.1.1 Characteristics of Muscle
    6.1.2 Why Are Muscles Essential for the Life of Animals?
    6.1.3 Why Are Muscles Especially Important for Domestic Animals?

6.2 Types of Muscle Cells (Myocytes)
    6.2.1 Overview of Types of Muscle Cells
    6.2.2 Skeletal Muscle
    6.2.3 Classification of Skeletal Muscles

- 6.2.4 Cardiac Muscle
- 6.2.5 Smooth Muscle
- 6.2.6 Types of Smooth Muscle
- 6.3 Functional Anatomy of Skeletal and Cardiac Muscle
  - 6.3.1 Overview
  - 6.3.2 What Are the Major Proteins in the Sarcomere?
- 6.4 Sarcoplasmic Reticulum
- 6.5 Myogenesis—The Development of Myofibers
  - 6.5.1 Overview
  - 6.5.2 Fetal (Mammals)/Embryonic (Poultry) Myogenesis
  - 6.5.3 Post-Natal Myogenesis
- 6.6 Stimulation of Contractions of Skeletal and Cardiac Muscle
  - 6.6.1 Definitions Related to Muscle Contractions
  - 6.6.2 How Does the Muscle Fiber Know When to Contract?
  - 6.6.3 Skeletal Action Potential
- 6.7 Mechanism of Skeletal and Cardiac Muscle Contraction
  - 6.7.1 Excitation
  - 6.7.2 Muscle Contraction
  - 6.7.3 Muscle Relaxation
  - 6.7.4 Steps in Muscle Contraction
  - 6.7.5 Muscle Contraction, Calcium, and the Sarcoplasmic Reticulum
  - 6.7.6 Special Features of Cardiac Muscle and Contraction
  - 6.7.7 ATP and Rigor Mortis
- 6.8 Functional Anatomy of Smooth Muscle
- 6.9 Mechanism of Smooth Muscle Contraction
  - 6.9.1 Overview of Smooth Muscle Contraction
  - 6.9.2 Activation of Smooth Muscle Myocytes
- 6.10 Energy Production for Muscle
  - 6.10.1 Overall
  - 6.10.2 Glycolysis
  - 6.10.3 Oxidative Respiration
  - 6.10.4 Creatine, Muscle, and ATP
  - 6.10.5 Why Is Creatine Important to Skeletal Muscle Functioning?
- 6.11 Muscle Contractions Leading to Twitches, Incremental Contractions, and Physiological Tetanus
- 6.12 Muscle as Meat
- 6.13 Double Muscling
- 6.14 Myopathy (Muscle Disease)
  - 6.14.1 Overview of Myopathy
  - 6.14.2 Myopathies in Livestock and Poultry
  - 6.14.3 Porcine Stress Syndrome
  - 6.14.4 Myopathies in Horses
  - 6.14.5 Fainting Goats

6.1 Introduction

6.1.1 Characteristics of Muscle

Muscles have the following characteristics:

- **Responsiveness** to internal stimuli (and consequently responding to external stimuli).
- **Excitable**.
- **Contractility** (muscles can contract).
- **Elasticity** (muscles when contracted or stretched return to their previous shape).
- **Extendibility** (muscles can be stretched).

Muscles are able to receive neural stimuli (responsiveness) and then conduct the messages down the muscle fibers (excitable). Muscles are able to respond through contraction or extension and are elastic such that they return to their resting state. Muscle or muscle

> **Textbox 6.1 Glossary Related to Muscle**
>
> - **Actin**: one of the proteins in the thin filaments in the sarcomeres.
> - **ATP** (adenosine triphosphate): the energy source for muscle contraction.
> - **Mitochondria**: the cell organelles generating ATP.
> - **Motor unit**: a single neuron and the muscle fibers.
> - **Myosin**: the protein in the thick filaments in the sarcomeres.
> - **Sarcolemma**: the plasma membrane of muscle fiber.
> - **Sarcomeres**: the functional units of muscle.
> - **Sarcoplasmic reticulum**: stores and releases calcium.
> - **T-tubules or transverse tubules**: conduct electrical signal from the sarcolemma to the terminal cisternae in the sarcoplasmic reticulum.
> - **Terminal cisternae**: part of sarcoplasmic reticulum. This organelle stores calcium, releases calcium, and takes up calcium.
> - **Tetanus (physiological)**: a muscle contraction that engages all muscle fibers in the muscle and achieves a maximum contraction.
> - **Troponin**: one of the proteins in the thin filaments in the sarcomeres.
> - **Tropomyosin**: one of the proteins in the thin filaments in the sarcomeres.
> - **Twitch**: a muscle contraction that consists of the latent period together with the period of contraction and relaxation.

layers are not the only tissues in muscles. In addition, there are connective tissue, blood vessels, nerves, and adipose tissue.

A glossary related to muscle is provided in Textbox 6.1.

6.1.2 Why Are Muscles Essential for the Life of Animals?

Muscles are essential for the following biological functions in domestic animals:

- Standing and maintaining posture (requiring bones and joints).
- Movement, including the following:
  - Contraction of skeletal muscle to walk, run, mate, etc. (skeletal muscle and requiring bones and joints).
  - Contraction of the diaphragm and intercostal muscles between the ribs, moving air and out of the lungs (skeletal muscle and requiring bones and joints).
  - The pumping action of the heart to move blood around the body (cardiac muscle and not requiring bones or joints).
  - **Peristalsis** of the gastrointestinal tract to move ingested food through the tract (smooth muscle and not requiring bones or joints).
  - **Galactokinesis** to move milk through the lobuloalveolar system of the mammary gland (smooth muscle and not requiring bones or joints).
  - Uterine contraction to move the fetus through the birth canal during parturition (smooth muscle and not requiring bones or joints).
- Heat generation as part of the homeostatic mechanism to keep the body temperature at the set-point.
- **Glucose homeostasis** with glucose stored in muscles as glycogen.

6.1.3 Why Are Muscles Especially Important for Domestic Animals?

The muscles of domestic animals are, or have been, used for the following two purposes by people:

- As meat (see Figure 6.1).
- Locomotion (historical uses indicated by *):
  - Plowing* (horses and cattle).
  - Herding (dogs).
  - Pulling sleds (dogs and horses).
  - Pulling carriages (horses).
  - Transportation* (horses and donkeys) (see Figure 1.1D).

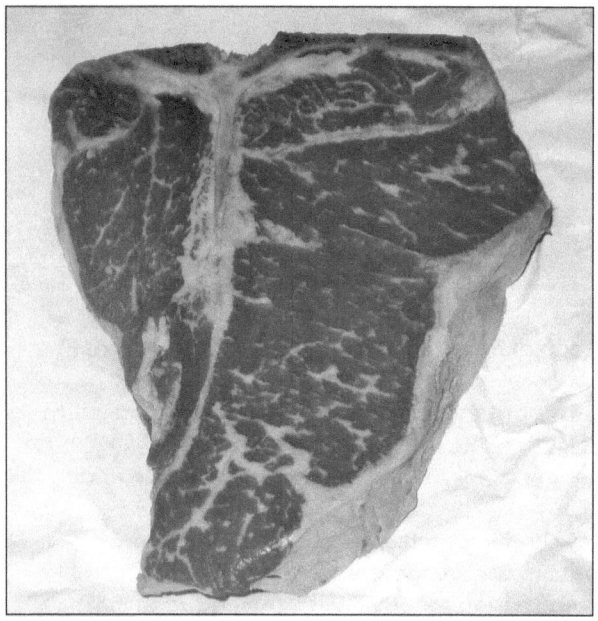

Figure 6.1 Meat from domestic animals is widely used. (Left) Raw well-marbled (cream colored) T-bone steak (beef) (Michael C. Berch, CC BY-SA 2.5). (Right) Raw chicken thighs with attached adipose tissue. (Gran, CC BY 3.0)

- Entertainment such as horse and dog races (see Figure 6.2), and cockfighting* (now illegal in the United States and many other countries).
- Military (e.g., horses and dogs).
- Search and rescue dogs.

Figure 6.2 Horses racing. (Courtesy Softeis, CC BY 3.0)

6.2 Types of Muscle Cells (Myocytes)

6.2.1 Overview of Types of Muscle Cells

Muscle is classified as (1) skeletal muscle, (2) cardiac muscle, and (3) smooth muscle.

6.2.2 Skeletal Muscle

Skeletal muscle is **voluntary, striated** (striped) muscle (see Figure 6.3). Skeletal muscles are attached to the skeleton. This attachment to bones provides an anchorage point for the muscle to contract against. At the macroscopic level, skeletal muscle consists of **long cylindrical muscle fibers** with the functional unit being the sarcomere. Skeletal muscle has multiple nuclei per cell arranged along the cell membrane (see Figure 6.3). The **satellite cells**, undifferentiated muscle cells, can get incorporated into the myofiber.

Skeletal muscles are surrounded by a connective tissue called the epimysium.

6.2.3 Classification of Skeletal Muscles

There are three categories of muscle fibers:

- **Slow oxidative fibers** (Type 1 slow-twitch muscle).
- **Fast oxidative fibers** (Type II-A fast-twitch muscle).
- **Fast glycolytic fibers** (Type II-B fast-twitch muscle).

Table 6.1 Differences between muscle types.

	Skeletal muscle cells	Cardiac muscle cells	Smooth muscle cells
Nuclei per cell	Multiple	1–2	1
Striated	Yes	Yes	No
Branched	No	Yes	No
Presence of desmosomes	No	Yes	No

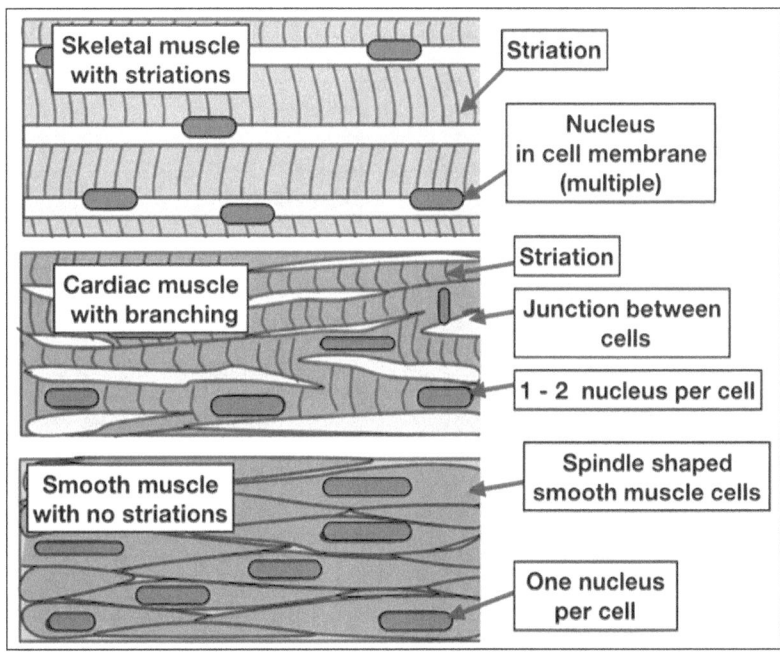

Figure 6.3 Schematic showing differences between smooth muscle, cardiac muscle, and striated muscle. (Adapted from Mdunning13, CC BY 3.0)

They are classified by color (red or white) or by the speed at which the muscle will contracts (fast or slow) or by whether their metabolism is primarily **oxidative** or **glycolytic**. There are additional differences between the three muscle types (see Table 6.2). Red muscle fibers are colored due to the presence of both the high concentrations of **myoglobin** and the high numbers of **capillaries**.

6.2.4 Cardiac Muscle

Cardiac muscle is **striated** (striped) muscle (see Figure 6.4). Cardiac muscle cells are found in the heart, specifically in the **myocardium**, the middle layer of the heart laying between endocardium and epicardium. The function of the cardiac muscle is to contract moving blood around the body of an animal.

The myocardium is composed of **cardiac muscle cells** or cardiomyocytes. There are chains of myofibrils within the cardiac muscle cells (myofibrils are discussed in Section 6.7 below). There are networks of branched cardiac myocytes with connective tissue between them. There are one or two nuclei per cell. The cardiac myocytes are striated. The sarcolemma surrounds muscle fiber. **Intercalated discs** are between the longitudinal ends of adjacent cardiac cells (see Textbox 6.2). Figures 6.3 and 6.4 illustrate the structure of cardiac muscle while Table 6.1 includes a summary of structural features.

Table 6.2 Types of muscle fibers in skeletal muscles.

	Slow oxidative fibers	Fast oxidative fibers	Fast glycolytic fibers
Alternative names	Type 1 slow-twitch muscle; Oxidative fibers	Type 2A fast-twitch muscle; Intermediate fibers	Type 2B fast-twitch muscle; Glycolytic
Speed of contraction	Low	Fast	Fast
Resistance to fatigue	High	Intermediate	Low
Metabolism	Aerobic	Aerobic producing ATP	Anaerobic
Number of capillaries	High	High	Low
Glycogen concentration	Low	Intermediate	High
Fiber diameter	Low	High	Intermediate
Number of mitochondria	High	High	Few
Myoglobin concentration	High	Moderate	Low
Color	Red	Pink to red	Pale or white

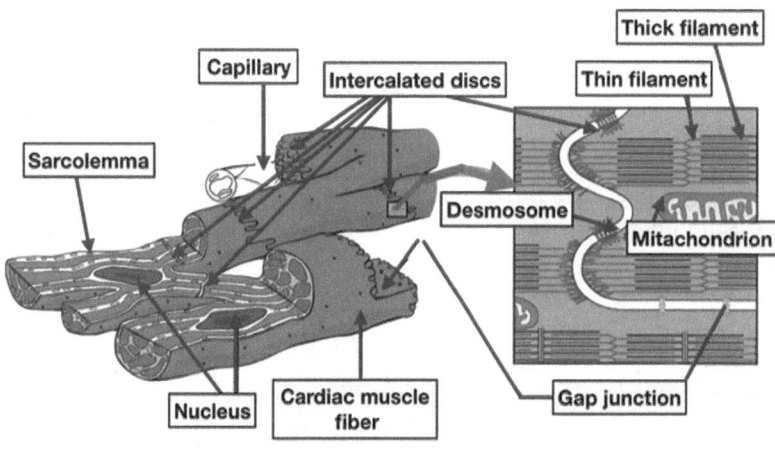

Figure 6.4 Structure of cardiac cells showing branching of cells and presence of nuclei. (Adapted from OpenStax, CC BY 4.0)

Textbox 6.2 Definitions About the Intercalated Discs

Intercalated Discs
There are **intercalated discs** between the longitudinal ends of neighboring cardiac myocytes (see Figures 6.3 and 6.4). They have two functions:

- Holding cells together in response to mechanical stress.
- Cell to cell communication.

The intercalated discs include the following:

- Desmosomes (for a deeper dive into desmosomes see Textbox 6.3).
- Adherens junctions.
- Gap junctions.
- Voltage-gated sodium channels.

(continued)

Textbox 6.2 (continued)

Desmosomes
There are desmosomes between cardiac muscle cells binding the cells together and providing the strength needed to resist mechanical stressors (see Textbox 6.2 and Textbox 6.3 for a deeper dive into desmosomes).

Adherens Junctions
Adherens junctions act like a "glue" holding cells together.

Gap Junctions
Gap junctions are groups of channels passing between cells. Gap junctions allow the passage of ions and other small molecules between cells. There are gap junctions between many other cell types also.

Voltage-Gated Sodium Channels
Voltage-gated sodium channels play a critical role in depolarization by causing cardiac cells to contract and action potentials in axons. Voltage gated sodium channels are, as their name implies, channels that are opened in response to the potential difference across a cell membrane.

Textbox 6.3 A Deeper Dive Into Desmosomes

There are desmosomes between cardiac muscle cells. They are part of a system that allows tissues to respond to mechanical stress by linking cells together in cell-to-cell adhesion. Desmosomes are also found in other tissues that are subjected to high mechanical stress including bladder smooth muscle and gastrointestinal muscle, together with skin (epidermis).

Desmosomes are linked to cytoskeleton proteins, specifically intermediate filaments forming a desmosome intermediate filament complex between two cells (see Figure 6.5). The linkage is due to intracellular plaque proteins linked to both intermediate filaments and to the desmosomal cadherin proteins.

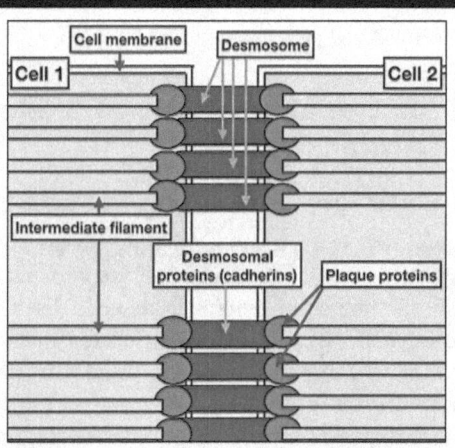

Figure 6.5 Structure of desmosome intermediate filament complexes.

6.2.5 Smooth Muscle

Smooth muscles are **involuntary muscles**. Smooth muscles do not have sarcomeres, giving a lack of the appearance of striations. There are filament bundles composed of thick filaments (myosin) and thin filaments (actin) attached to dense bodies and anchored to the attachment or dense plaques in the cell membrane.

Smooth muscles are predominantly found around hollow structures such as blood vessels, intestines, and uterus. Smooth muscles are found in the following:

- Blood vessels.
- Ciliary muscle of the eye.
- Erector pili of the skin.
- Gastrointestinal tract.
- Iris of the eye.
- Lymphatic vessels.
- Oviduct.
- Respiratory tract.
- Urinary bladder (not present in birds).
- Uterus.
- Vagina.

Smooth muscle cells are **spindle-shaped** cells rather like fibroblasts (see Table 6.1). They have one nucleus per cell but many mitochondria (see Figure 6.3 and

Table 6.1). Smooth muscle cells do not have striations (see Figure 6.4 and Table 6.1). The functional anatomy and contraction of smooth muscles are discussed in Sections 6.8 and 6.9.

6.2.6 Types of Smooth Muscle

There are two types of smooth muscle:

- **Single-unit smooth muscles** (or unitary or visceral smooth muscle) are found in the gastrointestinal tract. Contraction of single-unit smooth muscle is coordinated due to nerves supplying groups of muscle cells and gap junctions composed of **connexins** between adjacent myocytes. The gap junctions allow passage of action potentials and ions between the smooth muscle cells.
- **Multiple smooth muscles** are found in respiratory ducts, arteries, and ciliary muscles of the eye. Each myocyte has a nerve innervating it with a neuromuscular junction.

6.3 Functional Anatomy of Skeletal and Cardiac Muscle

6.3.1 Overview

Skeletal muscle is made up of multiple muscle fibers and connective tissue, together with nerves and blood vessels. Muscle fibers are multinucleate cells. Each muscle fiber is surrounded by connective tissue, the endomysium (see Figure 6.6). Groups of muscle fibers form **fascicles** which are surrounded by connective tissue, the **perimysium** (see Figure 6.6). A muscle comprises of groups of fascicles. These are surrounded by the **epimysium** (see Figure 6.6). The epimysium is composed of connective tissue with collagen fibers. In addition, the epimysium contains adipose cells, capillaries, and nerves.

> **Textbox 6.4 Interesting Factoid: Alternative Names in the Culinary Community**
>
> The epimysium is called the silver skin. The fascicles (bundles of muscle fibers) are known as the grain of the meat in the culinary world, with a fine grain being associated with tender cuts of meat. Meats with distinct grains are tough and include brisket, flank steak, and skirt steak (diaphragm).

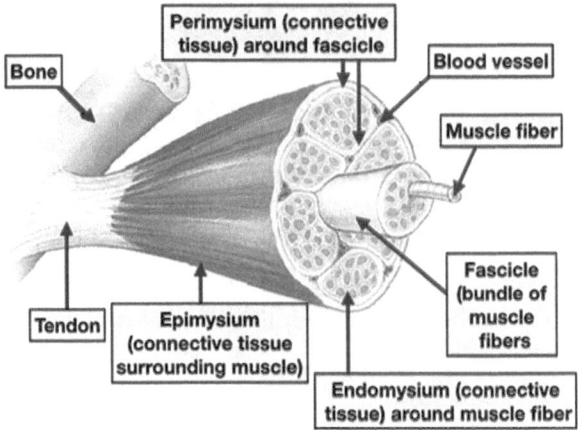

Figure 6.6 Structure of skeletal muscle.

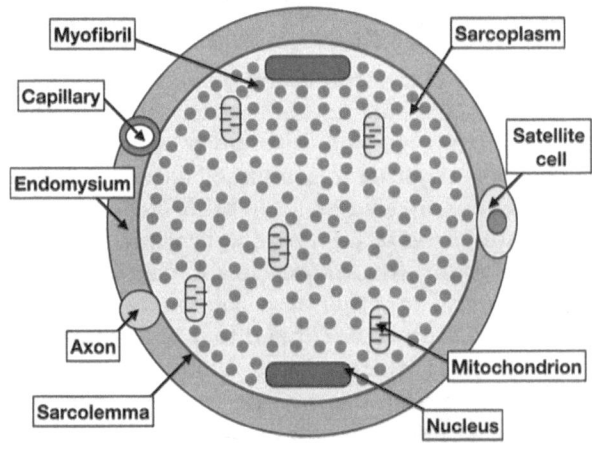

Figure 6.7 Idealized section of muscle fiber showing sarcoplasm; multiple myofibrils, nuclei, and mitochondria; blood vessels; and axons surrounded by the sarcolemma and then the connective tissue, endomysium.

Muscle fibers are made up of **multiple tubular myofibrils** (see Figures 6.7 to 6.9) and are responsible for the contraction of muscles. Myofibrils make up 80% of muscle fibers. The myofibrils are composed of **contractile units** called sarcomeres. Muscle fibers are surrounded by the sarcolemma; this being the cell membrane of the muscle fiber (Figure 6.7). The sarcolemma is a membrane that allows for selective movement of ions and other materials in and out of the muscle fiber. Additionally, the sarcolemma has specialized properties allowing it to carry action potentials to the transverse tubules, or T-tubules, in the sarcoplasmic reticulum (see Figure 6.9; see Section 6.4).

CHAPTER 6   Muscle   133

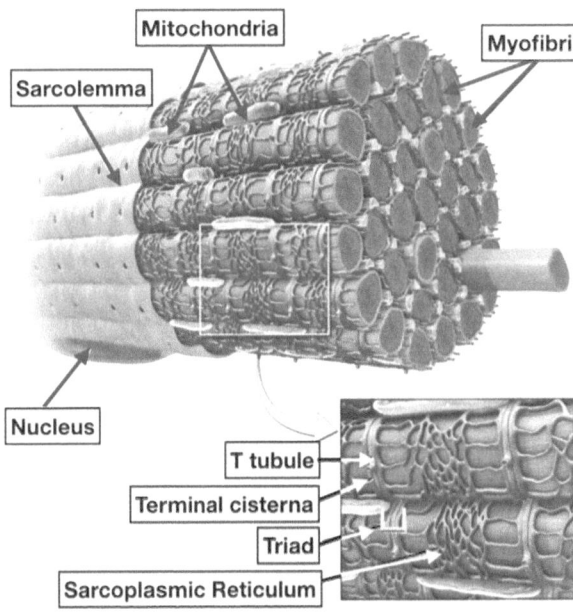

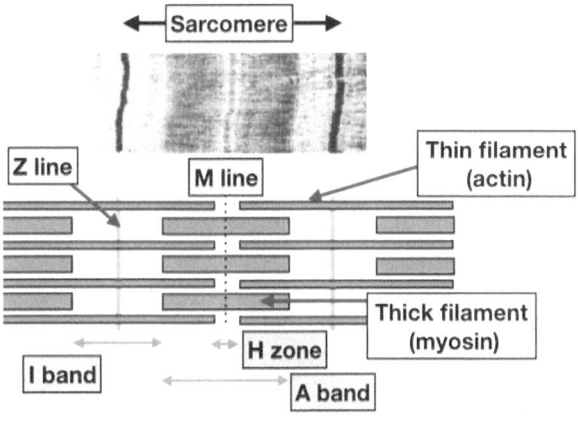

Figure 6.8   The sarcoplasmic reticulum surrounding skeletal muscle fibers. (Adapted from Blausen.com staff, CC BY 3.0)

Figure 6.10   Structure of sarcomere (electron micrograph [top]). During contraction, the thick filament is moved along the thin filament. This can be seen by the reduction of the widths of H zone and the I band. (Sarcomere image from Sameerb, CC BY-SA 4.0)

(see Figure 6.9 and 6.10). Myofibrils are tubular and are made up of a series of sarcomeres (see Figures 6.9 and 6.10). Myofibrils are the critical organelles within muscle fibers responsible for muscle contraction.

The sarcomeres are composed of thin and thick filaments that run from the Z line to the next Z line (see Figure 6.10). The thick filaments are composed of myosin while the Z lines separate sarcomeres. The A bands contain the thick filaments, including those overlapping the thin filaments (containing actin). The H zone contains thick filaments (containing actin, troponin, and tropomyosin) not overlapping with thin filaments. During muscle contraction, there are reduced widths of both the **A band** and the **I band** (see Figure 6.10). The reduction of width is due to actin sliding past myosin. This sliding reaction is known as the **sliding filament theory**.

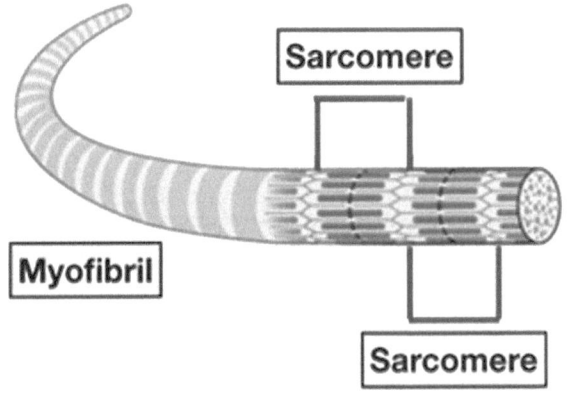

Figure 6.9   Structure of the myofibril. (Adapted from Shutterstock/Blamb)

6.3.2 What Are the Major Proteins in the Sarcomere?

Sarcomere proteins within the sarcomere can be placed into two major classes:

- **Contractile proteins.**
- **Regulatory proteins.**

Contractile proteins consist of the following:

There are also many nuclei in the sarcolemma. Multiple mitochondria are found in proximity to the myofibrils (see Figures 6.7 and 6.8); these being essential to muscle contraction as they supply the energy in the form of ATP. Within the sarcoplasm is the sarcoplasmic reticulum, mitochondria, myoglobin (binding oxygen), and **glycogen** (the storage form of glucose as glycogen) (see Figure 6.8).

The functional unit of the myofibril is the sarcomere. There are repeating sarcomeres along the myofibril

- Myosin in the thick filament with heads protruding (Figure 6.11).

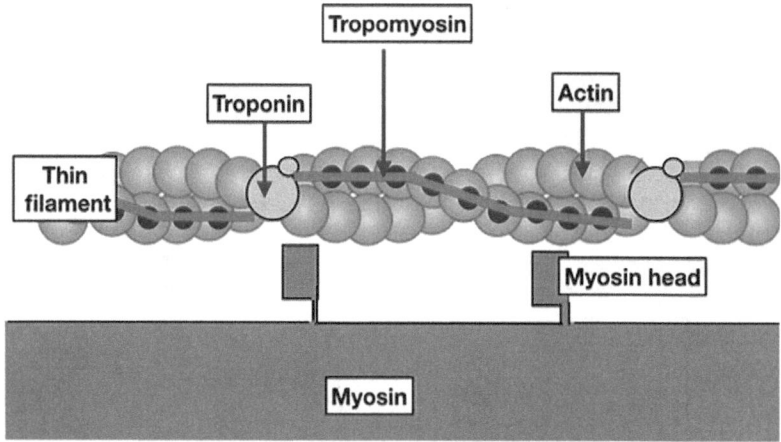

Figure 6.11 Detailed structure of thin and thick filaments in muscle. (Adapted from of Zlira'a, CC BY-SA 3.0)

- Actin, the major protein in the thin filaments with two intertwining strands composed of globular actin molecules (Figure 6.11).

These two proteins interact through the cross-bridge cycle and change the length of the sarcomere. Regulatory proteins control the interaction of myosin and actin. The primary regulatory proteins are the following:

- Tropomyosin is a thin protein that covers the myosin binding sites on actin.
- Troponin is a protein the moves tropomyosin away from the binding sites on actin. It is controlled by sarcoplasmic calcium concentrations.

6.4 Sarcoplasmic Reticulum

The sarcoplasmic reticulum is a system of tubes around the muscle fibers in skeletal muscles (see Figures 6.8 and 6.12). It is composed of membranes and is the equivalent of the smooth endoplasmic reticulum. The functions of sarcoplasmic reticulum are the following:

- To store calcium.
- To release calcium when the muscle fiber is stimulated.
- To pump calcium back into the sarcoplasmic reticulum to the end of the contraction.

The sarcoplasmic reticulum contains enlarged ends that connect with the T-tubule (or transverse tubule). These ends are called terminal cisternae. They function

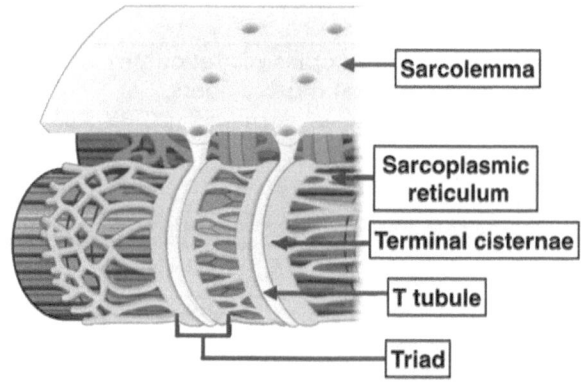

Figure 6.12 Structure of the sarcoplasmic reticulum together with transverse or T-tubules, terminal cisternae, and sarcolemma in skeletal muscle. (Adapted from OpenStax, CC BY 4.0)

to release stored calcium ions when stimulated. The combination of the terminal cisternae–T-tubule–terminal cisternae is referred to as the triad (see Figures 6.8 and 6.12).

6.5 Myogenesis—The Development of Myofibers

6.5.1 Overview

Textbox 6.5 provides a series of definitions related to myogenesis. Development of muscle occurs during fetal life in mammals and during embryonic life in birds. The dam can influence development of the fetus due to shifts in

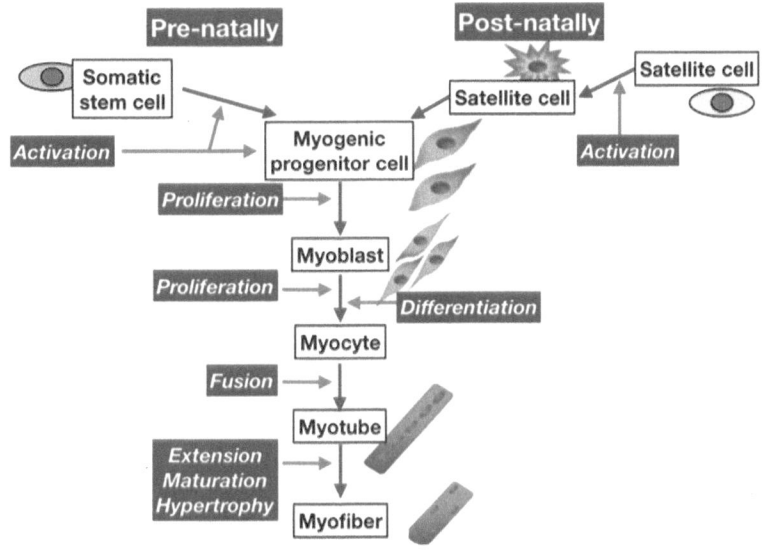

Figure 6.13 Myogenesis—development of muscle fibers during embryonic/fetal life (domestic mammals), embryonic development (poultry) and post-natal/post-hatching growth.

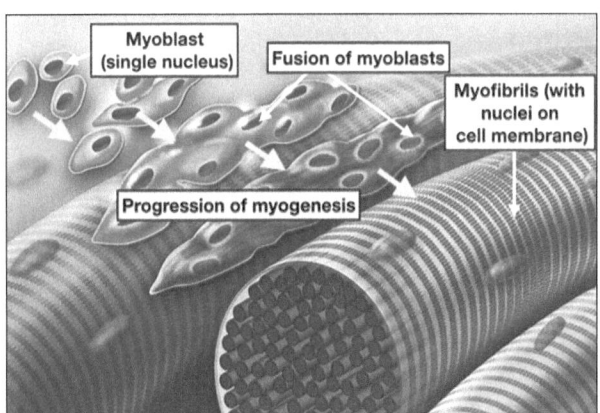

Figure 6.14 Myogenesis with myoblasts (each with a single nucleus) fusing to form myocytes (multinucleated) and ultimately myofibrils. (Courtesy of Darryl Leja, National Human Genome Research Institute, NIH)

the provision of nutrients and oxygen. In birds, the dam influences embryonic development by provision of different amounts and quality of yolk and egg white (albumen). Moreover, development is influenced by incubation temperature. There are a series of key regulatory genes affecting myogenesis. These include **myoblast determination protein** (MyoD) and paired box protein (Pax) 7.

6.5.2 Fetal (Mammals)/Embryonic (Poultry) Myogenesis

Embryonic somatic stem cells are activated to form **myogenic progenitor cells** (see Figure 6.13). These divide repeatedly and differentiate into myoblasts (proliferation) (see Figure 6.13). In turn, myoblasts further proliferate and differentiate to form **myocytes** (see Figure 6.13). The process of formation of the myotubes is started with nuclei moving out of the **myotubes** (see Figures 6.13 and 6.14).

Textbox 6.5 Definitions Related to Myogenesis and the Development of Other Organs

- **Differentiation** is the process where a cell becomes more specialized. This process is generally not reversible.
- **Embryo** is the stage of life between implantation and the formation of all major organs in mammals. In contrast, the word *embryo* refers to the entire period of development in the egg in poultry.
- **Fetal** is the life stage in mammals from the end of embryonic development (when all major organs are formed and true bone is being made) to birth.

(continued)

> **Textbox 6.5** *(continued)*
>
> - **Growth factors** are specific proteins that that modify a cellular process such as cell division or differentiation.
> - **Hypertrophy** is growth by cells increasing in size but without cell division.
> - **Hyperplasia** is growth by increases in cell numbers.
> - **Myogenesis** is the development of muscles.
> - **Myogenic progenitor cells** are cells that will become muscle cells.
> - **Pre-natal** is before birth.
> - **Proliferation** is multiple cell divisions.
> - **Stem cells** are undifferentiated cells.
> - **Transcription factors** are specific proteins that stimulate transcription of specific genes.

The muscle fibers or myofibers develop by lengthening of the myotubes, maturation, and hypertrophy (see Figures 6.13 and 6.14).

These processes are controlled by growth factors and transcription factors. For instance, myostatin modulates the early stages of myogenesis, resulting in decreased numbers of myoblasts (Figure 6.13).

6.5.3 Post-Natal Myogenesis

Post-natal myogenesis occurs during growth or in the adult, for instance, after injury to a muscle. Satellite cells are activated to become myogenic progenitor cells. Myogenesis continues as in embryonic/fetal myogenesis.

6.6 Stimulation of Contractions of Skeletal and Cardiac Muscle

6.6.1 Definitions Related to Muscle Contractions

Definitions related to muscle contraction are summarized in Textbox 6.6.

6.6.2 How Does the Muscle Fiber Know When to Contract?

Neurons **innervate** muscle fibers so that the muscles contract when needed (see Figure 6.15). Action potentials (waves of depolarization) pass along the axon

> **Textbox 6.6 Definitions Related to Muscle Contractions**
>
> - **Action potential**: a wave of depolarization passing along the muscle fiber (action potentials are discussed in Chapter 5).
> - **Acetylcholine**: binds to nicotinic cholinergic receptors on the motor plate and stimulates an action potential in the muscle.
> - **Depolarization**: the loss of potential difference across a cell membrane.
> - **Excitation of the muscle fiber**: due to the action potential initiating the muscle to contract.
> - **Motor endplate**: the post-synaptic membrane of the neuromuscular junction.
> - **Motor unit**: comprises the motor neuron and the muscle fibers innervated by the motor neuron.
> - **Neuromuscular junction**: the synapse between the motor neuron and muscle fiber.
> - **Neurotransmitter**: a chemical signal that travels across the neuromuscular junction. Examples include acetylcholine and norepinephrine.
> - **Potential difference**: the sum of charges on the inside and outside of a membrane.
> - **Relaxation** of the muscle fiber after contraction: due to a fall in the cytoplasmic concentrations of calcium; accomplished by a sarcoplasmic calcium pump.
> - **Ryanodine receptors**: proteins that facilitate calcium transit (calcium release channels).
> - **Single muscle contraction**: see twitch.
> - **Twitch** is defined by three phases: (1) latent or excitation coupling, (2) contraction, and (3) relaxation.
> - **Twitch Motor unit**: contraction of a single muscle fiber stimulated by an action potential passing along an axon.

CHAPTER 6 Muscle

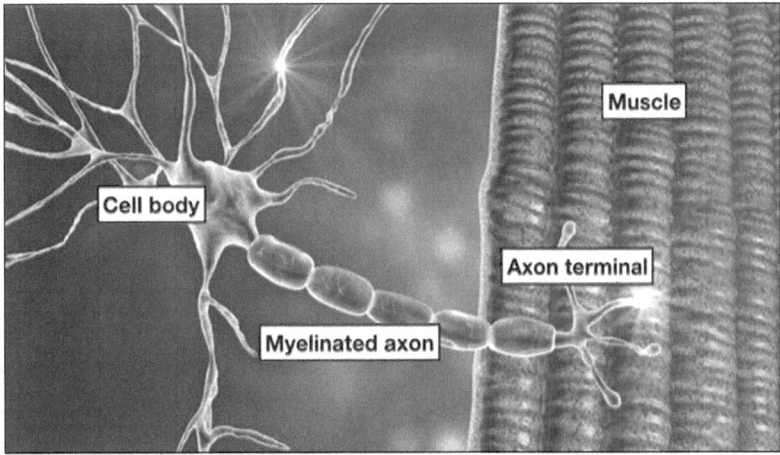

Figure 6.15 Nerve innervating a muscle. (Adapted from Shutterstock/Kateryna Kon)

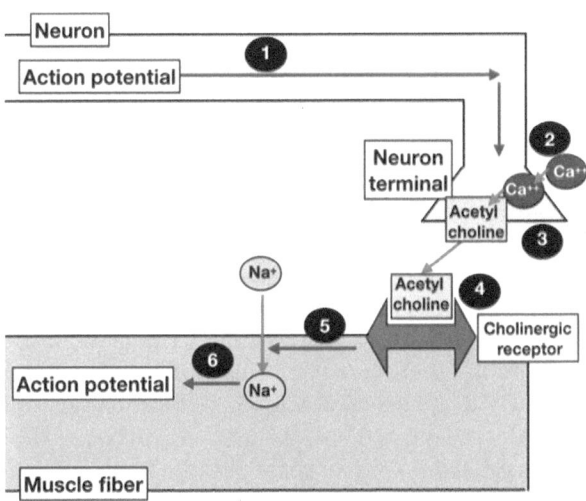

Figure 6.16 Nervous stimulation of muscle.

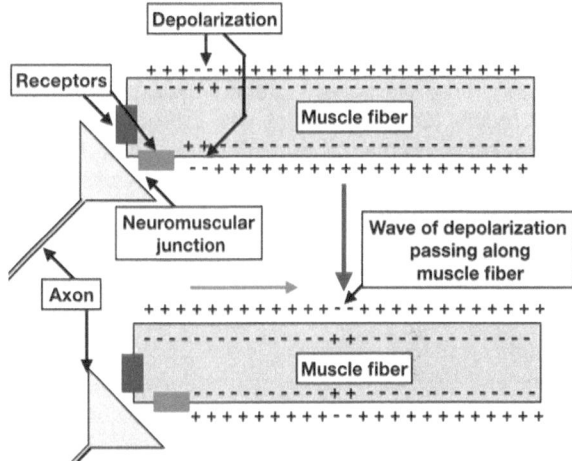

Figure 6.17 After neurotransmitters are released from a nerve terminal, they bind to receptors in the neuromuscular junction. This leads to depolarization of the muscle fiber. Subsequently, a wave of depolarization or action potential passes down the muscle fiber.

(Figure 6.15). At the neuromuscular junction, the depolarization leads to influx of calcium ions. This is followed by release of a neurotransmitter such as **acetylcholine** (Figure 6.15) or norepinephrine. In the case of acetylcholine, after transit across the synaptic cleft, the acetylcholine binds to **nicotinic cholinergic** receptors (Figure 6.16). This leads to an action potential (wave of depolarization) in the muscle fiber with sodium ions entering the muscle fiber (Figure 6.16).

6.6.3 Skeletal Action Potential

There is a potential difference across the muscle fiber membrane or sarcolemma (see Figures 6.17 and 6.18). This reflects the higher concentrations of sodium ions (Na$^+$) on the outside and high concentrations of chloride ions (Cl$^-$) in the cytoplasm. When the muscle is stimulated, a wave of depolarization passes along the muscle fiber. This depolarization occurs due to the opening of the voltage-gated sodium channels and influx sodium ions (Na$^+$) (see Figures 6.17 and 6.18). Depolarization is followed by repolarization. This being caused by the closure of the voltage-gated sodium channels and the influx of potassium

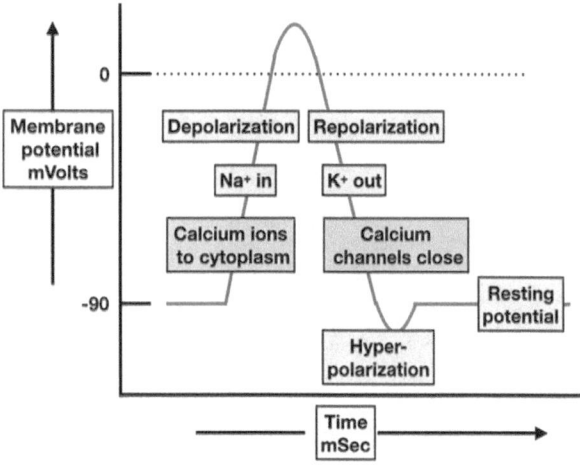

Figure 6.18 An action potential in a muscle fiber showing the following: the resting state due to high sodium concentrations outside the fiber. When stimulated, there is depolarization of the muscle fiber due to voltage-gated sodium channels opening with sodium ions (Na⁺) entering the cytoplasm and voltage-gated calcium channels opening in the sarcoplasmic reticulum with calcium ions (Ca²⁺) moving into the sarcoplasm. This is followed by repolarization with voltage-gated sodium and calcium channels closing. In addition, there is efflux of potassium ions (K⁺).

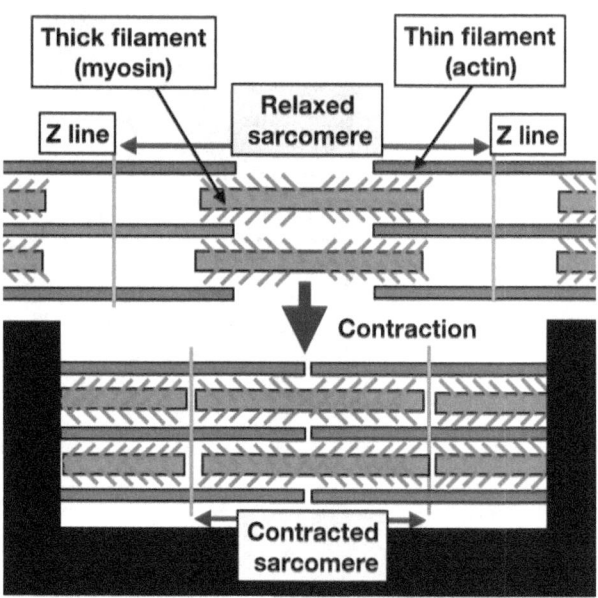

Figure 6.19 Movement of the thick filaments along the thin filaments in a muscle sarcomere during contraction.

ions (K⁺) (see Figures 6.17 and 6.18). After a brief hyperpolarization, the resting polarized state is restored with the higher concentrations of sodium ions (Na⁺) on the outside and high concentrations of chloride ions in the cytoplasm. This describes a single skeletal action potential, but many are required to transmit the signal down the down the sarcolemma to the T-tubule. The action potential alters the membrane potential in the surrounding sarcolemma. This leads to neighboring voltage-gated sodium channels to open. This results in an action potential at the neighboring section of the sarcolemma. Since the sarcolemma enters a state of hyperpolarized state following an action potential, it forces the action potential to move in a single direction down the sarcolemma.

6.7 Mechanism of Skeletal and Cardiac Muscle Contraction

6.7.1 Excitation

Excitation of the muscle fiber is the first step leading to contraction of the muscle fibers. During excitation, previously stored calcium ions are released across the inner membrane of the sarcoplasmic reticulum. There is a specific protein (**dihydropyridine receptor**) in the sarcolemma of the T-tubules (see Figure 6.10) in the region between the T-tubules and the **terminal cisternae** of the sarcoplasmic reticulum. This protein is the mediator between the depolarization to calcium release. The ryanodine receptors allow calcium release. With a high frequency of action potentials passing along the motor neuron, there is more calcium released and there is a greater contraction.

6.7.2 Muscle Contraction

During muscle contraction the thick filaments progress or slide along the thin filaments, shortening the length of the sarcomere (see Figure 6.19 and 6.20). During muscle contraction there are reduced lengths of the A band and the I band (see Figures 6.8 and 6.19).

6.7.3 Muscle Relaxation

During relaxation, calcium is pumped back into the sarcoplasmic reticulum by a calcium pump.

Textbox 6.7 provides a summary of binding sites on actin, troponin, and tropomyosin.

CHAPTER 6   Muscle   139

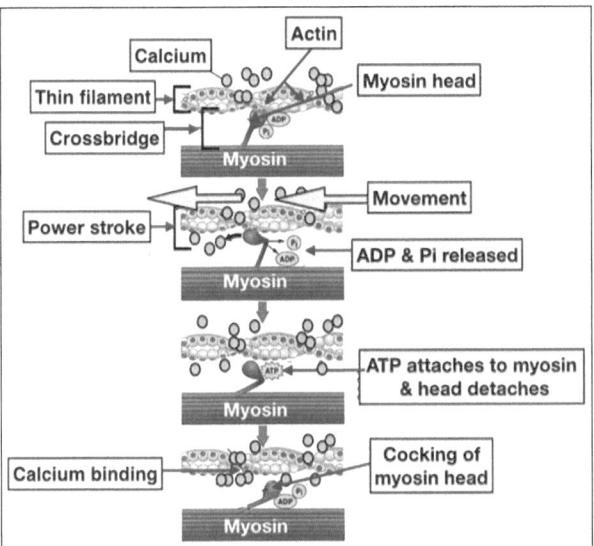

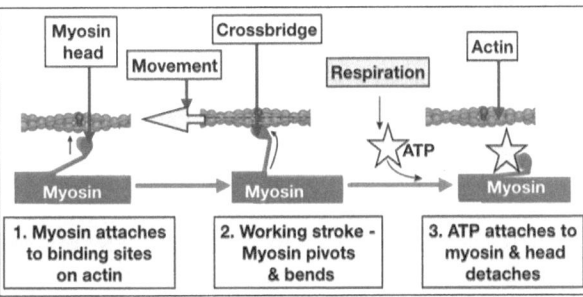

Figure 6.20 Mechanism of muscle contraction—the cross-bridge cycle. (Upper image adapted from OpenStax, CC BY 4.0; lower image adapted from Shutterstock/Blamb)

Textbox 6.7   Binding Sites on Actin, Troponin, and Tropomyosin

There are the following binding sites:

1. Actin molecules have binding sites for the head of myosin.
2. Troponin has a binding site for calcium ions (Ca^{2+}) (see Figure 6.21).
3. Troponin ~ calcium ion complex binds to tropomyosin causing a shift in the position of tropomyosin, unmasking myosin binding sites on the actin.

Consequences of Binding

1. Troponin binding to calcium ions
   • Binding of calcium ions (Ca^{2+}) to troponin unmasks the myosin binding sites on

(continued)

Textbox 6.7   (continued)

the actin. This is due to a shift in the position of tropomyosin (see Figure 6.21).
   • When calcium concentrations fall, the calcium ions (Ca^{2+}) dissociate from the troponin and myosin binding sites for actin to become masked again (see Figure 6.21).
2. Actin binding to myosin
   • The actin binding site for myosin is masked by tropomyosin until calcium ions (Ca^{2+}) binds to troponin (see Figure 6.21).
   • In the absence of sufficient calcium ions (Ca^{2+}), myosin cannot bind to actin (see Figure 6.21).
   • In the presence of sufficient concentrations of calcium ions (Ca^{2+}), myosin can bind to actin (see Figure 6.21).

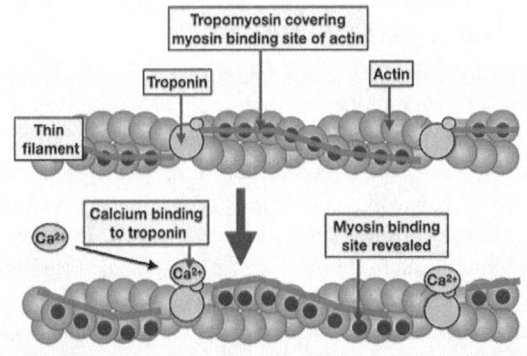

Figure 6.21 Role of calcium (Ca^{2+}) in the molecular mechanism of muscle contraction. (Actin structure image adapted from Zlir'a'a, CC BY 3.0)

6.7.4 Steps in Muscle Contraction

The following are steps in muscle contraction (see Figures 6.20 and 6.21):

1. The cycle can be viewed as starting when calcium ions (Ca^{2+}) are released from the sarcoplasmic reticulum flooding the sarcoplasm.

2. The calcium ions (Ca^{2+}) bind to troponin, causing a shift in the position of the tropomyosin molecules.
3. Tropomyosin moves away and, consequently, unmasks the myosin binding sites on actin molecules.
4. The myosin head binds to binding sites on actin molecules forming a **cross-bridge**.
5. ADP and phosphate are released from the myosin.
6. This causes a **conformational change** in the myosin head, which results in the power or working stroke and actin being pulled toward the M-line.
7. ATP binds to the myosin head.
8. The myosin head detaches from the binding sites on actin, breaking the cross-bridge (see Figures 6.20 and 6.21).
9. The ATP is hydrolyzed to ADP and the myosin resets or re-cocks.
10. The cycle continues as long as calcium ions are present.

6.7.5 Muscle Contraction, Calcium, and the Sarcoplasmic Reticulum

There are calcium pumps (sarcoplasmic reticulum calcium ATPase or SERCA) in the sarcoplasmic reticulum (see Figure 6.22). One ATP molecule is required to transport two calcium ions. Calcium ions (Ca^{2+}) are stored in the sarcoplasmic reticulum predominantly bound to the protein **calsequestrin** (see Figure 6.22). This protein can bind 50 calcium ions (Ca^{2+}) per molecule (see Figure 5.22). Calcium ions (Ca^{2+}) are released from the sarcoplasmic reticulum, specifically at the junction between the sarcoplasmic reticulum and the terminal cisternae, due to the ryanodine receptor (RyR) (see Figure 6.22).

There is a different RyR in skeletal muscle (RyR1) and cardiac muscle (RyR2).

6.7.6 Special Features of Cardiac Muscle and Contraction

The overall system of calcium stimulating contraction of cardiac muscle is similar to that of skeletal muscle (above) (see Figures 6.20, 6.21, and 6.23), however, they lack terminal cisternae. Uptake of calcium ions in the sarcoplasmic reticulum of cardiac myocyte leads to relaxation (see Figure 6.23).

6.7.7 ATP and Rigor Mortis

Calcium is released as the sarcoplasmic reticulum breaks down post-death causing rigor mortis (stiffening of muscles). In the absence of ATP after death, the myosin head remains bound to the actin in the thin filament (see Figure 6.21).

6.8 Functional Anatomy of Smooth Muscle

Unlike skeletal and cardiac muscle, smooth muscles do not contain striations as they do not contain sarcomeres. The smooth muscle layer of gastrointestinal tract contains the following:

- **Smooth muscle myocytes.**
- **Interstitial cells of Cajal.**
- **Nerve bundles.**

The interstitial cells of Cajal are located in close proximity with nerve bundles in the gastrointestinal muscle layers and serve as the pacemakers for peristaltic contractions in the gastrointestinal tract. These are also found in

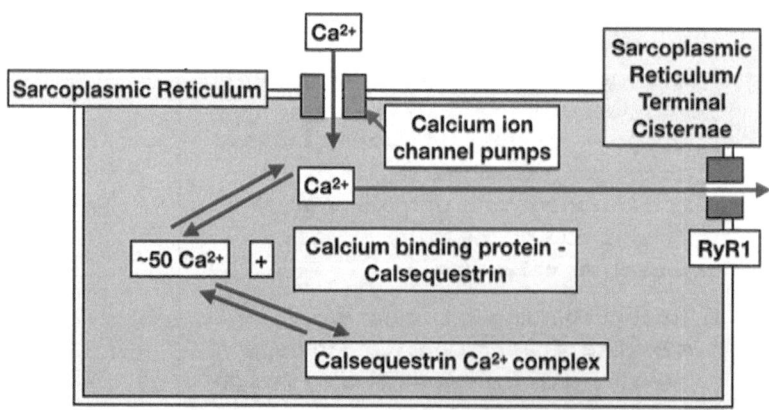

Figure 6.22 Schematic of the role of the sarcoplasmic reticulum showing calcium ions (Ca^{2+}) entering the sarcoplasmic reticulum via calcium ion channel pumps, binding to calsequestrin (50 Ca^{2+} per molecule) and, subsequently, released from the sarcoplasmic reticulum via RyR receptors/calcium channels.

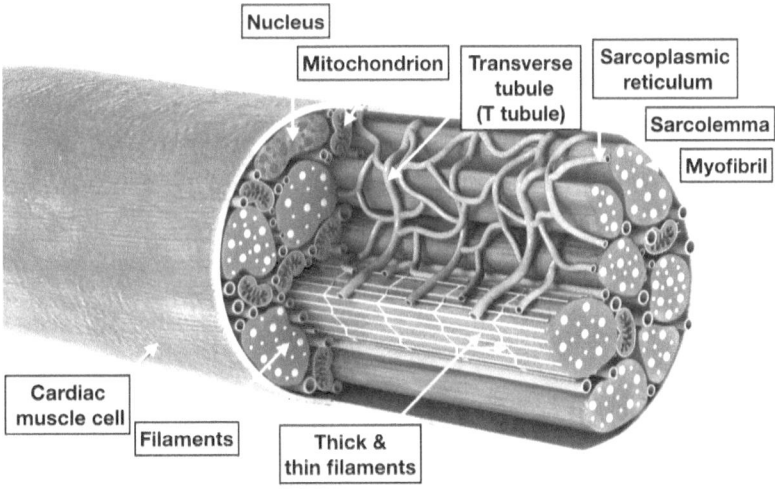

Figure 6.23 Structure of cardiac muscle showing cardiac muscle cell or fiber and the location of nuclei, mitochondria, thick and thin filaments together with transverse tubules and sarcoplasmic reticulum.
(Adapted from Shutterstock/stockshoppe)

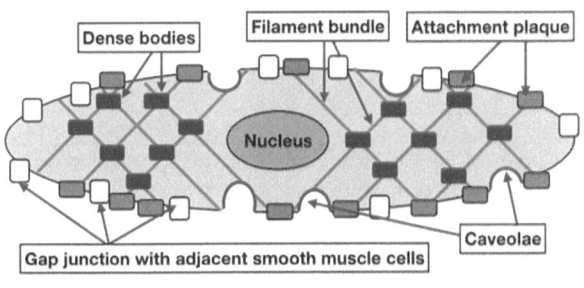

Figure 6.24 Schematic of smooth muscle cell (myocyte) showing the contractile elements, the filament bundles with central dense bodies, and attachment or dense plaques attaching the filament bundles to the cell membranes. In the sarcolemma (cell membrane), there are both gap (particularly in single-unit smooth muscles) and tight junctions between smooth muscle cells. In addition, there are caveolae. There are multiple mitochondria but these are not shown.

the submucosa and intermuscular layers. The interstitial cells of Cajal act on smooth muscle myocytes, at least in the gastrointestinal tract, playing a role post-junctionally. There are also gap junctions between the interstitial cells of Cajal and the smooth muscle myocytes.

There are spontaneous contractions of GI smooth muscle due to a spontaneous pacemaker current in the interstitial cells of Cajal. Figure 6.24 provides a schematic diagram of a smooth muscle cell. Smooth muscle cells do not contain myofibrils or sarcomeres. This is unlike the situation in skeletal muscles. They do, however, contain both actin and myosin. Textbox 6.8 summarizes the structure of smooth muscle cells.

Textbox 6.8 Structure of Smooth Muscle Cells

Smooth muscle cells consist of the following:

- A single nucleus (see Figure 6.24).
- Multiple mitochondria.
- **Filament bundles** anchored to the cell membrane at attachment or dense plaques and to other filament bundles at dense bodies (see Figure 6.24). The filament bundles contain thin filaments (containing actin) and thick filaments (containing myosin). The filament bundles are the contractile elements of muscle.
- The dense bodies in smooth muscle cells anchor the filament bundles and are equivalent to the Z lines in skeletal muscle.
- **Attachment or dense plaques** attach the filament bundles to the cell membranes.

(continued)

Textbox 6.8 (continued)

- A **sarcoplasmic reticulum** (see Figure 6.24).
- **Caveolae** are flask-like invaginations in the sarcolemma of smooth muscle myocytes. L-voltage-gated calcium channels are present in the caveolae. These are responsible for calcium ions (Ca^{2+}) passing into the sarcoplasm from the exterior of the myocytes.
- **Tight junctions** are present in the sarcolemma of smooth muscle myocytes. They provide mechanical strength. Tight junctions link with adjacent smooth muscle cells such that there is a network of cells that are mechanically linked (see Figure 6.24).
- **Gap junctions** are present in the sarcolemma of smooth muscle myocytes. They provide intimate connections between GI smooth muscle myocytes and other smooth muscle myocytes and interstitial cells of Cajal. They allow movement of ions from cell to cell. The gap junctions facilitate coordinated contraction of the smooth muscles with ions and, hence, potentials passing from myocyte to myocyte.

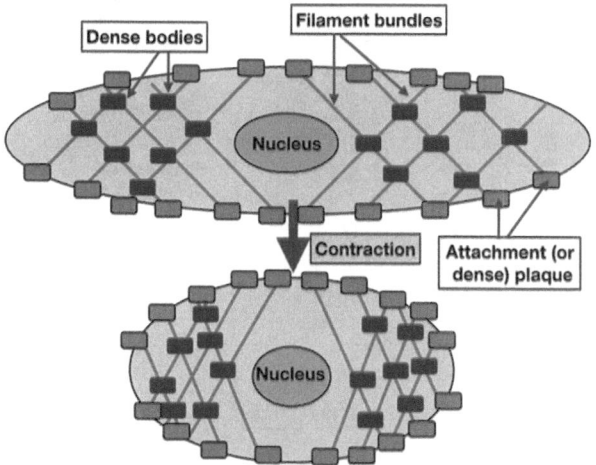

Figure 6.25 Contraction of smooth muscle cells.

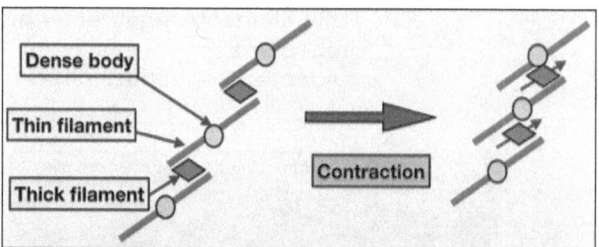

Figure 6.26 Contraction of smooth muscle cells due to shortening of the filament bundles, in turn, due to thick filaments (myosin) sliding along thin filaments (actin).

6.9 Mechanism of Smooth Muscle Contraction

6.9.1 Overview of Smooth Muscle Contraction

Contractions of smooth muscle are very different from that in skeletal or cardiac muscle. However, similar proteins control contraction.

To achieve contraction of the smooth muscle, the thick filaments move along the thin filaments in a manner analogous to that of the skeletal and cardiac muscle (see Figures 6.25 and 6.26). However, the net result of contractions of the filament bundles is a change in the shape, but not size, of the smooth muscle myocytes (see Figure 6.25). There is a decrease in the length of long axes of the smooth muscle myocytes (see Figure 6.25).

6.9.2 Activation of Smooth Muscle Myocytes

Smooth muscles are innervated by neurons that are located along the myocyte. This differs from the situation with skeletal and cardiac muscle where there are specific neuromuscular junctions. Neurotransmitters, such as acetylcholine and norepinephrine, bind to specific receptors on smooth muscles in neuromuscular junctions. Contraction of smooth muscle cells is the result of the following:

- Depolarization of the sarcolemma (muscle cell membrane) (see Figures 6.27 and 6.28).
- An influx of calcium ions (Ca^{2+}) into the sarcoplasm of the myocytes through the L-voltage-gated calcium channels in the caveolae. When activated by the increase in concentrations of calcium ions (Ca^{2+}) together with a signal molecule (**inositol trisphosphate**). The calcium ions (Ca^{2+}) pass from stores in the sarcoplasmic reticulum into the cytoplasm through the RyR receptors/calcium ion channels (Ca^{2+}) (see Figures 6.27) (discussed in Textbox 6.9).
- Smooth muscle cells are stimulated either by nerves or ion movements from adjacent cells via gap junctions.

- In the presence of calcium ions (Ca^{2+}), together with a signal molecule (calmodulin), **myosin** is **phosphorylated** (see Figure 6.28).
- Phosphorylated myosin binds to actin (see Figure 6.28).
- There is cross-bridge cycling (see Figure 6.28).
- **Dephosphorylation** of myosin stops the cycle and, thereby, stops contraction (see Figure 6.28).

6.10 Energy Production for Muscle

6.10.1 Overall

Muscles utilize a large amount of energy in the form of ATP (adenosine triphosphate) when contracting for the following:

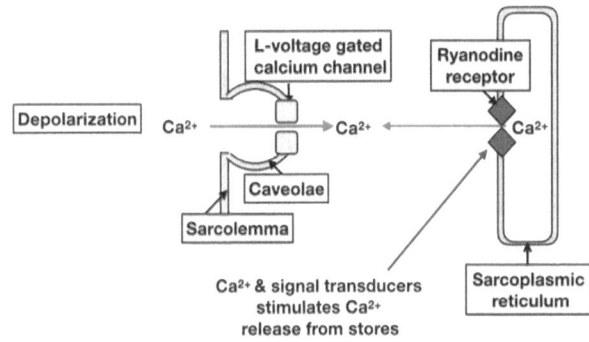

Figure 6.27 Initiation of contraction of smooth muscles.

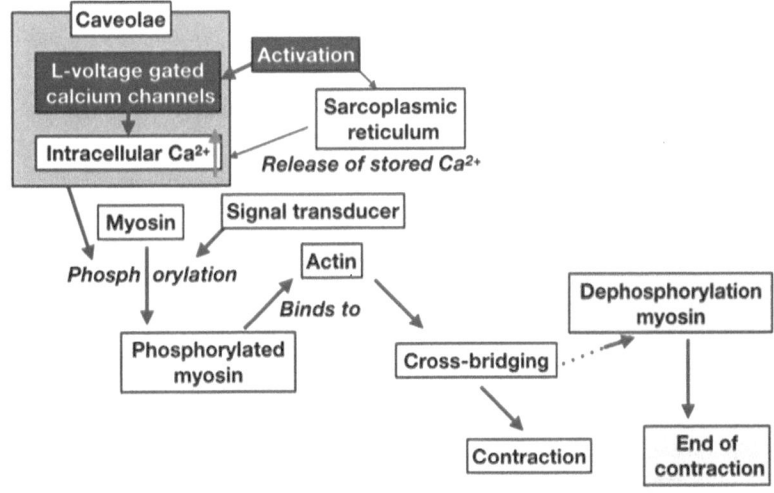

Figure 6.28 Mechanism of contraction of smooth muscles.

Textbox 6.9   A Deeper Dive Into Muscle Contraction

Other Neurotransmitters and Smooth Muscle Functioning
There are other neurotransmitters that influence smooth muscle functioning. Nitric oxide (NO) released from enteric inhibitory neurons. Opioids exert an inhibitory effect on contraction of GI smooth muscle myocytes acting on the enteric nervous system. γ-Aminobutyric acid (GABA) has an inhibitory effect on the smooth muscles of airways.

Important Proteins in Muscle Contraction
The ryanodine receptor (RyR) is an intracellular calcium channel named for its ability to bind the plant alkaloid, ryanodine. There are different RyR in different organs:

- RyR1 isoform in skeletal muscle.
- RyR2 isoform in cardiac muscle.
- RyR3 isoform in brain.

Calsequestrin is a protein that binds 50 calcium ions (Ca^{2+}). There are different calsequestrin isoforms in skeletal and cardiac muscle.

(continued)

Textbox 6.9 (continued)

Voltage-gated sodium channels (Ca^{++}) channels (a.k.a. dihydropyridine receptors). Skeletal and cardiac muscle have different voltage-gated Ca^{2+} channel isoforms.

Isoforms are proteins with similar functions but with similar, but not identical, amino acid sequences. The isoforms are encoded by separate genes or by different processing (splicing) of RNA. Isoform genes are derived by gene duplication during evolution.

- To correct the concentrations of sodium and potassium in the sarcoplasm after action potentials.
- To detach the cross-bridges.
- To pump calcium ions (Ca^{2+}) from the sarcoplasm into the terminal cisternae.

Muscle has only very short-term stores of ATP that can last for about five seconds. There must be mechanisms to generate ATP. There are three mechanisms to generate ATP in order of activation:

- **Phosphocreatine** (also called creatine phosphate).
- **Glycolysis**.
- **Oxidative respiration**.

6.10.2 Glycolysis

Glucose can be catabolized by the glycolytic pathway to generate pyruvate (see Figure 6.29). This pathway generates two molecules of ATP for every molecule of glucose.

In the absence of sufficient oxygen, the pyruvate cannot enter the **citric acid cycle**. Instead, the pyruvate is metabolized to lactate (see Figure 6.29). In turn, the lactate goes to the liver and kidneys via the blood stream. The lactate is converted to glucose in the process of gluconeogenesis. Post-death, there are increases in the concentrations of **lactate/lactic acid** in muscle/meat, and consequently a reduction in pH.

6.10.3 Oxidative Respiration

In oxidative respiration (or oxidation phosphorylation), much of the energy released by the oxidation of glucose is captured as ATP. Overall, one molecule of glucose generates 36 molecules of ATP (see Figure 6.30).

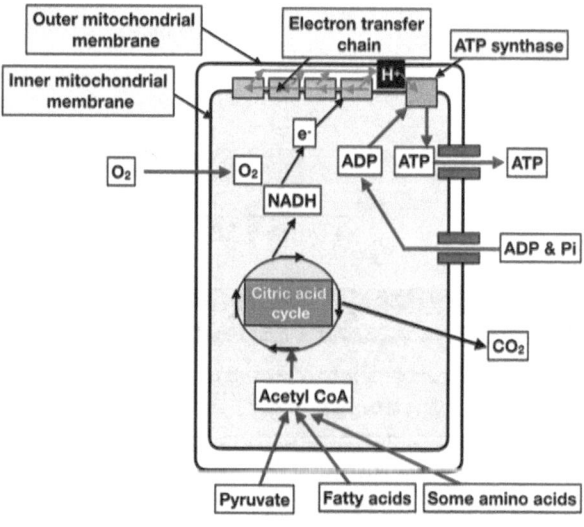

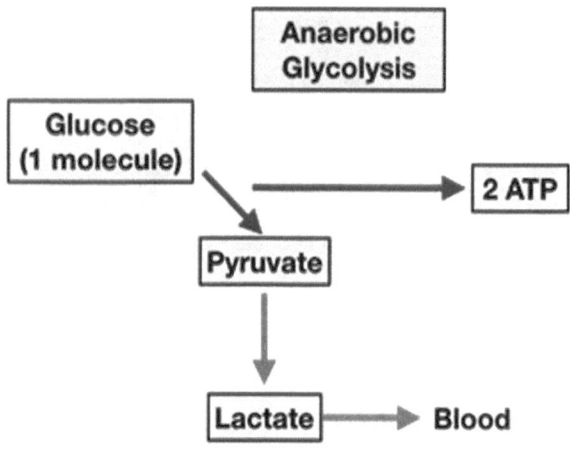

Figure 6.29 Anaerobic respiration generates 2 ATP molecules (adenosine triphosphate) for every molecule of glucose. This occurs in the absence of oxygen. The ATP is essential for muscle contraction. Lactate is generated. The lactate passes to the liver and kidneys via the bloodstream.

Figure 6.30 Oxidative respiration generates 36 ATP molecules (adenosine triphosphate) for every molecule of glucose. This occurs only in the presence of oxygen. The ATP is essential for muscle contraction. [Key: ADP = adenosine diphosphate; ATP = adenosine 5' triphosphate; CO_2 = carbon dioxide; e– = electron; NADH = reduced nicotinamide adenine dinucleotide; O_2 = oxygen]

Glucose enters the glycolytic pathway to generate pyruvate. This produces 2 ATP while generating a pyruvate molecule. In the presence of oxygen, pyruvate enters the citric acid cycle (or Kreb's cycle) to produce **acetyl-coenzyme A** (acetyl-CoA) in the mitochondria. Acetyl-CoA (two carbon atoms in the acetyl part of acetyl CoA) then enters the mitochondria to enter the electron transport chain to produce an additional 34 ATP. Muscles can also use fatty acids (see Textbox 6.10 for an example) and some amino acids to generate ATP through the electron transfer chain.

There is some storage of oxygen bound to the protein myoglobin. Like hemoglobin, myoglobin contains a heme subunit and ferrous ion (Fe^{2+}) and binds oxygen.

> **Textbox 6.10 Interesting Point: Volatile Fatty Acids**
>
> In ruminants, volatile fatty acids (VFAs) are produced in the rumen. VFAs include acetate (60%) together with propionate and butyrate. Acetate is a 2-carbon fatty acid that is used by ruminant animals to produce ATP by entering the electron transport chain directly.

6.10.4 Creatine, Muscle, and ATP

About 95% of the creatine in a domestic animal is in the skeletal muscles. The remainder is in other tissues including the brain and cardiac muscle. Creatine is synthesized by the liver and kidney and is transported into the muscle fibers by sodium- and chloride-dependent creatine transporters.

6.10.5 Why Is Creatine Important to Skeletal Muscle Functioning?

Creatine is important to muscle as it is readily phosphorylated by ATP (adenosine triphosphate) to phosphocreatine (creatine phosphate) in a reversible reaction (see Figures 6.31 and 6.32). When energy in the form of ATP is needed, for instance for muscle contractions, phosphocreatine rapidly phosphorylates ADP (adenosine diphosphate) (see Figure 6.31). Thus, phosphocreatine is a store of readily accessible energy and can donate phosphate to ADP at times of need without any need for oxygen. Phosphocreatine is generated when there are high concentrations of ATP (see Figure 5.32).

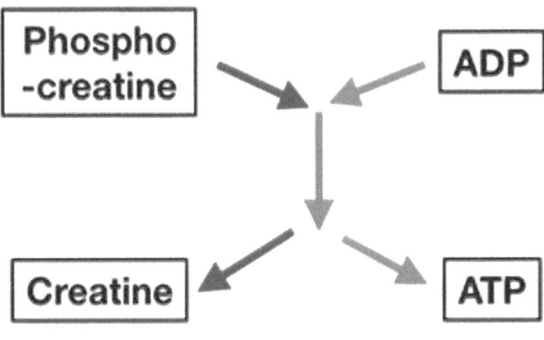

Figure 6.31 ADP (adenosine diphosphate) is directly phosphorylated to ATP (adenosine triphosphate), receiving phosphate from the donor molecule phosphocreatine. This occurs at times of low ATP in the muscle.

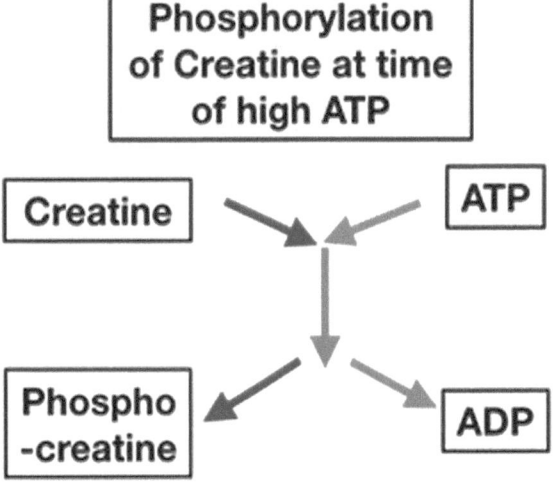

Figure 6.32 Creatine is directly phosphorylated to phosphocreatine by ATP (adenosine triphosphate). This occurs at times of high ATP in the muscle.

6.11 Muscle Contractions Leading to Twitches, Incremental Contractions, and Physiological Tetanus

When a muscle fiber receives a single stimulus above the threshold, it contracts in what is called a muscle twitch or twitch (see Figure 6.33A). The twitch consists of three phases:

- Latency period.
- Contraction.
- Relaxation.

The latency period is the phase where the neuron initiates a signal that is carried to the motor units and, ultimately, to the contractile units, the sarcomere. The contraction phase is when the sarcomeres are contracting with the cross-bridge cycles. The relaxation phase is when the muscle and the sarcomeres are returning to its **resting length**.

There are two ways that the tension of the contracting muscle can be changed:

1. **Frequency of the stimulus.**
2. **Strength of the stimulus.**

When the frequency of stimuli is such that a stimulus occurs before the muscle can complete relaxation, it allows a partial relaxation before a second stimulus arrives. This will increase the tension of the muscle above the initial or single twitch. With multiple stimuli/twitches, the tension can increase in a **step-wise manner** (see Figure 6.33B). If the muscle is allowed to only undergo partial relaxation, this is called **unfused** or incomplete tetanus. However, if the frequency is rapid enough, it will completely inhibit relaxation and result in fused or complete tetanus (see Figure 6.33C).

Increases in the strength of the response occurs through recruitment of additional motor units, allowing for changes in the tension of the muscle. This is considered the change in the "strength" of the muscle and is required for proper control of the muscle, such as the difference between the "strength" used by muscles in legs during a leisurely walk versus kicking down a door.

By increasing the strength of the stimulus, more muscle fibers are recruited. As more motor units are recruited, the tension of the contraction increases until a **maximal contraction** is reached (see Figure 6.33D).

There are predominantly two types of muscle contraction (see Textbox 6.11).

Textbox 6.11   Types of Muscle Contractions

Isotonic Contraction
Isotonic contraction is contraction under constant tension. Isotonic contraction allows movement of limbs, etc.

Isometric Contraction
Isometric contraction is contraction under constant length. During recovery from an injury, a muscle may be forced to contract in an isometric manner. A subset of isometric contraction is **isovolumetric contraction**. This occurs with contractions of the heart ventricles until a critical pressure is reached and valves open.

6.12 Muscle as Meat

The muscle from livestock and poultry that is used for meat is predominantly **skeletal muscle** (see Figure 6.34). In some countries, cardiac muscle is also consumed as meat. Intestines, including the smooth muscles, are used in human food, with small intestines used as the casings of sausages or consumed directly in chitlins (chitterlings). The rumen is also used as human food as tripe and, traditionally, in the Scottish dish haggis (sheep rumen). From a meat science technical viewpoint, muscle is converted to meat when it reaches the "ultimate pH". This occurs roughly 24 hours postmortem.

6.13 Double Muscling

There is a high frequency of **double muscling** in Belgian blue cattle (see Figure 6.35B), Piedmontese cattle

Chapter 6 Muscle

(see Figure 6.35A), and Texel sheep. As its name implies, animals with double muscling have increased muscle mass (see Figure 6.35A and number of myofibers). Double muscling is a heritable and autosomal trait. The cause of double muscling is mutations or deletions in the **myostatin gene** leading to inactivation of myostatin (an inhibitor of myogenesis).

Myostatin is one of the growth factors that control myogenesis (the development of muscle). Myostatin depresses myogenesis.

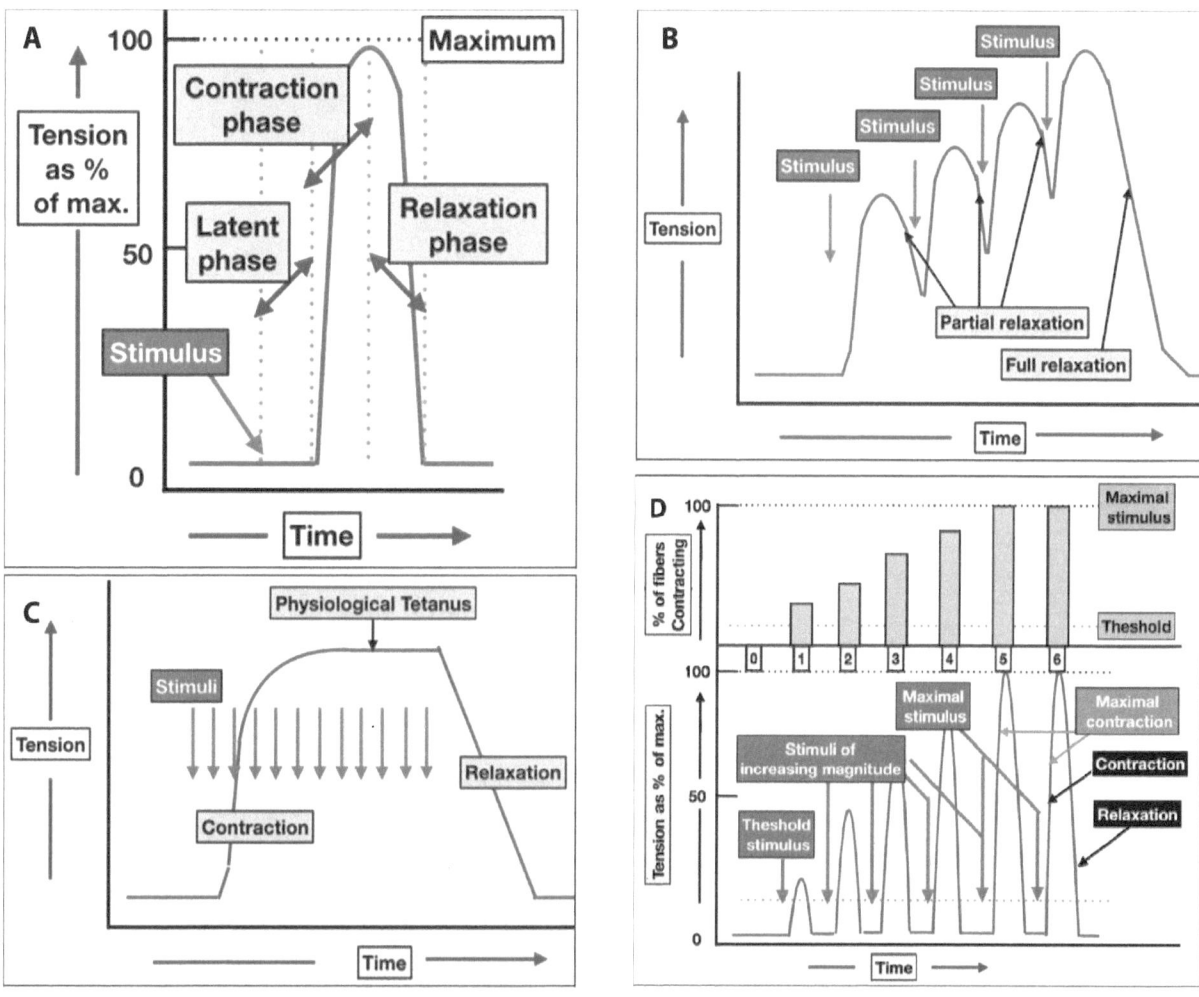

Figure 6.33 Muscle contractions. (A) Single-twitch contraction after stimulation. (B) Repeated stimulation of muscle by repeated action potentials. (C) Action potentials rapidly in succession lead to physiological tetanus. (D) With increasing magnitude of the stimulus, more muscle fibers will be contracting.

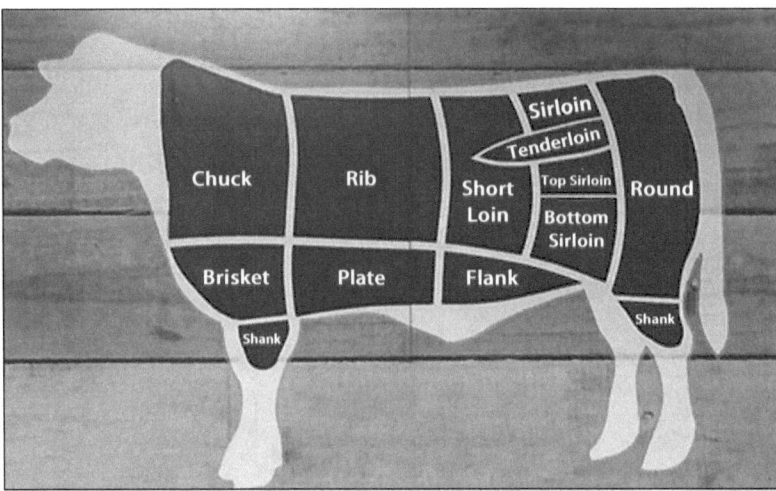

Figure 6.34 Cuts of meat from cattle. (Adapted from Luis Tamayo, CC BY-SA 2.0)

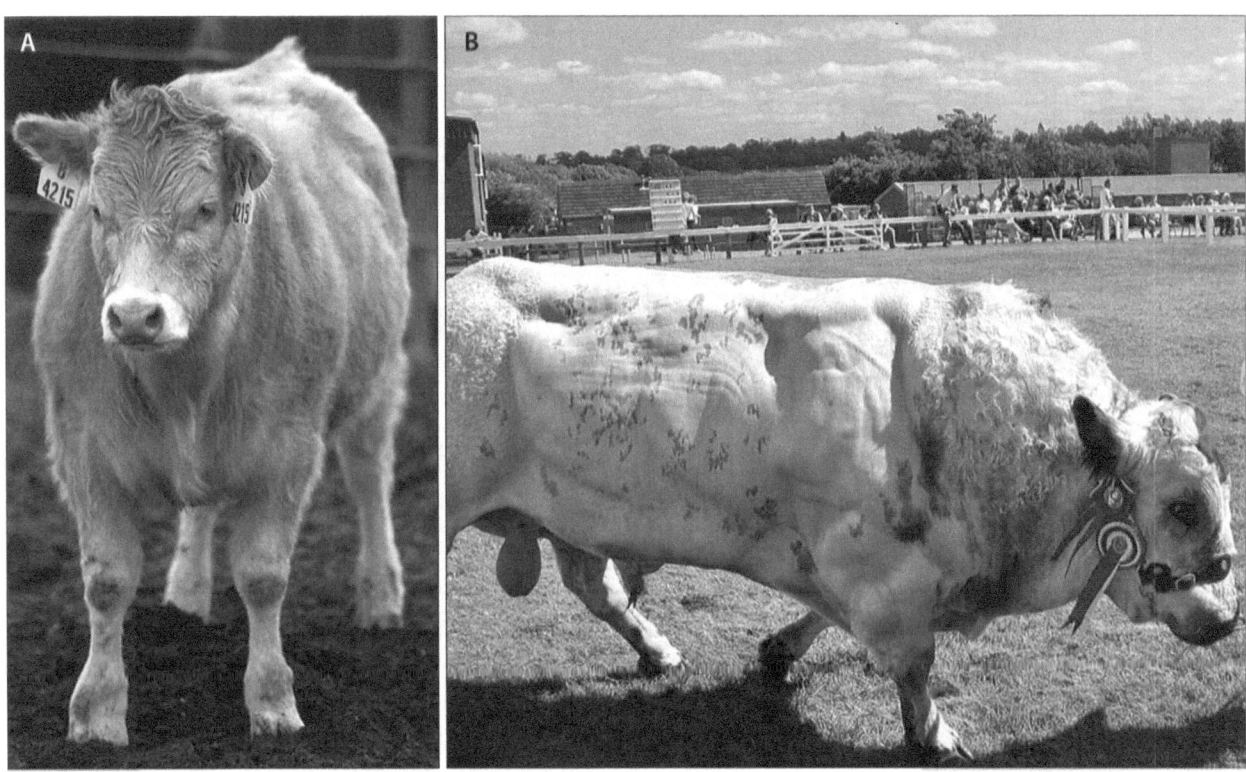

Figure 6.35 Double muscling in cattle. (A) Calf showing double muscling due to a defective myostatin gene (courtesy of USDA ARS/Weller). (B) Belgian Blue bull at the Great Yorkshire Show. (Adapted from Malcolmxl5, CC BY-SA 4.0)

6.14 Myopathy (Muscle Disease)

6.14.1 Overview of Myopathy

Myopathies are muscle diseases and there are numerous examples in domestic animals.

6.14.2 Myopathies in Livestock and Poultry

Examples of myopathies in livestock and poultry are summarized in Table 6.3.

6.14.3 Porcine Stress Syndrome

Porcine stress syndrome (a.k.a. porcine malignant hyperthermia or transportation myopathy) is genetically transmitted (see Textbox 6.12). Pigs with this mutation produce meat of poor quality. The meat is categorized as **pale, soft, and exudative (PSE)**. With molecular testing, carriers of the mutation can be identified and not used for breeding.

In stress susceptible pigs (particularly Pietrain pigs), there is a mutation in the calcium channel, RyR1 (also called the halothane gene); see Sections 6.7.5 and 6.9.2 for examples). This is seen after exercise or exposure to the anesthetic halothane. The mutation affects the uptake and release of calcium in muscle fibers.

6.14.4 Myopathies in Horses

Examples of myopathies in horses include the following:

- Tying up. This is a painful condition where horses become stiff, refuse to move, and sweat profusely. There is damage to the muscles. The syndrome has multiple causes.
- Sycamore poisoning is where horses consume a natural toxic chemical. This affects skeletal muscle and can result in the horse dying.

There are also injuries to the joints, tendons, and ligaments which are seen, for instance, in race horses.

6.14.5 Fainting Goats

Fainting goats (a.k.a. Tennessee goats or Tennessee wooden leg goats) are prone to fall over. These goats exhibit myotonia congenita. There is reduced chloride conductance in the muscles.

Table 6.3 Myopathies in livestock.

Species	Myopathy	Characteristics
Poultry	White striping	Infiltration of adipose tissue into the muscle fiber
Chicken	Wooden breast	Infiltration of collagen into the muscle fiber. This increases the toughness of the meat/lowers the quality of the meat.
Poultry, Swine, and Beef	Pale, Soft, and Exudate (PSE) [also see Textbox 6.12]	Abnormal pale or light color, a soft consistency, and decreased water holding capacity; resulting in an increase purge. This results in dry and unattractive meat.
Ruminants	Dark cutter	Increased stress peri-mortem decreases glycogen content and subsequent alters post-mortem pH. This results in decreased blooming of the meatA and a darker color.

ABlooming is the change of color of meat due to myoglobin binding oxygen from the atmosphere.

Metabolism

7

(in the Liver, Skeletal Muscle, and Adipose Tissue)

Learning Objectives

1. To understand the importance of maintaining glucose concentrations in the plasma at a set point (see Sections 7.1.3 and 7.1.4, together with Textbox 7.1).
2. To understand how glucose is stored as glycogen or glycogenesis (see Section 7.3.2).
3. To understand glycogenolysis (see Section 7.3.3).
4. To understand gluconeogenesis (synthesis of glucose from lactate, glycerol, and glucogenic amino acids) (see Section 7.4).
5. To understand the roles of glucose transporters (see Section 7.5).
6. To understand the roles of the liver in carbohydrate and lipid metabolism (see Section 7.6).
7. To understand the roles of skeletal muscle in carbohydrate and lipid metabolism (see Section 7.7).
8. To understand the roles of adipose tissue in carbohydrate and lipid metabolism (see Section 7.8).
9. To understand the development of adipose tissue (see Section 7.8.2).
10. To understand that there are many similarities in metabolism together with some differences in mammalian and avian domestic animals (see Section 7.9).
11. To understand examples of metabolic diseases in domestic animals (see Section 7.11).

Table of Contents

7.1 Introduction
    7.1.1 Overall Introduction
    7.1.2 Introduction to Carbohydrate Metabolism
    7.1.3 When Do Blood Concentrations of Glucose Change?
    7.1.4 Why Do Animals Attempt to Maintain Blood Concentrations

of Glucose at Around a Set Point?
7.2 Glucose and Its Metabolism
7.3 Glucose and Glycogen
  7.3.1 Introduction to Glucose
  7.3.2 Glycogen Synthesis (Glycogenesis)
  7.3.3 Glycogen Breakdown (Glycogenolysis)
7.4 Gluconeogenesis
7.5 Glucose Transporters
  7.5.1 Introduction to Glucose Transporters
  7.5.2 Glucose Transporters and Glucose Absorption in the Small Intestine
  7.5.3 Glucose Transporter 2 (GLUT2)
  7.5.4 Glucose Transporter 4 (GLUT4)
  7.5.5 Glucose Transporter 7 (GLUT7)
7.6 Liver and Metabolism
  7.6.1 Introduction to the Liver and Metabolism
  7.6.2 Glucose-6-Phosphatase
7.7 Skeletal Muscle and Metabolism
7.8 Adipose Tissue
  7.8.1 Adipocytes
  7.8.2 Adipogenesis
  7.8.3 Brown Adipose Tissue (BAT)
7.9 Differences Between Metabolism in Poultry Compared to Livestock/Companion Animals
7.10 Oxidative Stress and Reactive Oxygen Species
7.11 Metabolic Diseases
  7.11.1 Diabetes (Cats and Dogs)
  7.11.2 Equine Metabolic Syndrome (Horses)
  7.11.3 Ketosis (Cattle)
  7.11.4 Obesity (Cats, Dogs, and Horses)

7.1 Introduction

7.1.1 Overall Introduction

A glossary for terms used in metabolism is provided in Textbox 7.1.

7.1.2 Introduction to Carbohydrate Metabolism

Carbohydrate metabolism provides energy to animals. In a homeostatic manner, blood concentrations of glucose are maintained close to the set point (euglycemia) for the

---

Textbox 7.1 Glossary of Terms Related to Metabolism

- **Adipose tissue**: the tissue that stores energy in the form of triacylglycerol.
- **Adipogenesis**: the differentiation of adipocytes.
- **Aerobic**: requires adequate concentrations of oxygen.
- **Anaerobic**: the presence of inadequate concentrations of oxygen.
- **Euglycemia**: normal blood/plasma concentrations of glucose at a set point for the animal.
- **Fatty acid**: a carboxylic acid with an aliphatic (hydrocarbon) chain. Fatty acids can be used to generate energy in tissues or converted to triacylglycerol.
- **Glycogen**: storage form of glucose. Glycogen is a branched polymer of glucose.
- **Glycogenesis**: synthesis of glycogen from glucose.
- **Glycogenolysis**: breakdown of glycogen generating glucose.
- **Gluconeogenesis**: synthesis of glucose from lactate, pyruvate, some amino acids, and glycerol.
- **Glucose**: a monosaccharide (simple sugar) that is used to generate energy and important biological chemicals.
- **Glucose transporter**: molecule(s) that allow glucose to enter or leave a cell.
- **Glucose utilization**: use of glucose.
- **Hyperglycemia**: high blood/plasma concentrations of glucose.

(continued)

> **Textbox 7.1** *(continued)*
>
> - **Hypoglycemia**: low blood/plasma concentrations of glucose.
> - **Lipogenesis**: synthesis of fatty acids.
> - **Lipolysis**: breakdown of triacylglycerol generating fatty acids and glycerol in a ratio of 3:1.
> - **Triacylglycerol**: a molecule containing fatty acids and glycerol in a ratio of 3:1. This is the major storage form of energy in an animal's body. Triacylglycerol is also called triacylglyceride or triglyceride.
> - **Volatile fatty acid**: short chain fatty acids (containing 2, 3 or 4 carbon atoms). These are generated by microbial fermentation of carbohydrates through anaerobic metabolism.

individual species. Glucose is required for the functioning of the following organs: brain, erythrocytes, kidneys, and mammary glands.

Plasma concentrations of glucose are markedly higher in poultry than mammalian domestic animals.

7.1.3 When Do Blood Concentrations of Glucose Change?

Blood concentrations of glucose rise following a meal as starch is digested and glucose is absorbed in the small intestine in pigs, poultry, and dogs. There are differences with starch digestion and monosaccharide absorption in cats (for further details see Textbox 7.3) and some other species. **Insulin** is released to return plasma concentrations of glucose to normal by doing the following:

- Increasing utilization of glucose by tissues.
- Storing glucose as glycogen, primarily in the liver, but also muscle, heart, and other tissues.
- Increasing conversion of glucose to fatty acids (lipogenesis) and, consequently, triacylglycerol.

Blood concentrations of glucose decline after absorption of glucose is completed. To maintain blood concentrations of glucose, the hormone **glucagon** is released, and it acts to increase blood glucose concentrations in the following manner:

- Decreasing utilization of glucose.
- Mobilizing glucose from its storage form as glycogen in the liver (glycogenolysis).
- Decreasing lipogenesis.
- Increasing gluconeogenesis.

> **Textbox 7.2 Examples of Dynamics of Glucose**
>
> **Glucose Dynamics in a Dairy Cow**
> The cow uses 72 g glucose to produce one kg of milk.
>
> If a cow is producing 30 kg of milk per day, it will use 72×30 g glucose per day or 2.16 kg of glucose per day.
>
> Cattle digest their feed in the rumen, generating volatile fatty acids (acetate, propionate, and butyrate) and lactate (see Chapter 7). Glucose absorption is minimal. Therefore, the cow needs to synthesize over 2 kg of glucose from precursors (propionate and lactate). This is achieved in the process of gluconeogenesis.
>
> **Brain Utilization of Glucose**
> In fasted cats, the brain uses 30% of the glucose derived from gluconeogenesis.

> **Textbox 7.3 A Deeper Dive Into Carbohydrate Digestion in Cats**
>
> The ancestors of cats were obligate carnivores, as are feral cats today. The typical diet of feral cats contains the following:
>
> - Protein: 52%.
> - Fat: 46%.
> - Carbohydrate including glycogen: < 2%.
>
> Cats are less effective in digesting and absorbing carbohydrate due to the following:
>
> - Very low salivary and intestinal amylase activity.
> - Low disaccharidase activity in cat brush border.
> - Slow Na+/glucose co-transporter 1 (SGLT1) activity.

7.1.4 Why Do Animals Attempt to Maintain Blood Concentrations of Glucose at Around a Set Point?

Two scenarios will be discussed:

1. Following a carbohydrate meal, plasma concentrations of glucose rise as glucose is absorbed. Were it not for the actions of the hormone insulin, plasma concentrations of glucose would rise to dangerous levels. Glucose would not be reabsorbed in kidney tubules. Hence, glucose would be lost in the urine. This is wasteful. Moreover, due to osmotic effects of glucose, large amounts of water will also be lost in the urine. This would increase the need to drink large amounts of water. In addition to losing both glucose and water through the urine, the amount of glucose available to the body is limited or inadequate for proper body functions.
2. During periods of not consuming carbohydrate (fasting), plasma concentrations of glucose decline. Were it not for the actions of the hormone glucagon, plasma concentrations of glucose would decline to dangerous levels. This would imperil the functioning of the brain and other glucose dependent organs.

7.2 Glucose and Its Metabolism

Glucose is an energy source for an animal. The structure of the monosaccharide, glucose, is shown in Figure 7.1. Glucose is oxidized to carbon dioxide and water (H_2O) by three pathways to generate ATP:

- **Glycolysis.**
- **Citric acid cycle**.
- **Oxidative phosphorylation.**

Glycolysis is a process resulting in the formation of **pyruvate** that occurs under both aerobic and anaerobic conditions (Figure 7.2). When there is adequate oxygen

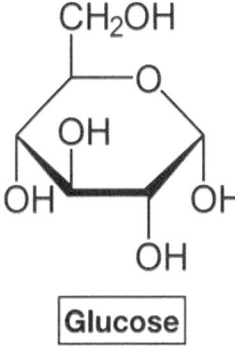

Figure 7.1 Structure of glucose.

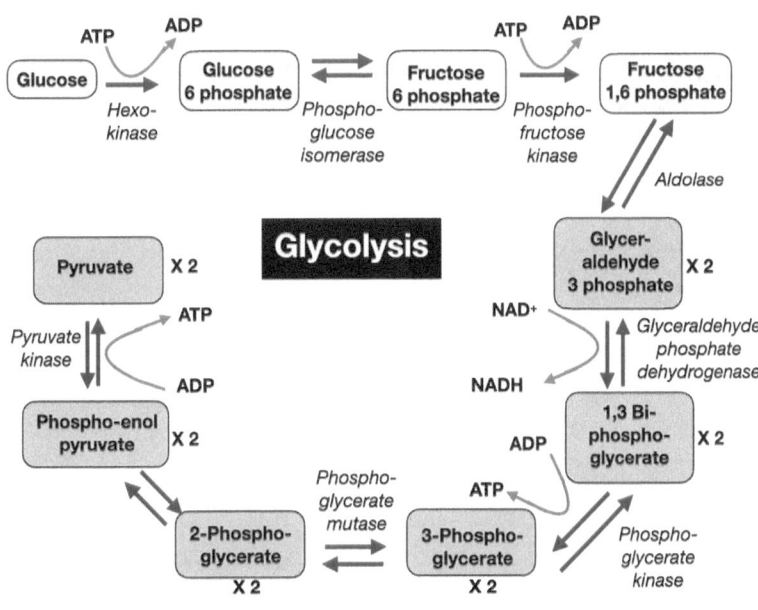

Figure 7.2 Glycolysis.

availability (aerobic), the pyruvate enters the citric acid cycle and is oxidized to carbon dioxide. At times of vigorous exercise, there is insufficient oxygen to allow for full oxidation (anaerobic conditions). Pyruvate is formed and can be converted to lactate (Figure 7.3). When there is sufficient oxygen availability, oxidation of pyruvate is completed in the citric acid cycle and ATP is produced (see Textbox 7.3 and Figure 7.4.).

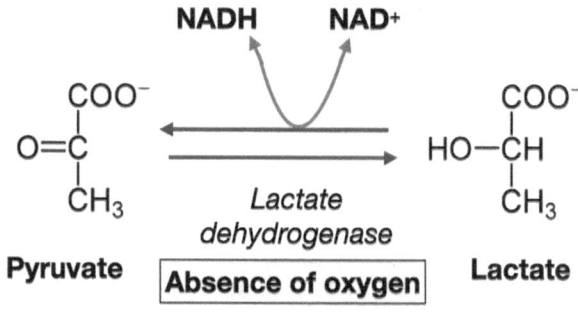

Figure 7.3 Conversion of pyruvate to lactate and vice versa.

Textbox 7.4 The Citric Acid Cycle

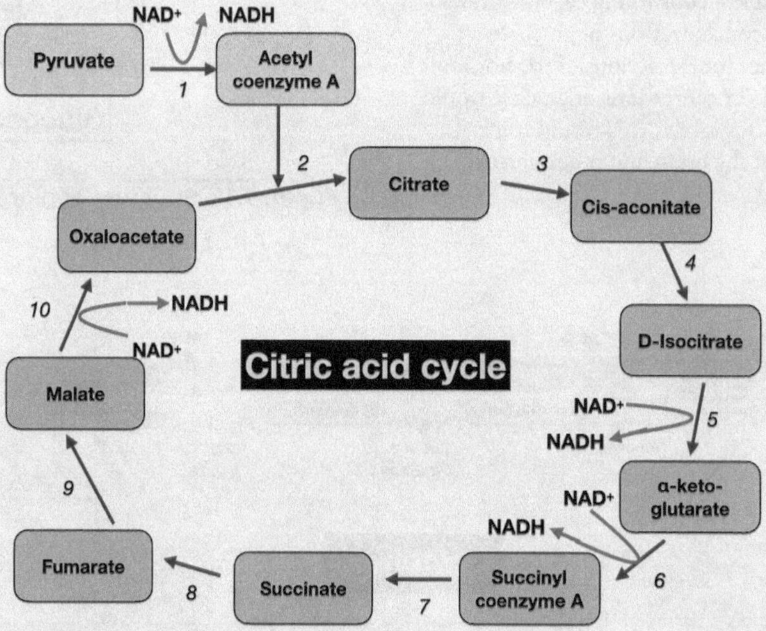

Figure 7.4 The citric acid (tricarboxylic acid) cycle. [Key: Enzymes (1) pyruvate dehydrogenase; (2) citrate synthase; (3) aconitase; (4) aconitase; (5) isocitrate dehydrogenase; (6) α-keto-glutarate dehydrogenase; (7) succinyl coenzyme A synthetase; (8) succinyl dehydrogenase (part of the electron transfer chain located on the inner membrane of the mitochondrion); (9) fumarase; (10) malic dehydrogenase, NAD$^+$ nicotinamide adenine dinucleotide, NADH reduced nicotinamide adenine dinucleotide.

7.3 Glucose and Glycogen

7.3.1 Introduction to Glucose

Glycogen is a **branched polysaccharide** [α(1→4) glycosidic polymer] of glucose. It is a readily mobilizable storage form of glucose.

7.3.2 Glycogen Synthesis (Glycogenesis)

Formation of glycogen is called glycogenesis. Glycogen is formed in the liver, skeletal muscle, and adipose tissue when both the plasma concentrations of glucose and plasma concentrations of insulin are high. Glucose is first phosphorylated to **glucose-6-phosphate** (Figure 7.5). This is converted to **glucose-1-phosphate** (Figure 7.5). Glucose-1-phosphate is combined with uridine triphosphate to form **uridine diphosphate glucose** and this is added to a glycogen chain by **glycogen synthase** (Figure 7.5.).

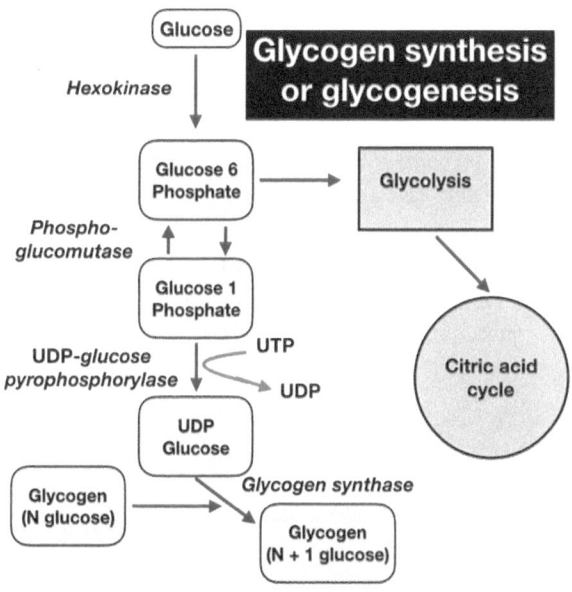

Figure 7.5 Metabolism of glucose at times of high glucose availability. Glucose is either oxidized by glycolysis and the citric acid cycle or stored as glycogen via glycogenesis. [Key: UDP = uridine diphosphate; UDP glucose = uridine diphosphate glucose; UTP = uridine triphosphate]

7.3.3 Glycogen Breakdown (Glycogenolysis)

Glycogenolysis is the breaking down of glycogen to generate glucose-1-phosphate (see Figure 7.5). The glucose-1-phosphate is changed to glucose-6-phosphate. In turn, the glucose-6-phosphate can be used as the substrate for glycolysis (see Figure 7.2). In the liver, the phosphate group is removed from glucose-6-phosphate by glucose-6-phosphase. The glucose is then exported from the **hepatocytes**.

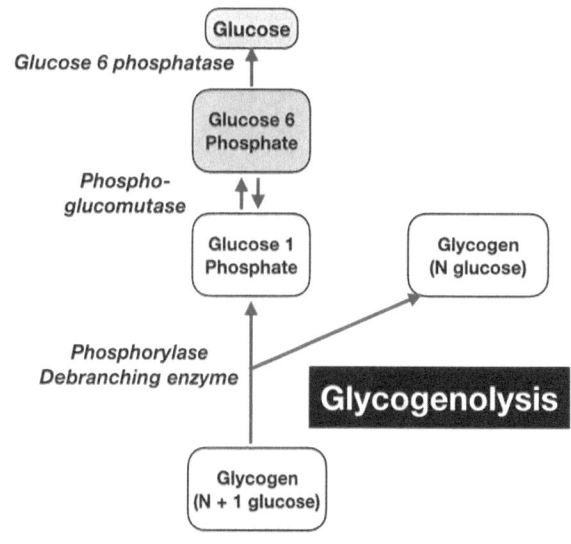

Figure 7.6 Breakdown of glycogen (glycogenolysis) to yield glucose occurring at times of low glucose availability.

7.4 Gluconeogenesis

Gluconeogenesis is generating glucose from glucogenic precursors such as lactate, pyruvate, glycerol, and glucogenic amino acids (see Figure 7.7). Gluconeogenesis is important to maintain blood concentrations of glucose, such as when the animal has not eaten for some time. Gluconeogenesis is particularly important in the following:

- Ruminants.
- Carnivorous companion animals (dogs and cats).

In ruminants, glucose has to synthesize principally from **propionate** and lactate. In turn, the propionate and lactate are generated by ruminal fermentation. In addition, there is some glucose produced from **alanine**, **valerate**, **isobutyrate**, and **glycerol**; these also come from fermentation in the rumen and digestion in the small intestine. However, ATP is primarily generated from acetate. In cats, there is little carbohydrate in the diet, little amylase, and consequently little glucose absorbed. Glucose needs to be generated by gluconeogenesis from **glucogenic amino** acids (e.g., alanine, arginine, cysteine, glycine, glutamate, glutamine, histidine, isoleucine, leucine, lysine, methionine, phenylalanine, serine, threonine, tryptophan, and valine in cats).

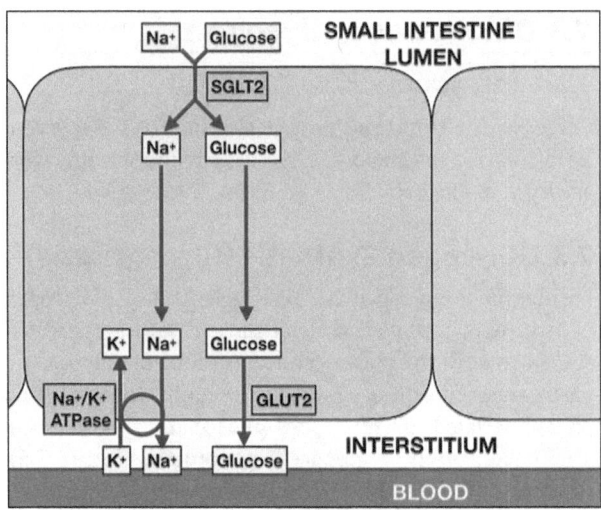

Figure 7.8 Glucose absorption from the lumen of the small intestine depends on two glucose transporters, SGLT2 (and in some species, SGLT1) and GLUT2.

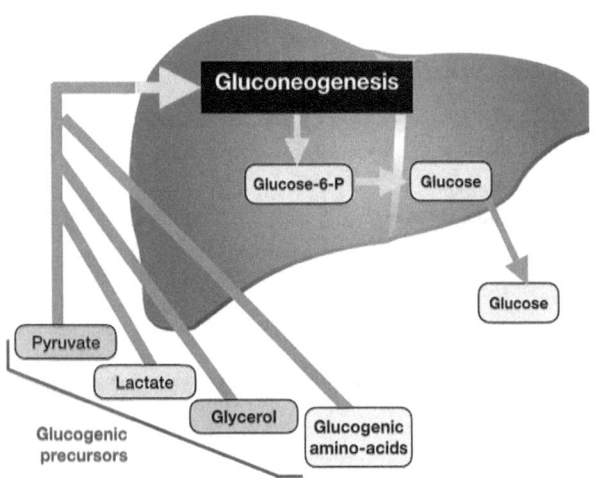

Figure 7.7 Gluconeogenesis occurs in the liver, synthesizing glucose from glucogenic compounds such as pyruvate, lactate, glycerol, and some amino acids such as alanine. (Outline of liver from DataBase Center for Life Science [DBCLS], CC BY 4.0)

7.5 Glucose Transporters

7.5.1 Introduction to Glucose Transporters

Glucose transporters (GLUT) or glucose channels allow facilitated diffusion of glucose. Glucose transits into or out of the cell in a passive manner through the GLUT down a **concentration gradient**. GLUT are members of the SLC2 (SoLute Carrier) **transmembrane** protein family. In addition, there is a series of sodium-dependent glucose cotransporters (SGLT).

7.5.2 Glucose Transporters and Glucose Absorption in the Small Intestine

Glucose transporters are critically important to absorption of glucose in the small intestine. At the luminal side, glucose is moved into the enterocyte up a concentration gradient by SGLT (See Figure 7.8). At the basal cell membrane, glucose transit into the blood is facilitated by GLUT2 (See Figure 7.8.).

7.5.3 Glucose Transporter 2 (GLUT2)

GLUT2 is the glucose channel responsible for glucose both entering or leaving hepatocytes. Net glucose transit depends on glucose moving down a concentration gradient.

7.5.4 Glucose Transporter 4 (GLUT4)

GLUT4 is an insulin-dependent GLUT. It is found in skeletal muscle, adipose tissue, and the heart of livestock and

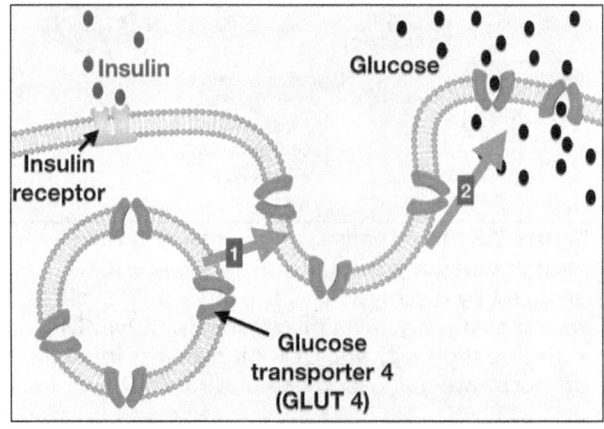

Figure 7.9 The glucose transmitter, GLUT4, in the membranes of vesicles in the cytoplasm. The vesicles are fused into the cell membrane in response to insulin binding to insulin receptors (Shutterstock/extender_01)

companion animals. GLUT4 is not found in poultry. In the unstimulated state, there are specialized vesicles containing GLUT4 in cytoplasm. In the presence of insulin binding to insulin receptors, the vesicles fuse with the cell membrane (see Figure 7.9). The more GLUT4 in the cell membrane, the more glucose transits through the glucose channel into the cell.

7.5.5 Glucose Transporter 7 (GLUT7)

GLUT7 is the glucose channel responsible for glucose leaving the **endoplasmic reticulum** (see Figure 7.10.).

7.6 Liver and Metabolism

7.6.1 Introduction to the Liver and Metabolism

The liver is considered to be the controller of metabolism, with the presence and regulation of the following pathways:

- Glycolysis.
- Citric acid cycle.
- Glycogen synthesis (glycogenesis) and breakdown (glycogenolysis).
- Gluconeogenesis.
- Fatty acid synthesis.

The shifts in hepatic metabolism following a meal or during fasting is summarized in Figure 7.11. Following a meal with the increase in blood concentrations of glucose (together with those of insulin), there are the following changes in metabolism (Figure 7.11):

- Glycolysis increased.
- Glycogenesis increased.
- Lipogenesis increased.
- Glycogenolysis decreased.
- Gluconeogenesis decreased.

Following fasting with the decrease in blood concentrations of glucose (and increase in those of glucagon), there are the following changes in metabolism (Figure 7.11):

- Glycolysis decreased.
- Glycogenesis decreased.
- Lipogenesis decreased.

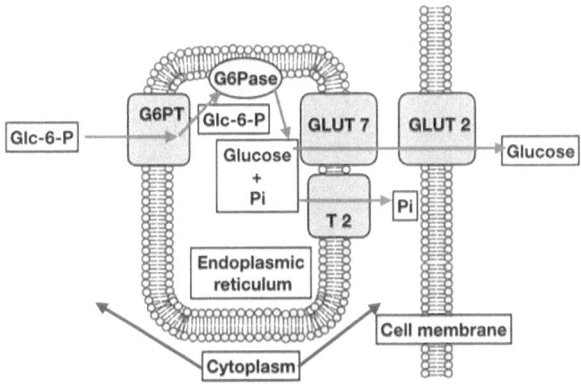

Figure 7.10 Importance of glucose transporters (GLUT) to release of glucose from the liver and, consequently, glucose homeostasis. [Key: Glc6P = glucose-6-phosphate; GLUT = glucose transporter; G6Pase = glucose-6-phosphatase; G6PT = glucose-6-phosphate transporter; Pi = phosphate; T2 = phosphate transporter]. (Adapted from Yikrazuul, CC BY-SA 3.0)

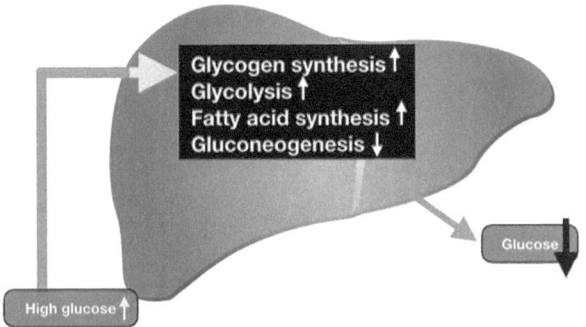 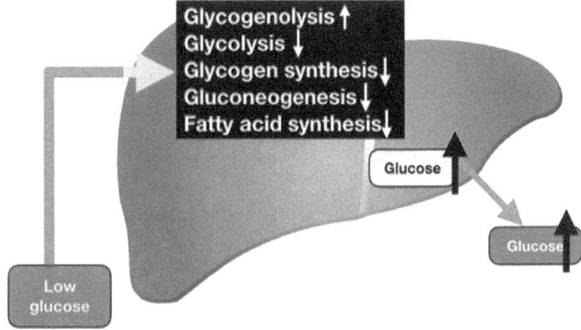

Figure 7.11 Changes in metabolism in the liver in fed animals (left: high blood glucose concentrations together with high concentrations of insulin) or fasted animals (right: low concentrations of glucose in the blood and high concentrations of glucagon plus low concentrations of insulin). (Outline of liver from the DataBase Center for Life Science [DBCLS], CC BY 4.0)

- Glycogenolysis increased.
- Gluconeogenesis increased.

7.6.2 Glucose-6-Phosphatase

Glucose-6-phosphatase is a critically important enzyme because without it the liver can't export glucose. Glucose-6-phosphatase is found in the liver, it is not found in skeletal muscle or adipose tissue or mammary tissue. It catalyzes the conversion of glucose-6-phosphate to glucose.

$$\text{Glucose-6-phosphate} \xrightarrow{\text{Glucose-6-phosphatase}} \text{Glucose}$$

Organs generate glucose-6-phosphate due to glycogenolysis (see Figure 11.6) or gluconeogenesis (see Figure 11.6). Glucose-6-phosphate does not cross the cell membrane. However, with the presence of both glucose-6-phosphatase and glucose transporters, the liver can and does export glucose into the bloodstream. This does not occur with either skeletal muscle or adipose tissue.

7.7 Skeletal Muscle and Metabolism

Skeletal muscle contains the following pathways:

- Glycolysis.
- Citric acid cycle.
- Glycogen synthesis and breakdown.

However, there is neither gluconeogenesis nor lipogenesis.

Following a meal, with the increase in blood concentrations of glucose (together with those of insulin), there are the following changes in metabolism (Figure 7.12):

- Glucose entry into the muscle fibers is facilitated by an insulin-induced increase in the number of GLUT4 in the plasma membrane.
- Glycolysis increased.
- Glycogenesis increased.
- Glycogenolysis decreased.

Following fasting, with the decrease in blood concentrations of glucose (and increase in those of glucagon), there are the following changes in metabolism (Figure 7.12):

- Glycolysis decreased.
- Glycogenesis decreased.
- Glycogenolysis increased.

7.8 Adipose Tissue

The predominant cell type in adipose tissue is the adipocyte. In addition, there are capillaries and macrophage.

7.8.1 Adipocytes

The role of white adipocytes is to store energy as triacylglycerol (triglyceride). At times of high glucose concentrations in the blood (and consequently high insulin

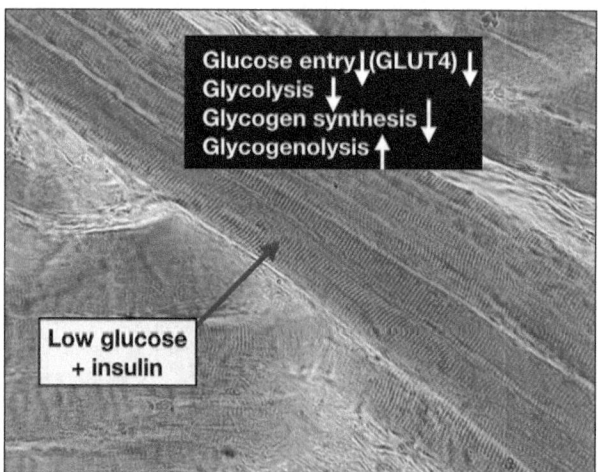

Figure 7.12 Changes in metabolism in skeletal muscle in fed animals (left: high blood glucose concentrations, together with high concentrations of insulin) or fasted animals (right: low concentrations of glucose in the blood and high concentrations of glucagon, plus low concentrations of insulin). (Background image of skeletal muscle from Reytan, CC BY-SA 4.0)

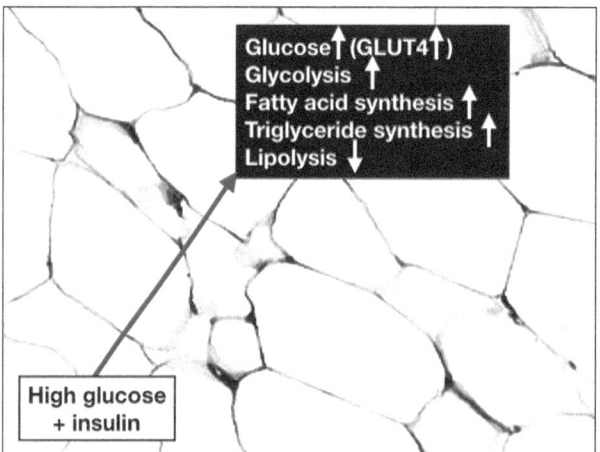 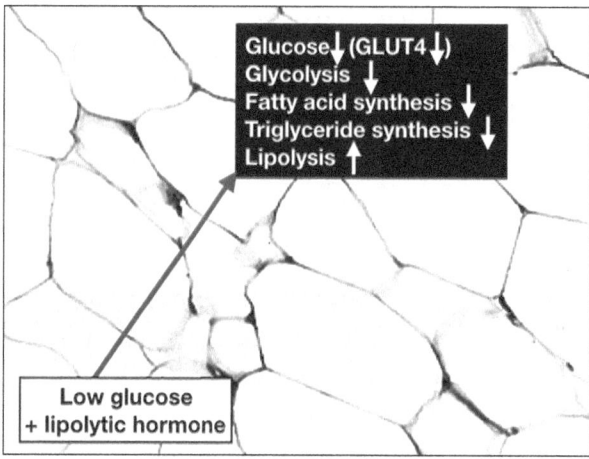

Figure 7.13 Changes in metabolism in the white adipose tissue in fed animals (left: high blood glucose concentrations together with high concentrations of insulin) or fasted animals (right: low concentrations of glucose in the blood and high concentrations of glucagon, plus low concentrations of insulin). (Background image from Berkshire Community College Image Library, CC0 1.0 Universal)

concentrations in the blood), the adipocytes synthesize fatty acids or obtain fatty acids from the circulation. The fatty acids are then combined with glycerol-3-phosphate to form triacylglycerol.

3 fatty acids (combined with coenzyme A) + glyceryl 3-phosphate → Triacylglycerol

In addition, there is movement of GLUT4 in vesicles into the cell membrane of the adipocyte. This increases the entry of glucose into the adipocyte. This glucose is metabolized in glycolysis to generate **glyceryl-3-phosphate** (Figure 7.13). At times of low concentrations of glucose in the blood and lipolytic hormones such as norepinephrine and glucagon, there is breakdown of triacylglycerol to fatty acids and glycerol in the process of lipolysis:

Triacylglycerol → 3 Fatty acids + glycerol

Adipocytes also synthesize hormones such as **leptin** (an energy homeostatic hormone) together with some immune-related proteins (known as adipokines). Adipocytes are generated from preadipocytes (see Figure 11.14) in the process of adipogenesis. In turn, preadipocytes are derived from mesenchymal precursor cells.

7.8.2 Adipogenesis

Adipogenesis is the process by which preadipocytes differentiate into adipocytes (see Figure 7.14). Adipogenesis is induced by a series of transcription factors including

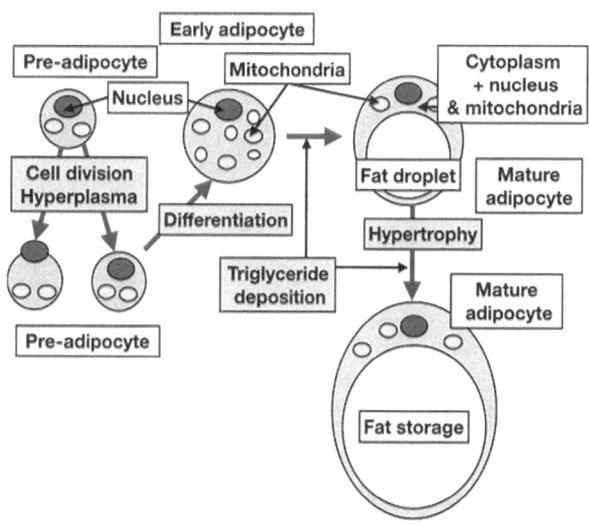

Figure 7.14 Adipogenesis.

peroxisome proliferator-activated receptor γ (PPARγ) and **CCAAT/enhancer-binding proteins** (C/EBPs). Once adipocytes are formed, there is synthesis of triacylglycerol and the presence of **lipid droplets**.

7.8.3 Brown Adipose Tissue (BAT)

Brown adipose tissue (BAT) is specialized adipose tissue that can generate heat when the animal is cold to restore body temperature to the set point. It has many more mitochondria than white adipose tissue, providing the color.

There is **uncoupling protein 1** (UCP1) in the mitochondria of brown adipose tissue. The implication of this is that **oxidative phosphorylation** is uncoupled from the generation of ATP and brown adipose tissue generates heat. The amount of BAT is higher in neonates. There is not BAT in pigs and poultry. White adipose tissue can be converted to BAT in some cases, such as long-term cold stress.

> **Textbox 7.5 Definitions**
>
> - **Adipogenesis**: the process by which pre-adipocytes differentiate into adipocytes.
> - **Hyperplasia** (increase in numbers of adipocytes): occurs as the result of proliferation of preadipocytes followed by their differentiation.
> - **Hypertrophy** of adipose tissue: occurs with deposition of triacylglycerol into the lipid droplet.
> - **Lipolysis:** the breakdown of triacylglycerol to fatty acids and glycerol.
> - **NEFA**: non-esterified fatty acids. They are also called free fatty acids. NEFA are found in plasma after lipolysis.
> - **Triglyceride synthesis**: the combination of fatty acids and glycerol.
> - **White adipose tissue** (WAT): functions to store energy.

7.9 Differences Between Metabolism in Poultry Compared to Livestock/Companion Animals

There are differences in detail in metabolism among the livestock and companion animals. For example, in ruminants, ruminal fermentation generates volatile fatty acids and lactate. There are also systematic differences between metabolism in poultry versus livestock and companion animals. These include the following:

- **Plasma glucose concentrations** are much higher in birds.
- **Fatty acid synthesis** is predominantly in the liver in poultry.
- There is no GLUT4 in poultry.
- The liver of female poultry produces **yolk precursors**.
- Absence of brown adipose tissue in poultry.

7.10 Oxidative Stress and Reactive Oxygen Species

Oxidative stress is the effect of **reactive oxygen species** (ROS). These ROS include the following:

- **Hydrogen peroxide.**
- **Superoxide radicals.**
- **Hydroxyl radicals.**
- **Singlet oxygen.**

These are by-products of metabolism and are mainly produced by **mitochondria**. ROS are capable of damaging or even destroying cells. To combat potential damage from ROS, cells have a series of **protective mechanisms**. These include:

- **Superoxide dismutase.**
- **Catalase (CAT).**
- **Glutathione peroxidase and glutathione.**

Glutathione is an antioxidant found at high concentrations in cells. It is a tripeptide consisting of three amino-acid residues:

$$\gamma\text{-Glutamate–Cysteine–Glycine.}$$

The cysteine residue has an SH (thiol) group. In the presence of a ROS, glutathione forms a dimer with a disulfide bridge.

$$2GSH + H_2O_2 \rightarrow GS\text{-}SG + 2H_2O$$
$$\text{glutathione peroxidase}$$

7.11 Metabolic Diseases

7.11.1 Diabetes (Cats and Dogs)

Diabetes is found in dogs and cats. Symptoms include excessive urination and drinking, together with increased appetite. In cats, diabetes is associated with obesity, inactivity versus physical activity (e.g., when cats are confined to the home), age, free-choice feeding, together with males and neutered cats.

7.11.2 Equine Metabolic Syndrome (Horses)

Equine metabolic syndrome (EMS) with insulin dysregulation (lack of regulation) is a risk factor for laminitis. The insulin dysregulation includes insulin resistance as evident from the prolonged hyperglycemia after a glucose challenge.

7.11.3 Ketosis (Cattle)

Ketosis is excess ketone bodies (acetone, acetoacetate, and beta-hydroxybutyrate) in the tissues including blood. Ketosis occurs when cattle are in a negative energy balance (i.e., consuming less energy than the animal needs). With a negative energy balance, the animal is mobilizing stored energy by breaking down triglyceride to fatty acids and glycerol. The fatty acids are metabolized by beta oxidation. If there are excessive amounts of fatty acids in the liver, there will be high concentrations of acetyl coenzyme A. If these are greater than the capability of these to enter the citric acid cycle, two acetyl coenzyme A molecules combine to form acetoacetyl coenzyme A. This is metabolized to acetone, acetoacetate and beta-hydroxybutyrate. The ketone bodies can be used as an energy source for skeletal muscle. Ketosis results when the liver is unable to use all of the fatty acids and there is insufficient metabolism of the ketone bodies.

7.11.4 Obesity (Cats, Dogs, and Horses)

Obesity is the excess accumulation of fat in adipose tissue. In cats, obesity is associated with inactivity versus physical activity, age together with males and neutered cats.

Skeleton

8

Learning Objectives

1. Understand the functions of bone (see Section 8.1.1).
2. Understand the composition of bones, including the following:
   a. Matrix (see Section 8.2.1).
   b. Calcium (see Section 8.2.1).
   c. Phosphate (see Section 8.2.1).
3. Understand the components of bones, including the following:
   a. Cartilage in joint (see Sections 8.4 and 8.5).
   b. Epiphyseal growth plate (see Section 8.2.1).
   c. Spongy bone (see Section 8.2).
   d. Compact bone (see Section 8.2).
   e. Bone marrow (see Section 8.9).
4. Understand the role of the cells of bone and cartilage (see Section 8.3).
   a. Osteocytes (see Section 8.3)
   b. Osteoblasts (see Section 8.3)
   c. Osteoclasts (see Section 8.3)
   d. Chondrocytes (see section 8.3)
5. Understand the functioning of joints (see Section 8.5).
6. Understand how bones develop and grow (see Section 8.6).
7. Understand what causes the production (laying down) of medullary bone in chickens, together with its the role (see Section 8.11).
8. Understand the roles of the marrow and pneumatic bones (see Sections 8.9 and 8.11).

Table of Contents

- 8.1 Introduction
  - 8.1.1 Introduction to the Skeleton
  - 8.1.2 What Are the Functions of the Bones of the Skeleton?
  - 8.1.3 Is Skeleton Dead Tissue?
- 8.2 Bone
  - 8.2.1 Composition
  - 8.2.2 Cortical or Compact Bone
  - 8.2.3 Spongy Bone
- 8.3 Bone Cells
  - 8.3.1 Osteoblasts
  - 8.3.2 Osteoclasts
  - 8.3.3 Osteocytes
- 8.4 Cartilage
  - 8.4.1 What Is Cartilage?
  - 8.4.2 Types of Cartilage
- 8.5 Joints
  - 8.5.1 Types of Joints
  - 8.5.2 Cartilage in Joints
  - 8.5.3 Ligaments
  - 8.5.4 Tendons
  - 8.5.5 Synovial Fluid
- 8.6 Bone Growth
  - 8.6.1 Prenatal Growth of Bones
  - 8.6.2 The Epiphyseal Plate and Growth of Long Bones via Endochondral Ossification
  - 8.6.3 Closure of the Epiphyseal Plate
- 8.7 Bone Remodeling or Turnover
- 8.8 Blood Supply
- 8.9 Marrow
- 8.10 Differences Between the Skeletons of Poultry and Mammalian Domestic Animals
  - 8.10.1 Avian Long Bone
  - 8.10.2 Why Is Having Medullary Bone Important to Hens?
  - 8.10.3 What Causes the Formation of the Medullary Bone?
  - 8.10.4 Air Sacs in Avian Bones
- 8.11 Diseases, Abnormalities, and Differences of Bones
  - 8.11.1 Bacterial Osteomyelitis
  - 8.11.2 Brachycephaly
  - 8.11.3 Bone Breakage
  - 8.11.4 Degenerative Joint Disease in Horses
  - 8.11.5 Equine Bone Fragility Syndrome (Osteoporosis, Silicosis) in Horses
  - 8.11.6 Hip Dysplasia in Dogs
  - 8.11.7 Manx Cat
  - 8.11.8 Osselets
  - 8.11.9 Osteoarthritis in Dogs and Cats
  - 8.11.10 Rotational and Angular Leg Deformities

8.1 Introduction

8.1.1 Introduction to the Skeleton

Textbox 8.1 provides a glossary of terms used relative to the skeleton. The skeleton consists of the following:

- **Axial skeleton** is composed of the bones of the central core of an animal's body (cranium, jaws, vertebrae, and associated structures such as ribs). It protects the brain and spinal cord.
- **Appendicular skeleton** is composed of the bones of the limbs and associated structures (mammals: skeleton of the front and hind limbs with associated structures in shoulders and hips; birds: skeleton of the

---

Textbox 8.1 Glossary Related to Bone

- **Bone**: consists of cells together with a matrix of proteins and hydroxylapatite.
- **Bones**: components of the skeleton.
- **Calcifying**: laying down of calcium phosphate.
- **Cartilage**: a flexible connective tissue.
- **Collagen**: a structural protein.
- **Chondrocytes**: cells embedded in cartilage that secrete cartilage matrix.
- **Haversian system**: contains blood vessels, nerves, and layers of bone.

(continued)

Textbox 8.1 (continued)

- **Osteoblasts**: cells that secrete the bone matrix (bone-building cells).
- **Osteoclasts**: cells that cut or chomp, breaking down bone.
- **Osteocytes**: cells buried in the bone that respond to mechanical stresses.
- **Osteoid**: unmineralized bone matrix that will be calcified.
- **Periosteum**: vascularized connective tissue surrounding bone except at joints.
- **Unmineralized bone**: bone with a matrix of proteins but prior to calcification.

wings and hind legs with associated structures). The appendicular bones form initially from the **limb buds** during embryonic development.

Figures 8.1 and 8.2 illustrate the skeletons of domestic animals, a dog and horse, respectively.

The structure of bones is shown in Figure 8.3. There are several tissues in the skeleton, including the following:

- Bone (or true bone).
  - **Spongy bone** (discussed in Section 8.2.2).
  - **Compact bone** (discussed in Section 8.2.3).
- Cartilage.
  - **Elastic cartilage** (discussed in Section 8.4).
  - **Fibrocartilage** (discussed in Section 8.4).
  - **Hyaline cartilage** (discussed in Section 8.4).

In addition, there are also ligaments and tendons in joints (discussed below).

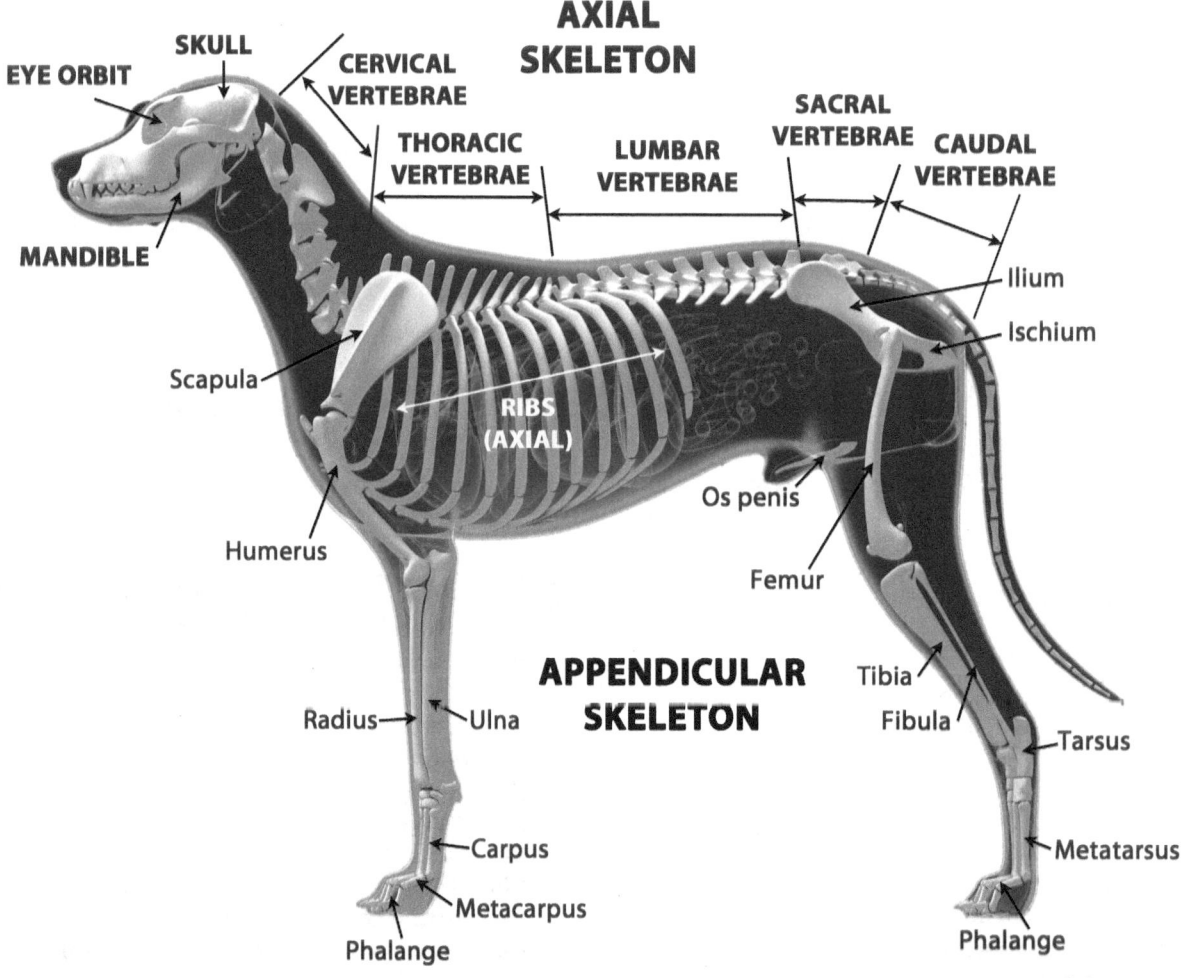

Figure 8.1 Skeleton of the dog. (Adapted from Shutterstock/decade3d–anatomy online)

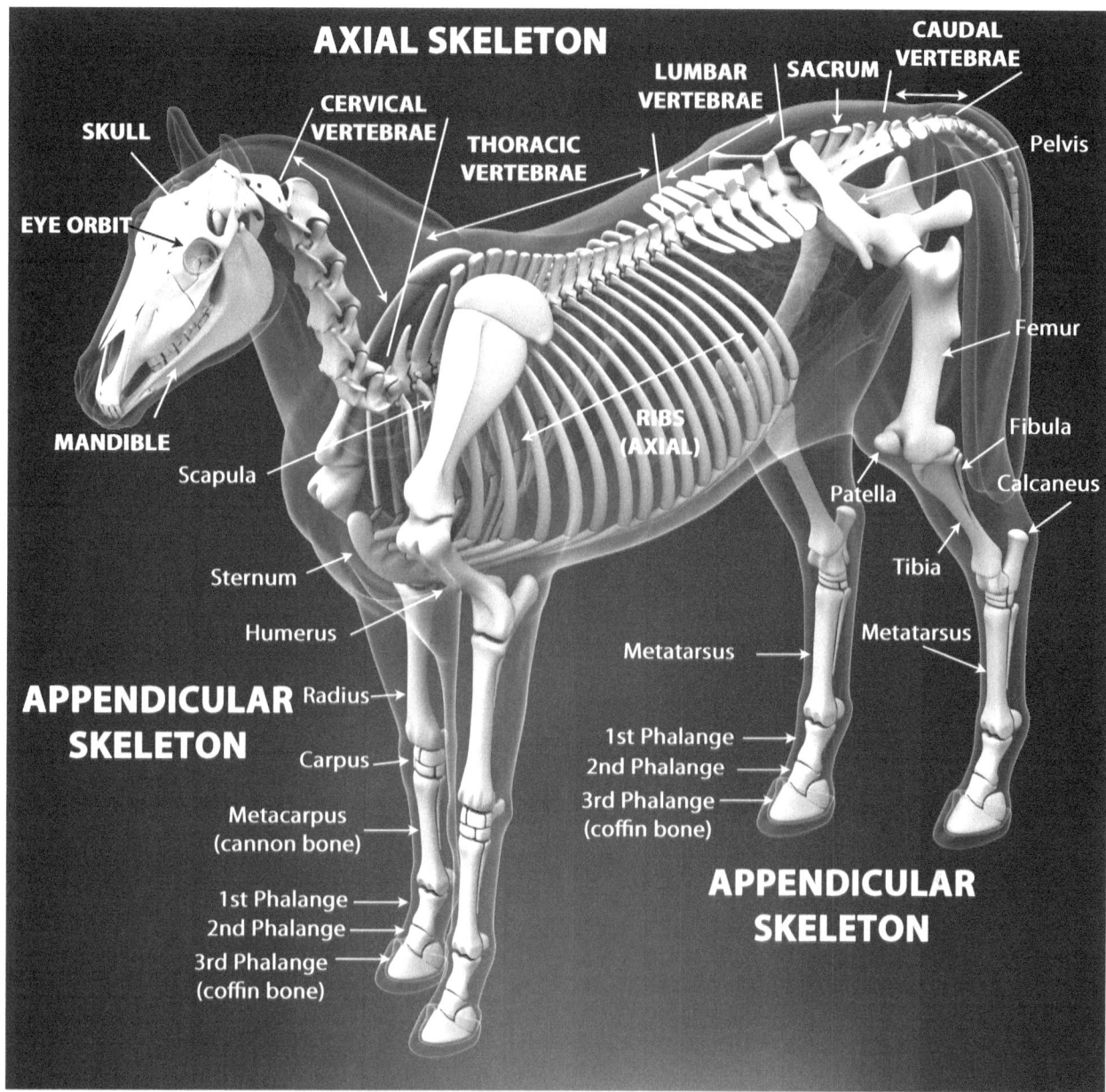

Figure 8.2 Skeleton of horse. (Adapted from Shutterstock/decade3d-anatomy online)

8.1.2 What Are the Functions of the Bones of the Skeleton?

The functions of the bones are the following:

1. Protection of brain, spinal cord, heart, and lungs.
2. Support (weight bearing).
3. Providing leverage points for muscles to contract against (anchorage).
4. Storage of calcium and phosphate and, consequently, assuring calcium and phosphate homeostasis.
5. Aiding maintenance of blood pH.
6. Production of erythrocytes, leukocytes, and platelets in bone marrow.
7. Production of hormones.

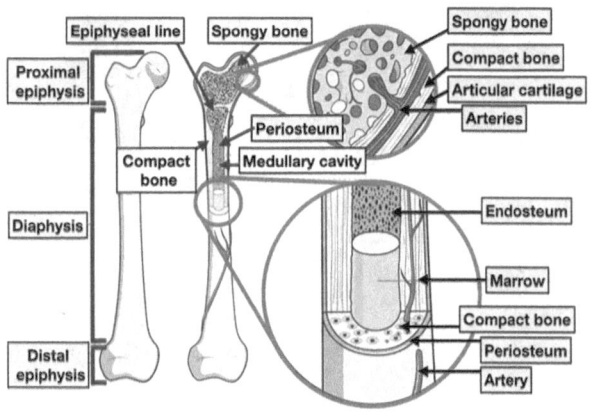

Figure 8.3 Bone structure.
(Adapted from Shutterstock/Alexander_P)

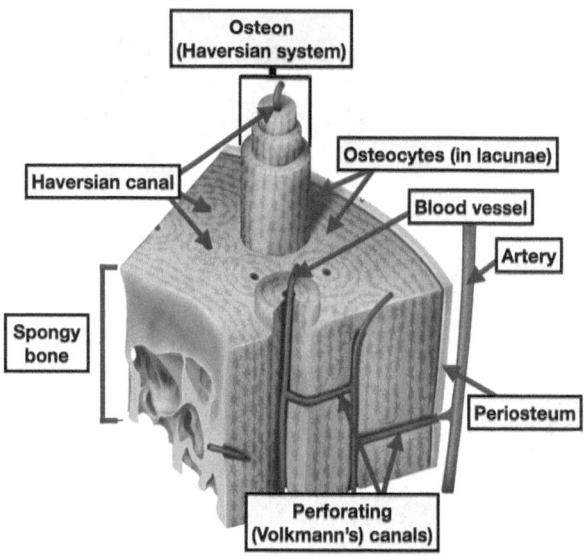

Figure 8.4 Cortical bone is made of concentric layers of bone (osteons), including osteocytes. Also shown are blood vessels in Haversian canals with associated osteocytes in bone.
(Adapted from Shutterstock/studiovin)

8.1.3 Is Skeleton Dead Tissue?

Based on its appearance, you might think that the skeleton is dead tissue. However, bone is a dynamic tissue with continuous bone turnover. Bone turnover is breakdown of bone by osteoclasts followed by new bone production by osteoblasts. (This is discussed in Section 3.3.1).

8.2 Bone

8.2.1 Composition

Bone is composed of **osteoid**, or bone matrix, together with minerals and cells. The osteoid consists of organic cross-linked **collagen fibers** interspaced by minerals. The major mineral in **inorganic bone** is **hydroxyapatite** (or hydroxyapatite) a form of **calcium phosphate** $[Ca_{10}(PO_4)_6(OH)_2]$. In addition to the calcium phosphate, there are the following other inorganic materials in bone: bicarbonate, sodium, potassium, citrate, magnesium, carbonate, fluorite, zinc, barium, and strontium.

8.2.2 Cortical or Compact Bone

Cortical bone (or compact or dense bone) makes up about 80% of bone. The material in the cortex of bones (see Figure 8.3) provides the skeleton with its strength. It is more resistant to mechanical stress than spongy bone. Compact bone has a structure of **concentric layers** (osteons) arranged around the **central canals** of the osteon or Haversian system. The central canals contain blood vessels supplying oxygen and nutrients to osteocytes, together with nerves running along the blood vessels (see Figure 8.4).

8.2.3 Spongy Bone

Spongy bone (or trabecular bone or cancellous bone) is a light, porous type of bone with greater elasticity than compact bone. It is highly vascularized and located at the ends of long bones (the epiphyses) and in the medullary spaces. It is less resistant to mechanical stress than compact bone. To remedy this, spongy bone in the ends of bone is surrounded by compact bone, increasing the overall strength of the bone (see Figure 8.3).

8.3 Bone Cells

There are five types of cells in bone:

- Osteoblasts.
- Osteoclasts.
- Osteocytes.
- Osteogenic stem cells (these can differentiate into osteoblasts).
- Bone lining cells.

Textbox 8.2   Mnemonic Device Using Alliteration

bbb osteo**b**lasts **b**uild **b**one
ccc osteo**c**lasts **c**ut or **c**homp on bone

8.3.1 Osteoblasts

While osteoblasts are derived from mesenchymal cells and comprise 5% of the cells in bone, they are essential as bone-forming cells. They are **polarized** cells with proteins released to the surface adjacent to the bone. The **single layer** of osteoblasts is forming bone. Osteoblasts are major producers of proteins with a well-developed endoplasmic reticulum and Golgi complex. They synthesize osteoid or bone matrix which contains collagen fibers (see Figures 8.5 and 8.6).

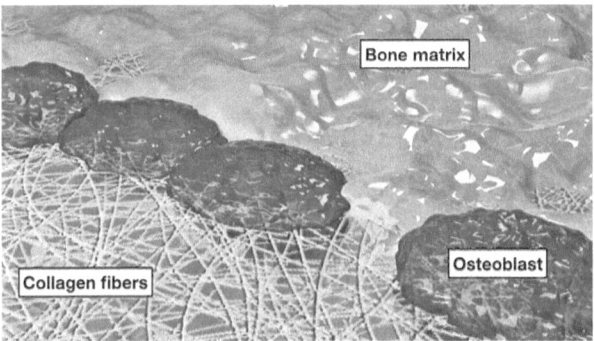

Figure 8.5 Osteoblasts synthesizing the collagen fibers into extracellular matrix of the bone. This plays an important role in bone modeling and remodeling.
(Adapted from Shutterstock/ALIOUI MA)

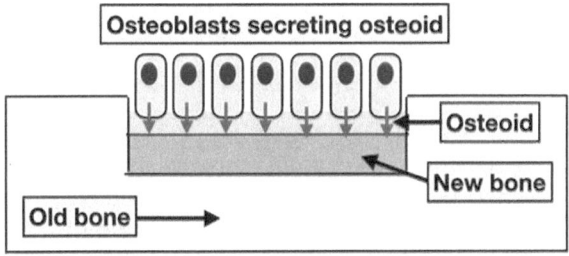

Figure 8.6 Polar osteoblasts secrete proteins including collagen, forming the osteoid. The osteoid is calcified, forming new bone.

8.3.2 Osteoclasts

Osteoclasts are **terminally differentiated multinucleate cells** that destroy (cut or chomp) bone. They are polarized cells with a **ruffled border** against bone resorption surface.

The osteoclasts have **carbonic anhydrase** and **proton pumps** to transport protons (H^+) facilitating solubilization of hydroxyapatite.

carbonic anhydrase

$$H_2O + CO_2 \rightarrow H_2CO_3 \rightarrow HCO_3^- + H^+ \rightarrow$$
exported via proton pump.

Protons pass through the cell membrane of the ruffled border onto the bone. The **hydroxyapatite** of the bone is **solubilized** in the acid. The osteoclasts also release **proteases** including **cathepsin**, matrix metalloproteinase-9, and tartrate-resistant acid phosphatase; these degrade the organic matrix of the bone.

8.3.3 Osteocytes

Osteocytes are derived from osteoblasts. Osteocytes are long-lived cells that are located in **lacunae of bones** (in the mineralized extracellular matrix) (see Figure 8.4). They comprise about 90% of the cells in bone. Their **dendritic processes** extend through bone with some extending to marrow and others to blood vessels. These processes pass through **canaliculi**.

Osteocytes are both **endocrine** and **sensory cells**. As endocrine cells, osteocytes in bones release **fibroblast growth factor** (FGF) 23. This hormone then plays an important role in phosphate homeostasis. Osteocytes, with their extended dendrites, act as **mechanosensory cells**. When there is a mechanical load on the bone, the osteocytes detect **shear stress** through the flow of fluid in the canaliculi and are activated. Osteocytes can also promote either bone formation by osteoblasts or bone destruction by osteoclasts.

8.4 Cartilage

8.4.1 What Is Cartilage?

Cartilage is a firm, resilient whitish, flexible connective tissue. Cartilage plays a pivotal role in joints being resistant to pressure and provides the elasticity of bones and joints. It is also important in the development of bones. It does not contain nerves. It is composed of the following:

- Chondrocytes in lacunae (see Figures 8.7 and 8.8). These make up less than 5% of cartilage. The chondrocytes synthesize the extracellular matrix.
- Extracellular matrix. Macromolecules, such as hyaluronan, endow the fluid of the joint with the following properties:
  o Viscous.
  o Semi-fluid.
  o Acting as a lubricant.

The fluid exhibits elasticity such that when deformed, it will return to its original state (shape and size) when

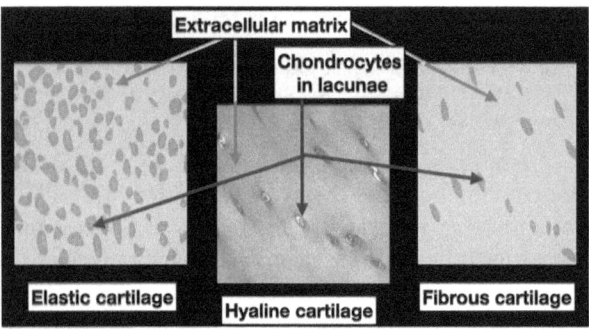

Figure 8.7 Structure of elastic, hyaline, and fibrous cartilage with chondrocytes present in lacunae (cavities) in the extracellular matrix of cartilage. Note: the darker extracellular matrix in hyaline cartilage is due to eosin (red) staining. (Adapted from Shiloh117981894 and Emmanuelm, CC BY 3.0)

the deforming agent is removed. The extracellular matrix includes collagen fibers and chondroitin sulfate (see Textbox 8.1 for a deeper dive into the constituents of bones). It provides a suitable locus for the survival of chondrocytes.

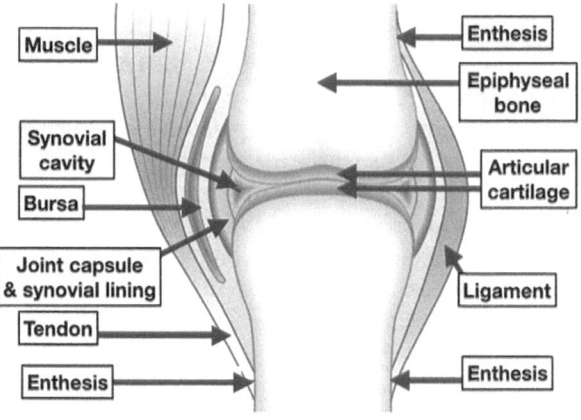

Figure 8.8 Synovial joint (Adapted from and courtesy Madhero88, Wikimedia Commons)

8.4.2 Types of Cartilage

There are three types of cartilage:

- Elastic cartilage has a high number of chondrocytes and a low amount of matrix (see Figure 8.7). Elastic cartilage contains a network of elastic fibers, making this type of

---

Textbox 8.3   A Deeper Dive Into the Composition of Components of the Skeleton

Cartilage
The **extracellular matrix** of cartilage is composed of the following:

- Collagen fibers.
- Elastin fibers.

A ground substance composed of the following:

- Water (65–80%).
- Chondroitin sulfate proteoglycans consisting of core proteins and a sulfated glucosaminoglycan, chondroitin sulfate. The latter is a water-soluble sulfated polysaccharide composed of repeated units of disaccharides; each being composed of N-acetylgalactosamine and glucuronic acid.

Tendons
Tendons have high tensile strength due to its composition of collagen type 1 fibers (arranged parallel to the long axis of tendon) together with elastin embedded in a proteoglycan matrix (mainly decorin and hyaluronan).

Ligaments
Ligaments need to be strong to connect bones and stabilize joints. Ligaments are composed of tough dense bundles of collagen fibers (type 1).

Synovial Fluid
Synovial fluid is composed of the following:

- An ultra-filtrate of plasma.
- Macromolecules that give synovial fluid its properties of high viscosity and elasticity.
  - **Hyaluronan** (or hyaluronic acid) is a high viscosity mucopolysaccharide, specifically a glycosaminoglycan, with repeated disaccharide units. Hyaluronan is synthesized by the synovial membrane.
  - **Lubricin** is a mucinous glycoprotein.
- Enzymes (collagenase and proteases) that maintain the integrity of the synovial fluid.

Osteocytes
Osteocytes express specific proteins in the SIBLING (small, integrin-binding ligand, N-linked glycoprotein) family of proteins. These proteins play a key role in mineralization.

cartilage the most flexible. It is found, for instance, in the pinna (external ear), epiglottis, and larynx.
- Fibrocartilage, or fibrous cartilage, has a low number of cells and high amount of matrix (see Figure 8.7). It is a very strong form of cartilage with type I collagen. It can distribute mechanical loads. This type of cartilage is found in the **intervertebral discs**, the equivalents to the knee, and **mandibular condyle** in the jaw.
- Hyaline cartilage is the most common type of cartilage. It is translucent and flexible. The matrix contains type II collagen and **proteoglycans** (see Figure 8.8). It is found in tracheal cartilage rings, bronchus, bronchioles, nose, and the ribs. **Articular cartilage** is found in joints, and it is a specialized form of hyaline cartilage.

8.5 Joints

Joints are where two or more bones meet (Figure 8.9). Joints allow bones to move. Joints contain the following:

- Articular cartilage (discussed in Section 8.5.2).
- A ligament stabilizing the joint (discussed in Section 8.5.3).
- Tendons (discussed in Section 8.5.4).
- A fluid filled synovial cavity (discussed in Section 8.5.5).

8.5.1 Types of Joints

There are six types of synovial joints.

- **Ball-and-socket joints**: These joints allow movements backward, forward, sideways, and rotation (see Figure 8.10).
- **Ellipsoidal joints**: These joints allow movement in two directions (see Figure 8.9).
- **Hinge joints**: These joints allow movement in one direction (see Figure 8.9).
- **Pivot joints**: These joints allow rotary movement around one axis.
- **Plane joints**: With these joints, two flat bones pass over each other (see Figure 8.9).
- **Saddle joints**: These resemble saddles and allow angular movement (see Figure 8.9).

Joints include the following:

- Bones with a cartilage on the articular surface.
- Ligaments.
- Tendons.
- Synovial fluid in bursa (see Figure 8.8).

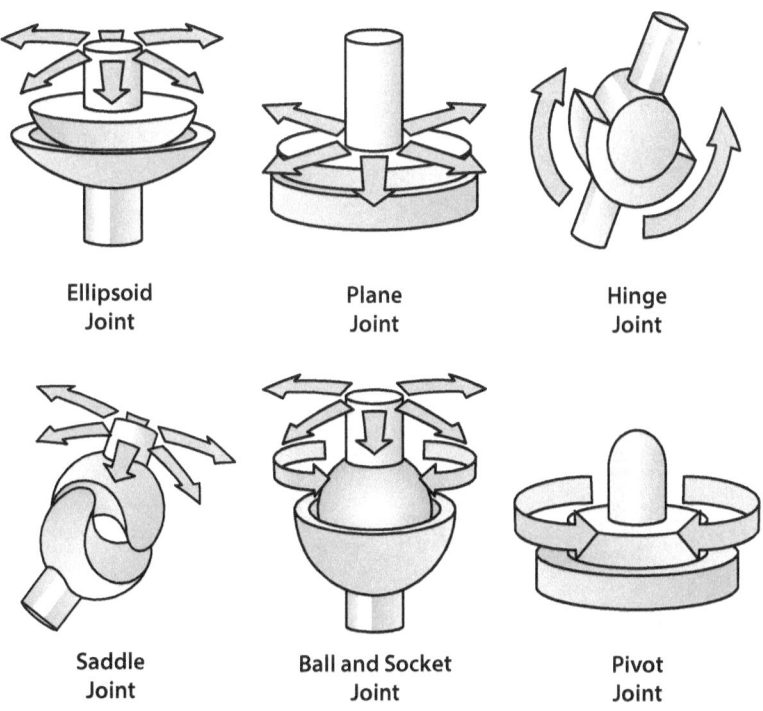

Figure 8.9 Types of synovial joints. (Adapted from Produnis, Wikimedia Commons)

8.5.2 Cartilage in Joints

There is articular cartilage (a subtype of hyaline cartilage) on those parts of bones that are in joints (see Figure 8.10). The collagen fibers of the superficial layer are aligned with the movement of the joint while those of the deep zone are aligned with the direction of the bone.

8.5.3 Ligaments

Ligaments connect bones and stabilize joints (see Figures 8.8 and 8.11). Ligaments are composed of tough dense bundles of collagen fibers (type 1). The stabilization may be supplemented by the muscles.

8.5.4 Tendons

Tendons connect bone with muscles. The tendons transfer mechanical force from the muscles to the bones, have high tensile strength, and are surrounded by a sheath (see Figures 8.8 and 8.11).

8.5.5 Synovial Fluid

Synovial fluid acts as **semifluid viscous lubricant** facilitating movement in the joint. It is synthesized by the synovial membrane.

8.6 Bone Growth

8.6.1 Prenatal Growth of Bones

Bones develop during embryonic development with, firstly, bone composed of cartilage as models for development including, later, formation of true bone or **ossification** (see Figure 8.12). In mammals, the cartilage model bones start to ossify at the end of embryonic/beginning of fetal development with the primary ossification center (see Figure 8.12). Ossification spreads in the cartilage in all directions by the process of **endochondral ossification**. Subsequently, there are secondary ossification centers towards the ends of the bones (see

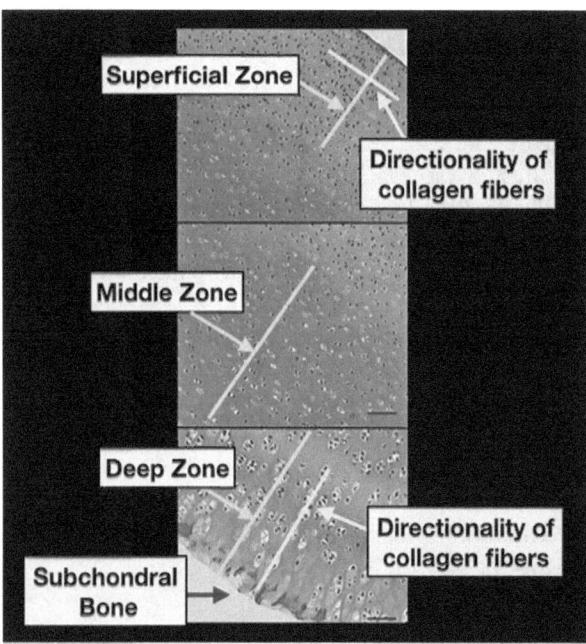

Figure 8.10 Articular cartilage showing zone with different directionality of collagen fibers. (Adapted from Catherine A. Bautista and colleagues, Wikimedia Commons)

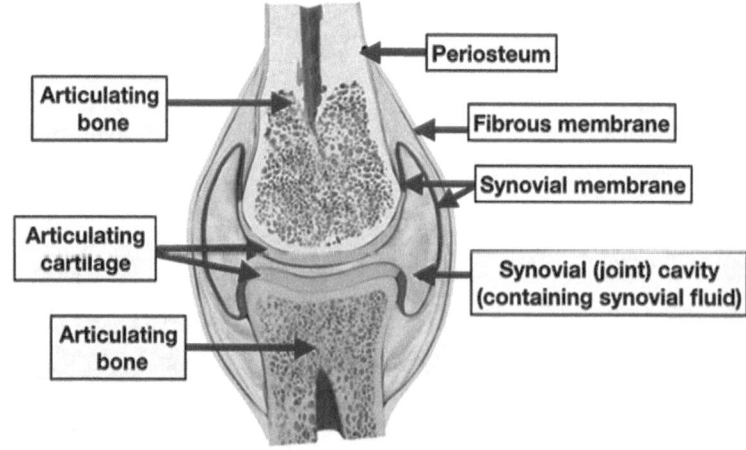

Figure 8.11 Joint showing the articulating cartilage on each bone with the synovial membranes enclosing a cavity filled with the viscous synovial fluid. The fluid-filled synovial cavity cushions the joint. (Adapted from Shutterstock/sciencepics)

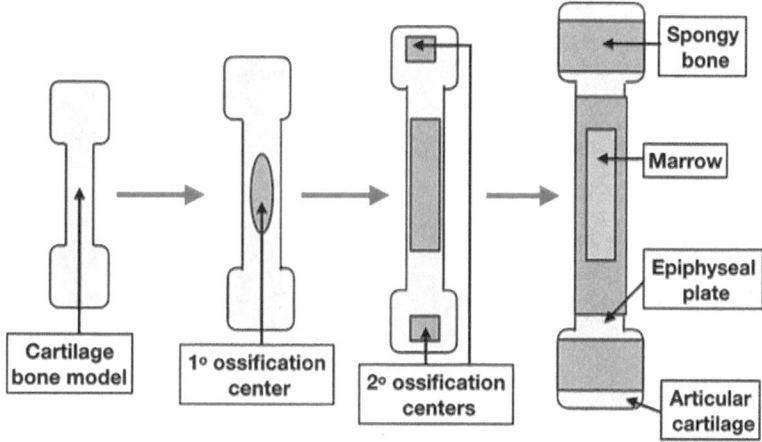

Figure 8.12 Embryonic and fetal development of bones. [Key: 1° = primary; 2° = secondary].

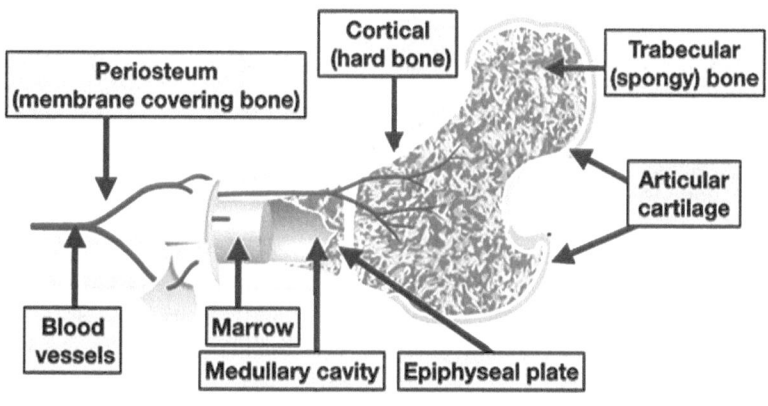

Figure 8.13 Bone structure showing epiphyseal plate. (Adapted from Pbroks13, CC BY 3.0)

Figure 8.12). This results in a bone composed of cortical bone, marrow in the medulla, articular cartilage, and a cartilaginous epiphyseal plate (see Figure 8.12). In chickens, most of ossification occurs during embryonic development.

8.6.2 The Epiphyseal Plate and Growth of Long Bones via Endochondral Ossification

The growth plate or epiphysis is a thin layer of cartilage. Growth of long bones occurs at the growth plate with first cartilage growth with a proliferation of chondrocytes (see Figures 8.12 to 8.14). Bone growth is the process by which the bone increases in length at the epiphyseal plate (or growth plate). In this region, chondrocytes expand both through hyperplasia and hypertrophy in the epiphyseal plate.

The epiphyseal plate has five zones (see Figure 8.14):

1. **Zone of reserve**—Quiescent chondrocytes.
2. **Zone of proliferation**—Hyperplasia (proliferation or mitosis) of chondrocytes due to insulin-like growth factor-1 (IGF-1) and growth hormone.
3. **Zone of maturation/hypertrophy**—Cell division of chondrocytes ceases and there are **increases in the size of cells** (hypertrophy).
4. **Zone of calcification**—Chondrocytes die due to **apoptosis**. Beginning of **calcification of matrix**.
5. **Zone of ossification**—Breakdown of calcified cartilage and replacement with bone tissue.

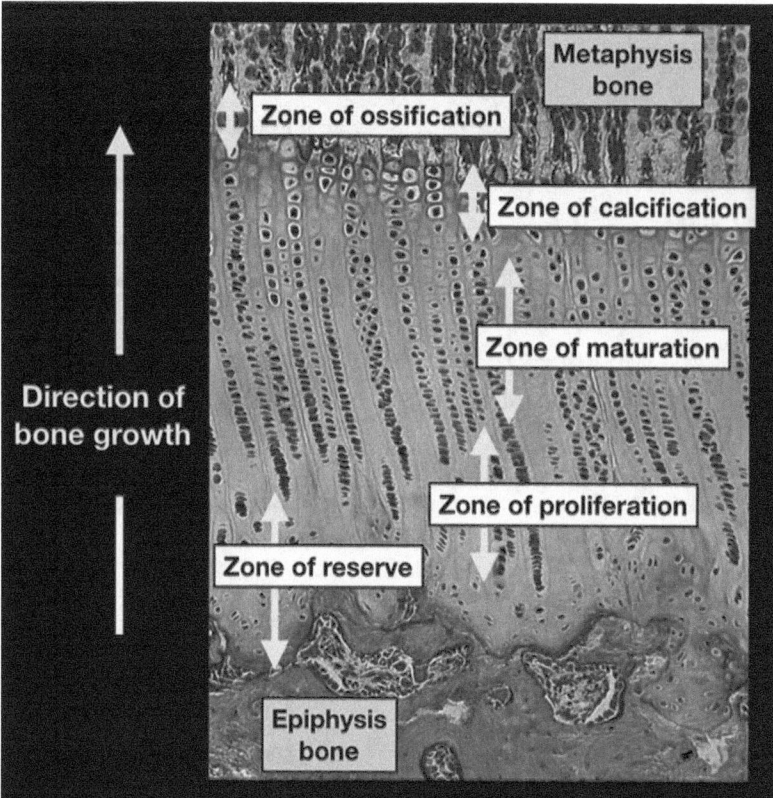

Figure 8.14 The epiphyseal plate is the site of the growth of long bones showing zone of reserve, zone of proliferation, zone of maturation/hypertrophy, zone of calcification, and zone of ossification. (Adapted from Jeppe Achton Nielsen, CC BY-SA 3.0)

There is a progression from zone 1 to 2 to 3 to 4 and, finally, to 5. Growth of the long bones is stimulated by growth hormone (GH) and insulin-like growth factor-1 (IGF-1). The cartilage in the growth plate is replaced by bone in the process of ossification. Osteoclasts destroy cartilage matrix. This allows blood vessels and osteoblasts to enter and the formation of new bone.

8.6.3 Closure of the Epiphyseal Plate

The **epiphyseal plate** closes at puberty due to the effects of sex steroids (estrogens and androgens), together with other hormones and growth factors. Closure of the growth plate results from a cessation of cartilage growth but continued ossification. When the cartilage in the growth plate is completely replaced with bone, the growth plate is no longer present.

Table 8.1 summarizes when the epiphyseal plate closes in domestic animals.

Table 8.1 Age when the epiphyseal plate closes in domestic animals.

Species	Age of Closure	
	Beginning	End
Cats	4 months	7–9 months
Chicken	4 months	23 weeks
Dogs	2.5 months	12 months*
Horses	8 months	40 months
Pigs	21 months	42 months
Rats	No closure	No closure
Sheep	4 months	30 months

*In large breeds, closure can be delayed to 18–20 months of age.

8.7 Bone Remodeling or Turnover

In adult mammals, the turnover rate for bone is 10% per year. There is breakdown of bone by the osteoclasts and production of new bone by the osteoblasts (see Figure 8.6). This is controlled by the osteocytes.

With increases in the loads from musculature and exercise, there are structural changes in bones leading to increases in bone strength. For example, in horses, exercise/training is associated with increases in bone density and tendon diameter. There are also effects of sympathetic nerves on bone remodeling. This example demonstrates a principle known as **Wolff's law**.

Wolff's law states that bones will adapt to the load placed on them. As the load on a bone increases, there is an increase in bone remodeling such that the bone can resist the load. If the load on a bone decreases, remodeling decreases the strength of the bone.

8.8 Blood Supply

Bone is well supplied with oxygen and nutrients via blood vessels in the Haversian canals at the centers of the osteoid bone and the perforating canals (see Figure 8.4).

8.9 Marrow

The bone marrow is a soft, spongy fatty tissue in the medulla of bones. Bone marrow is the site of production of the following blood cells:

- **Erythrocytes.**
- **Leukocytes.**
- **Platelets/Thrombocytes.**

8.10 Differences Between the Skeletons of Poultry and Mammalian Domestic Animals

8.10.1 Avian Long Bone

Many bones of laying hens and other reproductively active female poultry are composed of the following (see Figure 8.15):

- Cortical bone.
- Spongy or trabecular bone.
- Medullary bone.

The bones with medullary bone include tibia, femur, pubic bone, ribs, ulna, toe bones, and scapula.

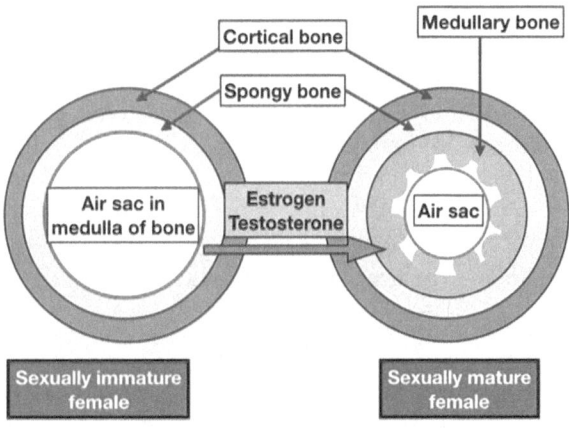

Figure 8.15 Transverse section through long bones of sexually immature and sexually mature female chickens. Estrogens and androgens stimulate the development of medullary bone.

Medullary bone provides a temporary storage or reservoir for calcium for egg shell formation (Figure 8.15). The reproductive hormones, estradiol, and testosterone, stimulate the formation of medullary bone in birds.

8.10.2 Why Is Having Medullary Bone Important to Hens?

Laying hens, as their name implies, produce one egg almost every day. Egg shells are composed principally of calcium carbonate (60% carbonate, 40% calcium). Each egg shell contains about 2.3 g of calcium. Primarily, this calcium comes from the feed. To produce one egg per day, a laying hen that is eating 100 g of feed per day, would need to consume 2.3 g of calcium to meet the needs of egg production, for homeostasis or a steady state. Therefore, the feed should contain at least 2.3% calcium. Chickens do not eat much at night. Therefore, the temporary storage in medullary bone provides a readily mobilizable source of calcium.

8.10.3 What Causes the Formation of the Medullary Bone?

Put simply, the medullary bone is formed by osteoblasts under stimulation by estrogens and androgens. Endosteal cells develop into pre-osteoblasts. In turn, these undergo mitosis and differentiate into osteoblasts. These produce woven bone progressively inwards to the medullar cavity.

174 THEME 2 Growth and Development

8.10.4 Air Sacs in Avian Bones

Some bones in birds contain air sacs. These are called **pneumatic bones**. Examples of bones with air sacs are the skull, humerus, keel, and some vertebrae. These air sacs play a role in respiration and also reduce the weight of the bones by depressing their density. Pneumatic bones are the mechanism by which the relative weights of avian bones are less than those of mammals. This is particularly important to reduce the energy needs for flight.

8.11 Diseases, Abnormalities, and Differences of Bones

8.11.1 Bacterial Osteomyelitis

Bacterial osteomyelitis is a disease of the bones and joints, particularly in the femoral head and proximal tibiotarsus. This leads to lameness in broiler chickens and turkeys.

8.11.2 Brachycephaly

Brachycephaly is foreshortening of the facial skeleton in dogs (see Figure 8.16). It is seen in short-muzzled dogs with flattened faces. Brachycephaly can cause problems with breathing, with increased airway resistance and increased effort being required to inhale. Dogs may exhibit snoring and snorting. Brachycephaly is seen in such breeds of dogs as the bulldog, pug, and French bulldog.

8.11.3 Bone Breakage

Long bones are subject to fractures. This is, for instance, seen in cattle, horses, pigs, and laying hens. In horses, there are fractures including chip fractures. These are seen in the following (see Figure 8.17):

- Long pastern bone (first phalanx).
- Short pastern bone (second phalanx).
- Sesamoid bone.

8.11.4 Degenerative Joint Disease in Horses

Degenerative joint disease is the equine equivalent of osteoarthritis. The cartilage is worn away in joints. This is a source of pain and restricted movement.

Figure 8.16 A pug dog showing foreshortening of the facial skeleton (with markedly flattening of the face). (Nancy Wong, CC BY-SA 4.0)

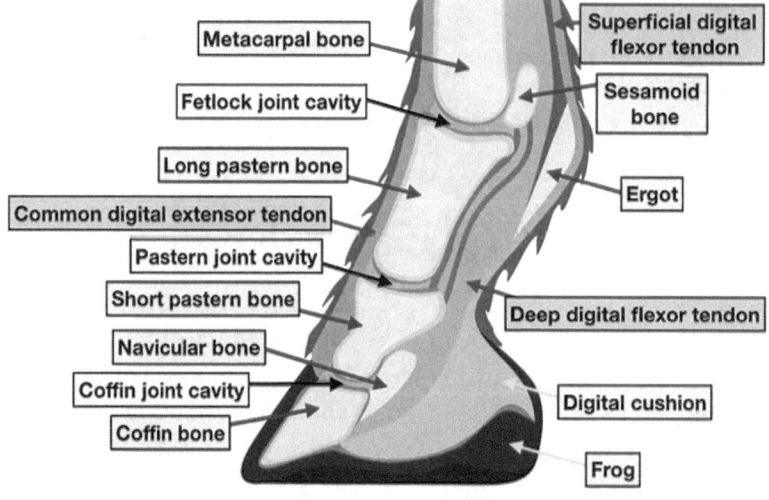

Figure 8.17 Foot of a horse.
(Adapted from Shutterstock/gritsalak karalak)

8.11.5 Equine Bone Fragility Syndrome (Osteoporosis, Silicosis) in Horses

This is a progressive disease in horses that is accompanied by lameness and exercise intolerance.

8.11.6 Hip Dysplasia in Dogs

Hip dysplasia is found in all breeds of dogs but is particularly prevalent in large dog breeds such as Great Danes, Labrador retrievers, and Saint Bernards. There is progressive damage to the joints of the hind legs. Dogs with hip dysplasia experience pain and lameness in their hind quarters.

8.11.7 Manx Cat

Manx is a cat breed with a bobbed tail. The lack of a tail is due to a natural mutation. Manx cats have weakened hindlimbs and problems related to the spine and spinal cord.

8.11.8 Osselets

Osselets is inflammation of connective tissue surrounding the canon bone and fetlock joint (see Figures 8.2 and 8.17).

8.11.9 Osteoarthritis in Dogs and Cats

Osteoarthritis is a disease found in one in five dogs older than one year. It is a progressive degenerative disease in which the cartilage in joints is worn away. Dogs and cats with osteoarthritis experience considerable pain and reduced mobility.

8.11.10 Rotational and Angular Leg Deformities

Rotational and angular leg deformities are seen in broiler chickens, pigs, cattle, and turkeys. This leads to lameness.

Theme Three

Interaction With the External Environment

Immune Function 9

Learning Objectives

1. Understand the roles of the primary and secondary immune organs (see Section 9.2).
2. Understand how an animal responds to pathogens (entire chapter).
3. Understand how the immune system can distinguish between "self" versus "non-self" (see Section 9.1.1)
4. Understand the primary and secondary immune organs (see Section 9.2).
5. Understand the immune cells (see Section 9.3).
6. Understand the process of phagocytosis and how phagocytes recognize pathogens (see Section 9.7).
7. Understand which cells are efficient at phagocytosis (see Section 9.7).
8. Understand what happens when a pathogen is subject to phagocytosis (see Section 9.7).
9. Understand the role of reactive oxygen species in the immune response (see Section 9.11).
10. Understand the structure and function of antibodies (immunoglobulins) (see Section 9.12).
11. Understand the structure and function of cytokines (interleukins, interferons, chemokines, etc.).
12. Understand the structure and function of PRRs and PAMPS.
13. Understand opsonin and opsonization (see Section 9.8).
14. Understand the complement cascade and three pathways (see Section 9.9.2).
15. Understand how complement aids defense against pathogens (see Section 9.9).
16. Understand innate immune responses (see Section 9.7).
17. Understand the types of innate immunity (see Section 9.7).
18. Understand adaptive immunity (see Section 9.10).
19. Understand the types of adaptive immunity (see Sections 9.10–9.12).
20. Understand the differences between B lymphocytes and plasma cells.
21. Understand how specific antibodies are produced by activation of B lymphocytes.
22. Understand activation of B cells (see Section 9.13).

23. Understand the utility and production of vaccines (see Section 9.14).
24. Understand what inflammation is and the role of mast cells (see Section 9.15).
25. Understand what allergies are (see Section 9.16).
26. Understand what cytokines are (see Section 9.5).
27. Understand how fever aids an animal fighting an infection (see Section 9.17).
28. Understand immune-related diseases in domestic animals (see Section 9.18).

Table of Contents

9.1 Introduction
   9.1.1 Overall Introduction and Key Concepts
   9.1.2 Introduction to Immune Organs
   9.1.3 Introduction to Immune-Related Molecules
9.2 Immune Organs
   9.2.1 Thymus
      9.2.1.1 What Does the Thymus Do?
   9.2.2 Bursa of Fabricius (Birds)/Bone Marrow (Mammals)
   9.2.3 Lymph Nodes
   9.2.4 Spleen
      9.2.4.1 What Does the Spleen Do?
9.3 Immune Cells
   9.3.1 Lymphocytes
      9.3.1.1 What Do Lymphocytes Do?
   9.3.2 Dendritic Cells
      9.3.2.1 What Do Dendritic Cells Do?
   9.3.3 Natural Killer Cells
      9.3.3.1 What Do Natural Killer Cells Do?
      9.3.3.2 How Do Natural Killer Cells Recognize Virally Infected or Tumor Cells?
9.4 Innate (Nonspecific) Immunity
   9.4.1 Introduction to Innate Immunity
   9.4.2 Barriers to Infection
      9.4.2.1 Introduction to Barriers to Infection
      9.4.2.2 Physical Barriers
      9.4.2.3 Hydrochloric Acid in the Stomach
   9.4.3 Antimicrobial Peptides
   9.4.4 Oxidative or Respiratory Burst
   9.4.5 Iron and Immunity
   9.4.6 Pattern Recognition Receptors (PRR)
9.5 Cytokines
   9.5.1 Introduction to Cytokines
   9.5.2 Interleukins
   9.5.3 Interferons
   9.5.4 Chemokines
9.6 Phagocytes, Platelets/Thrombocytes, and the Immune Response
   9.6.1 Overview of Phagocytes, Platelets/Thrombocytes, and the Immune Response
   9.6.2 Platelets and the Immune Response
   9.6.3 Movement of Leukocytes in Blood Vessels and Recruitment to Sites of Wounding or Infection
9.7 Phagocytosis (Cellular Innate Immunity)
   9.7.1 What Do Phagocytes Do?
   9.7.2 How Do Pseudopods on Phagocytes Discern That the Adjacent Cell or Organism Is a Pathogen?
   9.7.3 Do Antibodies or Other Factors Influence Phagocytosis?
9.8 Opsonization
   9.8.1 Introduction to Opsonization
   9.8.2 Opsonins
9.9 Complement

9.9.1 Introduction to Complement
9.9.2 Functions of Complement
9.9.3 Activation of Complement
9.10 Introduction to Adaptive (or Specific) Immunity
9.11 Cell-Mediated Immunity
9.12 Humoral (Antibody) Mediated Immunity—Antibodies/Immunoglobulin
9.12.1 What Is an Antibody?
9.12.2 Where Are Antibodies Produced?
9.12.3 Are There Multiple Types of Antibodies?
9.12.4 Why and How Are There Differences in the Antigen-Binding Site Allowing Binding Different Antigens?
9.12.5 How Does an Animal Produce Billions of Different Antibodies When There Are Limited Numbers of Genes?
9.12.6 What Do Antibodies Do?
9.13 B Lymphocytes and Antibody Production
9.14 Vaccines
9.15 Inflammation
9.15.1 What Is Inflammation?
9.16 Allergies
9.17 Fever
9.17.1 What Is Fever?
9.18 Immune Related Diseases in Domestic Animals

9.1 Introduction

9.1.1 Overall Introduction and Key Concepts

Immunity is the system that protects animals from infectious diseases and removes dead or dying cells and cell debris. Textbox 9.1 provides a brief overview of pathogens.

There are two overall lines of immune defense:

1. Innate immunity (not specific for an individual pathogen):

Textbox 9.1

Pathogens
Pathogens are bacteria, protozoa, fungi, and virions that cause diseases. Bacteria, protozoa, and fungi can be either of the following:

- Obligate pathogens that require a host(s) for their life cycle.
- Facultative pathogens—bacteria, protozoa, and fungi do not require a host but may infect an animal.

The degree of severity of the disease reflects the virulence of the pathogen.

Bacteria
The vast majority of bacteria are not pathogenic—they do not cause disease. There are billions of bacteria in the gastrointestinal tract. However, some bacteria are pathogenic. A pathogenic bacterium is *Yersinia pestis* (hosts: rodents and fleas) causing plague. Another pathogenic bacterium is *Bacillus anthracis*. This causes anthrax.

Viruses
Virions are the infectious viral particles. They contain a series of genes in the form of either RNA or DNA together with coat proteins making up the envelope and/or necessary for infection. An example of a viral disease is influenza.

Protozoa
Protozoa are single cells organisms. They can be parasites. Examples include trypanosomes causing sleeping sickness.

Fungi
Fungi can be pathogenic. Examples are *Aspergillus flavus* and *Aspergillus fumigatus* and these cause aspergillosis in poultry.

a Physical and chemical barriers.
b Internal cells and specific proteins.
2. Adaptive (specific) immune system:
a Humoral (antibodies) mediated immunity.
b Cell-mediated immunity.

Physical and chemical barriers, together with internal cells and specific proteins, are components of the **innate immune system**. These are the first lines of defense against pathogens. However, there is much that is common between innate and adaptive immunity.

It is important that the immune system can distinguish between "self" versus "non-self" to prevent it from destroying its own tissues. When this happens, it results in autoimmune diseases.

Adaptive immunity is either of the following:

- Active.
  o Natural in response to a pathogen.
  o Artificial—man-made, such as vaccination against a specific pathogen.
- Passive.
  o Natural, e.g., calves receiving antibodies from their mothers.
  o Artificial, e.g., administration of antivenom (antisera to a venom) after a venomous snake bite.

Textbox 9.2 provides a glossary of terms used relative to the immune system.

Textbox 9.2 Glossary and Definitions

- **Active immunity**: an immune response to a foreign substance such as pathogen or vaccine.
- **Adaptive (Specific) immunity**: immunity that an animal acquires during and after experiencing a foreign substance such as a pathogen or a vaccine.
- **Allergy**: overreaction or inappropriate reaction of the immune system.
- **Antimicrobial peptides/proteins**: proteins that kill pathogens.
- **Antibody**: a protein (gamma globulin) produced by lymphocytes that binds to a specific antigen.
- **Antigen**: a specific molecular signature that is recognized by the immune system. For instance, antigens bind to specific antibodies.
- **Antisera**: serum containing antibodies.
- **Apoptosis**: a type of programmed cell death.
- **Artificial immunity**: human-made immunity. For example, vaccines provoke an immune response (including antibody production).
- **B lymphocytes**: these develop in the bone marrow and mature in the bone marrow (mammals) and bursa of Fabricius (birds). When activated, they develop into antibody-producing plasma cells.
- **Cell-mediated immunity**: the non-antibody response to a pathogen. It includes activation of phagocytes.
- **Colostrum**: the first milk produced in a lactation. It contains antibodies that pass across the intestine of the newborn, providing natural passive immunity.
- **Cytokines**: proteins produced by immune cells that induce inflammation and impact immune function.
- **Dendritic cells**: a type of immune cell presenting antigen to other immune cells.
- **Gut-associated lymphoid tissue (GALT)**: the largest immune tissue; it contains both B- and T lymphocytes.
- **Humoral (antibody) mediated immunity**: production of antibodies that bind to antigens.
- **Immunocompetence**: the ability of B and T lymphocytes to react to antigens.
- **Immunogen**: an antigen that provokes the production of antibodies to be used against it, or production of cell-mediated immunity.
- **Innate (or non-specific) immunity**: a broad response that serves as the initial lines of defense to pathogens.
- **Interferons**: antiviral proteins produced by immune cells.
- **Interleukins**: proteins produced by immune and other cells that modulate immune function.
- **Leukocytes**: white blood cells that protect the animal from invading pathogens.

(continued)

Textbox 9.2 (continued)

- **Lymph nodes**: these play an important role in the immune response to pathogens and also filter lymph.
- **Lymphocytes**: types of leukocytes (also see B- and T- lymphocytes).
- **Maturation of B and T lymphocytes**: the development of immunocompetence and self-tolerance.
- **Natural immunity**: an immune response to a foreign substance.
- **Pathogen-associated molecular patterns (PAMPs)**: molecular patterns on bacteria and other pathogens.
- **Pattern-recognition receptors (PRRs)**: are found in the outer surface of immune cells. These recognize (bind to) pathogen-associated molecular patterns.
- **Passive immunity**: receiving antibodies from another animal. For instance, a calf will receive antibodies from its mother in colostrum and a chick embryo receives it in the yolk from the hen.
- **Pathogen**: an organism that causes disease. Examples of pathogens include bacteria, viruses, fungi, and protozoa.
- **Phagocytes**: an innate cell type that engulfs and "eats" pathogens or tissue debris in the process of phagocytosis. Examples of phagocytes are neutrophils, monocytes, and macrophage.
- **Phagocytosis**: the process by which phagocytes engulf pathogens and tissue debris. Subsequently, the pathogens and tissue debris are digested by the phagocyte.
- **Recognition of self**: the immune system ensures that it does not respond to the proteins in the body (self) but responses to "non-self," a foreign substance, such as a pathogen.
- **Self-tolerance**: the ability of B- and T- lymphocytes to *not* react to "self" proteins.
- **Spleen**: one of the secondary immune organs.
- **T lymphocytes**: these develop in the bone marrow differentiate but mature in the thymus and are important to cell-mediated immunity.
- **Thymus**: one of the primary immune organs. It plays the critical role in the maturation of T lymphocytes.
- **Vaccine**: a collection of immunogens (or the mRNA) related to a pathogen that produce an immune response and the development of memory cells so that an animal can more easily deal with an infection by the pathogen.

9.1.2 Introduction to Immune Organs

The organs responsible for immunity are considered as either primary or secondary immune organs. The primary immune organs are the following:

- Thymus (critical to the development of **T** lymphocytes).
- Bursa of Fabricius (in poultry) and **b**one marrow (in companion animals, livestock, and other mammals) (critical to the development of **B** lymphocytes).

The secondary immune organs are the following:

- Lymph nodes (not present in poultry).
- Spleen.
- Tonsils.
- **Gut-associated lymphoid tissue** (GALT) consisting of isolated lymphoid follicles and aggregates in Peyer's patches (associated with the small intestine).
- Blood containing leukocytes (white blood cells) and platelets (or nucleated thrombocytes in poultry) (also see Chapter 8).

In addition, the skin, bone marrow, and liver play roles in immunity.

9.1.3 Introduction to Immune-Related Molecules

The following are key proteins in immunity:

- Antibodies or immunoglobulin.
- Complement.
- Cytokines.
  - Interleukins.
  - Interferons.
  - Chemokines.
  - Tumor necrosis factors (TNF–α).
  - Growth factors.

- Opsonins.
- Pathogen-associated molecular patterns on/in pathogens.
- Pattern Recognition Receptors.

9.2 Immune Organs

9.2.1 Thymus

The thymus is one of the primary immune organs. It plays the critical role in the maturation and proliferation of T lymphocytes. The thymus is located in the thoracic/neck region. The structure of the thymus is shown in Figure 9.1. The thymus consists of a **cortex** and **medulla** surrounded by a **capsule** (see Figure 9.1). There are **multiple lobules** separated by **septa**.

The major cell types in the thymus are **thymic epithelial cells** (both cortical epithelial cells and medullary epithelial cells) and **thymocytes** (in the cortex) (see Figure 9.1C). In addition, there are monocytes, macrophage (MΦ), **mast cells**, and **dendritic cells** (see Figure 9.1C).

9.2.1.1 What Does the Thymus Do?

The thymus is critically important to the maturation of T lymphocytes. The T lymphocytes then leave the thymus to populate the secondary lymph organs, the spleen and lymph nodes. The thymus also produces a series of hormones called **thymosins**.

The size of the thymus declines with age; this being called **thymic involution**. This involution is due to the effects of sex steroid hormones, together with stress hormones (cortisol and corticosterone).

9.2.2 Bursa of Fabricius (Birds)/Bone Marrow (Mammals)

The bursa of Fabricius is located close to the cloaca in poultry species. The bursa of Fabricius was first recognized as being responsible for maturation of B lymphocytes in birds. This led to the concept of B lymphocytes producing antibodies and T lymphocytes responsible for cell-mediated immunity. Mammals lack a bursa of Fabricius, instead **B-lymphocytes** undergo maturation in the **bone marrow**.

9.2.3 Lymph Nodes

Lymph nodes play an important role in the immune response to pathogens and also filter lymph. The structure of a lymph node is shown in Figure 9.2. Each lymph node consists of a cortex and a medulla surrounded by capsule. There are multiple lymph nodes with approximately 900 in dogs and about 2,000 in horses.

Lymph nodes contain both B and T lymphocytes, together with **dendritic cells**. Dendritic cells present fragments of pathogens to initiate an immune response. There are both B and T lymphocytes in, respectively, the T and B lymphocyte zones.

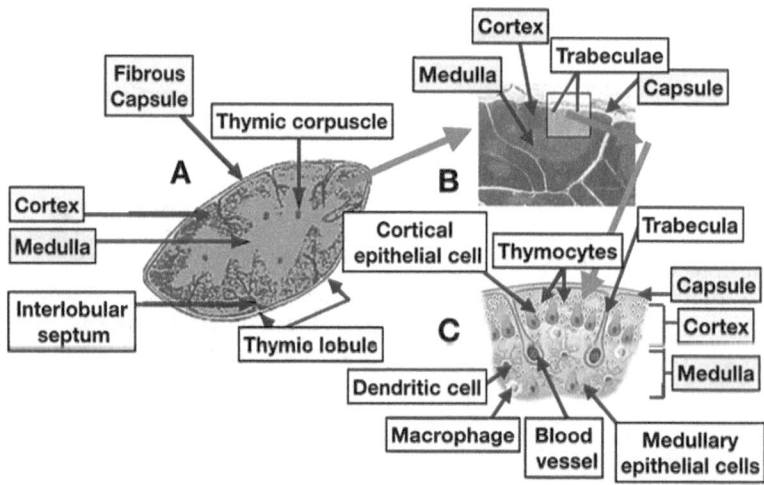

Figure 9.1 Structure of thymus. (A) Section through thymus; (B) Increased magnification of section of part of thymus; (C) Schematic diagram of structure of thymus.
(A) is adapted from the National Cancer Institute; (B) and (C) are adapted from OpenStax, CC BY 3.0)

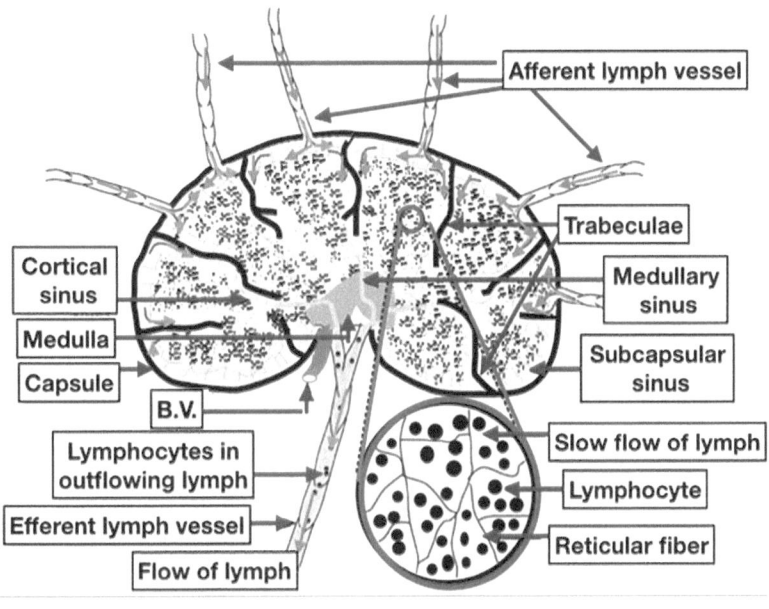

Figure 9.2 Structure of a lymph node [B.V. = blood vessel]. (Adapted from and courtesy of KC Panchai)

Lymph notes are not present in birds. There are, however, aggregates of lymphocytes in the lymph vessels of chickens and other birds. These are markedly enlarged when the birds are challenged with a foreign protein. The aggregates contain both B and T lymphocytes, together with dendritic cells.

9.2.4 Spleen

The spleen is the second largest immune organ and secondary lymphoid organ. It is composed of **white and red pulp**. Among the cells in the spleen are the following: macrophage, dendritic cells, lymphocytes, and antibody-producing plasma cells. As with lymph nodes, there are distinct zones, namely B and T lymphocyte zones, within the white pulp. There are large numbers of macrophage in the red pulp.

9.2.4.1 What Does the Spleen Do?

Not only is the spleen part of the immune system with functioning B and T lymphocytes but it also functions to filter blood, removing old erythrocytes. In addition, in some species such as dogs and horses, the spleen stores erythrocytes. These are then released during times when oxygen needs are very high, such as exercise. This occurs as a result of contraction of the spleen. The spleen also acts as a store or reservoir for monocytes.

9.3 Immune Cells

The cell types in the immune system include the following:

- Lymphocytes.
- Phagocytic cells.
  - Macrophage (MΦ).
  - Monocytes.
  - Neutrophils.
  - Dendrite cells.
- Other leukocytes.
  - Eosinophils.
  - Basophils.

9.3.1 Lymphocytes

Among the most important cells in the immune system are lymphocytes. These are found in the blood, lymph, lymph nodes, thymus, and bone marrow.

9.3.1.1 What Do Lymphocytes Do?

The types of lymphocytes are the following:

- B lymphocytes that develop into the antibody-producing plasma cells.
- T lymphocytes that play roles in cell immunity and stimulating B lymphocytes.
- Natural killer cells.

9.3.2 Dendritic Cells

9.3.2.1 What Do Dendritic Cells Do?

Dendritic cells are **antigen-presenting cells** (**APC**). They are characterized by their branched cytoplasmic projections (dendrites) (see Figure 9.3). These can be in contact with large numbers (potentially thousands) of surrounding immune cells, including epithelial cells, natural killer cells, neutrophils, and T lymphocytes. These APC have an important role in the **adaptive immune response**. Dendritic cells convert pathogen proteins to antigens (small peptides) that are presented. In the process of presenting the antigen, there is also binding to the **major histocompatibility protein** (MHC).

Dendritic cells are located in the blood, the lymph glands, the thymus (see Figure 9.3), and the skin (Langerhans cells), together with the lining of the respiratory and gastrointestinal tract organs.

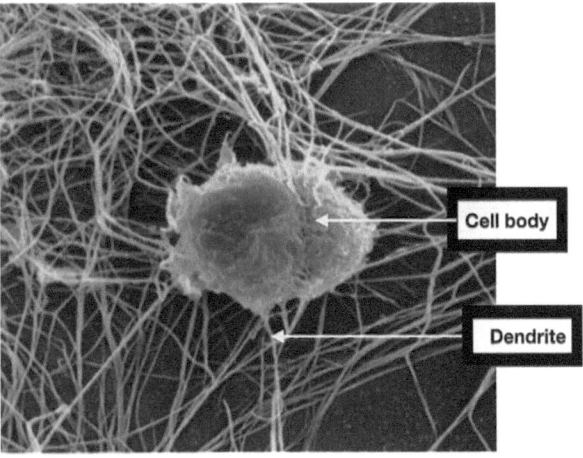

Figure 9.3 Dendritic cell with cell body and multiple dendritic cell processes (Aszakal Creative Commons Attribution-Share Alike 3.0)

9.3.3 Natural Killer Cells

9.3.3.1 What Do Natural Killer Cells Do?

Natural killer cells (NK cells) play a role in the early response to viruses and tumor cells. NK cells are **innate sentinels** and **cytotoxic lymphocytes**. As their name implies, NK cells kill:

- Tumor cells.
- Virus-infected cells.
- (Nonactivated) dendric cells, activated T-cells, and hyper-stimulated macrophage.

The mechanism by which NK cells kill distressed cells are the following:

- Being in close proximity of the target cell.
- The protein, perforin, forms holes in the plasma membrane of target cells, allowing the NK granules to enter the target cell.
- Inserting lysosomes into the target cell membrane.
- Inducing apoptosis in the target cells.

NK cells also produce cytokines. For instance, **interferon γ** is released in response to stimulation of the **toll-like receptors** (TLRs). Interferon γ is a cytokine (it is further discussed below). In addition, NK cells play a regulatory role in the immune system, activating macrophage, dendric cells, and T lymphocytes.

9.3.3.2 How Do Natural Killer Cells Recognize Virally Infected or Tumor Cells?

Normal cells have the **major histocompatibility protein class I** (MHC I) on the outside surface of the plasma membrane. In contrast, virally infected or tumor cells lack MCH I. The presence of MHC I indicates that the cell is "self" and healthy. NK cells have receptors to MHC I. When NK cells are in very close proximity to a healthy cell, MCH I on the heathy cell binds to the **MHC-I receptor** on the NK cell and the NK cell is suppressed. In the absence of MHC I, NK cells are activated. This allows NK cells to recognize healthy cells.

9.4 Innate (Nonspecific) Immunity

9.4.1 Introduction to Innate Immunity

Innate (or nonspecific) immunity is broad-spectrum and responds to pathogens upon exposure. Although innate immunity is present before invasion of a pathogen, it can be further activated. There are four major types of innate immunity:

- Barriers to infection.
- Physiological and chemical protection such as antimicrobial peptides and **reactive oxygen species** (ROS) that kill pathogens.
- Cellular protection, such as phagocytosis of pathogens by neutrophils, monocytes, macrophages, and natural killer cells.
- Pro-inflammatory cytokines, such as interleukin 1 (IL-1) and tumor necrosis factor-α (TNF-α) released by macrophages and other cells.
- Induction of fever by pyrogens.

9.4.2 Barriers to Infection

9.4.2.1 Introduction to Barriers to Infection

Barriers to infection can be physical, such as the skin or chemical barriers.

9.4.2.2 Physical Barriers

The skin is a physical barrier with tight junctions between the keratocytes. In addition, there is a layer of sebum that acts as a chemical barrier (also discussed in Chapter 10). Another physical barrier is the **mucosal membranes** in the gastrointestinal tract, the urogenital tracts, lungs, and trachea. The epithelial lining cells act as a physical barrier. This epithelium in the respiratory tract and gastrointestinal tract is overlaid by mucus and, thus, forms a chemical and physical barrier. The mucus can be moved by cilia, for instance, in the trachea. The blood−brain barrier represents another anatomical barrier protecting the brain from pathogens.

9.4.2.3 Hydrochloric Acid in the Stomach

There are high concentrations of hydrochloric acid in the gastric juice in the stomach. The parietal cells secrete the gastric juice in response to histamine and gastrin. This acidic environment kills bacteria, including many pathogenic bacteria.

9.4.3 Antimicrobial Peptides

Neutrophils and other phagocytic cells produce antimicrobial peptides such as defensins, kinocidins, and thrombocidins. These not only kill bacteria but also act as signaling molecules. Antimicrobial proteins are produced by natural killer cells, cytotoxic T lymphocytes, and platelets.

9.4.4 Oxidative or Respiratory Burst

Neutrophils and monocytes produce **respiratory bursts** (or **oxidative bursts**) of **reactive oxygen intermediates** (ROI; also known as reactive oxygen species, or ROS) to kill bacteria and viruses. Neutrophils and monocytes synthesize ROI/ROS from molecular oxygen (see Figure 9.4). They release superoxide anion radicals together with hydrogen peroxide and hypochlorous acid. This is part of the innate immune system. Excess production of the ROI/ROS can damage the neutrophils.

Figure 9.5 shows the duration of various aspects of the innate or nonspecific immunity.

9.4.5 Iron and Immunity

Bacterial proliferation requires free iron (ferrous ions). Transferrin in the blood plasma binds ferrous ions preventing bacterial growth. Bacteria can replicate in cells. One strategy used to reduce such replication is decreasing iron concentrations. For instance, IFN-γ decreases iron concentrations in macrophages suppressing bacterial multiplication.

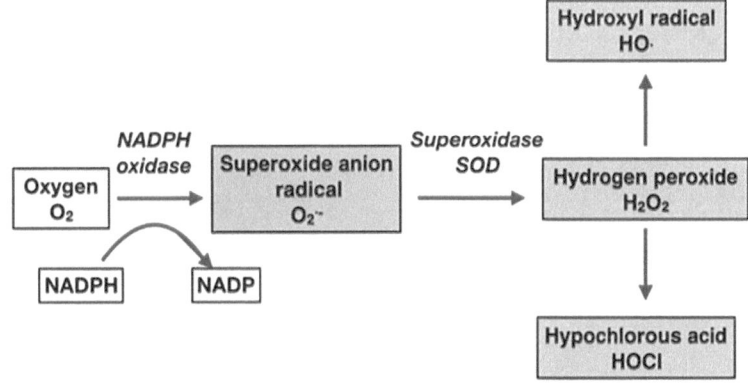

Figure 9.4 Production of reactive oxygen intermediates (ROI) (also known as reactive oxygen species [ROS]). The ROIs include superoxide anion radical (O_2^-), hydrogen peroxide hydroxyl radical, and hypochlorous acid. ROIs are released by activated neutrophils and they function to kill bacteria and viruses. [Key: NADP = nicotinamide adenine dinucleotide phosphate; NADPH = Reduced nicotinamide adenine dinucleotide phosphate.]

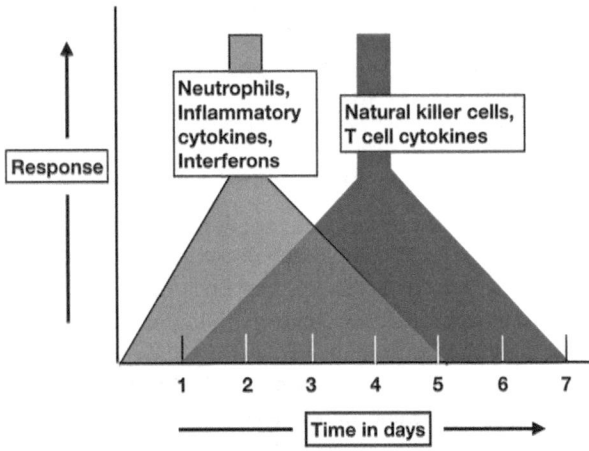

Figure 9.5 Timing of innate immune responses to invasion by a pathogenic organism.

9.4.6 Pattern Recognition Receptors (PRR)

Pattern Recognition Receptors (PRR) are expressed in neutrophils and other cell types. These recognize/bind to sites frequently found on/in pathogens. Binding of the pathogen to the PRR activates neutrophils or other cell types. Examples of PRR are the following:

- **C-type lectin receptors (CLRs)**: One type of CLR, Dectin-1, recognizes β-glucans; this being a carbohydrate on the cell surface of many fungi.
- Toll-like receptors (TLR).

Activation of phagocytosis is achieved by pathogen binding to PRR followed by signal transduction via integral immunoreceptor tyrosine-based activation (ITAM)-like motif in the cytoplasmic domain of the PRRs. The PRRs differentiate between self and non-self by binding the pathogens via pathogen-associated molecular patterns (PAMPs), or to damaged self with damage-associated molecular patterns (DAMPs), or to "altered self" via tumor-associated molecular patterns (TAMPs).

Toll-like receptors (TLR) are membrane associated. Different TLR recognize different nucleic acid PAMPs such as the following:

- TLR3—double stranded RNA.
- TLR7 and TLR8—single stranded RNA.
- TLR9 recognizes DNA.

Cytosolic DNA can be either of the following:

- Non-self—coming from pathogenic bacteria or viruses.
- Self—damaged DNA.

While TLRs do not activate phagocytosis alone, they do increase the efficiency of phagocytosis.

9.5 Cytokines

9.5.1 Introduction to Cytokines

Cytokines are chemical signals that control cells of the immune system. They also influence other tissues. They are soluble proteins or glycoproteins that do not cross the cell membrane and, therefore, bind to receptors on the surface of cells. They are produced by leukocytes (white blood cells) and other tissues. There are at least five subtypes of cytokines:

1. Interleukins (IL).
2. Interferons (IFN): IFN-α (alpha), IFN-β (beta), IFN-γ (gamma), IFN-δ (delta), IFN-ε (epsilon), IFN-κ (kappa), IFN-λ (lambda), IFN-τ (tau), and IFN-ω (omega).
3. Chemokines.
4. Tumor necrosis factors (TNF-α).
5. Growth factors, e.g., granulocyte colony stimulating factor (GCSF). This growth factor, GCSF, stimulates hematopoietic stem cells to become neutrophils.

Cytokines can cause **inflammation**. These are called **pro-inflammatory cytokines**. Examples of pro-inflammatory cytokines include the following:

- Interleukin: 1β (IL-1β).
- Interleukin: 6 (IL-6).
- Interferons: γ (IFN-γ).
- Tumor necrosis factor: α (TNF-α)
- Granulocyte-macrophage colony-stimulating factor (GM CSF).

There are also **anti-inflammatory cytokines** that decrease inflammation. They include interleukin-10 (IL-10), IL-11 and IL-13.

9.5.2 Interleukins

Interleukins are signaling proteins produced by specific leukocytes together with other cells. They are soluble and can be found in the circulation. Interleukins bind to specific receptors on the plasma membrane inducing the activation, migration, adhesion, and differentiation of other leukocytes. Interleukins can be either pro-inflammatory or anti-inflammatory. Frequently, interleukins act on the same type of leukocytes in an **autocrine** and **paracrine** manner.

9.5.3 Interferons

Interferons (IFN) are a family of proteins. They act to suppress replication of viruses and to act as signaling proteins in the immune system. IFN acts by binding to specific receptors on the cell membrane. IFN induces an antiviral environment in cells inhibiting viral replication. Some IFNs exert an antibacterial effect also (see Textbox 9.3 for a deeper dive into interferons). Interferons play both innate and adaptive immune roles. For instance, IFN-γ is part of the host response to intracellular bacteria.

9.5.4 Chemokines

There are about 50 chemokines. These are **chemotaxic** cytokines. They induce migration of leukocytes to where they are needed (chemoattraction) (see Figure 9.7). There are roles of chemokines in both innate and adaptive immunity.

Textbox 9.3   A Deeper Dive Into Interferons

There are three classes of this cytokines:

- **Type 1:** Type 1 IFN are produced by immune cells including neutrophils and monocytes and others cells. These IFN are a component of innate immunity due to their antiviral effects. They also participate in the initiation of adaptive immunity.
  - Interferon-alpha (IFN-α) encoded by about 12 genes.
  - Interferon-beta (IFN-β).
  - Interferon-tau (IFN-τ).
  - Interferon-omega (IFN-ω).
- **Type 2**: IFN-gamma (IFN-γ). Type 2 IFN are produced by immune cells (macrophages, natural killer cells and activated cytotoxic T lymphocytes) but not non-immune cells. This is produced by natural killer cells and T lymphocytes. IFN-γ evokes an anti-viral environment in cells.
- **Type 3**: Interferon-lambda (IFN-λ or IFNL). Type 3 IFN are produced by immune cells and others cells. Nucleic acids in the cytoplasm are detected by PRRs with the release of IFN-λs. IFN-γ induces reactive oxygen species killing intracellular bacteria, parasites, and fungi. It also has anti-tumor properties.

Double stranded RNA, a by-product of viral replication, induces the expression (production) of IFN. Binding of IFN to their receptors induces the expression of antiviral proteins inside the cell. These inhibit viral protein translation and destruction of virions by activating RNase.

IFN Expression in Response to Pathogens or Dead Cells
Toll-like receptors (TLR) are present on the plasma membrane and the endosome. TLR are pattern-recognition receptors (PRRs) recognize by binding to pathogen-associated molecular patterns (PAMPs). When PAMPs bind to TLR, this leads to expression (production) of type 1 IFNs.

Cyclic GMP–AMP Synthase (cGAS) Is a Stimulator of Interferon Genes (STING) Pathway
The **cGAS/STING** pathway stimulates INF expression (see Figure 9.6). Cyclic GMP–AMP synthase (cGAS) is activated by double stranded DNA in the cytoplasm. The double stranded

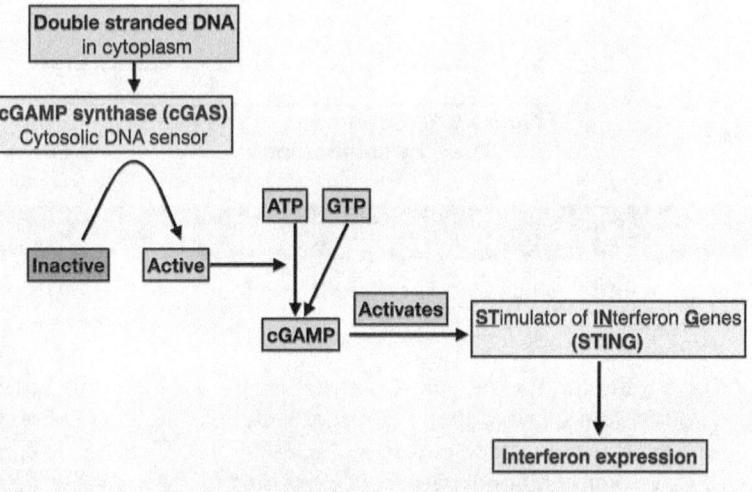

Figure 9.6   The Cyclic GMP–AMP synthase (cGAS)/Stimulator of Interferon Genes (STING) pathway resulting in interferon expression.

(continued)

Textbox 9.3 (continued)

DNA may be the result of cell death, bacteria, DNA viruses, and RNA viruses following reverse transcription. cGAS catalyzes the formation of the cyclic dinucleotide, cyclic 2'3'-GMP-AMP guanosine monophosphate-adenosine monophosphate (cGAMP) from GTP and ATP. In turn, cGAMP activates **St**imulator of **I**nterferon **G**enes (STING). This causes the expression of IFNs and consequently anti-viral effects.

Cyclic GMP–AMP synthase (cGAS) is activated by double stranded DNA in the cytoplasm. cGAS catalyzes the formation of the cyclic dinucleotide, cyclic 2'3'-GMP-AMP guanosine monophosphate-adenosine monophosphate (cGAMP) from GTP and ATP. cGAMP activates Stimulator of Interferon Genes (STING) which, in turn, provokes interferon expression, production, and release with these exerting antiviral effects.

IFN-τ is secreted by trophectoderm cells of the conceptus.

Immune Therapy
Immune therapy with IFNs is being used in people. In addition, they are used in domestic animals. For instance, IFN-ω is a treatment of canine parvoviral enteritis together with infections from feline leukemia virus, and feline immunodeficiency virus.

Other Roles for Interferons
IFN-τ acts as a **maternal recognition protein** for pregnancy in ruminants, supporting the establishment of pregnancy. In addition, IFN-τ exerts an anti-inflammatory effect.

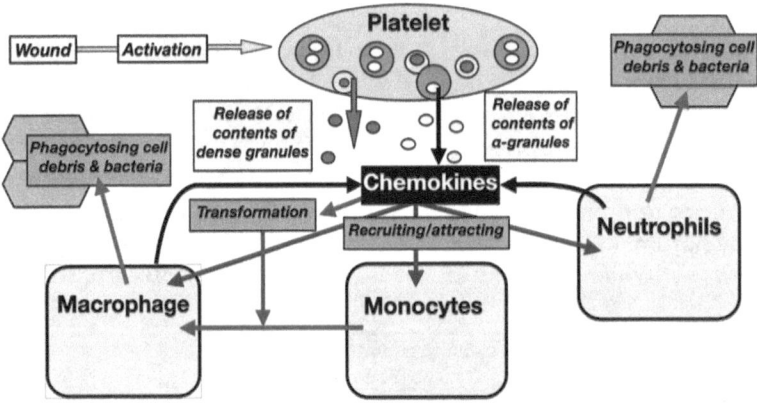

Figure 9.7 Schematic of the recruitment of phagocytes mediated by chemokines.

Textbox 9.4 A Deeper Dive Into Chemokines

There are about 50 chemokines within the four sub-classes:

- CCL containing two adjacent cysteine amino acid residues close to the N terminal. L stands for ligand that binds to receptors.
- CXCL containing almost adjacent cysteine amino acid residues with another amino acid residue between the cysteines (close to the N terminal). L stands for ligand that binds to receptors.
- CX3CL containing close cysteine amino acid residues with three other amino acid residues between the cysteines (close to the N terminal). L stands for ligand that binds to receptors.
- XCL containing one cysteine amino acid residue close to the N terminal. L stands for ligand that binds to receptors.

(continued)

> **Textbox 9.4 (continued)**
>
> The chemokine system is unique to chordates being found in fish, mammals and birds. Chemokines are synthesized and subjected to post-translational modification and released into the extracellular space when needed.
>
> **Receptors for Chemokines**
> There are about 20 receptors for chemokines depending on species. Chemokines bind to both conventional chemokine receptors (CCKRs) and atypical chemokine receptors (ACKR). CCRs and ACKRs are found on the cell membrane of leukocytes but also on other cells.
>
> **Activity of Chemokines**
> Binding of a chemokine to its receptor on leukocytes can lead to migration to where they are needed, adhesion (see Figure 9.7) and other effects, including the formation of new blood vessels (angiogenesis). For instance, CCR1 and CCR2 are expressed by both monocytes and macrophage. While CCR3 is expressed by eosinophils.
>
> ACKR2 acts as a scavenger by binding **chemokine ligands** and, thus, enabling their internalization and breakdown. Malaria parasites (some species of *Plasmodium*) enter erythrocytes using the ACKR1.

9.6 Phagocytes, Platelets/Thrombocytes, and the Immune Response

Platelets play a role in the immune response releasing histamine and serotonin, adhesion molecules such as fibrinogen, chemokines attracting and recruiting neutrophils to the specific site, and stimulating differentiation of monocytes into macrophage. In birds, the nucleated thrombocytes, when activated, also exhibit increased expression of immune-related proteins and peptides such as interleukin (IL)-6. Thrombocytes are activated via toll-like receptors (TLR).

9.6.1 Overview of Phagocytes, Platelets/Thrombocytes, and the Immune Response

Phagocytes play an important role in **host innate immune defense** response. They act as sentinels for infection or wounding. Subsequently, phagocytes engulf pathogens. Phagocytes are recruited to a site of inflammation (wounding and/or infection). They are attracted by chemokines binding to specific receptors on the cell surface. There are multiple mechanisms:

- Platelets and multiple cells around a wound induce the release of chemokines.
- Phagocytes are first captured by **endothelial cells** (lining blood vessels) close to a site of infection. After rolling, the phagocytes stop and move between the endothelial cells to the site of wounding/infection due to chemokines.
- Phagocytes release chemokines that attract other phagocytes either of the same or a different type.

Phagocytosis is induced by pattern-recognition receptors (PPRs) on phagocytes binding pathogen-associated molecular patterns (PAMPs). Monocytes transform into macrophage and dendric cells in a non-reversable manner.

9.6.2 Platelets and the Immune Response

Platelets play a role in the immune response releasing histamine and serotonin, adhesion molecules such as fibrinogen, chemokines attracting and recruiting neutrophils to the specific site, and stimulating differentiation of monocytes into macrophage. In birds, the nucleated thrombocytes, when activated, also exhibit increased expression of immune related proteins and peptides such as interleukin (IL)-6. Thrombocytes are activated via toll-like receptors (TLR).

Platelets contain the following:

- Mitochondria generating ATP.
- Three types of granules:
  - α-granules containing proteins including blood clotting factors V, XI, and XIII, growth factors and chemokines, e.g., CXCL1, CXCL4. These are the most common granules in platelets.
  - Dense granules containing small molecules including serotonin, ATP, ADP, and calcium.
  - Lysosomes with proteolytic enzymes.
- RNA enabling protein synthesis.

Both the platelets and granules are surrounded by membranes. Platelets release granules containing chemokines when activated by the process of exocytosis.

Platelets initiate the four phases in response to wounding and healing:

- Hemostasis (blood clot formation).
- Inflammation with recruitment of phagocytic cells—neutrophils, monocytes, and macrophage.
- Proliferation.
- Healing with fibroblasts laying down a new extracellular collagen matrix.

Platelets release cytokines, e.g., CXL1 and CXCL8 with CXCL8 attracting neutrophils to the wound site.

9.6.3 Movement of Leukocytes in Blood Vessels and Recruitment to Sites of Wounding or Infection

When activated leukocytes move to the site of wounding or infection. The phases of leukocytes movement in blood vessels and recruitment to sites of wounding or infection are the following:

- Leukocytes are tethered to the surface of endothelial cells lining blood vessels. This is due to selectins.
- Leukocytes roll along the surface of endothelial cells lining blood vessels. This is due to selectins.
- Leukocyte rolling slows, then stops. This is due to chemokine and selectin signaling.
- Leukocytes polarize and move between the endothelial cells to the site of wounding or infection.

Textbox 9.5   A Deeper Dive Into Recruiting and Differentiating Phagocytes

Neutrophils Passage to Sites of Wounding or Infection

The phases of passage to sites of wounding or infection are the following:

- Neutrophils flowing in the circulation are first captured by endothelial cells on the inner surface of blood vessels close to a site of infection/wounding. The endothelial cells have **selectins** on their cell membranes and neutrophils have selectin ligands.
- Neutrophils are tethered to the surface of endothelial cells.
- Neutrophils roll along the surface of endothelial cells lining blood vessels.
- Neutrophils rolling slows, then stops. This is due to chemokine signaling.
- Neutrophils polarize and move between the endothelial cells to the site of wounding or infection.
- Neutrophils are attracted to a wound by the platelet derived chemokine, CXCL8.

Neutrophils are attracted to a wound by the platelet derived chemokine, CXCL8.

Monocytes and Infection

The spleen is a reservoir for monocytes with more than half in the body present in the spleen. The rest are present in the circulation. The hormone **angiotensin II** mobilizes monocytes to exit the spleen. Binding of a chemokine to its receptor, CCR2, leads to out-migration of monocytes from the bone marrow.

There are two subsets of monocytes:

- Inflammatory monocytes with high concentrations of CCR2 on their surface. These monocytes can transform into macrophage.
- Monocytes with low concentrations of CCR2 on their surface. These can transform into dendritic cells.

Monocytes are present in the blood and are in close association with endothelial cells. Circulatory monocytes are a biomarker for inflammatory diseases. Those sites that are infected with bacteria release CCL7 and this induce monocyte migration. Migration of monocytes from the bloodstream to the site of wounding or infection is induced by chemokines (CCL2 and CCL7) binding to a chemokine receptor (CCR2).

Inflammatory monocytes have PPRs including TLR that bind PAMPs on/in pathogens. Such binding induces phagocytosis. Some inflammatory monocytes produce CCL3 to recruit natural killer cells.

9.7 Phagocytosis (Cellular Innate Immunity)

The word *phagocytosis* is derived from ancient Greek, with *phago* meaning to "eat or devour," *cytos* meaning "a cell," and *osis* meaning "a process." Phagocytosis is the process of engulfing pathogenic bacterium, dead cells, or tumor cells (see Figure 9.8 and 9.9 and Textbox 9.6).

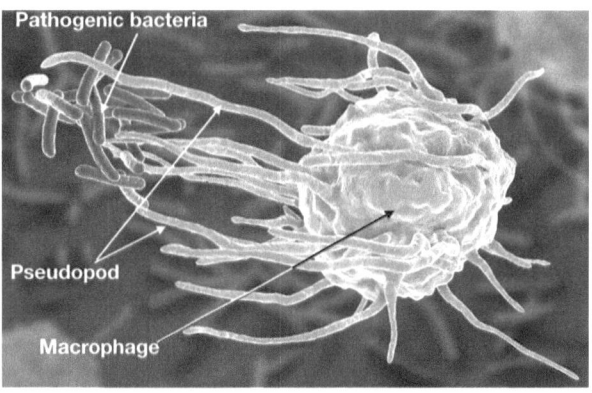

Figure 9.9 Macrophage engulfing tuberculosis bacteria *Mycobacterium tuberculosi*. (Shutterstock/Kateryna Kon)

Phagocytosis is a component of both the innate and specific immune system.

Examples of phagocytes are neutrophils (see Figure 9.8), monocytes, macrophage, and dendritic cells. Neutrophils and macrophages have a higher capacity for phagocytic than monocytes. Therefore, differentiation of monocytes into macrophage includes increasing their capacity for phagocytosis (see Figure 9.9).

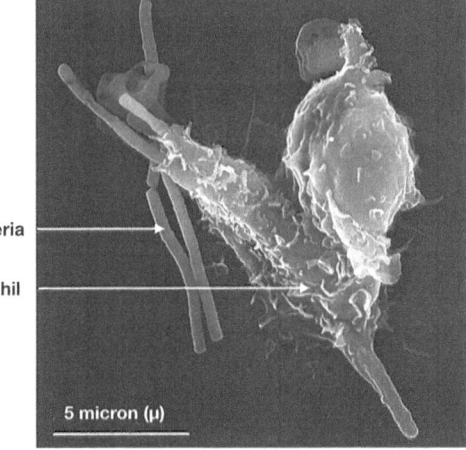

Figure 9.8 Neutrophil engulfing an anthrax bacterium in the process of phagocytosis. The image is a scanning electron micrograph with colorization. (Volker Brinkmann, CC BY 2.5)

9.7.1 What Do Phagocytes Do?

The process starts with the **extension of pseudopods** (see Figure 9.9). A pseudopod or series of pseudopodia

---

Textbox 9.6   Process of Phagocytosis

Figure 9.10 summarizes phagocytosis. The process can be viewed as the following:

1. Pathogen binds to receptors on the surface of the phagocyte plasma membrane (on as pseudopodia).
2. Pathogen is engulfed by phagocyte and placed in phagosome.
3. The pathogen containing phagosome fuses with a lysosome forming a phagolysosome.
4. The pathogen in the phagolysosome is broken down by lysosomal enzymes.
5. Debris from the phagolysosome is voided from the phagocyte by exocytosis.

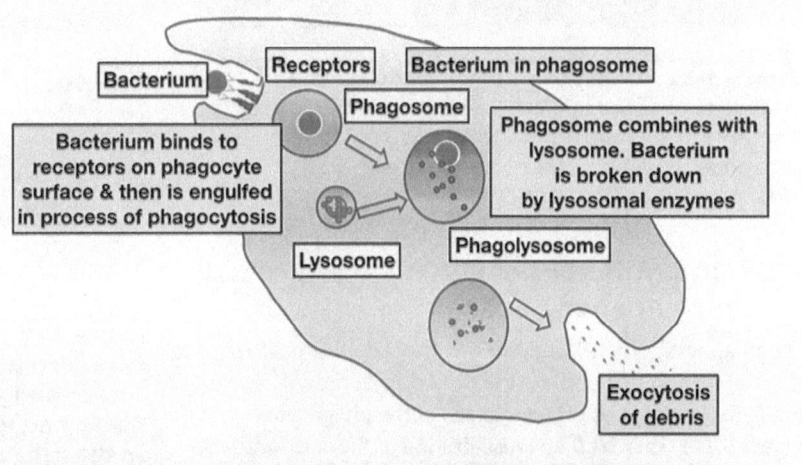

Figure 9.10 Schematic of the process of phagocytosis with a bacterium engulfed by a phagocyte. (Courtesy Graham Colm Creative Commons Attribution-Share Alike 3.0)

make contact with the target. The target is engulfed by the phagocyte, resulting in phagosome containing the pathogen or senescent cell or tumor cell (see Figure 9.10). The next step is the fusing of the destructive intracellular organelle, the lysosome, with the phagosome, resulting in a phagolysosome. The bacterium is killed and destroyed by a respiratory burst and broken down by lysosomal enzymes, including lysozyme and proteases (see Figure 9.10).

9.7.2 How Do Pseudopods on Phagocytes Discern That the Adjacent Cell or Organism Is a Pathogen?

There are receptors called pattern recognition receptors (PRR) that bind pathogens (see Section 9.4.5).

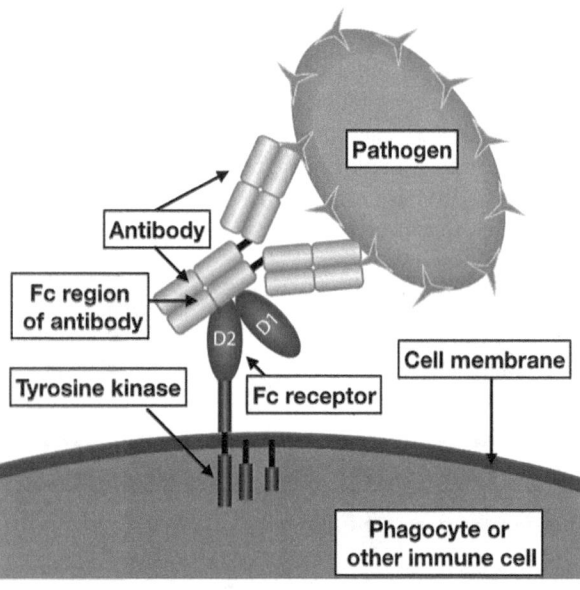

Figure 9.11 When there is an antibody to a pathogen antigen, Fc receptors on the surface of the immune system bind to the Fc region of the antibody. This facilitates phagocytosis. (Adapted from Ciar, public domain)

9.7.3 Do Antibodies or Other Factors Influence Phagocytosis?

When an antibody is bound to a pathogen, phagocytosis is facilitated. This is because the Fc region of the antibody binds to Fc receptors on the surface of the phagocyte (see Figure 9.11). There is then cross-linking of the antibodies, increases in intracellular calcium concentrations, and further activating of the phagocytes. This is discussed below in Section 9.9 and Section 9.10.

9.8 Opsonization

9.8.1 Introduction to Opsonization

Opsonization is the mechanism by which the immune system "labels" or "marks" pathogens to facilitate their elimination. The opsonized pathogen binds to pathogen recognizing receptors and ultimately increases the susceptibility of the pathogen to phagocytosis by both macrophage and natural killer cells. This coating of the pathogen can be viewed as making the pathogen more "appetizing" to the phagocytes, rather like "sprinkles on ice cream."

9.8.2 Opsonins

Opsonins are soluble factors that are part of the innate immune system. The word *opsonin* is derived from the Latin "to buy provisions." Opsonins label pathogens or dead cells to be targets of phagocytosis. Opsonins include **immunoglobulin G (IgG), fibronectin, complement,** and **mannose-binding lectins** (see Figure 9.12). There

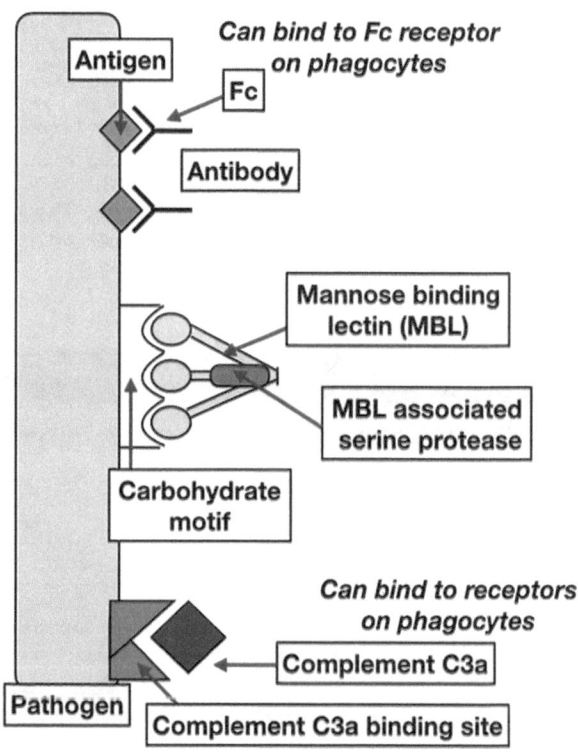

Figure 9.12 Opsonins binding to pathogens increasing phagocytosis. Specific antibodies (IgG) bind to antigens on the surface of pathogens. The Fc part of the antibodies bind to Fc receptors on the cell wall of phagocytes. Mannose-binding lectin binds to carbohydrate motifs on the surface of pathogens. Complement C3a binds to both the pathogen and to receptors on phagocytes.

are opsonin receptors on the surface of phagocytes. The binding of opsonin to opsonin receptors increases the efficiency of phagocytosis.

9.9 Complement

9.9.1 Introduction to Complement

Complement represents an ancient protective system that is found not only throughout the vertebrates but also in other chordates and, even, in arthropods such as horseshoe crabs. Complement consists of over 30 different proteins and peptides that are found in the blood and on cell surfaces. Complement has traditionally been viewed as part of the innate immune system complementing antibodies to combat pathogens. However, recent research places complement in both the innate and adaptive immune systems or bridging between the systems.

9.9.2 Functions of Complement

The functions of complement include the following:

- **Opsonization/phagocytosis**: facilitating phagocytosis of pathogens, apoptotic cells (cells undergoing apoptosis or programmed cell death), and cell debris.
- **Cell lysis**: killing cells or pathogens by creating holes in their plasma membranes. Complement protein (C3b) initiates the cascade, forming membrane attack complex (consisting of multiple activated complement proteins) that "punches a hole" in the cell walls of pathogens, killing them in the process of cytolysis.
- **Inflammation**: producing pro-inflammatory responses and, consequently, inducing inflammation that promotes **vascular permeability**, **leukocyte recruitment**, **chemotaxis of leukocytes**, **migration off leukocytes**, and release of other agents that promote inflammation (e.g., histamine).
- **Regulation**: inhibition of complement cascade (see Figure 9.11).
- Other functions include:
  o Cell differentiation and recruitment.
  o Modulation of immune cell migration.
  o Regulation of B- and T-cell functioning.
  o Leukocyte adhesion.
  o Synaptic pruning.

9.9.3 Activation of Complement

There is a cascade of complement with **precursor zymogens** subject to **proteolytic cleavage** by convertase generating the following (see Figure 9.13):

- **Active complement proteins (enzymes)**.
- **Anaphylatoxins**—proinflammatory complement peptides.

There is an analogous cascade with blood clotting (see Section 2.15.2).

There are three complement pathways. These converge with the activation of C3 complement:

1. **Classical**—antibody dependent and, consequently, a part of the adaptive immune system. The formation of antibody–antigen complexes induce a series of complement proteins to be activated, leading to the cleavage of complement protein (C3) to C3a and C3b (see Figure 9.13).
2. **Alternative**—antibody independent and, consequently, a part of the innate immune system. In the presence of the lipid carbohydrate complexes

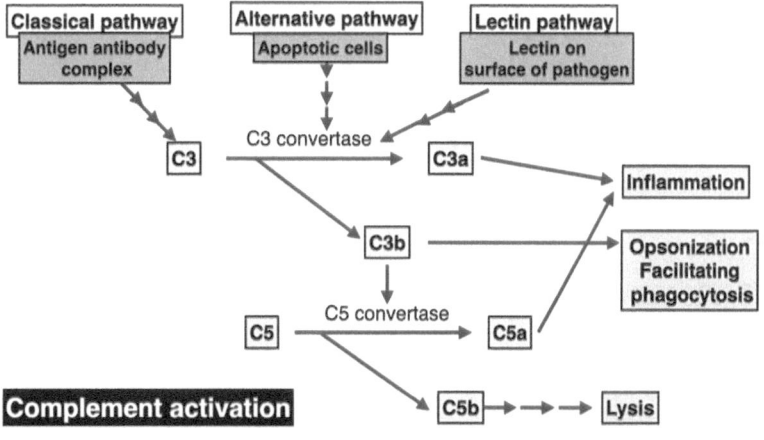

Figure 9.13 Pathways of complement activation.

(lipopolysaccharide) on the cell wall of pathogens, complement protein (C3) splits to form complement proteins, C3a and C3b (see Figure 9.13).
3. **Lectin pathway**—antibody independent and, consequently, a part of the innate immune system. The liver produces proteins that recognize and bind to pathogens and, thereby, are activated. There are then a series of complement proteins that are activated as proteases. Complement proteins, C3a and C3b, are generated (see Figure 9.13).

Textbox 9.7 A Deeper Dive Into Complement

1. **Classical pathway**
   Antibody-antigen complexes induce the activation of complement component protein C1. In turn, activated C1 cleaves C4 to C4a and C4b. Moreover, activated C1, together with C4b, acts as a protease to activate C2 to C2a and C2b. A combination of these activated complement proteins act as C3 convertase, generating to active C3a and C3b.
2. **Alternative pathway**
   In the presence of the lipid carbohydrate complexes on the cell wall of pathogens, complement protein (C3) splits to form complement proteins, C3a and C3b.
3. **Lectin pathway**
   Mannose-binding lectin (MBL) is released from the liver in response to the presence of pathogens (see Figure 9.14). It plays an important role in pattern recognition, specifically carbohydrate recognition, of pathogens particularly initially during invasion by pathogens (Figures 9.14 and 9.15).

   MBL binds a series of carbohydrate sequences (or motifs including mannose and N-acetyl-D-glucosamine) on the surface of pathogens (see Figure 9.15). This allows interaction with bacteria, viruses, fungi, and protozoa. MBL circulates in the blood. When MBL presence binds to a pathogen, there are conformational changes in MBL, activating MBL-associated serine proteases (MASP) (see Figures 9.14 and 9.15). This is followed by activation of complement proteins. Specifically, MASP cleaves inactive complement proteins, C2 and C4, generating active forms—C2a, C2b, C4a, and C4b. Proteins C2a and C4b combine to form C3 convertase, a proteolytic enzyme. This activates complement C3, cleaving it to C3a and C3b. C3a C3b binds to the surface of pathogens and, ultimately, facilitates phagocytosis (see Figure 9.16).

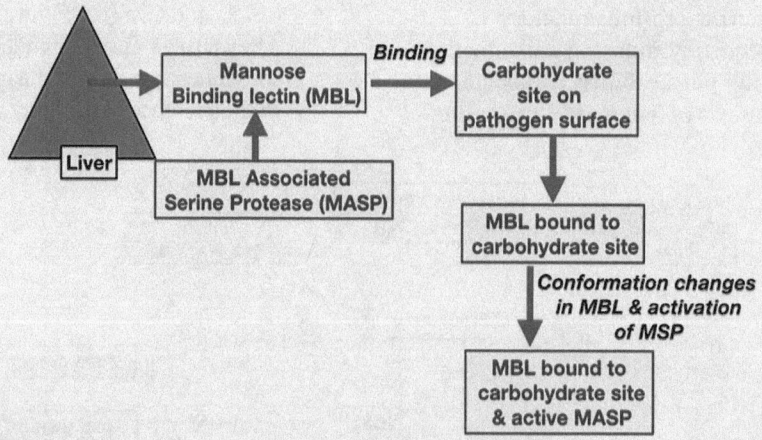

Figure 9.14 Mannose-binding lectin (MBL) is part of the complement system. MBL is released from the liver into the circulation. MBL binds a carbohydrate motif on the surface of pathogens. This causes confirmation changes in MBL and MBL-associated protein (MASP). The shift in the structure of MASP activates it.

(continued)

Textbox 9.7 (*continued*)

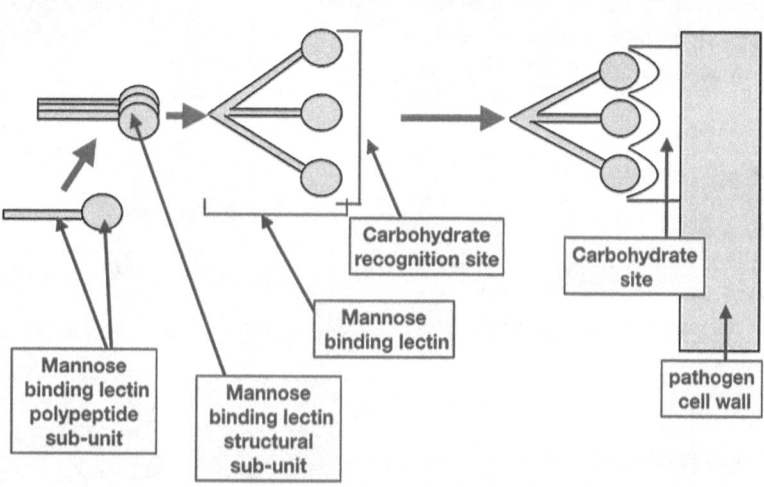

Figure 9.15 Mannose-binding lectin (MBL) is produced as a peptide subunit. Subunits come together to form structural subunits. The structural subunits come together to form mannose-binding lectin. In turn, this binds to a carbohydrate motif on the surface of pathogens.

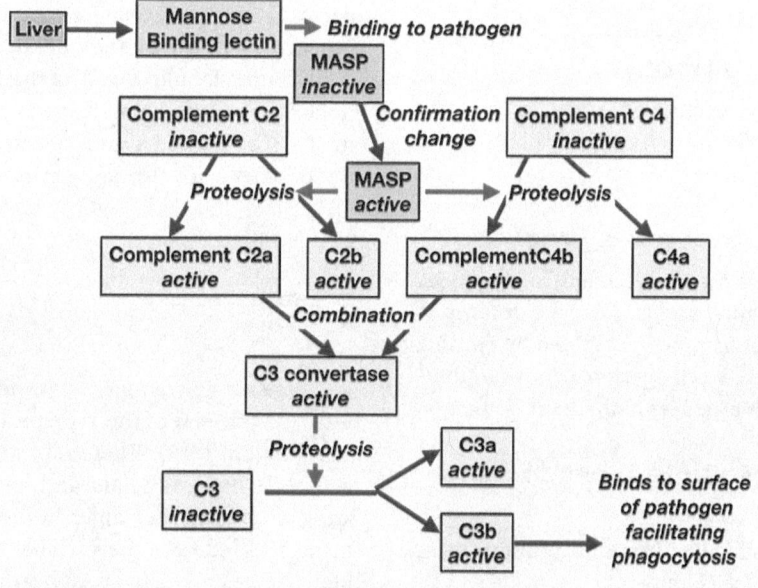

Figure 9.16 The lectin pathway of complement. [Key: MASP = MBL-associated serine proteases.]

9.10 Introduction to Adaptive (or Specific) Immunity

Adaptive (specific) immunity is immunity that is a defense to a specific antigen on a pathogen or non-health cell.

9.11 Cell-Mediated Immunity

Cell-mediated immunity does not involve antibodies. Instead, it includes activation of phagocytes (see above) and antigen-specific cytotoxic T lymphocytes. There are three types of cells involved in cell-mediated immunity:

- Cytotoxic T cells (or killer T cells) kill (lyse) and destroy cells with abnormal surfaces, such as virus-infected cells and tumor cells.
- Dendritic cells induce the activity of cytotoxic T cells.
- Regulatory T cells modulate a coordinated immune response.
- T helper cells are directly required for a coordinated immune response.

9.12 Humoral (Antibody) Mediated Immunity—Antibodies/Immunoglobulin

9.12.1 What Is an Antibody?

Antibodies or immunoglobulins are Y-shaped proteins with at least two binding sites. Each antibody binds only a single antigenic site or epitope.

9.12.2 Where Are Antibodies Produced?

Antibodies are produced by and secreted from plasma cells in the spleen and lymph nodes. These are **differentiated B lymphocytes** that have been activated by binding of a specific antigen. There is cell division forming a clone. The clone of plasma cells produces a single antibody.

9.12.3 Are There Multiple Types of Antibodies?

There are five types of antibodies: **Immunoglobulin (Ig)A, IgD, IgE, IgG, and IgM** (see Figure 9.17).

Immunoglobulin (Ig)A is found in saliva, tears, and milk, together with mucus membranes such as respiratory, gastrointestinal, and urinary-genital secretions. IgA is considered a **monomer** with two antigen-binding sites.

Immunoglobulin (IgD) is found on the surface of B lymphocytes. IgD is considered a dimer with four antigen-binding sites.

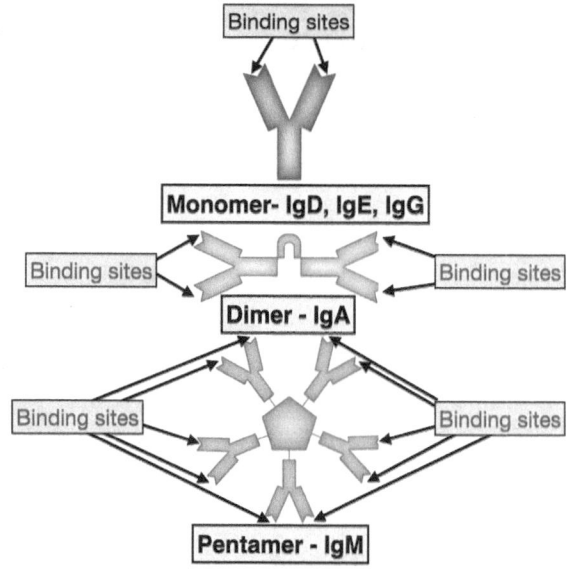

Figure 9.17 Schematic showing the types of immunoglobulins and sites that bind antigen. (Adapted from Martin Brändli, CC BY-SA 3.0)

Immunoglobulin (IgE) is found on the surfaces of mast cells. Binding of an allergen to IgE causes an inflammation response. IgE plays a role in the defense against parasites and insect/spider venom. IgE is considered a monomer with two antigen-binding sites.

Immunoglobulin (IgG) is the most abundant immunoglobulin neutralizing bacteria and viruses. It is found in the blood plasma and colostrum. IgG is considered a monomer with two antigen-binding sites (see Figures 9.17, 9.18, and 9.19). IgG is composed of four protein chains with a Y shape:

- Two heavy chains
- Two light chains.

These are linked together by **disulfide bridges** (Figure 9.18). At one end of the Y shape are two antigen-binding sites (Figure 9.19). These are specific to a single antigenic site. Both the light and heavy chains close to the antigen-binding sites differ in different antibodies; these being referred to as the **variable regions**. The rest of the light and heavy chains are nonvariable or constant (Figure 9.18). The heavy chains distant from the antigen-binding site are called the Fc (Figure 9.18). Binding of the Fc to Fc receptors plays an important role in immunity.

In poultry, the equivalent to IgG is called IgY; this being named IgY based on its presence in yolk. IgY in egg yolk is transferred into the blood of the embryo, providing the chick with passive immunity.

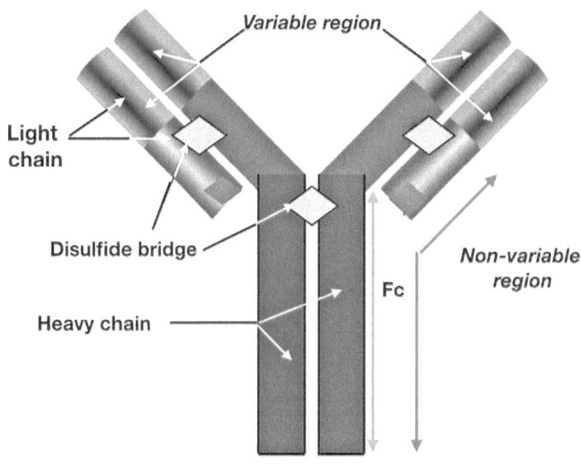

Figure 9.18 Schematic of immunoglobulin G (IgG) antibody molecule. (Adapted from DigitalShuttermonkey, CC BY-SA 3.0)

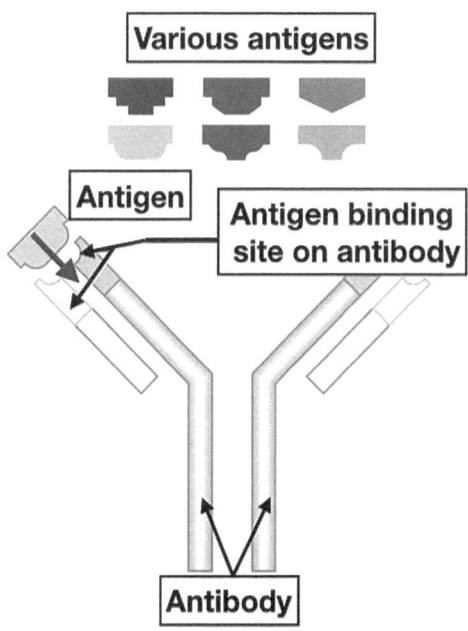

Figure 9.19 Schematic of an antibody (IgG) molecule binding to one specific antigen but not to others. (Adapted from Fvasconcellos, public domain)

Immunoglobulin (IgM) is the first antibody produced by plasma cells. It is found in blood plasma. IgM is considered a pentamer with ten antigen-binding sites (see Figure 9.20). It is the most effective immunoglobulin in opsonizing bacteria. This is due to cooperativity in the binding of IgM to bacteria due to the ten binding sites.

CHAPTER 9 Immune Function

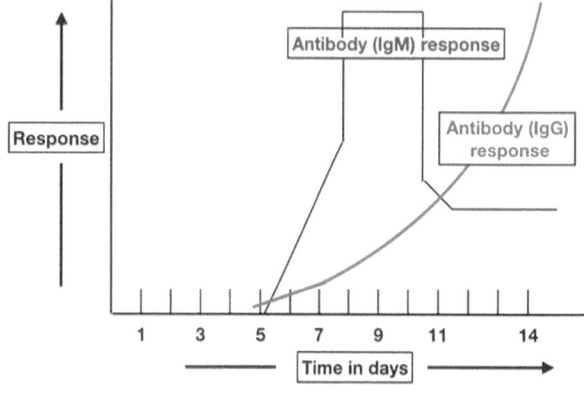

Figure 9.20 Timelines of specific or adaptive immunoglobulin (antibody) response to pathogens.

The five types of antibodies (IgA, IgD, IgE, IgG, and IgM) are produced by plasma cells. The first antibodies produced and released into the blood are IgM, peaking at about 9 days after exposure to the pathogen or challenge with a vaccine (see Figure 9.20). The more specific IgG antibodies are produced somewhat later.

9.12.4 Why and How Are There Differences in the Antigen-Binding Site Allowing Binding Different Antigens?

The answer is differences in the RNA transcripts for the light and heavy chains and, hence, the structures of the antibodies. There are two identical heavy chains and two identical light chains. There are multiple genes that encode the variable region of the heavy and light chains of immunoglobulin.

- **Five heavy chain isotypes** in IgA, IgD, IgE, IgG, and IgM.
- **Two light chain isotypes**, lambda (λ) and kappa (κ); the latter being absent in birds.

The regions of the genes for the light and heavy chains responsible for the variable region contain the following:

- **Multiple variable** or **V segments**.
- **Multiple joining** or **J segments**.
- **Constant** or **C segments**.

9.12.5 How Does an Animal Produce Billions of Different Antibodies When There Are Limited Numbers of Genes?

The simple answer is there are recombinations (rearrangements) and mutations within the variable region of the genes for both the light and heavy chains. This leads to huge numbers of different binding sites for antigens.

There are multiple mechanisms to generate huge numbers of different antibodies, each with different binding abilities including the following:

1. Recombination of the variable or V segments in the DNA of genes for both the heavy and light chains During the development of B cells. This is due to joining, or J, segments within the variable region of the antibodies.
2. Combination of different heavy and light chains during the development of B cells.
3. Somatic hypermutation during the development of B cells.
4. Unique recombinations and mutations in each B lymphocyte.

This might be viewed as analogous to combination locks with multiple possibilities but only one right answer.

9.12.6 What Do Antibodies Do?

The actions of the antibodies can be summarized by the acronym P.L.A.N. as follows:

- Precipitate the antigen and, hence, the pathogen.
- Lysis. Activation of the complement cascade by an antibody can result in the lysis of organisms or of infected cells.
- Agglutination—binding antigens and, hence, the pathogen, together.
- Neutralization. Intracellular pathogens can be neutralized by antibodies at post-attachment steps in their lifecycle or by blocking their capability to invade cells or depresses their ability to invade a cell by steric hinderance or by destabilizing the virus.

Much of the biological activity of an antibody is mediated through interactions between Fc and Fc receptors found on a number of cells important for host defense, including phagocytes.

9.13 B Lymphocytes and Antibody Production

B lymphocytes are activated by a specific antigen that binds to immunoglobulin molecules on the surface of the B lymphocytes (Figure 9.21). In addition, T helper cells

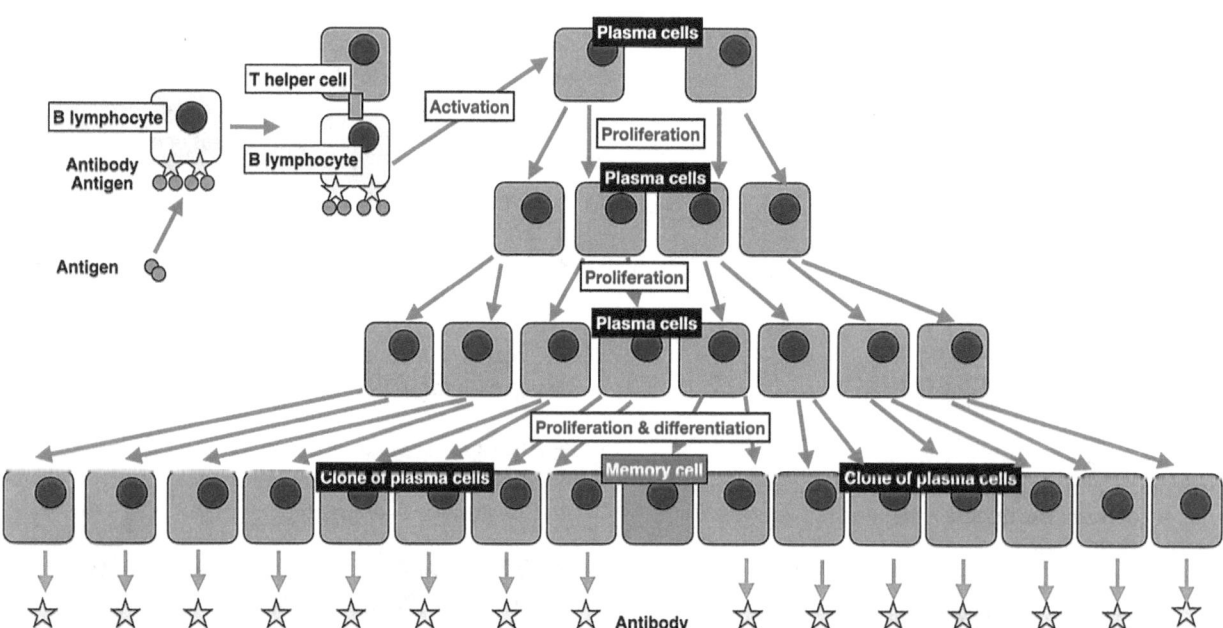

Figure 9.21 B lymphocytes are activated by a specific antigen binding to the respective immunoglobulin on the surface of the B cell. This leads to proliferation of the B cells. This generates a clone of plasma cells producing much of a single antibody (monoclonal antibody). In addition, memory cells are formed. These are ready to respond to a new exposure to the antigen, say on the surface of a pathogen.

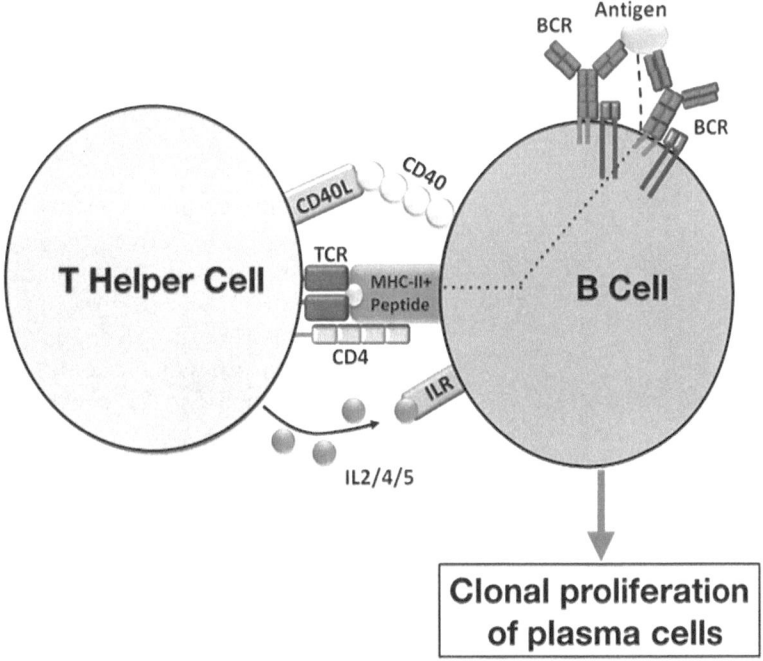

Figure 9.22 T-dependent B cell activation. T helper cells play a critical role in the activation of B lymphocytes. [Key: BCR = B cell receptor; CD4 = cluster of differentiation 4; CD40L = CD40 ligand; IL2 = interleukin 2; ILR = interleukin receptor; MHC-II = Major histocompatibility complex II; TCR = T-cell receptor.] (Adapted from Altaileopard, public domain)

are required for activation (Figure 9.22). The T helper cells bind to the B lymphocyte via CD40/CD40 ligands, cytokines [interleukins (IL) 2/4 and 5], IL receptor, T cell receptors, and major histocompatibility complex II (Figure 9.22). The activated B lymphocytes undergo proliferation, generating clones of plasma cells that all produce the same antibody (Figure 9.21).

Each B lymphocyte, and consequently the plasma cells derived from it, only produce one type of antibody with the same structures. In the presence of specific antigens, the specific B lymphocyte produces antibodies that bind it, and only these are activated. There is proliferation (multiplication) of the specific B cells and activation of these to the following:

- Plasma cells producing antibodies.
- Memory cells, such that the secondary response to the antigen is much larger than the first or primary response.

During development, most of the B cells expressing antibodies to the proteins of the animal or "self" are removed.

> **Textbox 9.8 Immunogens and Antigens**
>
> An **antigen** is a molecule that is recognized by the immune system and actives an immune response. An **immunogen** is an antigen that provokes the production of antibodies to be used against it, or the production of cell-mediated immunity.
>
> Not all antigens are immunogens. This is the case with small molecules such as hormones like testosterone. It is possible to "fool" the immune system by getting antibodies to an antigen that is not an immunogen by chemically binding the small molecule to a large protein. The bound antigen is called a **hapten**.

9.14 Vaccines

Vaccines and extensive vaccination have been invaluable in improving the well-being and quality of life of domestic animals. Vaccination increases productivity of livestock and poultry by greatly reducing mortalities and morbidities, together with reducing negative effects of pathogens on the growth and production of milk and eggs. Good management includes implementing a vaccination schedule as vaccinations are effective in provoking an immune response (see Figure 9.23).

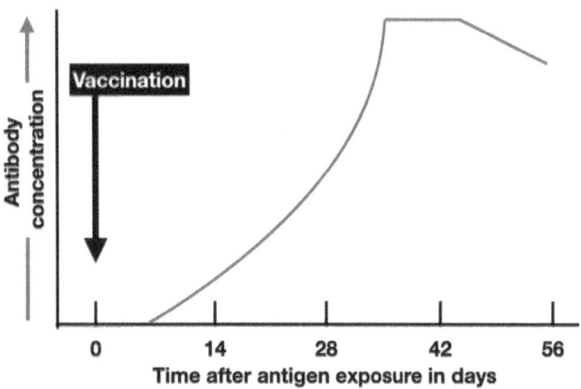

Figure 9.23 Example of an antibody response to a vaccination.

Why is the secondary response to an antigen much greater than the first response (such as a vaccination)? This is because of the memory cells produced in response to the first exposure are activated and develop into plasma cells that produce antibodies.

The key to the effectiveness of vaccines is that, in addition to their provoking an immune response, a set of memory cells are developed against antigens in the vaccine. Exposure to the pathogen activates the memory cells to divide; differentiate, e.g., into plasma cells that produce antibodies; and display a full immune response.

Not only do vaccinations prevent specific infectious diseases in companion animals but also vaccination against rabies can be required by law. This is as a public health measure. In poultry, embryos or newly hatched chickens are vaccinated in addition to later in life in breeding birds.

Textbox 9.9 Passive Immunity Via Artificial Means

Cows, sows, and ewes are typically vaccinated 30 days prior to parturition. This is a way to provide the offspring with passive immunity via artificial means.

Textbox 9.10 Interesting Factoids: History of Vaccination

Vaccines have been around for about 200 years.

- Edward Jenner (1749–1823) employed animal pox from cattle (cattle or horse pox) to protect against small pox in humans.

  The word "vaccination" is derived from the Latin word *vaccinus*, which comes from *vacca*, meaning "cow."

- Louis Pasteur (1822–1895) developed vaccines against the following:
  - Fowl cholera (the causative being the bacterium, *Pasteurella multocida*) in poultry.
  - Anthrax (the causative being the bacterium, *Bacillus anthracis*) in sheep and cattle.
  - Rabies (the causative being the RNA virus, *Rabies lyssavirus*) in many species.

Vaccines can be the following:

- Live bacteria or viruses that have been attenuated (greatly weakened) by growing them in a different animal (e.g., flu vaccines in chick embryos).
- Killed organisms.
- Bioengineered proteins.
- DNA for surface protein.
- mRNA for surface proteins. (Note: mRNA vaccines are fragile and need to be stored at very low temperatures.)

In addition, there may be adjuvants, agents that increase the immune response. The first adjuvants were aluminum compounds, or an oil water emulsion and killed bacteria, or **lipopolysaccharide** (LPS), or TLR agonists. Adjuvants are effective in increasing the immune response to a vaccine be the following:

- A depot effect releasing the immunogen sustainably over a prolonged time.
- Increased cytokines production.
- Increased recruitment of antigen-presenting cells.
- Increased uptake of antigen-presenting cells.
- Activation of antigen-presenting cells.

Not only do vaccines protect with antibodies (see Figures 9.23 and 9.24), but also, with priming, the cell-mediated immunity. This is such that the animal is much more capable of responding to a pathogen. Following vaccination, there are rapid increases in cytokines such as IL-8, IL-10, IFN-α, and IFN-γ.

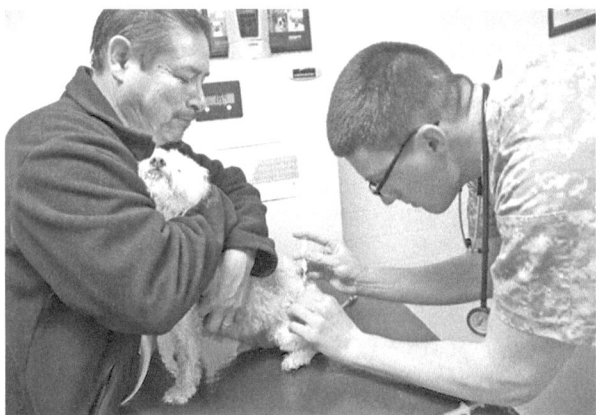

Figure 9.24 A dog being vaccinated with rabies vaccine at Joint Base San Antonio-Randolph. (Courtesy of United States Air Force, photo by Joel Martinez)

> **Textbox 9.11 Interesting Factoid: Antivenoms**
>
> Horse and sheep are used to produce antisera against snake venom. The antivenoms can be used to save the life of a person or domestic animal bitten by a venomous snake. Frequently, the antivenom will be **polyvalent** or **multivalent**, with antibodies to venom of all the snakes in a particular locality. This is an example of passive immunity by artificial means.

9.15 Inflammation

9.15.1 What Is Inflammation?

Inflammation is one of an animal's innate immune responses to a wound or pathogens (viruses or bacteria). The word "inflammation" comes from Latin, meaning "to set on fire." Inflammation promotes killing pathogens by phagocytosis and also healing. It is not surprising that signs of inflammation are the following:

- Swelling with edema due to leakage of capillaries.
- Warmness/heat due to dilation of arterioles.
- Redness due to dilation of arterioles.
- Pain due to all of the above.
- Sometimes immobility.

Figure 9.25A and B give examples of inflammation in dogs.

Inflammation includes the following:

- Mast cells that release histamine and other vasoactive peptides, such as bradykinin and prostaglandins. These increase blood flow due to dilation of arterioles and constriction of the bronchioles.
- Increased permeability of blood vessels, allowing leukocytes to pass to the site of infection or cell damage.
- Recruitment of leukocytes: attraction (chemotaxis) of leukocytes to the site of inflammation, e.g., monocytes

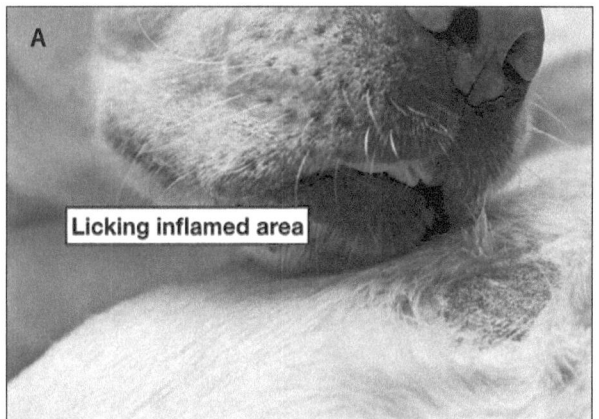

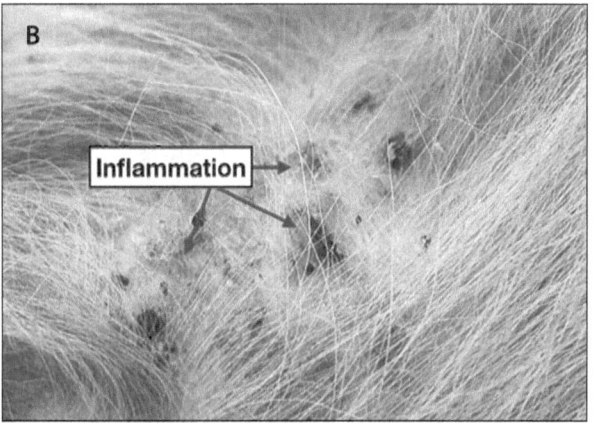

Figure 9.25 Inflamed sites in dog skin. (A) Inflammation in skin provoking itching and licking (Shutterstock/fetrinka). (B) Dog skin showing inflammation and breaks in the skin from scratching. (Shutterstock/February_Love)

that are transformed to macrophage, together with phagocytotic neutrophils.
- Blood clotting due to chemicals from the platelets.
- Release of pro-inflammatory cytokines (interleukin-1β [IL-1β], interleukin-6 [IL-6], and tumor necrosis factor-α [TNF-α]) from leukocytes.
- Activation of monocytes (transformed to macrophage) and neutrophils to release cytokines and phagocytose bacteria, viruses, and damaged cells.
- Release of antimicrobial peptides, defensins, together with antitrypsin (AAT) protein, c-reactive protein (CRP), and ferritin. These can be used as nonspecific indicators of inflammation.
- Systemic effects of proinflammatory cytokines include fever and anorexia (reduction of feeding).
- Release of acute phase proteins from liver; acute phase proteins include alpha-1.

Inflammation can be either acute (short-term) or chronic (long-term or prolonged) and either localized or systemic.

- **Acute inflammation**: burns, cuts and lacerations, infection, frostbite, chemical irritants, and allergic response.
- **Chronic inflammation**: cardiovascular disease; rheumatoid arthritis; autoimmune disease; and cancer, aging, and periodontitis.

Systemic inflammation is often accompanied by increased body temperatures, malaise (depression) and

---

Textbox 9.12 Mast Cells and Inflammation

Where are mast cells found?
Mast cells are located in mucosal and epithelial tissues, together with below the epithelium. Organs with multiple mast cells include the respiratory epithelium, gastrointestinal tract, and skin. Mast cells are frequently located close to blood vessels.

What do mast cells do?
Mast cells function as part of both innate and specific immunity.

How are mast cells activated?
Figure 9.26 summarizes the activation of mast cells. There are high affinity IgE receptors

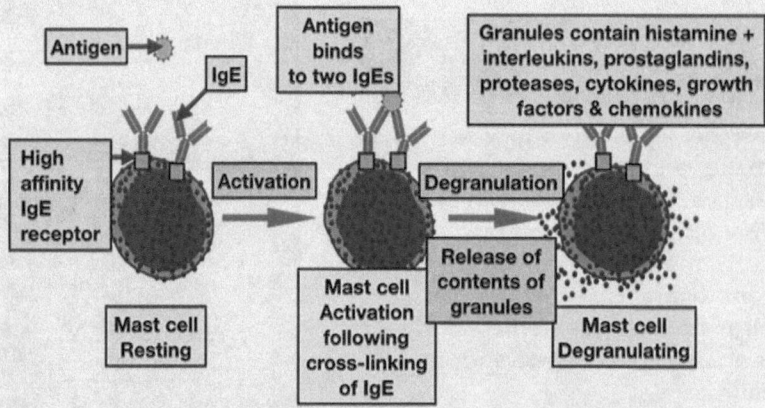

Figure 9.26 Mast cells are activated when an antigen binds to a specific immunoglobulin E (IgE). The antigen can be a pathogen or an allergen. IgE is on the surface of mast cells. IgE is anchored to high affinity IgE receptors. Antigens can bind to two IgE molecules leading to cross-linking of IgE, increased intracellular calcium ion concentration and activation of the mast cells. The activated mast cell degranulates releasing the contents of the mast cell granules including histamine together with other mediators. In addition, the activated mast cell synthesizes new biological mediators. (Adapted from Shutterstock/Soleil Nordic)

(continued)

Textbox 9.12 (continued)

spanning the plasma membrane of mast cells. These IgE receptors bind to IgE. When an antigen binds to the IgE, there is cross linking of IgE, followed by an increase in intracellular calcium ion concentrations. This leads to degranulation or the release of the contents of the granules, together with newly synthesized mediators. There are also binding sites for sites on the surface of many pathogens, these being the pathogen-associated molecular patterns (PAMPS).

What do mast cells release?
Masts cells release the following:

- Proteases.
- Biogenic amines—histamine and serotonin.
- Prostaglandins and leukotrienes.
- Heparin.
- The phospholipid platelet-activating factor.
- Nitric oxide.
- Numerous proteins, including:
  - Cytokines such as interleukins [IL-4 and IL-14].
  - Chemokines.
  - Growth factors (e.g., transforming growth factor-β).
- These were either released from granules during degranulation or newly synthesized.

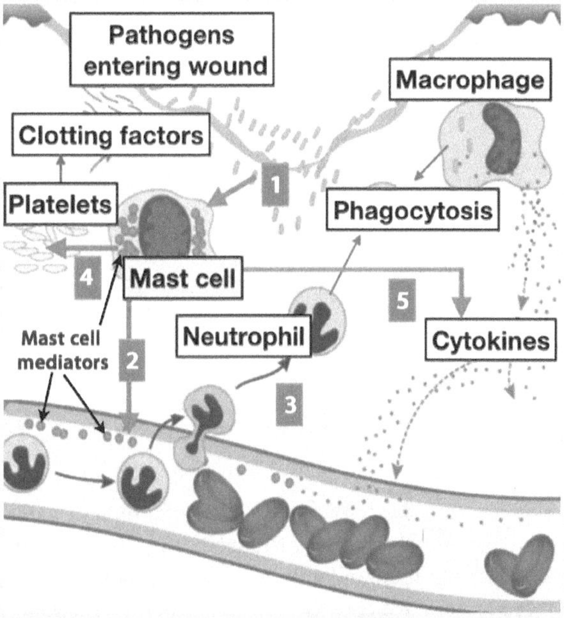

What are the effects of histamine and other mediators released from mast cells?
Effects of histamine include the following:

- Increasing vascular permeability, allowing neutrophils and dendritic cells to pass out of the blood vessels.
- Increasing the adhesiveness of the blood vessel walls.
- Activation of neutrophils, eosinophils, and platelets.
- Recruiting or attracting neutrophils, eosinophils, basophils, dendritic cells, and some T cells.
- Growth factors also act to promote the formation of new blood vessels and the repair of tissues.

Figure 9.27 Response of mast cells to the intoduction of pathogens into a wound. (Adapted from Shutterstock/Blamb)

1. Pathogens activate mast cells.
2. Mast cells release histamine and other mediators to induce dilation of blood vessels and permeability.
3. Mast cell mediators attract neutrophils, increase neutrophils adhering to the blood vessel wall, and stimulate neutrophils passing through the blood vessel wall to the site of infection.
4. Mast cell mediators increase the release of clotting factors from the platelets.
5. Mast cells release cytokines and growth factors.

---

decreased productivity (e.g., decreased milk production). Examples of systemic inflammation in domestic animals are the following:

- Infection of the mammary gland(s) in mastitis in dairy cattle.
- Infection of the uterus following calving in dairy cattle.

9.16 Allergies

Allergies are overreactions or inappropriate responses of the immune system to specific proteins or other biological chemicals. Allergic responses are due to a release of inflammatory mediators from the mast cells due to the binding of the allergen to immunoglobin E (IgE) molecules

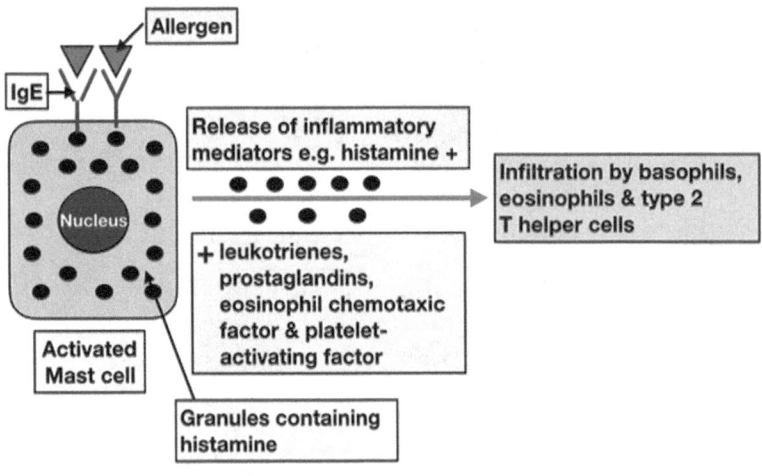

Figure 9.28 Schematic of the role of mast cells in an allergic response.

on the surface of the mast cells. The inflammatory mediators include histamine, prostaglandins, leukotrienes, platelet-activating factor, and eosinophil chemotactic factor (see Figure 9.28). These act to change blood flow, attract eosinophils (a type of leukocytes), and activate platelets.

Just like allergies to dog or cat dander are common in people, allergies are relatively common in dogs and other companion animals, as well as with livestock. Allergic responses include the following:

- Allergies to a food cause diarrhea and/or vomiting. This is seen, for instance, in dogs, calves, and piglets.
- Allergies to insect bites (e.g., fleas or horse flies) or arachnid bites (e.g., spiders and ticks) lead to dermatitis (inflammation of the skin) with intense itching. Scratching can lead to breaks in the skin and then secondary bacterial infections.
- Systemic allergies to allergens:
  - Relatively mild with bronchitis (inflammation of the bronchi) and coughing, nasal congestion, and trouble breathing, together with watery eyes. This is seen, for instance, with pollen.
  - Anaphylaxis, a life-threatening situation with profound narrowing of the respiratory bronchioles (blocking breathing), falling blood pressure, and even heart failure. The tongue and gums show cyanosis with a bluish color.

9.17 Fever

9.17.1 What Is Fever?

Fever is when an animal has a body temperature markedly higher than the normal set-point or homeostasis temperature. It is thought that fever improves the chances of survival of an infected animal, as bacteria do not operate as well at these elevated temperatures. In response to the elevated temperatures, the animal produces protective proteins called **heat shock proteins**. It should be stressed that it is important to prevent the increase in temperature going too far, as this will damage the animal.

Pyrogens induce fever by provoking sentinel cells to release **pyrogenic cytokines** (e.g., interleukin-1, interleukin -6, tumor necrosis factor [TNF]-α) and ceramide (a lipid derivative) into the bloodstream. These pass to the **organum vasculosum** in the brain, specifically in the anterior hypothalamus. This then affects the temperature control center in the pre-optic area in the hypothalamus by releasing prostaglandin E_2 (see Figure 9.29). In addition, Kupffer cells in the liver produce prostaglandin E_2 in response to lipopolysaccharide (LPS); LPS being produced by *E. coli* and other gram-negative bacteria (see Figure 9.29). Neural circuits also induce fever.

Fevers are systematic as opposed to local. A mild fever sequesters both ferrous ions (Fe^{2+}) and zinc ions (Zn^{2+}) in the liver. This inhibits bacterial replication and both Fe^{2+} and Zn^{2+} are required for DNA replication. Fever also increases the basal metabolic rate and, thus, aids repair mechanisms.

Domestic animals experience fever or pyrexia in response to infections from gram-negative bacteria, such as the following:

- Acinetobacter species, such as *Acinetobacter baumannii*.

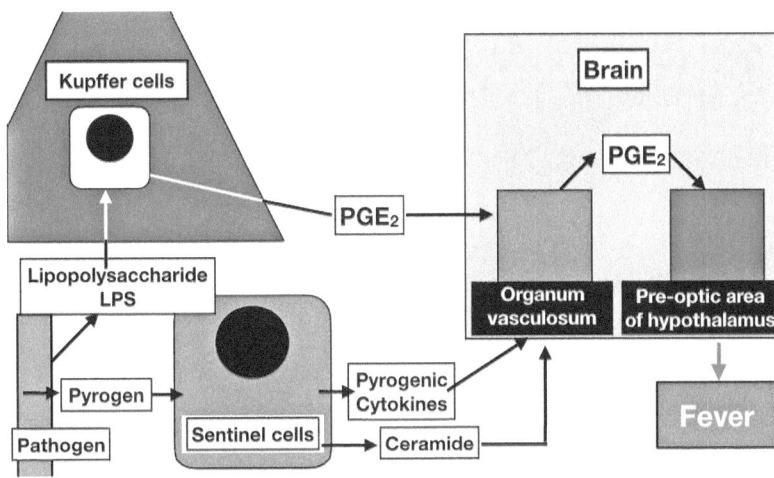

Figure 9.29 A schematic showing the mechanism by which pyrogenic bacteria induce fever. [Key: PGE$_2$ = prostaglandin E$_2$.] (Adapted from Walter et al., 2016)

- Types of *Escherichia coli* (*E. coli*).
- Klebsiella species, such as *Klebsiella pneumoniae* causing infections in dogs and cats.
- *Pseudomonas aeruginosa*, causing urinary tract infections in dogs and mastitis in cattle.

9.18 Immune Related Diseases in Domestic Animals

Textbox 9.13 summarizes immune-related diseases in domestic animals.

Textbox 9.13 Immune Related Diseases in Domestic Animals

- Allergies are found in domestic animals and livestock. For example, they are relatively common in dogs. For example, dogs may be allergic to fleas. The skin becomes red/inflamed and itches intensely (also see Chapter 10).
- Equine asthma is found in horses. This is a form of an allergy. In this case, it is an allergy to small particles in the air. There is constriction of the bronchioles and, consequently, decreased ability to breathe.
- Immune deficiency is found in horses that are not producing antibodies (immunoglobulins). It can be congenital or the consequence of another disease. There may also be a failure to passively transfer antibodies to the foal.
- Immune-Mediated Hemolytic Anemia in dogs is low blood erythrocytes and hemoglobin due to abnormal destruction of erythrocytes.
- Immune-Mediated Thrombocytopenia is abnormally low platelet concentration in blood. It is due to an abnormally high destruction of platelets. The result of Immune-Mediated Thrombocytopenia is lowered ability of the blood to clot.

Skin

(Integument)

10

Learning Objectives

1. Understand the structure of the skin (see Section 10.3).
2. Understand the functions of the skin (see Section 10.2).
3. Understand the functions of the protein keratin (see Section 10.4).
4. Understand production and role of sweat and sebum (see Sections 10.7 and 10.8).
5. Understand growth of hairs (see Section 10.6).
6. Understand the role of melanocytes in coloration of the skin, hair, and feathers (see Section 10.9)
7. Understand how ultraviolet light causes the conversion of 7-dehydrocholesterol to Vitamin D_3 (see Section 10.13).
8. Understand the role of skin in horns, hooves, and claws (see Section 10.11)
9. Understand the role of skin in temperature homeostasis (see Section 10.15).
10. Understand the role of skin as a barrier that reduces water loss (see Section 10.15.3).
11. Understand the role of skin as a barrier against pathogens (see Section 10.14)
12. Understand keratinization and its utility (see Section 10.4.3)
13. Understand skin as a secondary sexual characteristic (see Section 10.12).
14. Understand differences in skin in poultry and livestock/companion animals (see Section 10.17).
15. Understand diseases of the skin (see Section 10.18).

Table of Contents

10.1 Introduction
10.2 Functions of Skin
10.3 Functional Anatomy of Skin
10.4 Keratin, Keratinocytes, and Keratinization
    10.4.1 Overview
    10.4.2 Keratocytes
    10.4.3 Keratinization
10.5 Functions of Hair, Wool, Fur, and Feathers
10.6 Hair Structure and Growth
10.7 Sweat Glands and Sweating
10.8 Sebaceous Glands and Sebum
10.9 Skin, Hair, and Feathers Colors
    10.9.1 Overview of Skin, Hair, and Feathers Colors
    10.9.2 How Is Color of Skin, Hair, and Feathers Controlled?
10.10 Horns
10.11 Hooves and Claws
    10.11.1 What Are the Functions of Hooves and Claws?
10.12 Skin Appendages as Secondary Sexual Characteristics
10.13 Vitamin D and Skin
10.14 Skin and Bacteria
10.15 Skin and Temperature Control
    10.15.1 Overview
    10.15.2 How Is Heat Lost From Skin?
    10.15.3 How Is There an Aqueous Layer on the Skin Enabling Evaporative Heat Loss?
    10.15.4 Blood Flow to the Skin
    10.15.5 What Is the Impact of Insulation?
    10.15.6 Behavioral Responses Involving the Skin to Increased Temperatures
10.16 The Skin of Chickens and Other Poultry
    10.16.1 Overview
    10.16.2 Keratocytes
    10.16.3 Water Loss Across the Skin
    10.16.4 Glands in Chicken Skin
    10.16.5 Uropygial Gland in Birds
    10.16.6 Other Glands
    10.16.7 Pigmentation
10.17 Comparison of Skin in Poultry With Domestic Mammals
10.18 Diseases and the Skin
    10.18.1 Adult Blow Flies
    10.18.2 Alopecia or Hair Loss
    10.18.3 Anhidrosis
    10.18.4 Cushing's Disease
    10.18.5 Feather Pecking
    10.18.6 Fleas
    10.18.7 Hyperhidrosis or Excess Sweating
    10.18.8 Lameness in Cattle
    10.18.9 Laminitis or Founder
    10.18.10 Mite Infestation and Mange in Dogs
    10.18.11 Ringworm
    10.18.12 Ticks
    10.18.13 Trypanosomiasis

10.1 Introduction

Skin or skin-related materials of livestock and poultry have multiple uses, including the following:

- **Hair.**
  - Wool sheep (see Figure 10.1), cashmere wool from cashmere goats (Kashmir), angora wool from angora rabbits, llama wool, and alpaca wool.
- Horse hair has had multiple uses (e.g., stuffing upholstery, horse hair plaster).
- **Leather** is tanned skin after hair removal (e.g., from cattle, sheep, pigs, etc. plus alligators and snakes) used for shoes, purses, clothing etc.
- **Skins with hairs** used as fur clothing (e.g., from mink, sable, rabbits, etc.).

- **Feathers**, e.g., from geese (see Figure 10.1), are used as insulation in winter clothing and in bedding (pillows, mattresses, and duvets).
- **Horns** used in decorative items.

Textbox 10.1 provides a glossary of terms used relative to the skin.

10.2 Functions of Skin

Skin has multiple important functions:

1. Protective Barrier.
   a. Protection against pathogens by serving as a barrier and by producing antibacterial peptides, such as defensins.
   b. Acting as a repairable barrier protecting against predators and biting insects.
   c. Continual renewal of keratinocytes with dead keratinocytes lost and new keratinocytes formed close by the basement membrane (see Figure 10.3).
   d. A waterproof barrier preventing/reducing loss of water.
2. Temperature control (thermoregulation).
   a. At times of excess heat (high environmental temperature), removal of heat by radiating it together with either trans-epidermal water loss or to sweating.
   b. At times of heat deficits (low environmental temperature), reducing heat loss by decreased blood flow and insulation by hair (mammals) or feathers (birds).
3. Sensory input (touch, pressure, temperature, pain, etc.).
4. Vitamin D synthesis.
5. Coloration for camouflage or ornamentation.
6. Sexual signaling.

Figure 10.1 Woolly sheep. (Gavin Schaefer, CC BY-SA 3.0)

Figure 10.2 Goose feathers seen through a microscope. The feathers trap air and act as insulation. (Kateryna Martyniuk, CC BY 4.0)

Textbox 10.1 Glossary Related to Skin

- **Adipocytes**: fat cells.
- **Arrector pili muscles**: muscles associated with the base of hair follicles
- **Dermis**: a thick layer of the skin containing hair follicles, sweat glands, sebaceous glands, capillaries, and nerves.
- **Dermal papillae/dermal papilla cells**: the hair follicle.
- **Epidermis**: the exterior layer of the skin.
- **Eumelanin**: a brown/black pigment found in the skin, hair, and feathers.
- **Horns**: large structures on the heads of cattle, sheep, and goats.
- **Hypodermis**: the lowest (deepest) layer of the skin.
- **Keratin**: a structural fibrous protein.
- **Keratocytes**: the majority of the cells in the epidermis.
- **Langerhans cells**: immune cells in the epidermis.
- **Melanocytes**: skin cells that produce the pigment melanin.
- **α-Melanocyte-stimulating hormone (α-MSH)**: a neuropeptide.
- **Melanin**: a pigment.
- **Melanocortin 1 receptor (MC1R)**: found in the cell membrane and it binds α-MSH.
- **Phaeomelanin**: a red/yellow pigment found in the skin and hairs.
- **Proopiomelanocortin** (POMC): the precursor molecule for MSH.
- **Sebaceous glands**: found in the skin secreting an oily/waxy substance (sebum).
- **Sebum**: the oily secretion of the sebaceous glands.
- **Skin**: composed of three layers: epidermis, dermis, and hypodermis.
- **Sweat glands**: produce a watery secretion (sweat).
- **Uropygial gland** (or preen or oil gland): produces the oily secretion.

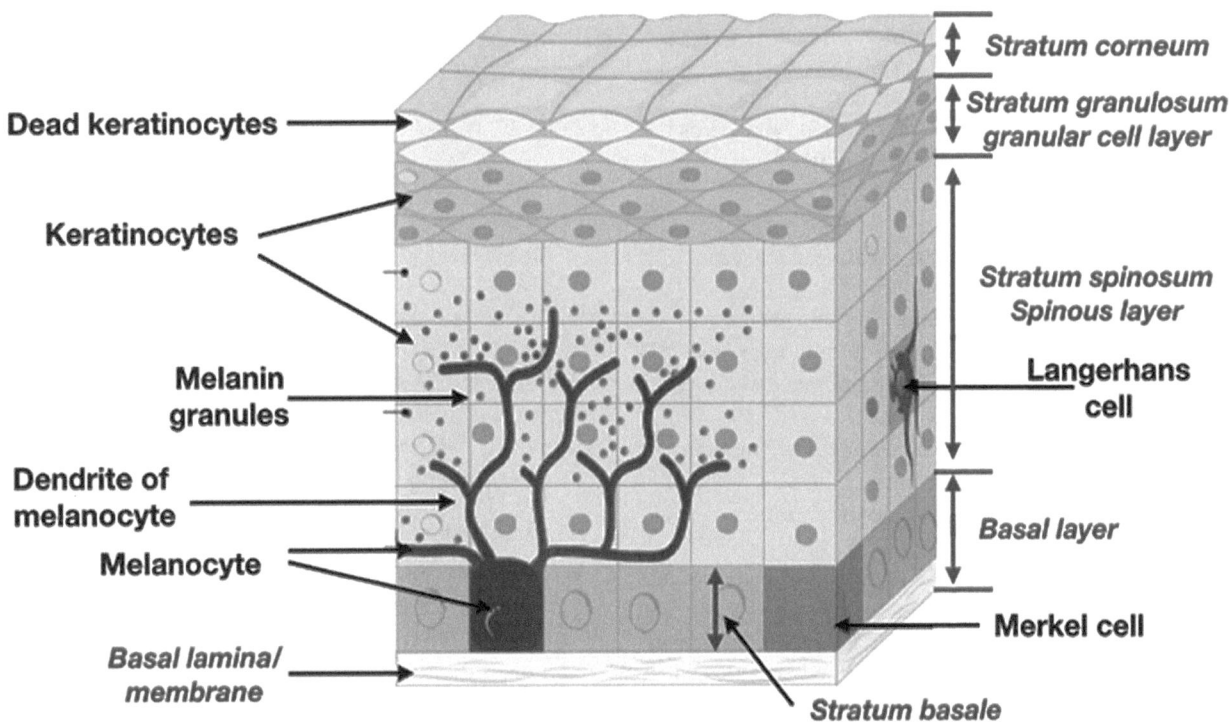

Figure 10.3 Schematic of the structure of the epidermis showing how melanocytes cause the color of the skin. Note: Langerhans cells are dendritic cells which are part of the immune systems (see Chapter 9 for a discussion of the dendritic cells). (Adapted from Stutterstock/Designua)

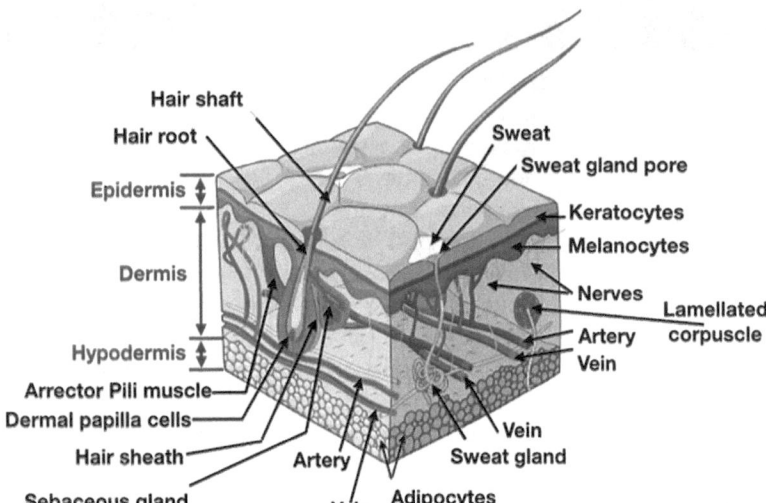

Figure 10.4 Structure of skin. (Adapted from Shutterstock/Alexander_P)

10.3 Functional Anatomy of Skin

Figures 10.3 and 10.4 show the structure of the skin. The skin of domestic mammals and birds is composed of the three layers:

1. Epidermis is the outer layer of the skin. It is derived from the **ectoderm** during embryonic development. The major protein in the epidermis is keratin. Figure 10.3 shows the structure of the epidermis. It does not have blood vessels and instead relies on nutrients and oxygen diffusing from blood vessels in the dermis. The epidermis is composed of the following cell types:
   a. **Keratinocytes**: the major cell type in the epidermis (see Figure 10.3). The outer keratinocytes form the **stratum corneum**. These keratinocytes are dead and slough off. The layer under the stratum corneum is the **stratum granulosum**. In this layer, the keratocytes undergo programmed cell death (apoptosis).
   b. Melanocytes giving the color of the skin, hair and feathers.
   c. Langerhans cells are involved in immune functioning.
   d. **Merkel cells** are involved with the perception of touch.
2. Dermis is the thickest and middle layer of the skin. It is largely connective tissue composed of extracellular protein including mainly **collagen** but also **elastins** and **proteoglycans**. In addition, it contains hair follicles (or feather follicles in birds), sweat and sebaceous glands, together with blood vessels.
3. Hypodermis (inner layer also called the subdermis or subcutaneous layer) consists of a connective tissue and adipose tissue.

10.4 Keratin, Keratinocytes, and Keratinization

10.4.1 Overview

Keratin is an important structural protein in the epidermis of the skin. In addition, keratin is an important constituent in hair, feathers, hooves, claws, and horns. There are multiple types of the protein keratin. The two major types are α-keratin and β-keratin:

- α-keratin is found in the skin (livestock and companion animals), wool (sheep), hooves (livestock and horses), and horns (cattle, sheep, and goats).
- β-keratin is found in the feathers, claws, scales, and beaks of poultry and other birds.

10.4.2 Keratocytes

Keratinocytes are the major cell type in the epidermis. Keratinocytes are responsible for the skin's role as a barrier to water loss and bacterial invasion.

10.4.3 Keratinization

Basal keratinocytes are next to the **basal membrane** between the dermis and epidermis (Figure 10.5). These undergo **proliferation** with the resultant keratinocytes moving into the **suprabasal** region of the epidermal layer (see Figure 10.5). The suprabasal keratocytes undergo differentiation with the final stage of differentiation being **apoptosis** (programmed cell death) (see Figure 10.5).

Textbox 10.2 A Deeper Dive Into Keratin

Keratin is synthesized from amino acids derived from adjacent capillaries and transported into the keratinocytes by specific amino acid transporters. Keratin has a high content of sulfhydryl or thiol (-SH). This results from its high content of the amino acid cysteine (see Figure 1.11). Subsequently, the sulfhydryl groups are oxidized, forming cross-linking disulfide bridges (see Figure 1.12).

10.5 Functions of Hair, Wool, Fur, and Feathers

Major functions of hair, wool, fur, and feathers are the following:

- Aiding thermoregulation by providing insulation (see Section 10.14).
- Protection against damage from trauma, ectoparasites, and predators.
- Providing some degree of waterproofing.

Feathers have additional functions:

- Aiding flight.
- Providing waterproofing.
- Coloration as sexual attractants, e.g., in male mallard ducks (drakes).

10.6 Hair Structure and Growth

Hairs are threads composed of keratin growing from hair follicles in the dermis of the skin. The presence of hairs is one of the characteristics of mammals. Hair, fur, and wool are fundamentally the same chemically. They are composed largely of keratin fibers but there are differences (for discussion of wool, see Textbox 10.3). An example of long-haired

Textbox 10.3 Definitions Related to Hair

- Hairs are threads composed of keratin growing from hair follicles in the dermis of the skin. The presence of hairs is one of the characteristics of mammals.
- Wool is a specialized type of hair that is fine, soft, and curly that can be shorn for making cloth or yarn. Wool is specialized hair from both primary and secondary follicles of sheep together with some breeds of goats (cashmere and mohair) and angora rabbits. Wool fibers are hairs that are fine (low diameters) and that are crimped or folded (up to 40 crimps per cm in merino sheep). The crimp in wool fibers makes the wool soft and springy. Crimping is due to oxidation of sulfhydryl groups in the keratins forming crosslinks (disulfide bridges).
- Fur is the coat of soft hairs covering the body of some animals. An example of a species of domestic animals with fur is the domestic cat.
- Pelage is the entire coat of hair, wool, or fur.

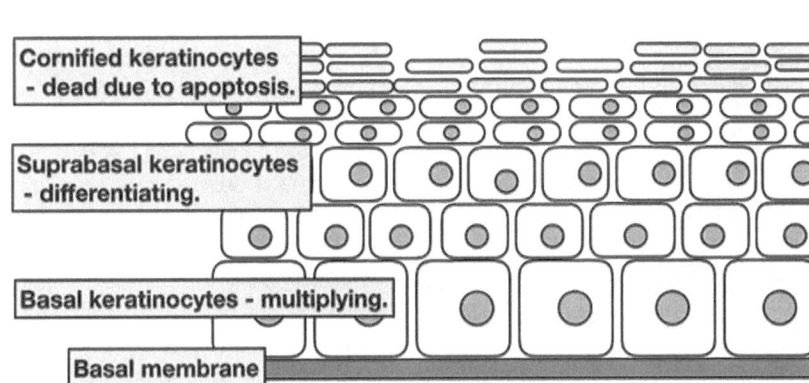

Figure 10.5 The process of keratinization.

> **Textbox 10.4 A Deeper Dive Into Wool**
>
> There are two types of hair follicles:
>
> - Primary follicles.
> - Secondary follicles (multiple around a single prim).
>
> Primary follicles produce three types of hair fiber:
>
> - Kemp (coarse fibers that do not dye evenly).
> - Medullated fibers with a central cavity.
> - Wool fibers.
>
> Secondary follicles only produce wool fibers. The amount of wool depends on the number of wool follicles. This varies considerably between different breeds of sheep (see Table 10.1).

Table 10.1 Differences in the number of follicles in different breeds of sheep.

Breed	Number of primary (1°) follicles per mm^2	Number of follicles (both primary [1°] and secondary [2°]) per mm^2	Secondary: primary ratio (2°: 1°)
Fine Merino	3.6	71.7	19.1
Medium Merino	2.9	64.4	21.0
Strong Merino	3.3	57.1	16.5
Dorset, Suffolk, & Romney	3.3	20.3	5.3
Border Leicester, & Cheviot	2.8	15.0	4.7

Data from www.infovets.com/books/smrm/a/a965.htm

Figure 10.6 Highland cattle showing a coat of very long hair. (Gordon Leggett, CC BY-SA 4.0)

cattle is shown in Figure 10.6 and of a sheep with a considerable accumulation of wool is shown in Figure 10.1.

Hairs grow in the **hair bulb** above the **vascular dermal papilla** (see Figure 10.7). There are multiple **mitotic divisions** of the matrix cells such that the base of the hair is composed of live cells. As more cells are incorporated into the hair, the hair moves up the **hair sheath**. While the hair close to the dermal papilla contains live cells, as the hair is moved up the hair sheath the cells die.

Hair follicles are surrounded by a **cuticle** consisting of overlapping dead cells (see Figure 10.7). Within the cuticle, the hair follicle consists of the **cortex** and **medulla**. The cortex is composed of keratinocytes and melanocytes. The keratinocytes become keratinized as the hair develops.

Hair growth can continue for many months or longer during the phase of hair growth called **anagen** (see Figure 10.8). At some point, there is cell death in the area of the dermal papilla and the hair bulb becomes detached (see Figure 4.8). This is the second phase of the hair cycle called the **catagen**. There is then a resting phase, the **telogen** (see Figure 10.8).

In people, there is a single hair shaft. However, in some animals, there are secondary hairs (see Figure 10.9).

CHAPTER 10 Skin

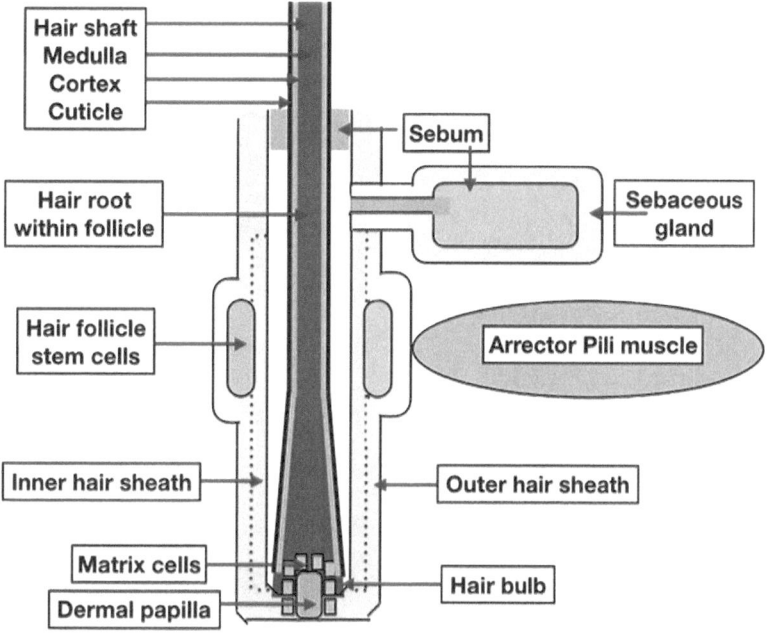

Figure 10.7 Schematic diagram showing the hair follicle (root and shaft) surrounded by a hair sheath, a sebaceous gland adding sebum to the hair root and shaft. The hair grows in the hair bulb from the dermal papilla and the matrix cells. The arrector pili muscle (APM) shifts the hair to a standing position.

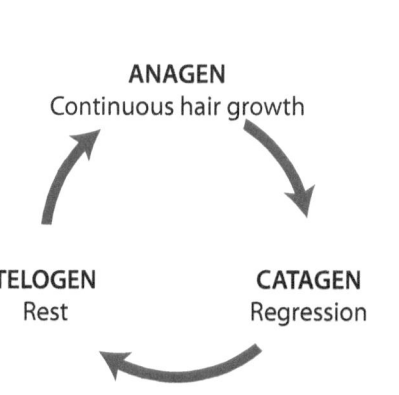

Figure 10.8 The hair cycle consists of three phases: anagen when hair is growing, catagen when the hair bulb (root) becomes separated from the dermal papilla and ultimately the hair is lost. This is followed by a resting period, telogen.

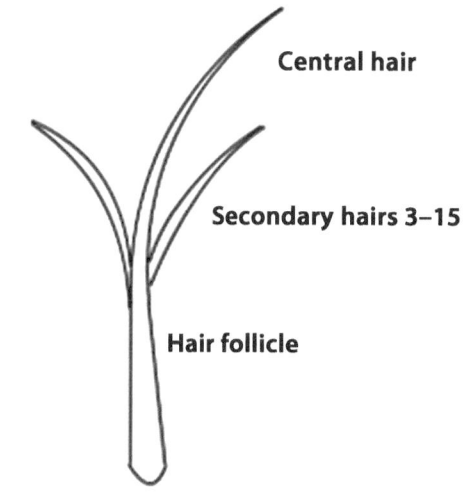

Figure 10.9 Compound hair follicle showing the central hair shaft (or guard hair) surrounded, in some species such as cats, by secondary hairs.

> **Textbox 10.5   Hair in Dogs**
>
> **Double or single coats**: Some breeds of dogs have a double coat with guard hairs and down hairs. This is found in Newfoundland, Great Pyrenees, and other guardian dogs. Other breeds, such as poodles, have a single coat.
>
> **Shedding**: In some breeds of dogs, hair (from the undercoat) is shed during specific seasons.

10.7 Sweat Glands and Sweating

Sweat glands are **exocrine glands** with ducts (tubes) leading to the outer surface of the skin. Only mammals sweat, but not all mammals. There are marked increases in sweating in horses with elevated temperature and/or exercise. Sweating is a very effective way of losing heat due to the sweat evaporating (Table 10.1).

> **Textbox 10.6   Interesting Factoids: Sweating Like a Pig**
>
> The expression "sweating like a pig" is strange, as pigs don't sweat. The expression does not come from pigs as animals. Instead, it comes from pig iron, molten iron that has cooled sufficiently so that it can be handled.
>
> **The Slick Gene**
> The "slick" phenotype in cattle was identified on St. Croix (one of the US Virgin Islands), the heat-tolerant Senepol breed of cattle. The slick hair coat is caused by a variation in the prolactin receptor gene which segregates in Holstein cattle. There are differences in body temperatures in cattle with the slick phenotype:
>
> - Slick hair coat cattle—vaginal temperature 38.5°C
> - Non-slick hair coat cattle—vaginal temperature 39.1°C

Apocrine sweat glands are found in horses, cattle, sheep, goats, and dogs. In contrast, in primates including people, sweat glands are **eccrine** (unbranched). In addition, there are **merocrine sweat glands** (unbranched) in the paws of dogs.

10.8 Sebaceous Glands and Sebum

Sebaceous (oil) glands produce sebum, an oily substance that provides waterproofing for the skin as a lipid layer and either the coat (of hair or wool) or feathers. Sebaceous glands are associated with hair follicles (see Figure 10.5). In sheep, the oily secretion is composed of **waxy esters** and **cholesterol**. It is used as **lanolin** or wool wax. In cattle, sebum is composed of triglyceride and phospholipids.

10.9 Skin, Hair, and Feathers Colors

10.9.1 Overview of Skin, Hair, and Feathers Colors

The major pigments in the skin, hair, and feathers of mammals and birds are the following:

- **Eumelanin** (brown/black).
- **Phaeomelanin** (red/yellow).

Both eumelanin and phaeomelanin are synthesized from the amino acid **tyrosine**. Examples of cattle and dogs with eumelanin (black) or phaeomelanin (red/yellow) coats are shown in Figure 10.10.

10.9.2 How Is Color of Skin, Hair, and Feathers Controlled?

The color of coat or feathers or skin is due predominantly to the following genes:

- **Tyrosinase.**
- **Tyrosinase-related protein 1.**
- **Agouti signaling protein** (ASIP).
- **Melanocortin 1 receptor** (MC1R).

together with a locus (SILV gene) responsible for **dilution of coloration** (see Figure 10.11).

Textboxes 10.7 and 10.8, respectively, provide a deeper dive into how the color of skin, hair, feathers are controlled at a cellular level, and to provide a deeper dive into epistasis and hair color.

A. Black angus cattle. (Shutterstock/Judy Drietz)

B. Red angus cattle. (Shutterstock/Steve Oehlenschlager)

C. Black Labrador retriever. (Peter Wadsworth, CC BY 2.0)

D. Irish red setter. (Canarian, CC BY-SA 4.0)

Figure 10.10 Examples of domestic animals with marked production of eumelanin (A and C) and phaeomelanin (B and D)

10.10 Horns

Horn in cattle (see Figure 10.13), sheep, and goats is composed of a sheath or shell of keratin overlying a bony core (see Figure 10.14).

Horns function as a weapon for the following purposes:
- Defense against predators.
- Offense against other males of the same species (conspecifics) causing injury or protecting by holding the opponent's head.

The horns can also injure farmers and ranchers. It is, therefore, desirable to have livestock without horns. This can be achieved by polling (removing horns) or having genetically polled cattle; the latter being due to an autosomal recessive (polled) gene.

10.11 Hooves and Claws

Other skin associated appendages are the hooves of livestock; the claws of dogs, cats, and poultry; and the finger and toe nails of a person (Table 10.2). These are all composed of keratin. An example of a hoof is shown in Figure 10.15 with a cross section through the hoof shown in Figure 10.16.

Textbox 10.7 A Deeper Dive Into How the Color of Hair, Feathers or Skin Is Controlled?

α-Melanocyte-stimulating hormone (α-MSH) is released by keratinocytes. α-MSH is produced by proteolytic cleavage of the proopiomelanocortin (POMC) (Figure 4.11). POMC is also the precursor for adrenocorticotropic hormone (ACTH) and β-endorphin (see Chapter 4). The melanocortin 1 receptor (MC1R) is a receptor for α-MSH. MC1R can be activated by α-MSH and thereby increase the synthesis of eumelanin (Figure 10.11). Agouti-signaling protein (ASIP) blocks the effect of α-MSH on the MC1R (Figure 10.10). The protein, ASIP, stops the synthesis of eumelanin, allowing the production of phaeomelanin as the default position (Figure 10.11). Agouti-signaling protein (ASIP) is a paracrine factor released from dermal papillae cells acting on nearby melanocytes (Figure 10.11). There are mutations in the MC1R gene that either effectively switch it on without stimulation or switch it off completely.

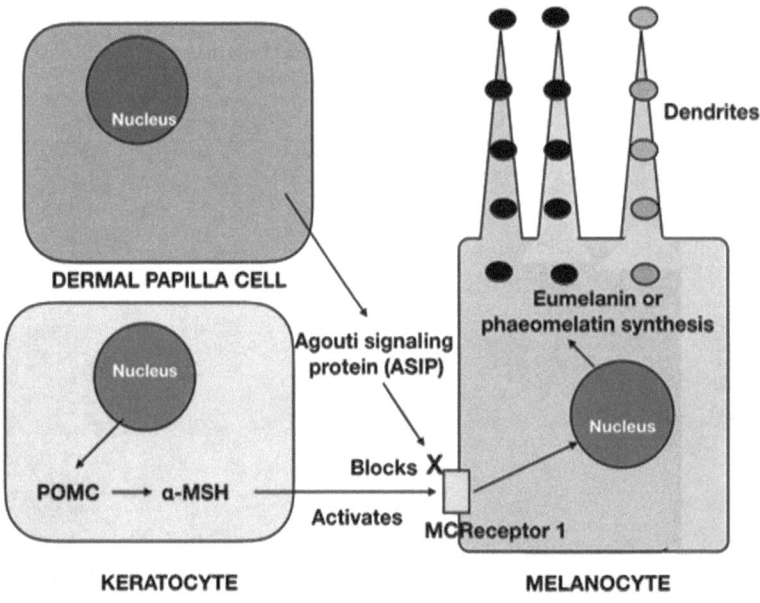

Figure 10.11 A melanocyte producing eumelanin or phaeomelanin in granules. These are exported into the skin or hair follicles or feathers. A keratocyte stimulating melanin production by synthesizing POMC (pro-opiomelanocortin). This is cleaved to produce α-MSH (α-melanocyte-stimulating hormone). This binds to MC-receptor-1 (melanocortin receptor 1 or MCR1) and stimulates eumelanin synthesis. In contrast, agouti-signaling protein (ASIP) blocks the MCR1 and the melanocyte produces phaeomelanin.

Textbox 10.8 A Deeper Dive Into the Color of Hair—Epistasis

Epistasis is when one gene regulates the expression of a second gene. A prime example of this is coat color in Labrador retrievers (see Figure 10.12). The "E" gene regulates the expression of the color gene "B". If the "E" gene is homozygous recessive ("ee") color is not expressed and the Labrador retriever will be yellow. If the "E" gene is homozygous dominant ("EE") or heterozygous ("Ee"), the color gene is allowed to be expressed and the dog will be black ("BB or Bb") or chocolate ("bb").

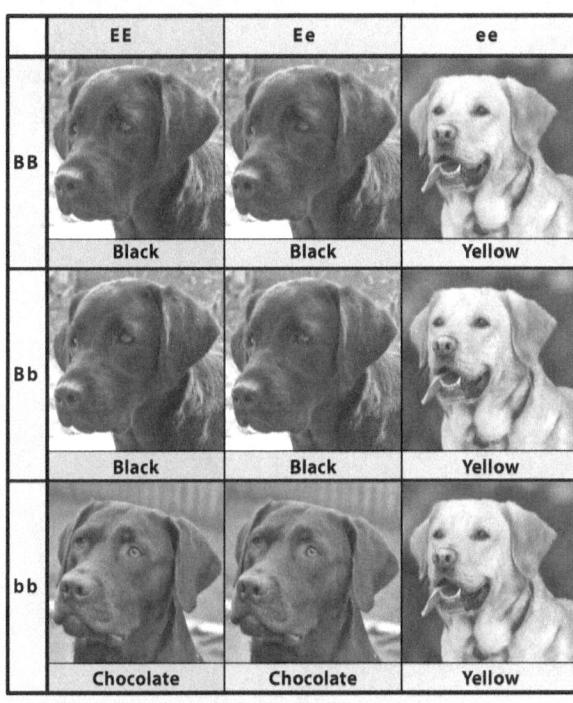

Figure 10.12 Epistasis is where one gene regulates the expression of a second gene. A prime example of this is coat color in Labrador retrievers. (Yellow lab from Pixabay; chocolate lab courtesy of Rob Hanson, CC BY 2.0; black lab courtesy of Kreuzschnabel, CC BY-SA 3.0)

Figure 10.13 Texas Longhorn cattle showing intermediate coat coloration and large horns. (Ed Schipul, CC BY-SA 2.0)

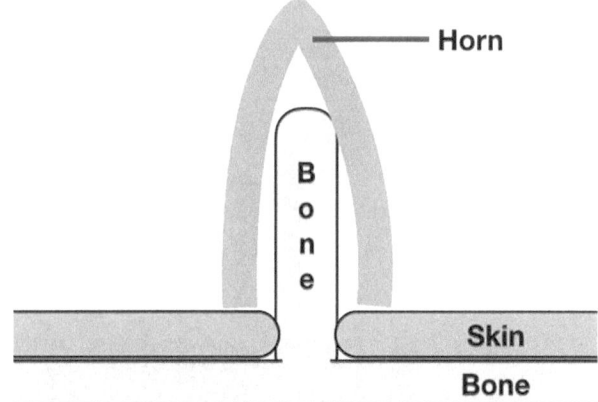

Figure 10.14 Schematic diagram showing a horn composed of keratin (or horn) surrounding a bone.

Table 10.2 Comparison of skin and skin appendages in livestock/mammalian companion animals with those in poultry and humans.

	Horns	Spurs	Sexual decoration/secondary sexual characteristics
Livestock & companion mammals	Yes in cattle (see Figure 4.6), sheep, and goats, but not in other domestic mammals or birds	No	No
Poultry	No	Yes, in chicken	Comb and wattle in male chickens. Snood and blue coloration of head and neck in male turkeys. Coloration in other avian species
Humans	No	No	Beards, Adam's apple, and increased body hair in men. Widened hips and breasts in women.

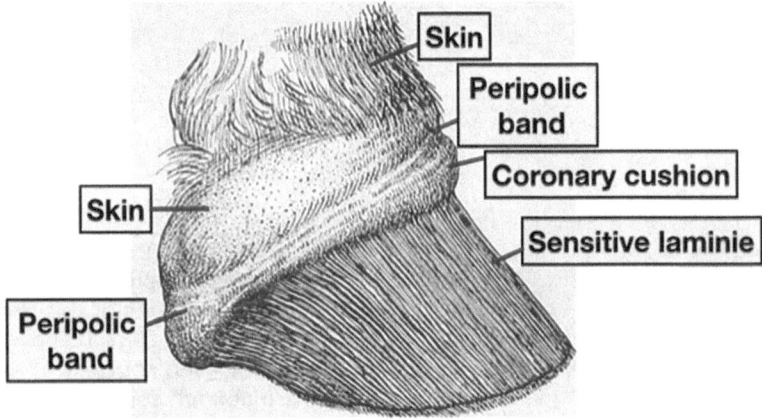

Figure 10.15 Side view of a horse's hoof. (Adapted from A.J. Wortley, public domain)

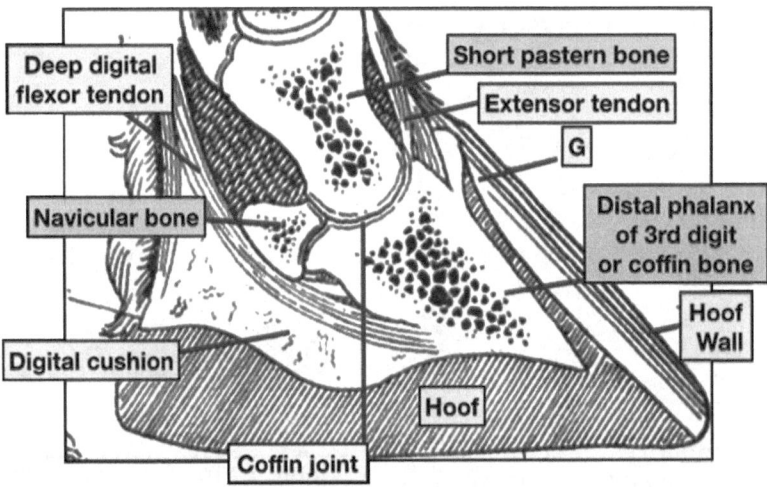

Figure 10.16 Section through the hoof of a horse. (Based on Henry Thompson, public domain)

10.11.1 What Are the Functions of Hooves and Claws?

- Hooves provide a hard surface for walking.
- Claws can be protective or offensive in killing prey.

10.12 Skin Appendages as Secondary Sexual Characteristics

Figure 10.17 shows skin appendages (comb and wattle) in roosters and male (tom) turkeys (wattle and snood). These grow under the influence of the male sex hormone testosterone. They may be viewed as displays to warn off other males or, possibly, to attract females.

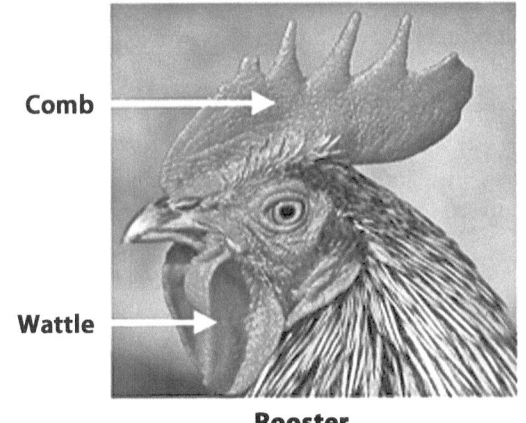

Rooster

Tom Turkey

Figure 10.17 Secondary sexual characteristics in male poultry. Figure shows skin ornamentations that are induced by the male sex steroid testosterone.
(Rooster photo from Clément Bardot, CC BY-SA 4.0; male turkey from Xuaxo, CC BY-SA 3.0)

10.13 Vitamin D and Skin

Vitamin D_3 is produced in the skin (epidermis) when exposed to sunlight (ultraviolet B), for instance, in pigs, cattle, and chickens.

$$7\text{-Dehydrocholesterol} \rightarrow \text{Vitamin } D_3 \text{ (cholecalciferol)}$$

Ultraviolet B

Light

Vitamin D_3 is hydroxylated first in the liver (by 25-hydroxylase) and then kidneys (by 1-hydroxylase) to generate the hormone 1,25 vitamin D_3 (1,25-dihydroxycholecalciferol). The activity of 1-hydroxylase is increased at times of low calcium. The 1,25 vitamin D_3 binds to the vitamin D receptor and initiates a series of changes that ultimately result in increased absorption of calcium in the small intestine.

$$\text{Vitamin } D_3 \text{ (cholecalciferol)} \rightarrow \text{25-hydroxy vitamin } D_3$$

25-hydroxylase

Liver

$$\text{25-hydroxy vitamin } D_3 \rightarrow \text{1,25 dihydroxy vitamin } D_3$$

1-hydroxylase (increased at times of low calcium)

Kidney

10.14 Skin and Bacteria

The skin of mammals and birds is covered with bacteria; most of these not being problematic, if not benign. An exception is Staphylococcus aureus, an organism responsible for skin infections in, for instance, dogs, cattle, and sheep. When the skin is damaged, opportunist pathogens such as Staphylococcus aureus invade and proliferate. This points to the importance that the skin has a physical barrier (see Section 10.15).

Antimicrobial peptides are found in the skin and in sweat. For instance, three **beta-defensins** are expressed in the skin of dogs. Defensins are antimicrobial peptides against pathogens (bacteria, fungi). Examples include α-, β-, and θ- defensins. Moreover, the antibacterial enzyme **lysozyme** and β-defensin are in sweat from the apocrine sweat glands of horses. While dermcidin is found in human sweat, it is not in sweat from horses or cattle.

10.15 Skin and Temperature Control

10.15.1 Overview

Domestic animals, irrespective of whether mammals or birds, are homeothermic with a set-point for temperature. Moreover, they produce heat as a by-product of metabolism. There are mechanisms in place to maintain the body temperature in a homeostatic manner and efficiently dissipate metabolic heat. The skin plays an important role in the temperature homeostatic system. Heat is transferred by three mechanisms:

- Conduction: transfer of heat through a solid, liquid, or gas.
- Convection: transfer of heat due to movement of air.
- Radiation: transfer of heat as electromagnetic waves, specifically as infrared light.

The skin has a very large surface area. Heat can be transferred by conduction by either of the following:

- From capillaries in the dermis, though the dermis and epidermis, to the exterior.
- From the exterior, through the dermis and epidermis, to the capillaries of the dermis.

The skin plays an important role in temperature homeostasis (see Table 10.3).

10.15.2 How Is Heat Lost From Skin?

There is some heat loss by radiation, conduction through the feet/claws/paws, and convection. A major method of losing heat is evaporation of water. Evaporative cooling involves the transition of liquid water into a gaseous state in an energy-requiring process. The energy is heat from metabolism or the environment. For every gram of water evaporated, 580 calories are lost.

10.15.3 How Is There an Aqueous Layer on the Skin Enabling Evaporative Heat Loss?

In all domestic animals, there is **transdermal water loss** across the skin due to it being an incomplete barrier for water. In many species, there is sweating that provides additional water to evaporate.

10.15.4 Blood Flow to the Skin

Blood flow to skin varies with environmental temperature:

- At high environmental temperatures there is high blood flow in the skin capillaries. This allows increased transdermal water loss (by sweating and/or transdermal water loss).
- At low environmental temperature there is low blood flow to the skin. This reduces heat loss.

10.15.5 What Is the Impact of Insulation?

Fur/hair (domestic mammals) and feathers (poultry) act as insulation due to the air trapped in them (see Figure 10.18). This is the same principle as when houses are insulated by using foam, fiberglass, or cellulose. There can be shifts in the ability of the hair or feathers to insulate, with the hair or feathers moved away from the skin trapping air as an insulator.

10.15.6 Behavioral Responses Involving the Skin to Increased Temperatures

There are some species-specific responses to elevated environmental temperatures. Given the opportunity, pigs wallow in mud. This gives rise to the expression "Happy

Table 10.3 Role of skin in temperature control

	Livestock & companion mammals	Poultry
Metabolic heat loss	Yes	Yes
Water lost across skin (transdermal water loss and loss of heat by latent heat of vaporization)	Yes	Yes
Insulation	Fur	Feathers
Sweating	Yes, in horses, cattle, sheep, and goats, but not in other domestic mammals or birds	No
Wallowing in mud	Yes, in pigs	No

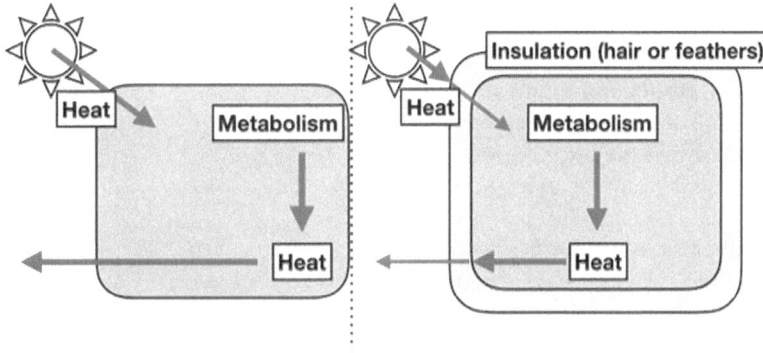

Figure 10.18 Effect of insulation on heat transfer.

as a pig in mud". The evaporation of water in the mud reduces the skin temperature and rids the body of some heat. Cats lick their fur. This wets the fur and the water in saliva will evaporate, leading to evaporative heat loss.

10.16 The Skin of Chickens and Other Poultry

10.16.1 Overview

The skin of chickens and other poultry contains layers of dermis and epidermis. There are no sweat glands and sebaceous glands. It has the following functions:

- Acting as a barrier, preventing invasion of pathogens.
- Acting as a barrier, albeit leaky, reducing water loss.
- Acting as an anchoring site for feathers and muscles to adjust the position of feathers.
- The site for the production of feathers.
- Controlling temperature.

The bodies of birds are covered with feathers except for the beak (chickens and turkeys) or bill (ducks and geese), lower legs, and feet. Feathers are light and strong; this is important for flight. Feathers are composed of proteins (~85%), predominantly β-keratin.

The skin is thinner where there are feathers and thicker where there are scales but no feathers. The skin is even thicker on the under surface of the feet.

10.16.2 Keratocytes

As in mammals, the keratocytes in the avian epidermis undergo proliferation and differentiation followed by apoptosis and loss of dead cells (see Figure 10.5). Avian keratocytes are very capable of synthesizing lipids. The cornified keratocytes lose water and have a high lipid content. The latter includes a **lipid envelope** (rather than a plasma membrane) around the dead cornified keratocytes (rather than a plasma membrane). The layer of lipid-filled cornified keratocytes reduces epidermal permeability to water or transepidermal water loss. There is not a layer of sebum on the surface of the skin, although there may be some uropygial secretions (important for waterproofing).

10.16.3 Water Loss Across the Skin

Despite the high lipid-containing cornified keratinocytes, there is some water loss across skin. This evaporative cooling reduces the temperature of the bird. The amount of water lost will be affected by the closeness of the feathers to the body. At high environmental temperatures the feathers are moved away from the body, increasing water loss.

10.16.4 Glands in Chicken Skin

There are not sebaceous glands in avian skin, with the exception of the uropygial glands that are considered modified sebaceous glands.

10.16.5 Uropygial Gland in Birds

The uropygial gland (also known as the preen or oil gland) is a two-lobed sebaceous gland near the base of the tail in birds. Preen oil is a thick oil that is spread over the feathers of birds. It is important for waterproofing (particularly in waterfowl—ducks and geese) and for feather integrity. The uropygial gland secretions consist of aliphatic fatty acids esters/waxes. The uropygial secretions are distributed on the feathers by the beak (chickens) or bill (ducks and geese).

10.16.6 Other Glands

Other glands in the skin are the following:

- **Ear glands**—secreting ear wax.
- **Vent gland**—secreting mucoprotein.

10.16.7 Pigmentation

In chickens, the skin of the lower legs and feet may be pigmented yellow. This is due to **carotenoid pigments** from the feed being deposited into the skin. This carotenoid can be reabsorbed by the body. As the hen progresses in her laying cycle, her legs/shanks can become lighter as the pigment is depleted. Feathers and skin can also be pigmented with carotenoid and other plant-derived pigments or melanin from melanocytes in the same manner as in mammals.

10.17 Comparison of Skin in Poultry With Domestic Mammals

Table 10.4 shows a comparison of skin between livestock/companion animals and poultry.

10.18 Diseases and the Skin

10.18.1 Adult Blow Flies

Adult blow flies lay their eggs on damp area on the skin sheep or cattle (see Figure 10.19). The larvae or maggots hatch on the skin, attach to it, and breach the skin barrier to feed on it. There can be a secondary invasion of the lesions by bacteria.

10.18.2 Alopecia or Hair Loss

Alopecia or hair loss is found in dogs and can be due to a variety of underlying causes (e.g., genetic, excessive licking, allergies, mange, etc.).

Figure 10.19 The Australian blowfly lays eggs on damp areas of the skin of sheep. The eggs develop into maggots that feed on the skin and blood of sheep.
(Photo by Fir0002/Flagstaffotos, GFDL v1.2).

10.18.3 Anhidrosis

Anhidrosis, the lack of the ability to sweat, is found in horses.

10.18.4 Cushing's Disease

Cushing's disease is found in horses and dogs. This is associated with long curly haired coats.

Table 10.4 Comparison of skin between livestock/companion animals and poultry.

	Livestock and companion animals	Poultry
Keratinized epidermis	Yes	Yes
Dermis	Yes	Yes
Barrier to pathogens	Yes	Yes
Barrier to water loss	Yes	Yes
Hairs	Yes	No
Feathers	No	Yes
Pigmentation	Yes	Yes
Scales on lower legs and feet	No	Yes
Sweat glands	Yes (cattle, goats, horses, and sheep) No (dogs, cats, and pigs)	No
Sebaceous glands	Yes	No

10.18.5 Feather Pecking

Individual chickens and turkeys peck at another of the same species and, thereby, injure it. With loss of feathers, the red tissue is an even more tempting target for pecking. This results in cannibalism in chickens and turkeys. To prevent this, beaks are trimmed.

10.18.6 Fleas

Fleas and other ectoparasites can cause the host to itch due to allergies and inflammation (see Figure 10.20). The host may then themselves cause breaks in the skin from scratching, or the ectoparasite may break the skin.

10.18.7 Hyperhidrosis or Excess Sweating

Hyperhidrosis or excess sweating is found in dogs.

10.18.8 Lameness in Cattle

In feedlot cattle, there can be a problem of lameness. This is caused by necrosis of the distal phalanx.

10.18.9 Laminitis or Founder

Laminitis or Founder in horses is inflammation of the laminae of the feet. A high rate of fermentation in the ceca and colon (e.g., following feeding on fructan) is associated with laminitis.

10.18.10 Mite Infestation and Mange in Dogs

The itch mite *Sarcoptes scabiei var canis* is an ectoparasite that lives on the skin and coat of dogs. The female burrows into the skin to lay its eggs. There is

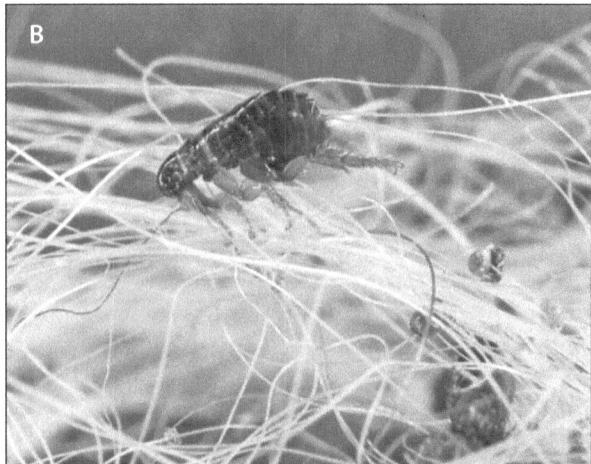

Figure 10.20 Dog fleas. (A) Dog flea showing its head and its biting ability. (Luis Fernández García, Creative Commons Attribution 2.5); (B) Dog fleas in dog hair. (Shutterstock/photowind); (C) Dog scratching because of fleas. (Shutterstock/kobkik)

inflammation (also see Chapter 9) in response to the mite's excreta and the dog biting the itchy area. Cats can also be infected with Sarcoptes scabiei (see Figure 10.21). There will then be loss of hair. There can be secondary bacterial or fungal infection. These mites can be easily spread.

10.18.11 Ringworm

Ringworm is caused by a fungal infection. There can be a round ring around the fungal infection caused by inflammation. The fungi feed on the keratin from dead cells. It is found in livestock, companion animals, and people.

10.18.12 Ticks

Ticks are another ectoparasite, for instance in dogs (see Figure 10.22).

10.18.13 Trypanosomiasis

Trypanosomiasis or sleeping sickness (humans) and nagana (cattle) are diseases found in Sub-Saharan Africa and both are caused by the trypanosome protozoa, Trypanosoma brucei. The protozoa are transmitted to people or cattle via bites from infected tsetse flies.

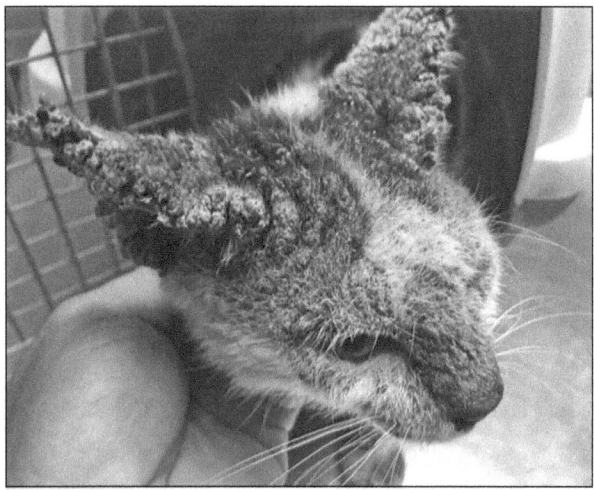

Figure 10.21 A cat suffering from a mange infection. (Shutterstock/M. Sam)

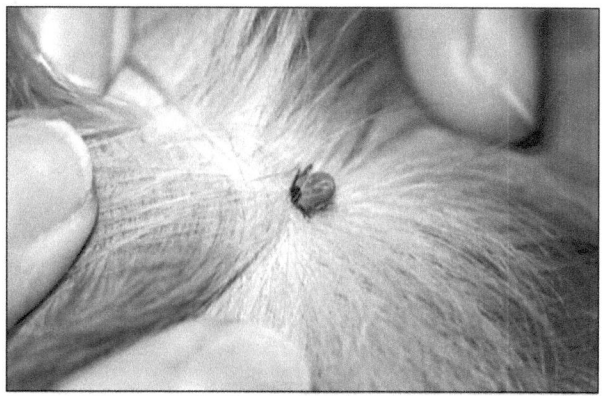

Figure 10.22 A tick imbedded in the skin of a dog. (Shutterstock/Igor Chus)

Digestion and the Gastrointestinal Tract   11

Learning Objectives

1. To understand what digestion is and what its stages are (see Section 11.1).
2. To understand why animals need to digest their food (see Section 11.1).
3. To understand types of digestion (see Textbox 11.1).
4. To understand the functions of the following regions of the gastrointestinal tract:
   a. Mouth (see Section 11.2).
   b. Esophagus (see Section 11.3).
   c. Stomach (see Section 11.4).
   d. Small intestine (see Section 11.5).
   e. Large intestine (colon) (see Section 11.7).
   f. Cecum/ceca (Section 11.8).
5. To understand the functions of the accessory organs associated with the gastrointestinal tract, namely the liver and pancreas (see Sections 11.10 and 11.11).
6. To understand the differences in gastrointestinal functioning between the following:
   a. Foregut fermenters (cattle, sheep, and goats) verses non-fermenters (pigs, dogs, and cats) (see Section 11.16).
   b. Hindgut fermenters (horses and rabbits) verses non-fermenters (pigs, dogs, and cats) (see Sections 11.17, 11.18, and 11.19).
7. To understand the process of digestion including:
   a. Mechanical breakdown of foods (see Section 11.16.1).
   b. Enzymic breakdown of constituents of proteins in foods to amino acids that can be absorbed (see Section 11.12).
   c. Enzymic breakdown of constituents of starch in foods to monosaccharides, such as glucose, that can be absorbed (see Section 11.13).

d. Enzymic breakdown of constituents of lipids in foods to fatty acids that can be absorbed.
   e. Absorption of simple compounds (monosaccharides such as glucose, amino acids, and fatty acids).
8. To understand the functioning of the rumen in ruminants (foregut fermenters) and the colon and cecum in hindgut fermenters, such as horses (see Sections 11.16 and 11.17).
9. To understand the differences between gastrointestinal physiology between nonruminant mammals and that in poultry (see Section 11.20).
10. To understand diseases of the gastrointestinal tract (see Section 11.21).

Table of Contents

11.1 Introduction
   11.1.1 What Is Digestion?
   11.1.2 What Are the Types of Digestion?
   11.1.3 Why Do Animals Need to Digest Food?
   11.1.4 What Are the Types of Nutrition?
   11.1.5 Overall Structure of the Gastrointestinal Tract
   11.1.6 What Are the Stages of Digestion in Livestock, Companion Animals, and Poultry?
   11.1.7 How Is Ingesta Moved Through the Gastrointestinal Tract?
   11.1.8 What Are the Functions of the Gastrointestinal Tract?
11.2 Mouth
   11.2.1 What Are the Functions of the Mouth in Digestion?
11.3 Esophagus
11.4 Stomach
   11.4.1 Functional Anatomy of the Stomach
   11.4.2 Parietal Cells
   11.4.3 Enteroendocrine Cells
   11.4.4 Lifespan of Stomach Epithelial Cells
11.5 Small Intestine
   11.5.1 Overall Introduction to the Small Intestine
   11.5.2 Cell Types in the Villi
   11.5.3 Functioning of the Small Intestine
   11.5.4 Functioning of the Microvilli
   11.5.5 Transporters in Microvilli
11.6 Cecum
11.7 Colon
11.8 Rectum
11.9 Accessory Organs
11.10 Liver
   11.10.1 Introduction to the Liver
   11.10.2 What Does Bile Contain?
   11.10.3 Are There Enzymes in Bile?
   11.10.4 Is the Bile Important to Digestion?
   11.10.5 What Does the Gallbladder Do?
11.11 Pancreas
   11.11.1 Introduction to the Pancreas
   11.11.2 Pancreatic Secretion
   11.11.3 Why Does the Pancreas Secrete Bicarbonate?
   11.11.4 Are There Enzymes in Pancreatic Fluid?
11.12 Digestion of Proteins
11.13 Carbohydrate Digestion
11.14 Lipid (Fat) Digestion
11.15 Control of Gastrointestinal Functioning

Chapter 11 Digestion and the Gastrointestinal Tract

- 11.16 The Gastrointestinal Tract and Digestion in Ruminants
  - 11.16.1 Mouth
  - 11.16.2 Ruminant Stomach
  - 11.16.3 Rumen
  - 11.16.4 Reticulum
  - 11.16.5 Omasum
  - 11.16.6 Abomasum
  - 11.16.7 Rumination
  - 11.16.8 Small Intestine
  - 11.16.9 Colon and Cecum
- 11.17 Digestion in Horses (Pseudoruminants)
  - 11.17.1 Introduction to Digestion in Horses
  - 11.17.2 Fermentation in the Foregut (Stomach and Small Intestine)
  - 11.17.3 Functioning of the Stomach
  - 11.17.4 Functions of the Small Intestine
  - 11.17.5 Fermentation in the Hindgut
- 11.18 Digestion in Rabbits
- 11.19 Digestion in Cats and Dogs
  - 11.19.1 Protein Digestion
  - 11.19.2 Carbohydrate Digestion
  - 11.19.3 Taste Receptors
- 11.20 Comparison of the Gastrointestinal Tract and Digestion in Livestock, Companion Animals, and Poultry
- 11.21 Diseases of Gastrointestinal System in Domestic Animals

11.1 Introduction

11.1.1 What Is Digestion?

Digestion is breaking large molecules in the food into smaller molecules that can be absorbed. This occurs in the gastrointestinal tract. Textbox 11.1 provides a glossary of terms in digestion.

11.1.2 What Are the Types of Digestion?

There are two types of digestion:

- **Mechanical digestion**. This includes chewing to break up food and the muscular contractions of the stomach. This makes pieces of food smaller and,

---

Textbox 11.1 Glossary of Terms Related to Digestion

- **Absorption**: movement of simple nutrients such as glucose, amino acids, and fatty acids from the lumen of the gastrointestinal tract into the bloodstream.
- **Accessory organs**: organs associated with the gastrointestinal tract and contributing to digestion. These are the liver and exocrine pancreas.
- **Alimentary canal**: another name for the gastrointestinal tract.
- **Amylase**: enzyme that breaks the polysaccharide starch, to the disaccharide maltose.
- **Bile salts**: emulsify fats to facilitate digestion.
- **Digestion**: breaking of large molecules in food into smaller molecules that can be absorbed. This occurs in the gastrointestinal tract.
- **Enzymes**: proteins that catalyze (speed up) chemical processes.
- **Gastrointestinal tract**: the intestines from the mouth to the anus.
- **Hindgut fermenter or pseudoruminant**: an animal, such as a horse, where the majority of the digestion occurs in the colon and ceca.
- **Hydrolysis**: addition of water molecules. This can be part of the chemical digestive process.
- **Ingesta (or chyme)**: what food is called after it is eaten (after ingestion) and as it passes down the gastrointestinal tract.
- **Lactase**: enzyme that digests lactose to glucose and galactose.
- **Lipase**: enzyme that digests triacylglycerol to fatty acids and glycerol.
- **Maltase**: enzyme that digests maltose to glucose.
- **Mastication**: chewing of food using the molar teeth (in mammals).
- **Monogastric**: having one simple stomach.

(continued)

> **Textbox 11.1 (continued)**
>
> - **Microbiome or microflora or gut flora**: These are the microorganisms, bacteria, fungi, protozoa (single-celled organisms), and archaea (single celled organisms formally known as archaebacteria), found in the intestine. Ruminal microorganisms are predominantly anaerobic (not requiring oxygen).
> - **Peptidase**: enzyme that digests peptides to amino acids.
> - **Protease**: enzyme that digests proteins to peptides and amino acids.
> - **Ruminant**: animal with a rumen consisting of four chambers where plant matter is fermented.
> - **Zymogen**: inactive precursor of enzymes.

therefore, have a much increased surface area to facilitate enzymic digestion.
- **Enzymic or chemical digestion**. This occurs in the mouth, stomach, and small intestine.

11.1.3 Why Do Animals Need to Digest Food?

Digestion is needed to convert foods into **simple molecules** so that they can be absorbed.

11.1.4 What Are the Types of Nutrition?

Textbox 11.2 summarizes types of nutrition.

11.1.5 Overall Structure of the Gastrointestinal Tract

The overall structure of the gastrointestinal tract of domestic animals is shown in Figure 11.1. There are the following major components which form the alimental tract:

- **Mouth** ingesting food. It starts mechanical digestion through chewing and releases saliva with an amylase to digest carbohydrate in non-ruminants.
- **Esophagus** transits food from the mouth to the stomach.

> **Textbox 11.2 Types of Animals Based on Their Nutrition**
>
> - **Herbivores** are animals that consume only plants. Among domestic animals, herbivores include cattle, sheep, and horses. There are consequences to gastrointestinal functioning, with herbivores being either of the following:
>   - **Foregut fermenters** in ruminants such as cattle, sheep, and goats.
>   - **Hindgut fermenters** in horses and rabbits.
>   - **Fermentation** allows breakdown of non-digestible plant materials by microbial population in the intestine. The breakdown allows the host to absorb additional nutrients.
> - **Carnivores** are animals that consume meat from animals. Examples include cats and dogs.
> - **Omnivores** are animals that consume both plants and animals, including insects. Examples of omnivores are pigs and chickens. There may be some microbial fermentation in the gastrointestinal tract of omnivores.

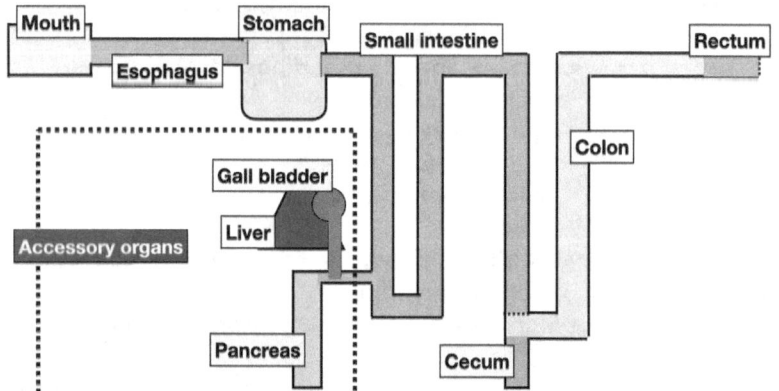

Figure 11.1 Generalized schematic diagram of the gastrointestinal tract and associated accessory organs.

Chapter 11 Digestion and the Gastrointestinal Tract

- **Stomach** provides mechanical digestion of food together with release of the zymogen **pepsinogen**, which is converted to the protease **pepsin**, in the presence of acid which is released into the stomach.
- **Small intestine** where foods are enzymically digested and the nutrients absorbed (see Textbox 11.3 for structure of the small intestine).
- **Cecum** (plural ceca) where there are bacteria and bacterial fermentation.
- **Colon** (large intestine) where there is bacterial fermentation and absorption of water and some nutrients.
- **Rectum** where water is removed from the stored feces.

> **Textbox 11.3 Structure of Small Intestine** 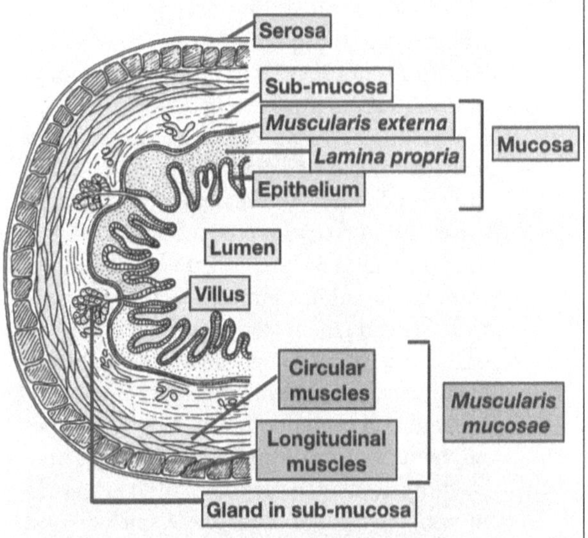
>
> **Figure 11.2** Overall structure of the small intestine shown in cross-section. (Adapted from the National Cancer Institute)
>
> Figure 11.2 shows the structure of intestine in cross section. There are four layers as follows:
>
> - Mucosa-absorbing nutrients.
> - Sub-mucosa releasing enzymes from glands.
> - *Muscularis mucosae* (muscle layer) with both circular and longitudinal muscles that move the undigested food along the intestine by peristalsis.
> - Serosa layer (connective tissue) maintaining the integrity of the gastrointestinal tract.

11.1.6 What Are the Stages of Digestion in Livestock, Companion Animals, and Poultry?

The stages of digestion in livestock, companion animals, and poultry are the following:

- Ingestion.
- Digestion.
  - Mechanical digestion.
  - Chemical digestion.
- Regurgitation in ruminants.
- Absorption of nutrients.
- Fermentation in the rumen of ruminants (cattle, sheep, and goats) and the caeca (horses).
- Consolidation (water absorption).
- Defecation.

11.1.7 How Is Ingesta Moved Through the Gastrointestinal Tract?

Peristalsis is the movement of ingesta along the gastrointestinal tract by a wave of contractions of the **circular smooth muscles** (see Figure 11.3). Peristalsis is controlled by an elaborate neural network referred to as the **enteric nervous system**.

11.1.8 What Are the Functions of the Gastrointestinal Tract?

The gastrointestinal tract is essential to life. Its functions are the following:

1. Digestion of food:
   a. Mechanical (chewing and churning in stomach) breaking down.

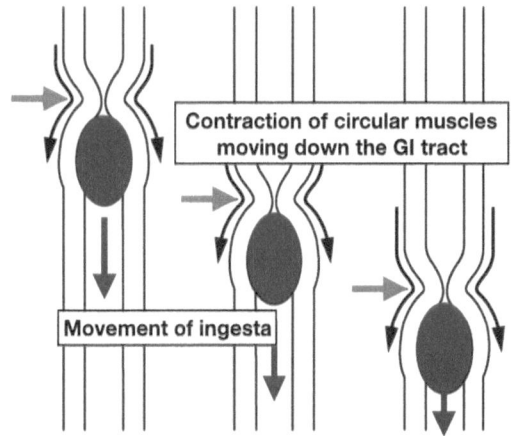

Figure 11.3 Diagrammatic representation of peristalsis.
(Adapted from OpenStax College, CC BY 4.0)

b. Chemical (enzymatic) breaking down insoluble and non-absorbable constituents of the food with starch to simple sugars (e.g., glucose), proteins to amino acids, and triglycerides to glycerol and free fatty acids.
c. Absorption of nutrients (e.g., simple sugars, amino acids, glycerol, fatty acids, minerals, and vitamins).
d. Fermentation of components of the ingesta and absorption of the products of fermentation with marked differences between species (see below).
2. Protection:
a. Stomach-killing pathogens.
b. The gastrointestinal tract is a major immune tissue. The gastrointestinal tract is protected through the gut-associated lymphoid tissue (GALT).
c. Teeth protecting the animal.
3. Mobility: transporting food (by peristalsis), then ingesta, from the mouth to the environment finally as feces.
4. Communication—nervous and endocrine.
a. The quality by taste and smell.
b. Playing a role in determining the adequacy of the quantities of food eaten to meet the requirements of the animal and, therefore, the need to consume more feed (appetite—eating more; satiety—stopping eating).
c. Playing a role in the control of the release of insulin and other hormones.
5. Excretion of bile pigments (waste products from the breakdown of hemoglobin).

The functioning of the gastrointestinal tract will be considered first in a generalized domestic animal. Subsequently, gut functioning will be considered for specific species or groups of species: (1) ruminants (cattle, sheep, and goats), (2) horses, (3) cats and dogs (carnivorous species), and (4) chickens and turkeys.

11.2 Mouth

11.2.1 What Are the Functions of the Mouth in Digestion?

The functions of the mouth include the following:

1. **Prehention**: Biting off food (by the incisor teeth in mammals and the beaks of birds) (for discussion of teeth see Section 11.2.2).
2. Starting digestive process with the following:
a. Mastication: Food is chewed using the premolar and molar teeth in mammals (see Figures 11.4 and 11.5); for discussion of teeth see Section 11.2.2). There is tearing and grinding to break the food into smaller pieces with a consequent greater surface area to mass ratio. There is also breakage of plant walls in herbivores and connective tissue in carnivores. This mechanical digestion facilitates enzymic digestion later in the gastrointestinal tract. Mastication mixes the food with saliva, making it softer and more easily swallowed. The cheeks, lips, and tongue are used to keep the feed between the teeth and develop a **bolus** of masticated ingesta. This is then swallowed. There is no chewing in birds as they lack teeth.
b. Insalvation: The lingual glands in the mouth produce saliva and this is added to the food. This contains mucus (for lubrication), together with the starch-digesting enzyme amylase. Amylase is not present in saliva from ruminant or dogs. In some species, such as pigs, there is also **lingual lipase** which is activated in the stomach. Saliva also contains lysozyme and immunoglobulin A to maintain the health of the mouth. **Lysozyme** is a bactericidal enzyme killing bacteria by breaking their cell walls. Immunoglobulin A inhibits bacterial growth by preventing them from adhering to the oral cavity. Secretion of saliva by the salivary glands is regulated by the **salivary nuclei** in the higher brain center and, locally, by the vagus nerve. Activation of the salivary nuclei can be stimulated by the presence of food in the mouth, the sight of food, or the smell of food.
c. Deglutition or swallowing: Moving ingesta into the esophagus via the pharynx. The muscle contractions that regulate swallowing consist of two different phases: (1) the **buccal phase** with voluntary contractions of the tongue initially, and (2) the pharyngeal-esophageal phase, which is under involuntary control in the brain.
d. In ruminants, but not other species, ingesta is regurgitated and the cud is re-chewed (discussed in more detail under ruminant gastrointestinal tract and digestion in a section below).
e. Tasting by the taste buds, for instance, on the dorsal side of the tongue, together with smelling food (rejecting contaminated food). (See Section 5.16.13 for more details on tasting.)

3. Defense/aggression, e.g., teeth in dogs, cats, and horses biting other animals or even people (see Figure 11.4).
4. Vocalization. While the larynx is where sounds are produced, the mouth affects the loudness of the sound by being either open or closed.

Teeth are composed of the following:

- **Enamel**: a very hard mineralized tissue that protects the dentine of the tooth. It is composed of an extracellular matrix embedded with mineral crystals of calcium phosphate and, specifically, hydroxyapatite. Minerals represent over 95% of enamel. Enamel is produced by epithelial derived cells (ameloblasts) (see Figure 11.5).
- **Dentine**: a hard tissue composed of minerals, representing over 70% of dentine, and extracellular matrix (20%) (see Figure 11.5).
- **Pulp**: is a soft connective tissue containing blood vessels (see Figure 11.5).

There are four types of teeth (see Figures 11.4 and 11.6):

- **Incisor teeth** (front teeth). These bite off chunks of food.
- **Canine teeth** (cucsids or fangs). These are used to kill prey in carnivores.
- **Premolar teeth**. These crush and grind food.
- **Molar teeth**. These crush and grind food.

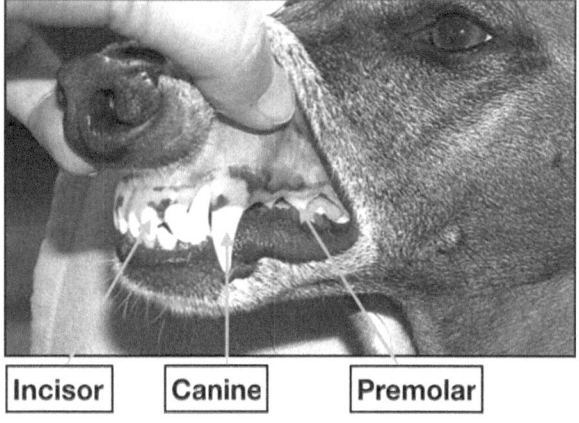

Figure 11.4 Teeth of a dog showing very large canine teeth, incisor teeth, and small pre-molar teeth. (Adapted from Jimmy Vivier, CC BY-SA 3.0)

The basic pattern of teeth on both upper and lower jaws (left or right sides) are as follows (see Figures 11.4, 11.6, and 11.7, and Textbox 11.4):

- Incisor teeth: 3 or 4 teeth.
- Canine teeth: 1 or 0 teeth.
- Premolar teeth: 3 or 4 teeth.
- Molar teeth: 3 or 4 teeth.

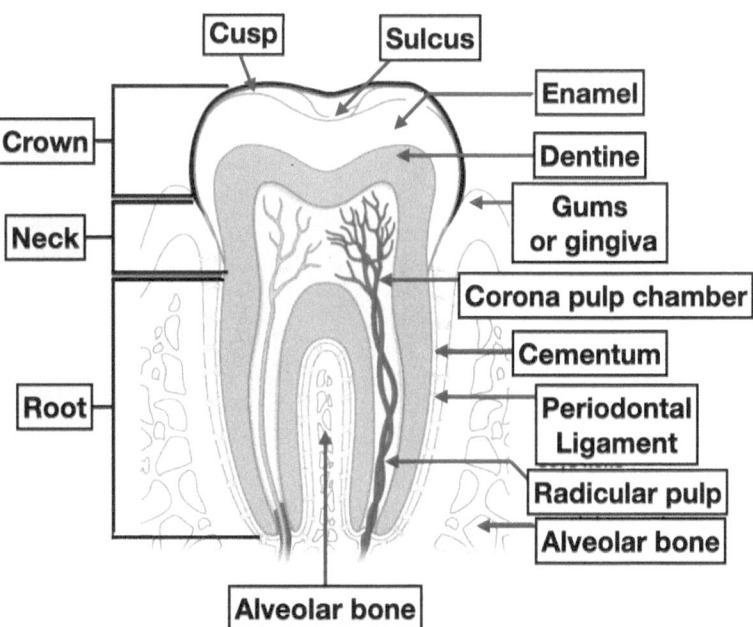

Figure 11.5 Section through tooth showing its structure. (Adapted from Openstax, CC BY 4.0)

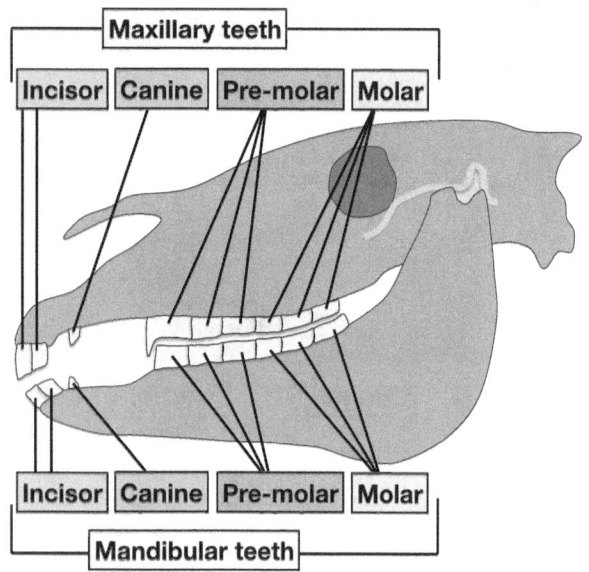

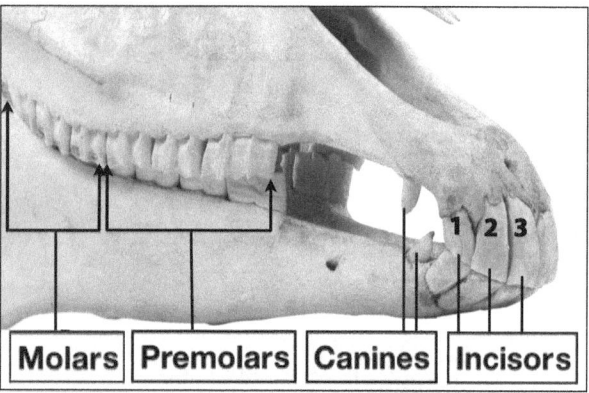

Figure 11.6 Types of teeth in horses shown as a schematic (Top) and on a skeleton (above). (Left photo adapted from Jayci, CC BY-SA 4.0; photo above adapted from Shutterstock/Wallenrock)

Textbox 11.4 Comparison of Dentition (Teeth) In Horses and Dogs

Comparison of dentition (teeth) in horses (Figure 11.6) and dogs (Figure 11.4) is as follows:

1. The three pairs of incisors in both, these being labelled 1 to 3 in the horse.
2. The large incisors in both horses and dogs. These are for grabbing food.
3. The large canine teeth in dogs but very small canine teeth in horses.
4. The large premolar and molar teeth in horses but small molar teeth in dogs.

Textbox 11.5 Interesting Factoid: Baby Teeth

In young domestic mammals, there are deciduous or baby teeth. These are shed during development and replaced by adult teeth (see Table 11.1 for age of eruption of adult teeth).

Table 11.1 Age of eruption of adult teeth.

	Cattle	Horses	Dogs
Incisor	1–3 years	2–4 years	~4 months
Canine	Not present	~4 years	~5 months
Premolar	2–3 years	½–4 years	~5 months
Molar	½–2 years	1–4 years	~6 months

11.3 Esophagus

The esophagus is a long tube from the pharynx to the stomach. It is a **distensible organ** with the same overall structure as shown in Figure 11.2, with the exception of a lack of glands in the submucosa. The esophagus is located in the thoracic cavity passing through the diaphragm to terminate at the stomach with the esophageal sphincter (a muscular valve) between the esophagus and stomach.

The functions of the esophagus are the following:

1. Transporting ingesta via peristalsis to the stomach (see Figure 11.3).
2. Gases are transported to the mouth to be lost. The gases can be produced during stomach digestion or can be air that is swallowed along with food.
3. In ruminants (cattle, sheep, and goats), ingesta is regurgitated (by reverse peristalsis) to the mouth from the rumen. The cud is then chewed and returned to the rumen. (This is discussed in more detail in Section 11.16 below).

4. There is an out-pocketing of the lower esophagus in many birds (chickens and turkeys) but not mammals. This is extensible and called the **crop**. Crop predominantly functions to store ingesta. In addition, there is some microbial fermentation generating lactate from starch. In ducks, there is a rudimentary crop able to store feed. In pigeons, crop milk is produced by the crop. This is regurgitated to feed the offspring.

11.4 Stomach

11.4.1 Functional Anatomy of the Stomach

The stomach is both a muscular and a glandular organ within the gastrointestinal tract. The stomach is located in the abdominal cavity. Ingesta pass into the stomach from the esophagus via the lower **esophageal sphincter (cardiac sphincter)**. This is a muscular valve that impedes backflow of ingesta. After ingesta has been in the stomach, it passes to the first part of the small intestine, the duodenum, via the **pyloric sphincter**. This is another muscular valve.

In nonruminants such as the pig and dog, the stomach has four regions:

- **Pars esophagea**, the region after the lower esophageal sphincter (nonglandular).
- **Cardiac region** (glandular).
- **Fundus or fundic region** (glandular).
- **Pyloric region** leading to the pyloric sphincter (nonglandular).

Unlike other regions of the gastrointestinal tract, there are three layers of muscles in the stomach, namely, the following:

- **Longitudinal.**
- **Circular.**
- **Oblique.**

The stomach musculature churns the ingesta when both sphincters are closed. This mechanical digestion breaks the ingesta into smaller and smaller units. The smaller the size of the pieces of ingesta, the larger the total surface area for chemical (enzymic) digestion. The glandular component of the stomach releases mucus, hydrochloric acid, and the zymogen pepsinogen. The acid kills pathogens in the ingesta, activates pepsinogen to pepsin, and denatures/opens up proteins (see Figures 11.5).

> **Textbox 11.6 What Is a Zymogen**
>
> A zymogen is an inactive protein that is converted to an active enzyme by enzymes or acidity. As many enzymes are needed quickly during digestion, enzymes are stored as zymogens to allow quick release and activation. Moreover, this storage system prevents active enzymes from damaging the cells that produce them.

The functions of the stomach are the following:

1. Mechanical digestion breaks up pieces of ingesta.
2. Enzymatic digestion starts due to gastric secretions of the following:
   a. The zymogen pepsinogen is secreted. It is activated by stomach acid to the protease pepsin (see Figure 11.7).
   b. Enzymatic digestion of triglyceride starts due to gastric lipase, together with lingual lipase in some species.
3. Bacteria are killed in acidic fluid (pH 1.5–2.5).
4. Absorption of a few nutrients including vitamin B_{12} (with stomach produced intrinsic factor) together with volatile fatty acids in ruminants.
5. Proteins are denatured/opened up to facilitate protein digestion by proteases and peptidases.
6. Transporting ingesta from the esophagus to the duodenum.
7. Protection
   a. Secretes a bicarbonate rich mucus to protect the epithelial cells of the stomach.
   b. The epithelial cells have tight junctions that prevent gastric acid from seeping between cells and damaging other tissue of the stomach.
   c. The epithelial cells have a short life and are replaced rapidly.
8. Messaging to influence the following:
   a. Appetite/satiety (eating) by the hormone **ghrelin**.
   b. Release of gastric juices by the hormone **gastrin**.
9. Storage of ingesta.
10. Absorption of some pharmaceutical drugs such as nonsteroidal anti-inflammatory drugs (NSAID). (In people, alcohol is absorbed in the stomach.)

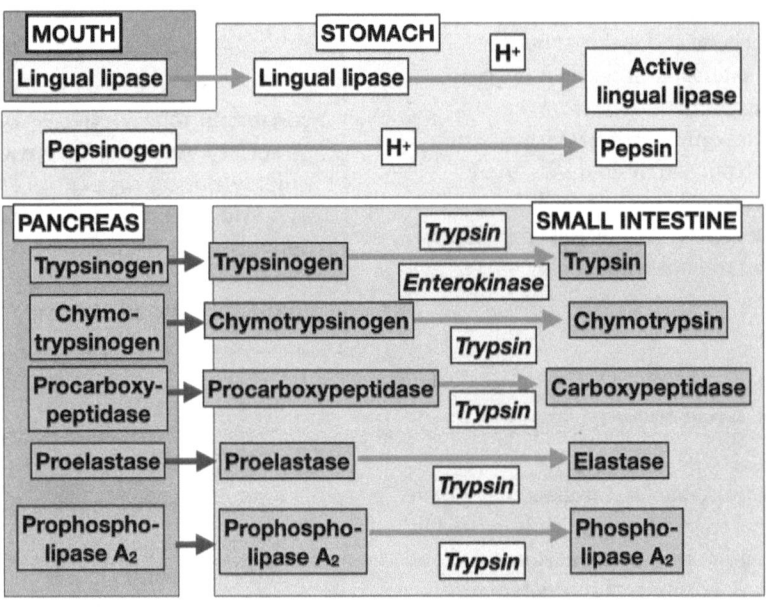

Figure 11.7 Activation of zymogens to form active enzymes.

There are differences in the anatomy of the functioning of the stomach in ruminants (discussed under the digestive system of ruminants below). In addition, in poultry, the glandular and muscular aspects of the stomach are in separate organs: respectively, the proventriculus and the gizzard.

The vagus nerve stimulates both gastric secretion and peristalsis.

There are four major secretory cells in the gastric pits of the stomach (see Figure 11.8).

- **Mucous neck cells** producing bicarbonate rich mucus to protect the cells of the stomach from the stomach acid.
- **Parietal** or **oxyntic** cells secreting hydrogen and chloric ions (hydrochloride acid) and **intrinsic factor**; the latter being critical to the absorption of vitamin B$_{12}$.
- **Chief or zymogenic cells** secreting pepsinogen. This is converted to the protease pepsin by the stomach's acidity. Pepsin starts the digestion of proteins (also see Textbox 11.7).
- Multiple **enteroendocrine cells** such as the G cells that produce gastrin.

11.4.2 Parietal Cells

Parietal cells release hydrogen ions (protons) and chloride ions into the lumen of the stomach (see Figure 11.9).

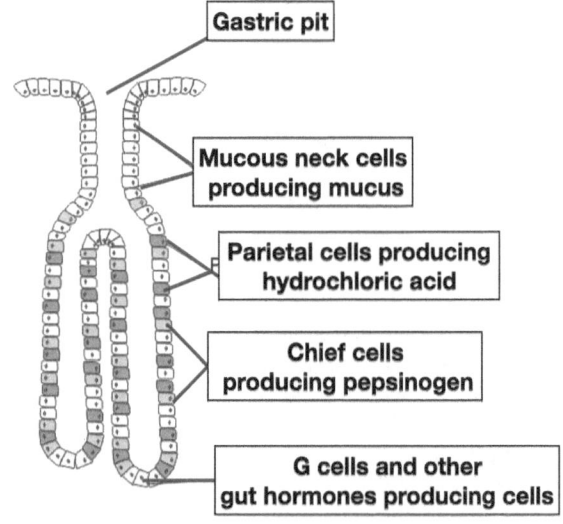

Figure 11.8 Gastric pit in stomach of a nonruminant. (Adapted from Frank Boumphrey, Creative Commons)

Textbox 11.7 Possible Way to Remember What Chief Cells Do

You go to a Chiefs game and have a Pepsi. Chief cells produce pepsinogen.

Together, these ions are **hydrochloric acid**. The hydrogen ions are responsible for reducing the pH on the stomach content to approximately 1.0.

The hydrogen ions are then exported into the lumen of the stomach by **proton pumps** (H^+ K^+ ATPase or H^+ K^+ ATPase antiporter) (see Figure 11.9). There is reciprocal movement of potassium ions into the parietal cells via H^+ K^+ ATPase (see Figure 11.9). Luminal potassium ions are derived from release of potassium ions from the parietal cells via the potassium channels (see Figure 11.9). To maintain **homeostasis** (electrical neutrality), chloride ions are pumped through **chloride channels** in the apical surface of the parietal cell into the stomach of the lumen (see Figure 11.9).

What is the source of the hydrogen ions? The parietal cells contain **carbonic anhydrase**. This catalyzes the combination of carbon dioxide and water to carbonic acid which in turn ionizes to bicarbonate and hydrogen ions (protons) (see Figure 11.10).

$$CO_2 + H_2O \rightarrow H_2CO_3 \rightarrow HCO_3^- + H^+$$

Carbon dioxide is either generated by metabolism of the parietal cells or diffuses from the blood (see Figure 11.10). Bicarbonate ions pass into the blood.

What controls the release of hydrogen ions? Stomach acid is released when food is present in the stomach. There are a series of chemicals that influence H^+ K^+ ATPase activity (see Figure 11.11). Stimulators of proton pumps include the neurotransmitter **acetyl choline**, the neuropeptide gastrin, and histamine (see Figure 11.11). In addition, the neuropeptide somatostatin inhibits H^+ K^+ ATPase activity (see Figure 11.11).

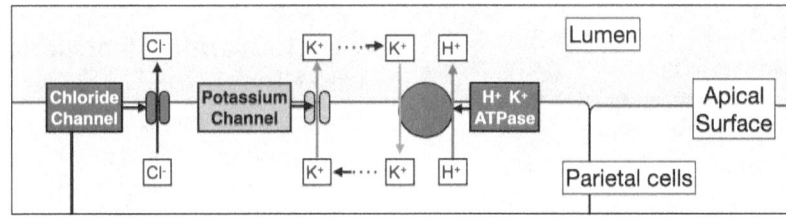

Figure 11.9 Mechanism of release of hydrochloric acid from parietal cells via proton pumps (H+ K+ ATPase) and a series of ion channels.

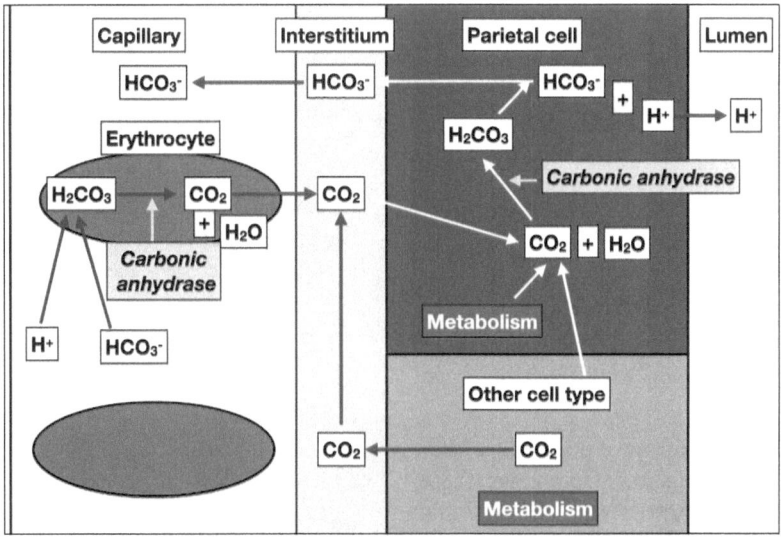

Figure 11.10 Role of carbon dioxide and carbonic anhydrase in the generation of acid (H^+) by parietal cells.

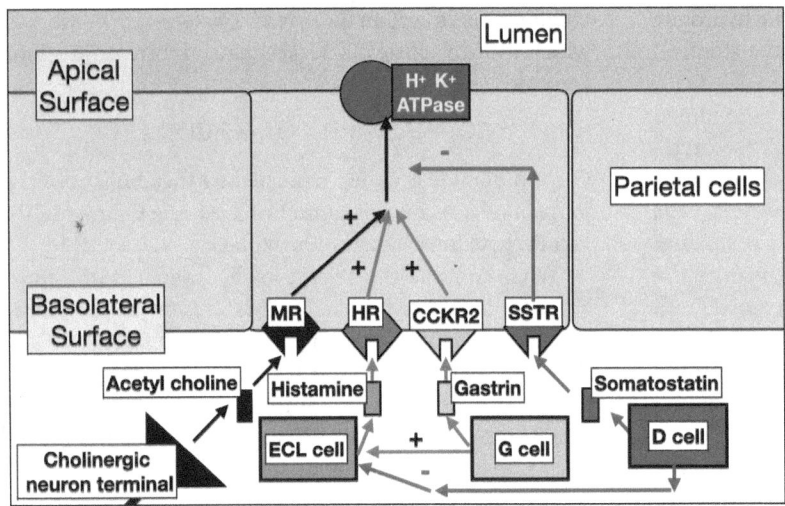

Figure 11.11 Control of the release of hydrochloric acid from parietal cells by increasing proton pumps. [Key:
$H^+ K^+$ ATPase = proton pump;
MR = muscarinic receptor;
HR = histamine receptor,
CCKR2 = gastrin receptor (a.k.a. cholecystokinin receptor 2);
SSTR = somatostatin receptor
ECL cell = enterochromaffin-like cells;
G cell = produce gastrin;
D cell = producing somatostatin.]

11.4.3 Enteroendocrine Cells

Enteroendocrine cells secrete chemical signals (hormones) into the lamina propria to regulate gastric digestion. The chemical signals that are secreted in a paracrine fashion to regulate intestinal functioning. This encompasses different cell groups such as the following:

- **EC cells (Enterochromaffin cells)** in the small intestine release serotonin, a.k.a. *5-hydroxytryptamine (5 HT)*.
- **ECL cells (Enterochromaffin-like cells)** in the stomach secrete histamine, which stimulates release of gastric acid (see Figure 11.11).
- **D cells** in the stomach release somatostatin, which inhibits release of gastric acid (see Figure 11.11).
- **G cells** in the stomach release gastrin which stimulates release of gastric acid by increasing the intracellular concentration of calcium ions (see Figure 11.11).
- **I cells** in the small intestine secrete cholecystokinin. This induces contraction of the gall bladder and the secretion of pancreatic enzymes.
- **S cells** in the duodenum secrete secretin, which stimulates release of pancreatic bicarbonate.

11.4.4 Lifespan of Stomach Epithelial Cells

The epithelial cells in the stomach have a short life. They die and are sloughed off to be digested. On average, these surface epithelial cells are replaced every 3–6 days. The cells are replaced very quickly through division of stem cells in the gastric pit. This serves as a form of protection for the stomach.

11.5 Small Intestine

11.5.1 Overall Introduction to the Small Intestine

The small intestine is a long tube with the same overall structure as shown in Figure 11.2, again without glands in the submucosa. It consists of the following three regions:

- **Duodenum** (the first region after the stomach with the pancreas lying in the loop of the duodenum).
- **Jejunum** (the second region of the small intestine).
- **Ileum** (the third region of the small intestine).

The small intestine is located in the abdominal cavity.

The small intestine is the site of digestion of ingesta and absorption of the products of digestion. When the ingesta leaves the small intestine, digestion and absorption is complete in nonfermenters and ruminants. An exception to this statement is horses with hindgut fermentation. To facilitate both digestion and absorption of nutrients, the small intestine has a huge surface area. This is achieved by the presence of villi (singular villus) and microvilli on the surface of enterocyte (see Figure 11.12).

11.5.2 Cell Types in the Villi

The villus contains several different cell types. These are the following:

1. **Enterocytes** or columnar absorptive cells.
2. **Goblet cells** producing mucus.
3. **Enteroendocrine cells** producing peptide hormones.
4. **Exocrine cells** producing digestive enzymes.

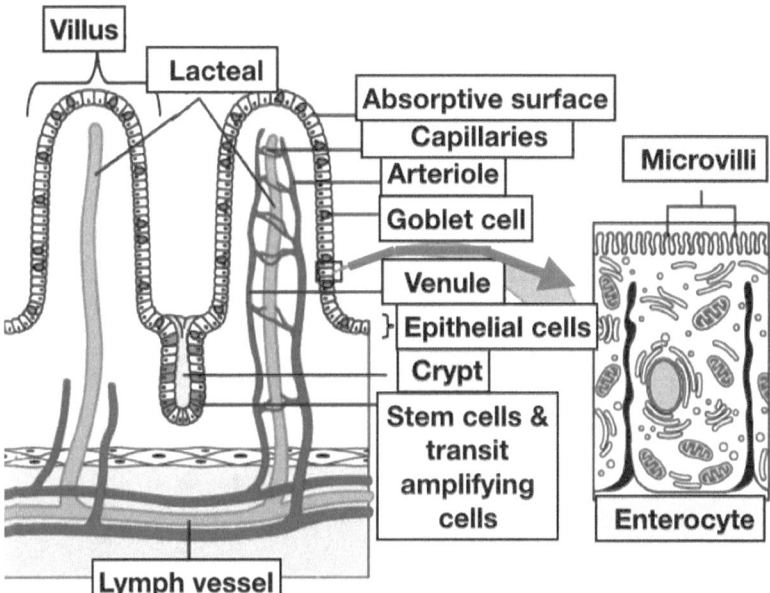

Figure 11.12 Villi of small intestine showing blood vessels, lymph vessels (including lacteals), crypts, and the microvilli. The villi and microvilli greatly expand the surface area of the small intestine facilitating digestion and absorption of nutrients.
(Adapted from OpenStax, CC BY 4.0)

5. **Subepithelial fibroblast cells** functioning as mechanoreceptors.
6. **Smooth muscle cells**.
7. **Mucosal endothelial cells** of the blood vessels.
8. Stem cells.

11.5.3 Functioning of the Small Intestine

The functions of the small intestine (duodenum, jejunum, and ileum) are the following:

1. Synthesizing digestive enzymes.
2. Receiving secretions from the liver (bile stored in gall bladder to aid digestion of fats) and pancreas (exocrine pancreatic secretions containing zymogens, enzymes, and bicarbonate).
3. Enzymatic digestion as follows.
   a. **Proteases** (**trypsin** and **chymotrypsin**) and **peptidases** breaking down proteins to amino acids.
   b. **Amylases** breaking down starch and other **polysaccharides** to **disaccharides**.
   c. **Disaccharidases**, enzymes digesting disaccharides such as maltase and lactase. These are found on the **brush border** and cleave disaccharides to monosaccharides (e.g., glucose, fructose, and galactose).
   d. **Lipases** that (with bile salts) digest **triacylglyceride**.
4. Absorption of nutrients including simple sugars (monosaccharides), amino acids, and glycerol together with minerals and vitamins into the bloodstream.
5. Absorption of fatty acids into the lymph.
6. Messaging to influence the following:
   a. Appetite/satiety (eating).
   b. Release of pancreatic (from the pancreas) and liver juices (or bile from the gall bladder).
   c. Releasing intestinal hormones that increase the secretion of insulin and other hormones.
7. Transporting ingesta or chyme by peristalsis (see Figure 11.3) from the stomach to the colon and/or cecum.
8. Acting as a barrier to prevent pathogens invading the body.

There are tight junctions between epithelial cells in the gastrointestinal tract creating a barrier between the lumen of the intestine and the interstitial space and then to the circulation (also see Section 1.6.6 in Chapter 1). The **epithelial barrier** is dynamic with permeability to some ions being shifted by nutritional state. Epithelial layers with tight junctions in the gastrointestinal tract can become so leaky that they allow pathogens to enter the body from the gut lumen.

11.5.4 Functioning of the Microvilli

Not only is the ingesta being converted into absorbable nutrients in the lumen of the small intestine but also the brush border of the microvilli is critically important to digestion. Enzymes tethered to the brush border include what are referred to as brush border enzymes:

- Maltase (catalyzing splitting the disaccharide maltose to two monosaccharides, glucose, molecules).
- Sucrase/isomaltase (catalyzing splitting the disaccharide sucrose to two monosaccharides, respectively, a glucose and a fructose molecule).
- Lactase (catalyzing splitting the disaccharide lactose to two monosaccharide molecules, a glucose and a galactose molecule). This is very important to newborn mammals that are receiving milk as their food.
- A series of peptidases (catalyzing the splitting of the peptide bonds to yield amino acids).
- Several lipases (catalyzing splitting the triglyceride to a glycerol molecule and three fatty acid molecules or two fatty acids and one monoglyceride).

11.5.5 Transporters in Microvilli

There are a series of transporters in the cell membranes of the microvilli. These are responsible for the transport of nutrients into the enterocytes (also known as absorptive cells). The nutrients pass through the luminal cell membrane into the interstitial fluid (and then into the blood) via the basal cell membranes, including the following:

- Glucose and other monosaccharides.
- Basic, neutral, and acidic amino acids.
- Ions such as sodium, calcium, etc.

An example of absorption in the small intestine is that of glucose. There is a flux of glucose from the small intestine lumen into the hepatic portal vein and hence to the liver. Figure 11.13 summarizes glucose transport from the lumen of the small intestine through the enterocyte to the

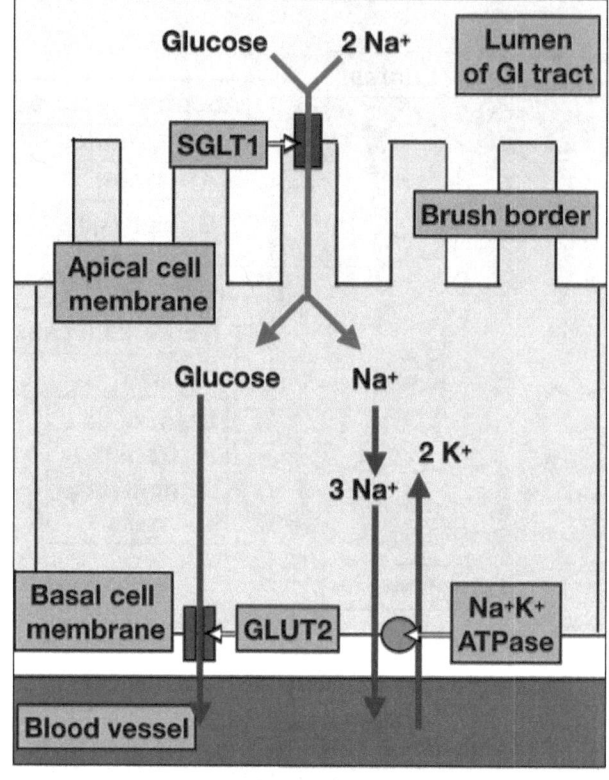

Figure 11.13 Transport of glucose from the lumen of the small intestine through the enterocytes into the blood.
[Key: GLUT2 = glucose transporter 2; SGLT1 = sodium-dependent glucose cotransporters, Na^+ sodium ions]

blood vessels leading to the hepatic portal veins. There are both **sodium-dependent glucose cotransporters** (e.g., SGLT1) and **glucose transporters** (GLUT).

Textbox 11.8 A Deeper Dive Into Absorption—Calcium Absorption

Absorption of calcium ions is summarized in Figure 11.14.

Concentrations of calcium ions are maintained within tight limits in both the plasma and within the cell. Concentrations of calcium ions are 12,500 times higher in plasma than within the cell; the calcium ion concentrations being the following, respectively:

- Plasma 1.25 mmoles L^{-1}.
- Intracellular 100 nmoles L^{-1}.

Within the enterocytes of the small intestine, calcium ions are bound to calbindin-D. This prevents the emergence of high concentrations of ionized calcium and consequent damage to the cells. Calcium absorption is greatly increased by the hormone 1,25-dihydroxyvitamin D (see Sections 4.1.3.7, 4.4.1.2, and 4.10.3). 1,25-dihydroxyvitamin D binds to the vitamin D receptor which binds to the retinoid X receptor (RXR) forming a heterodimer. This leads to increased expression of plasma membrane

(continued)

Textbox 11.8 (continued)

ATPase (PMCA), apical calcium channel (TRPV6), and calbindin-D (see Figure 11.14).

Calcium ions pass from the lumen to the interstitial space and then into the blood due to the action of a series of transporters including plasma membrane ATPase (PMCA), apical calcium channel (TRPV6).

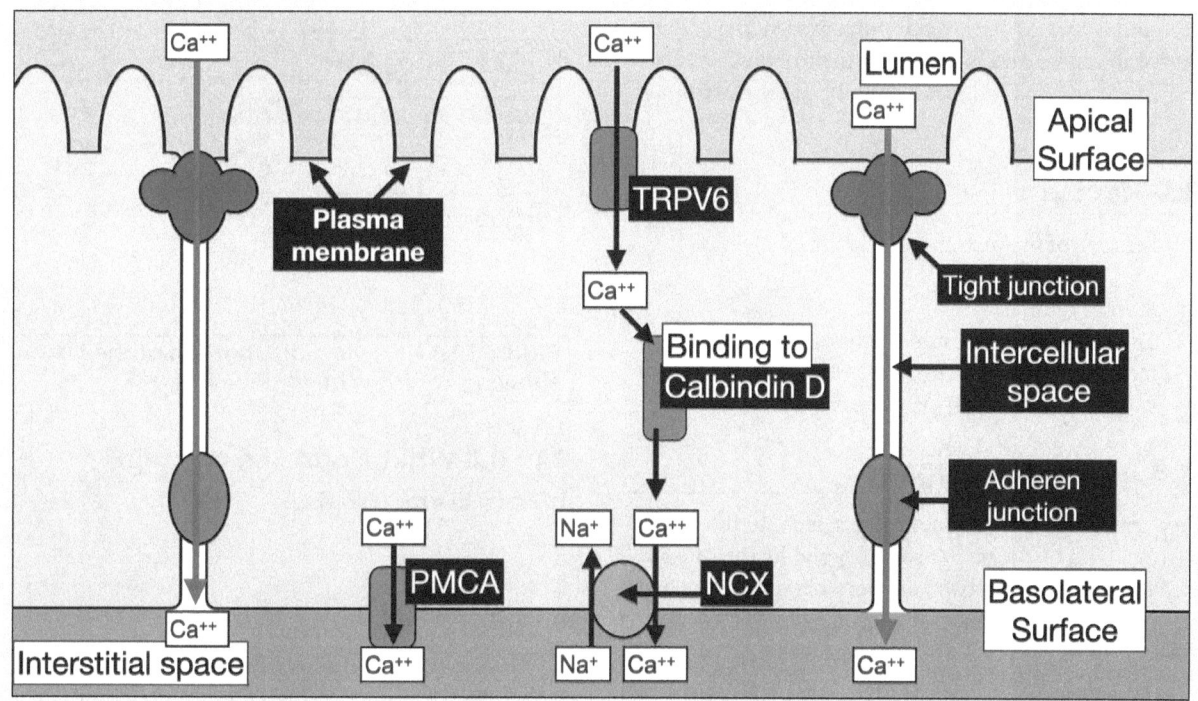

Figure 11.14 Mechanism of calcium absorption in the epithelium of the small intestine. Calbindin D aids intracellular movement of calcium ions.
[Key:
Ca^{++} = calcium ions;
TRPV6 = Transient Receptor Potential Vanilloid Channel Type 6;
PMCA = Plasma Membrane Calcium ATPase (an ATP driven calcium pump);
Na^+ = sodium ions;
NCX = sodium calcium exchanger.]

11.6 Cecum

The functions of cecum (singular, one in mammals) or ceca (plural, two in birds) include the following:

1. Fermentation by microbiome.
2. Absorption of water and electrolytes.
3. Absorption of volatile fatty acids and other products of fermentation.

The cecum is massive in horses, being over a meter long and capable of holding about 30 liters. It is critical to their obtaining nutrients. This is covered in more detail under the gastrointestinal tract of horses.

11.7 Colon

The functions of the colon or large intestine include the following:

1. Absorption of water and electrolytes.
2. Transporting ingesta (peristalsis) from the small intestine to the rectum.

3. Particularly important in hindgut fermenters representing 40–50% of the gastrointestinal tract in horses.
4. Fermentation by bacteria.
5. Absorption of fermentation products, e.g., vitamin (biotin) and vitamin K.

The colon is located in the abdominal cavity. The colon is massive in horses and is critical to their obtaining nutrients. This is covered in more detail under the gastrointestinal tract of horses.

11.8 Rectum

The functions of the rectum include the following:

1. Storage of feces.
2. Transporting indigestible components of ingesta/feces (by peristalsis) from the rectum to the exterior (defecation) via the anus.

11.9 Accessory Organs

There are two major accessory organs tied into the gastrointestinal tract by ducts, by proximity and by their importance to digestion. The two accessory organs are the liver with the gallbladder, and the pancreas.

11.10 Liver

11.10.1 Introduction to the Liver

The liver is found in the abdominal cavity. It is a critical organ having multiple functions including the following:

- Making bile (important for the digestion of fats).
- Removing products of the degradation of hemoglobin, namely the bile pigments bilirubin and biliverdin.
- Controlling metabolism (discussed in more detail in Chapter 7).
- Producing **erythrocytes** (red blood cells) during fetal development.
- Producing hormones, such as insulin-like growth factor 1 (IGF1) or hormone precursors such as **angiotensinogen**.

The hepatocytes of the liver (Figure 11.15) produce bile which is then stored in the gallbladder. There is a duct (the bile duct) leading ultimately into the duodenum (small intestine). Bile is released when there is ingesta in the small intestine.

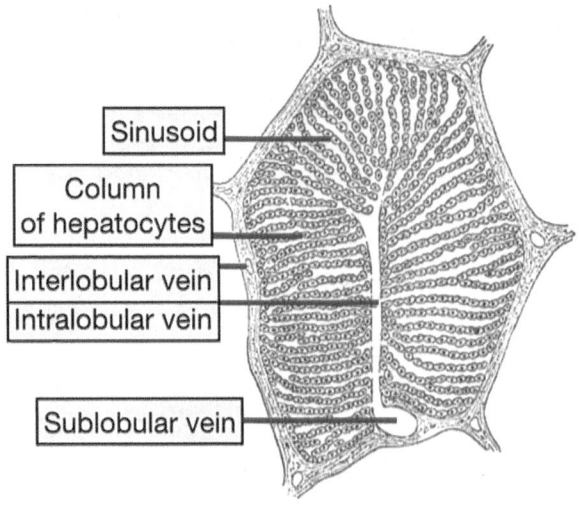

Figure 11.15 A view of a portion of the liver showing columns of hepatocytes.

11.10.2 What Does Bile Contain?

Bile contains the following:

- Water.
- Bile acids or bile salts.
- Bile pigments (bilirubin).
- Phospholipids and cholesterol.
- Some drugs and environmental contaminants.

Bile acids (also known as bile salts) are chemically related to cholesterol. They act as detergents to emulsify fat globules into small droplets. The bile acids are the following:

- Chenodeoxycholic acid in horses.
- Cholic acid in cattle and dogs.

The conjugated bile salts are reabsorbed into the enterohepatic circulation (see Figure 11.16) such that there is little bile acid in the feces.

11.10.3 Are There Enzymes in Bile?

The simple answer is no.

11.10.4 Is the Bile Important to Digestion?

The simple answer is yes. The bile salts play an important role in digestion of fats but do not directly digest fats. Bile salts emulsify large globules of fat (triacyglyceride) to droplets that are readily digested.

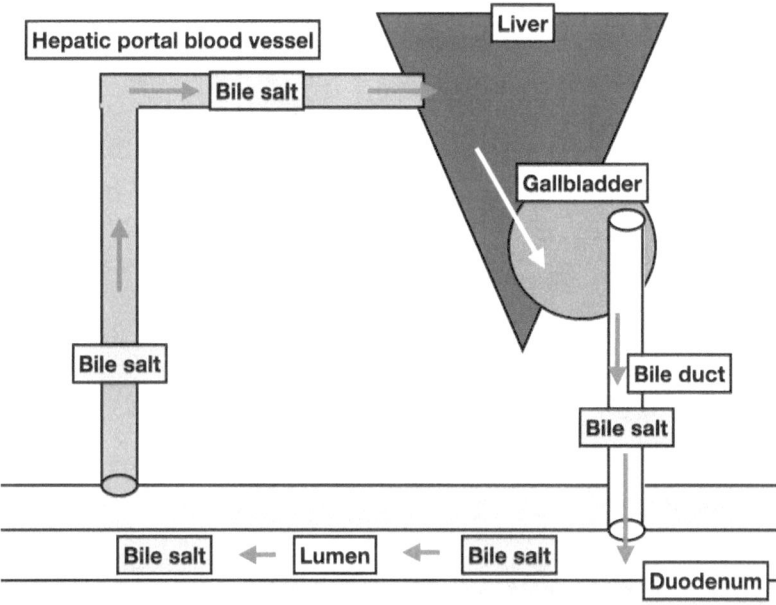

Figure 11.16 Bile salts (also known as bile acids) are released into the small intestine when the gallbladder contracts due to stimulation by cholecystokinin (CCK). The bile salts emulsify fat globules in the small intestine lumen, then are reabsorbed into the enterohepatic circulation.

11.10.5 What Does the Gallbladder Do?

The gallbladder stores bile. This is released by contraction of the gallbladder when there is ingesta in the duodenum. The contraction is induced by the hormone cholecystokinin (CCK).

11.11 Pancreas

11.11.1 Introduction to the Pancreas

The pancreas consists of two distinct tissues:

- The **exocrine pancreas** produces **pancreatic fluid**, which is secreted into the duodenum when there is ingesta in the small intestine. Pancreatic fluid plays an important role in digestion.
- The **endocrine pancreas** produces hormones including insulin and glucagon (discussed in Chapter 4).

Figure 11.17 shows the structure of the exocrine pancreas.

11.11.2 Pancreatic Secretion

The pancreas secretes large quantities of a fluid with high concentrations of sodium bicarbonate ions (HCO_3^-) in response to the small intestine hormone secretin and/or the neurotransmitter acetylcholine. The bicarbonate ions are derived from the following: (1) bicarbonate entering the cells by the sodium bicarbonate cotransporter (in the basolateral membrane), (2) carbonic anhydrase catalyzing carbon dioxide combining with water to form carbonic acid, which rapidly dissociates to form bicarbonate ions (intracellular), and (3) chloride bicarbonate exchangers (luminal membrane).

The small intestine produces two hormones when ingesta is present:

- Cholecystokinin, which stimulates the release of pancreatic enzymes.
- Secretin, which stimulates the release of pancreatic bicarbonate.

11.11.3 Why Does the Pancreas Secrete Bicarbonate?

The sodium bicarbonate acts to neutralize the acid ingesta coming from the stomach.

11.11.4 Are There Enzymes in Pancreatic Fluid?

There are multiple enzymes and zymogens in pancreatic fluid. Zymogens are inactive but become active enzymes when activated after cleavage.

For example, the pancreatic fluid contains the following zymogens:

- Trypsinogen.
- Chymotrypsinogen.

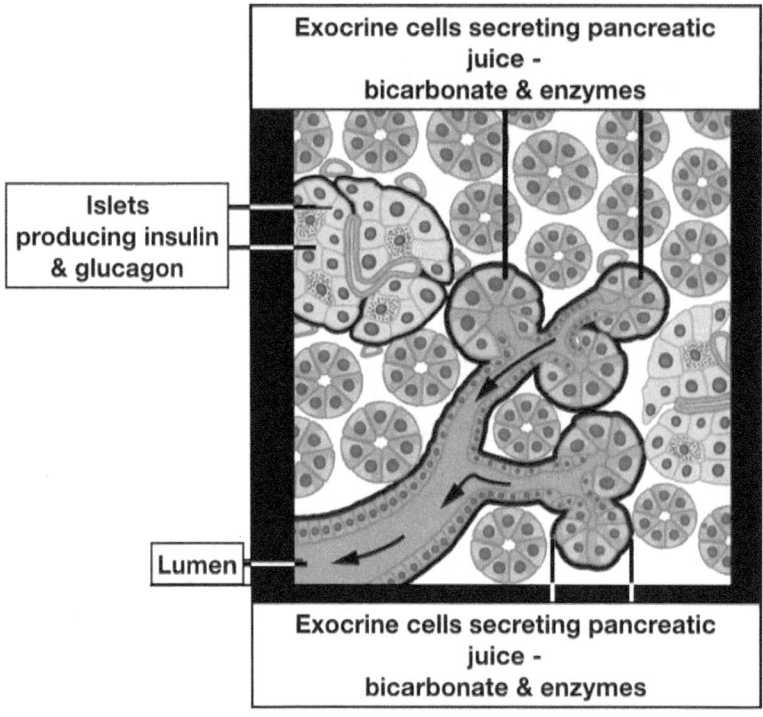

Figure 11.17 Schematic diagram of pancreas. (Adapted from OpenStax, CC BY 4.0)

- Procarboxypeptidase.
- Phospholipase A_2.
- Proelastase.

Trypsinogen is activated to trypsin by the enzyme **enterokinase** (enteropeptidase) in the brush border of the duodenum (part of the small intestine). In turn, trypsin activates more trypsinogen (autocatalysis), chymotrypsinogen to chymotrypsin, and procarboxypeptidase (see Figure 11.5). Trypsin, chymotrypsin, and carboxypeptidase are proteases that break down proteins.

Pancreatic fluid also contains digestive enzymes including the following:

- Amylase (breaking down starch to glucose).
- Lipase (breaking down triglyceride to fatty acids and glycerol).
- Nuclease (breaking down RNA and DNA to generate nucleotides).

11.12 Digestion of Proteins

Protein digestion can be viewed as starting in the mouth with chewing of the food. This starts mechanical digestion breaking protein containing ingesta to smaller sizes to aid later enzymic digestion. This mechanical digestion continues in the muscular stomach as it churns the ingesta. Proteins are subjected to chemical digestion in the stomach with some denaturing of the protein in the stomach acid.

Enzymic digestion of proteins is summarized in Figure 11.18. Enzymic digestion of proteins in the ingesta starts in the stomach, with the proteolytic enzyme pepsin cleaving peptide bonds. This is after pepsin have been activated from pepsinogen due to the stomach acid. Enzymic digestion of ingesta proteins continues in the small intestine. The proteases trypsin and chymotrypsin are generated from the zymogens trypsinogen and chymotrypsinogen in the pancreatic fluid (see Figure 11.7). Similarly, carboxypeptidase is activated by trypsin (see Figure 11.7). The proteases cleave peptide bonds creating peptide fragments (Figure 11.15). In turn, the peptides are digested to amino acids by peptidases in the pancreatic fluid and in the brush border (Figure 11.18).

There is a flux of amino acids from the small intestine lumen into the hepatic portal vein and hence to the liver. Amino acids are absorbed entering the enterocytes by a series of transporters for acid, neutral, and basic amino acids. By the end of the passage of ingesta through the small intestine, over 80% of protein has been digested, and the amino acids absorbed.

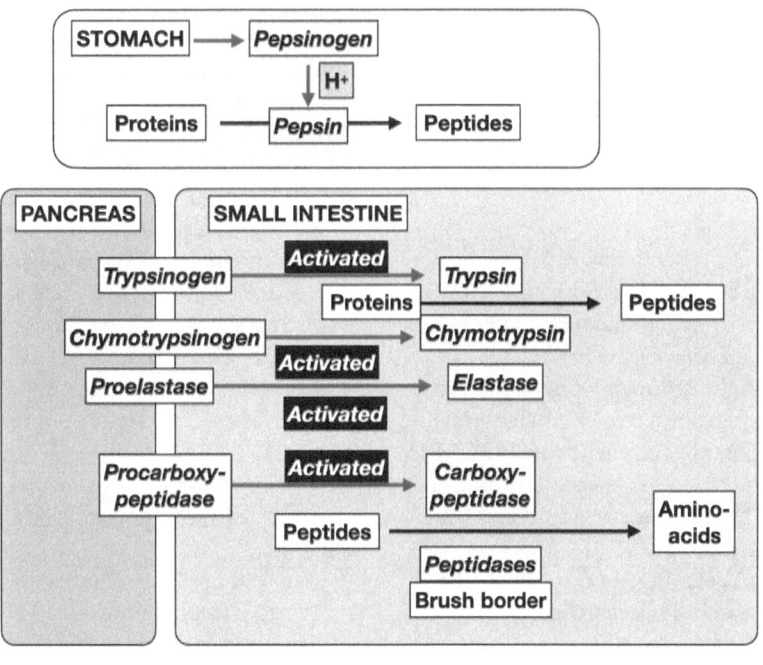

Figure 11.18 Digestion of proteins.

11.13 Carbohydrate Digestion

Discussion of carbohydrate digestion will be limited to the nonruminants (ruminant digestion is covered below in Section 11.16 below). Mechanical digestion of carbohydrates is similar to that of proteins (see above). Specific carbohydrate digestion starts in the mouth with the release of the saliva containing amylase (the enzyme breaking down starch) (Figure 11.19). Digestion of starch and other carbohydrates resumes in the small intestine.

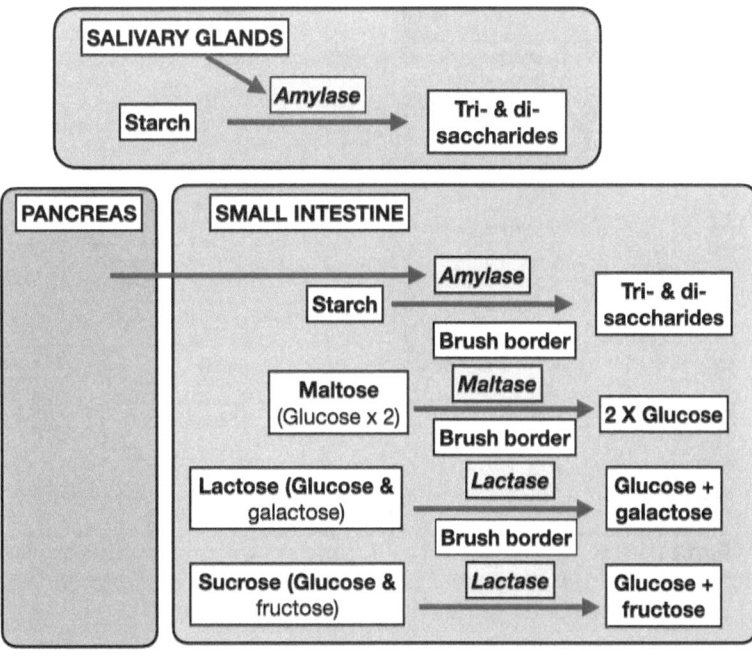

Figure 11.19 Carbohydrate digestion.

This is achieved by amylase in the pancreatic fluid, along with a series of brush-border enzymes cleaving disaccharides (Figure 11.19). The disaccharides and the enzymes that cleave them are the following:

- **Maltose** cleaved by **maltase**.
- **Sucrose** cleaved by **sucrase**.
- **Lactose** cleaved by **lactase**.

It is important to realize lactose is the principal carbohydrate received by the pre-weaning mammal, irrespective of whether it is a ruminant or not.

Glucose passes across the cell membrane of the brush border of the intestinal epithelium cells or enterocytes due to **sodium-dependent glucose cotransporter 1** (**SGLT1**) and then exported to the blood by glucose transporter 2 (GLUT2) (see Figure 11.14). There are additional transporters for galactose and fructose.

By the end of the passage of ingesta through the small intestine, over 90% of starch has been digested and the glucose generated absorbed.

11.14 Lipid (Fat) Digestion

Mechanical digestion of fats in the mouth and stomach is similar to that of proteins (see Sections 11.2 and 11.4).

The liver plays an important role in the digestion of fats. The liver produces bile, which contains bile salts/acids, and this acts as a detergent reducing the size of fat globules to tiny droplets (see Figure 11.20). This provides a much larger surface area for digestion of triacylglyceride to either free fatty acids or monoglycerides by lipase. While most of the lipase is from the pancreatic secretion, there is also **lingual lipase** (from the salivary glands in the mouth) and **gastric lipase** (from the stomach). These are combined with bile salts to produce **micelles**. These pass across the brush border into the enterocytes. The fatty acids can be re-esterified to triacylglyceride or phospholipids and placed into chylomicrons (see Figure 11.21). These are secreted into the lymph vessels (see Figure 11.21), except in poultry as they lack lacteal ducts. In poultry, lipids are packaged into protomicrons in the enterocytes and exported to the liver via the hepatic portal system.

Phospholipids are digested in the small intestine by the enzyme, phospholipase A_2. This enzyme is produced in the pancreas as a zymogen (see Figure 11.23). It is activated by trypsin in the small intestine (see Figure 11.24). It breaks each phospholipid into one fatty acid and one lysophospholipid. These are transported into the enterocyte as micelles (see Figure 11.21). Cholesterol also enters the enterocytes in the micelles.

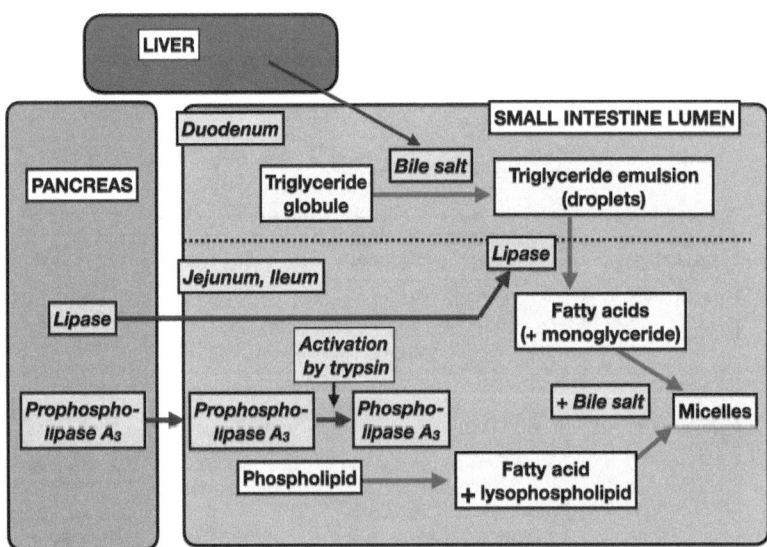

Figure 11.20 Digestion of triglycerides and phospholipids. Note in addition to pancreatic lipase, there is also some lingual lipase from the saliva and gastric lipase from the stomach.

11.15 Control of Gastrointestinal Functioning

Eating, digestion, and movement of ingesta through the intestinal tract is controlled by the nervous system and gastrointestinal hormones. Biting and chewing are due to voluntary muscles controlled consciously by the brain. However, the autonomic nervous system, together with enteric nervous system and enteric hormones, control the digestion and movement of ingesta through the gastrointestinal tract. The enteric nervous system is part of the autonomic nervous systems, which is responsible primarily for involuntary actions, and communicates with the central nervous system (brain) through the parasympathetic and sympathetic nervous system via the vagus nerve and prevertebral ganglia, respectively.

The enteric hormones regulate appetite (the desire to eat), satiety (feeling "full"), and gastric motility. In the initial or **cephalic phase** of gastric digestion, the sight, smell, and taste of food results in the stimulation of the vagus nerve to begin secretion of gastric juice and contractions of the stomach. During the second or **gastric phase** of gastric digestion, the stomach senses stretch, the change in pH from food and gastric juices entering, or endocrine signals to stimulate the production of gastric juice.

11.16 The Gastrointestinal Tract and Digestion in Ruminants

Textbox 11.9 provides some key points about ruminants.

Figure 11.22 shows the passage of ingesta in the gastrointestinal tract of cattle.

Textbox 11.9 Key Points About Ruminants

How Can Ruminants Live on Plants That Human Can't Digest?
Micro-organisms ferment the ingesta, chemically breaking the ingesta down, and supplying the animal with volatile fatty acids.

What Is a Volatile Fatty Acid (VFA)?
Volatile fatty acids (VFA) are very short chain fatty acids (SCFA) with the major VFAs in ruminants being acetic, propionic, and butyric acids. These are discussed as salt or ester forms, e.g., acetate, propionate, and butyrate.

Ruminal Fermentation
The microbiome in the rumen metabolizes cellulose, hemicellulose, starch, and simple sugars to VFAs, carbon dioxide, and methane. Proteins are also metabolized to VFAs, ammonia, and other gases. Triglyceride can be digested to fatty acids and glycerol.

Types of Ruminants
There are three types of ruminants:

- Forage grazers, such as cattle and bison.
- Intermediate, such as sheep and goats.
- Concentrate feeders, such as deer.

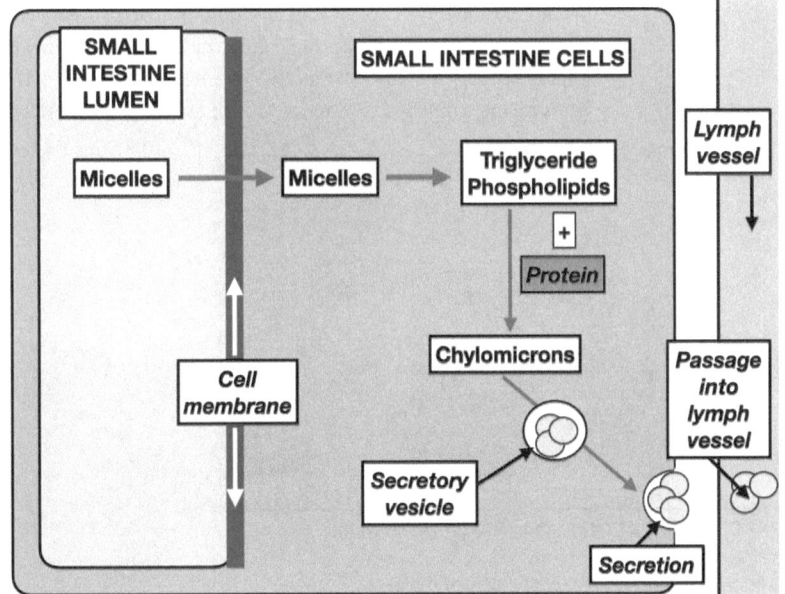

Figure 11.21 Absorption and processing of triglyceride and phospholipids.

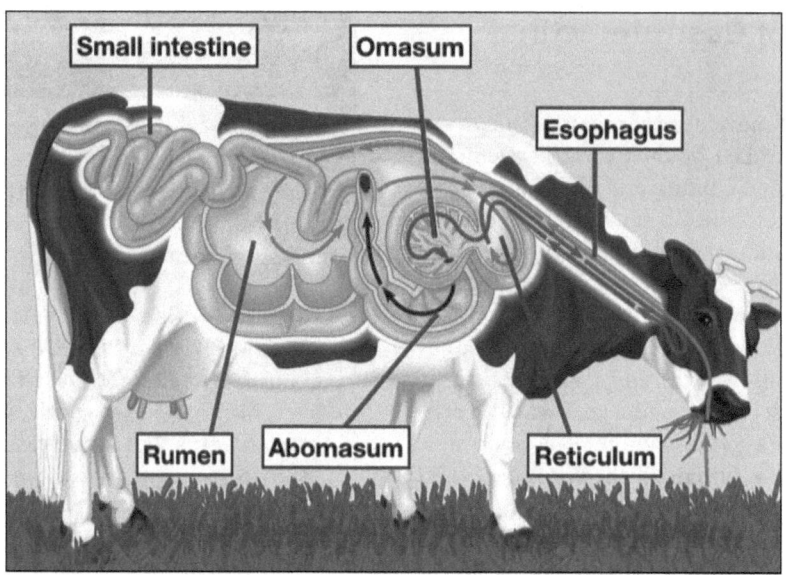

Figure 11.22 Flow of ingesta through the rumen, reticulum, omasum, and abomasum in order of passage.
(Adapted from N B Shridhar, CC BY-SA 4.0)

11.16.1 Mouth

Ruminants eat forages and high fiber plants. There is little chewing of the food before swallowing. Ruminants produce large qualities of saliva. For instance, cattle produce 150 liters (37 gallons) of saliva per day. The saliva contains sodium and potassium ions that bring ingesta to optimal pH for fermentation in the rumen. Unlike monogastric animals, ruminant saliva does not contain amylase.

11.16.2 Ruminant Stomach

There are four chambers to the ruminant stomach.

- **Reticulum** (or honeycomb).
- **Rumen** (or paunch).
- **Omasum**.
- **Abomasum** (or true stomach).

The four chambers take up most of abdominal cavity (see Figures 11.19 and 11.20). In cattle, the rumen plus reticulum contains about 200 liters (50 gallons) of fluid, and partially digested ingesta.

11.16.3 Rumen

Ingesta pass from esophagus into rumen (also see Figures 11.22 and 11.223). The rumen is a **muscular fermentation vat**. The cellulose and hemicellulose in the plant-based ingesta is subject to **anaerobic fermentation** generating volatile fatty acids (VFA) also known as **short**

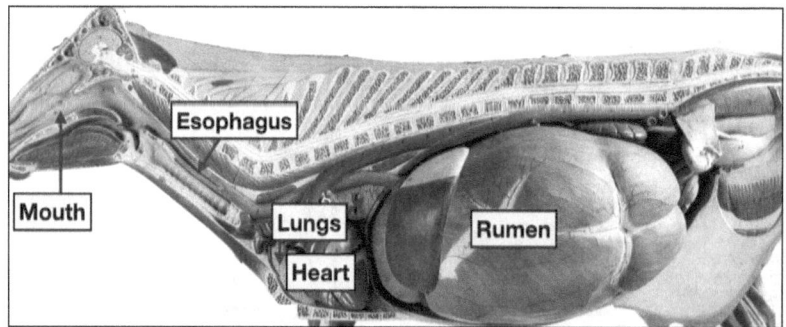

Figure 11.23 Model of cow's abdominal cavity showing rumen taking up most of space.
(Adapted from the Museum of Veterinary Anatomy)

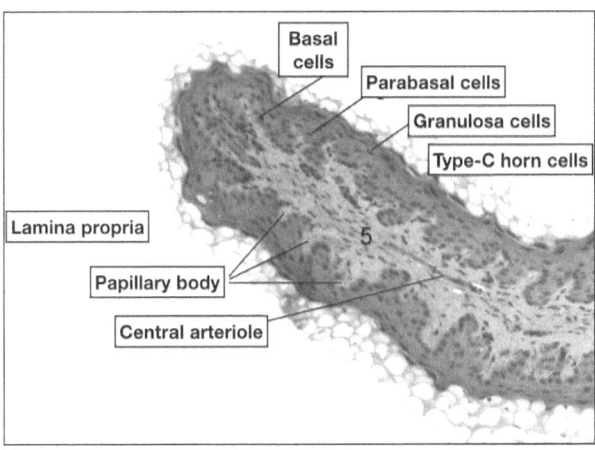

Figure 11.24 Ruminal papilla with squamous epithelium (type-C horn cells). (Adapted from Uwe Gille, CC BY-SA 3.0)

chain fatty acids (SCFA), together with lactate. The VFA generated by microbial fermentation are **acetate**, **propionate**, **butyrate**, and **isobutyrate**, and these are absorbed. In addition, to fermentation and absorption of VFA, the rumen churns moving ingesta, acting as mechanical digestion. In addition to generating VFA, there is generation of methane and carbon dioxide. These are lost from the rumen by belching (eructation). In cattle, there is about ~40 liters (10 gallons) of gas produced each day. The microbial population of the rumen also ferment protein from the food, together with ammonia and urea, to generate microbial proteins; these become a protein source for the animal.

The surface of the **rumen papillae** (Figure 11.24) is covered with **squamous epithelium**. This is the site of absorption of VFA [short chain fatty acids (SCFA)] and lactate. VFA are absorbed through the rumen wall. There are transporters that move from the lumen of the rumen into, through and out of the epithelial cells. The VFA are transported across the cell membrane by diffusion, SCFA bicarbonate exchangers, anion exchangers, monocarboxylate transporters, and anion channels.

11.16.4 Reticulum

The role of the reticulum is to sort particles by size. Small particles of ingesta and microbes are collected and pass through the reticulum to the omasum. In contrast, large particles are returned to the rumen.

11.16.5 Omasum

The omasum is the penultimate chamber of the stomach (also see Figures 11.22 and 11.23). It is a site for water absorption and mechanical digestion. In addition, it moves ingesta to the abomasum.

11.16.6 Abomasum

The **abomasum** is the final part of the stomach (also see Figures 11.22 and 11.23). It functions as the true stomach with the release of the zymogen pepsinogen and hydrochloric acid. Pepsinogen is activated by the acid to pepsin (Figure 11.5) and this starts digestion of protein, including microbial protein.

There is a major difference in the size of abomasum between pre-weaning and post-weaning ruminants. For instance, the abomasum is the largest part of stomach in calves prior to weaning. Milk bypasses the rumen via the esophageal groove.

11.16.7 Rumination

Ruminants spend as much as a third of their time on rumination. Rumination consists of the following:

1. **Regurgitation**: moving the ingesta, also known as the cud, from the rumen to the mouth by reverse peristalsis.
2. **Remastification**: chewing the cud.
3. **Reinsalvation**: addition of saliva.
4. **Redeglutition**: re-swallowing the chewed cud.

Rumination is illustrated in Figure 11.19. Chewing the cud reduces the particle size of the ingesta to allow passage of the ingesta through reticulum and, hence, aid fermentation. In addition, the addition of saliva optimizes the pH for fermentation. There is little rumination when cattle are consuming a high-grain feed.

Textbox 11.10 Interesting Factoid: Methane from Ruminants

Is the Global Warming Effect of Cattle, Sheep, and Goats Due to Methane in Their Farts (Flatus)?
Largely, the answer is no! Methane is generated by the microbiome in the rumen. It is estimated that 95% of the methane goes to the environment in the breath of cattle, sheep, and goats. This is from the ruminant belching and out-gassing when the cud is being chewed. However, the methane is still a problem as it contributes to global release of greenhouse gases (GHG).

11.16.8 Small Intestine

As in other animals, amino acids and fats are digested and absorbed in small intestine.

11.16.9 Colon and Cecum

There is secondary fermentation in the colon and cecum generating 10–15% of the animal's energy.

11.17 Digestion in Horses (Pseudoruminants)

11.17.1 Introduction to Digestion in Horses

Horses are grazing herbivores consuming grass, haylage, and hay (see Figure 11.25). Figures 11.26 and 11.27 show the structures of the gastrointestinal tract in horses and their relative sizes.

Horses are hindgut fermenters. The cecum (a blind ended sac) and colon (ventral and distal) is the site for hindgut fermentation of the cell walls of plants (including fiber cellulose), together with any starch and proteins remaining after digestion in the stomach and small intestine.

11.17.2 Fermentation in the Foregut (Stomach and Small Intestine)

There is some fermentation of feed ingredients such as starch in the stomach and small intestine. This is indicated by the presence of hydrogen on the breath of horses.

Figure 11.25 Horses eating hay.

11.17.3 Functioning of the Stomach

The stomach is a site of mechanical digestion. In addition, there is addition of acid and proteases.

11.17.4 Functions of the Small Intestine

The small intestine is the site for chemical (enzymic) digestion of starch, proteins, and lipids. In addition, it is the site for absorption of simple sugars, amino acids, fatty acids, minerals, and vitamins. Food passes through the small intestine very rapidly moving at 30 cm per minute (or 12 inches per minute).

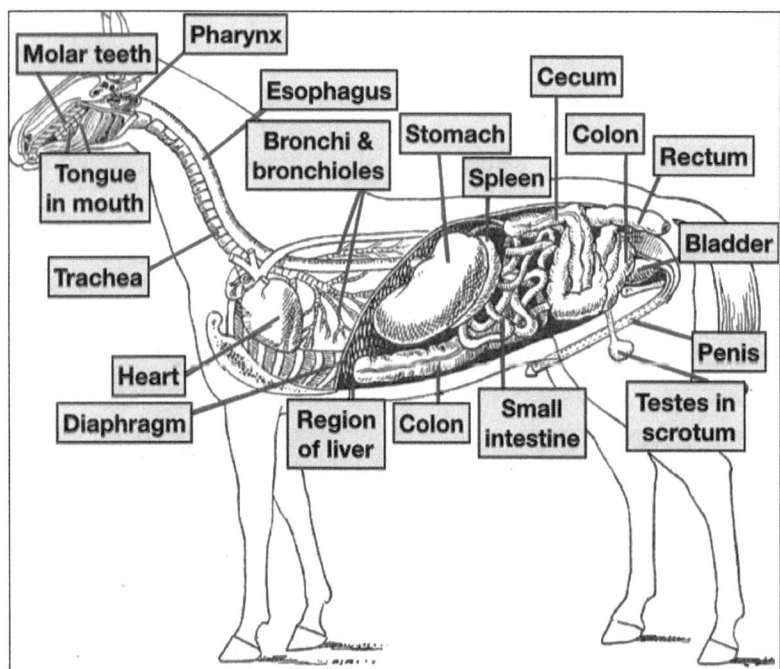

Figure 11.26 Structure of the gastrointestinal tract of the horse. (Adapted from Henry Thompson)

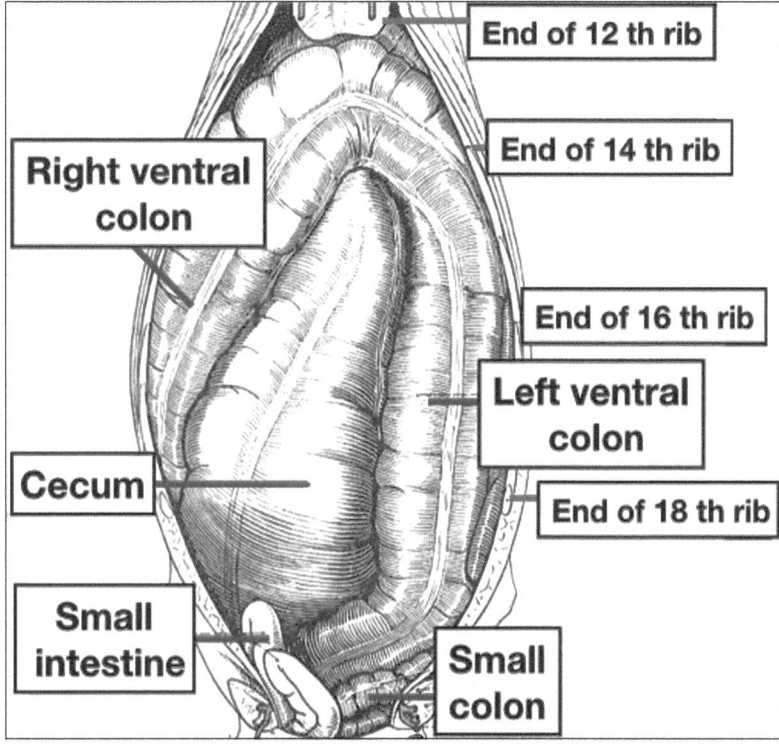

Figure 11.27 The colon and cecum make up most of the abdominal cavity in horses. (Adapted from O. C. Bradley)

11.17.5 Fermentation in the Hindgut

The microbial population in the cecum and colon ferment undigested food to generate volatile fatty acids (acetate, propionate, and butyrate), together with lactate. These are absorbed and provide nutrients to the horse. In addition, methane is generated. The microbial population uses urea to synthesize amino acids. There is some absorption of amino acids into the horse. Food spends about 35 hours in the cecum and colon, allowing considerable time for fermentation.

11.18 Digestion in Rabbits

The rabbit is included as it is another hindgut fermenter. In addition, rabbits have the peculiar practice of **coprophagy**, or ingestion of feces. In rabbits, the feces ingested are soft fecal pellets. These nutrient-dense fecal pellets are called cecotropes. After a further round of digestion, hard fecal pellets are released.

11.19 Digestion in Cats and Dogs

11.19.1 Protein Digestion

Cats and dogs were domesticated from ancestral carnivores. They are superbly able to digest proteins: cats 95%, dogs 88%.

11.19.2 Carbohydrate Digestion

Cats have a limited ability to digest starch with the following:

- Low levels of amylase in the saliva and the small intestine.
- Low levels of disaccharidase (the enzyme that breaks down disaccharide to monosaccharides, such as glucose) in the small intestine.

11.19.3 Taste Receptors

Cats also appear to lack the ability to taste sweetness.

11.20 Comparation of the Gastrointestinal Tract and Digestion in Livestock, Companion Animals, and Poultry

Similarities between in the gastrointestinal tract between livestock, companion animals, poultry, and humans are shown in Table 11.2, while differences are shown in Table 11.3.

Table 11.2 Examples of similarities in gastrointestinal tract between livestock, companion animals, poultry, and humans.

GI region	Livestock, cats, dogs, poultry & other birds
Mouth	Salivary glands, taste buds.
Esophagus	Duct through which ingesta pass from mouth to stomach.
Stomach	Mechanical digestion; Releases acid gastric secretion containing proteolytic enzyme.
Small intestine	Three regions—duodenum, jejunum, and ileum, where digestion is completed and nutrients are absorbed.
Ceca	Site of microbial fermentation.
Colon	Absorption of water and ions.
Liver	Produces bile containing bile salts that facilitate fat digestion.
Pancreas	Produces pancreatic fluid containing digestive enzymes and bicarbonate to neutralize pH in ingesta.

Table 11.3 Examples of differences in gastrointestinal tract between livestock, companion animals, and poultry.

	Livestock, cats & dogs			Poultry & other birds
	Pigs, dogs, and cats	Sheep, goats, and cattle	Horses	
Mouth	Taste buds			Few taste buds
Expanded end of esophagus	No crop			Crop present in chickens and turkeys but not ducks—storing feed and fermentation
Stomach chambers	1	4	1	Two (proventriculus producing gastrin juice & gizzard mechanically digesting ingesta
Major energy nutrient absorbed	Glucose and fatty acids	Volatile fatty acids	Glucose and fatty acids	Glucose and fatty acids
Fat flow after absorption	As chylomicrons into the lymph blood vessels of the lacteals			As protomicrons into the hepatic portal blood vessel in the absence of lacteals
Cecum/ceca	Modest	Modest	Large hind gut fermenter	Moderate size of 2 ceca. Some fermentation

11.21 Diseases of Gastrointestinal System in Domestic Animals

Textbox 11.11 summarizes diseases of gastrointestinal system in domestic animals.

Textbox 11.11   Examples of Diseases of Gastrointestinal System in Domestic Animals

Bloat in Cattle
There is a rapid build-up of gas in the rumen and reticulum. In this painful condition, the excess gas exerts pressure such that breathing is difficult. Bloat can be a cause of death.

Colic in Horses
Colic in Horses (abdominal pain) has multiple causes:

- Impaction colic due to coarse and/or dry feed
- Gas colic due to excess gas production by the microbes in the gastrointestinal tract
- Colon shift or twisted colon
- Inadequate blood supply to the intestines due to a build-up of fatty deposits around the intestines.
- Poor intestinal mobility due to an infection in the gastrointestinal tract.

Colitis
Colitis (inflammation of the colon and cecum) is a serious problem in horses. Symptoms include diarrhea, lethargy, edema in the abdomen, and reduced appetite. Death can result from fluid loss.

Intestinal Parasites
Intestinal parasites compete with the gastrointestinal tissues of animals for nutrients and damage gastrointestinal tissues. Examples of the intestinal parasites are the following:

- Nematode or roundworms (see Figure 11.28), such as *Toxocara cati* (cats), *Toxocara canis* (dogs), and *Strongylus Vulgaris* (horses).
- Cestodes (tapeworms)
- Single cellular parasites, e.g., *Eimeria* species and *Giardia duodenalis*.

Megaesophagus
Megaesophagus is a condition in which the esophagus becomes enlarged. However, instead of allowing for increased feed to be passed through the esophagus, motility is either prevented or greatly reduced. As a result, animals have a reduced ability to ingest food and water. This is a condition that is either congenital or acquired and is observed in dogs more than cats.

Rumen Acidosis
Rumen acidosis occurs when there is rapid fermentation of readily digestible carbohydrate such as ground corn. With the generation of acids, there is a decrease in the ruminal pH and a shift in the microbial populations with increases in lactic acid producers and decreases in lactic acid users and protozoa. There is an increase in the osmotic pressure in the rumen drawing water into it. This can result in dehydration.

Teeth Problems in Horses
If untreated, the horse will experience pain, reduced feed intake, and loss of condition. Examples of teeth problems is horses are the following:

- There can be sharp edges cutting (damaging) the cheeks and tongue.
- Excessively worn teeth.

Ulcers
Ulcers (erosion of the mucosa of the stomach) are commonly found in the stomach lining of horses and pigs.

Figure 11.28   Nematode worms are intestinal parasites. (Shutterstock/Rattiya Thongdumhyu)

Urinary System

12

Learning Objectives

1. Understand the functioning of the components of the urinary system (entire chapter).
2. Understand the functioning of the nephron (see Sections 12.1–12.5).
3. Understand the roles of the blood vessels in the glomerulus and surrounding the tubules and nephron loop (a.k.a. loop of Henle) (see Figure 12.4).
4. Understand the importance of filtration, reabsorption, and secretion in kidney functioning (see Figures 12.2–12.5).
5. Understand why mammals use urea as their nitrogenous waste while birds use uric acid (see Section 12.1.4).
6. Understand how glucose is reabsorbed in the proximal convoluted tubule (see Section 12.3.1).
7. Understand how sodium is reabsorbed in the proximal convoluted tubule (see Section 12.3.3).
8. Understand how water is reabsorbed in the descending loop of Henle and the collecting ducts (see Section 12.4).
9. Understand how sodium is reabsorbed in the ascending nephron loop (loop of Henle) (see Section 12.4).
10. Understand how chloride is reabsorbed (see Section 12.3.5).
11. Understand what happened to small peptides in filtrate (see Section 12.3.6).
12. Understand the roles of the hormones produced by the kidneys (see Section 12.7).
13. Understand the roles of the ureters, the bladder, and the urethra (see Sections 12.8 and 12.9).
14. Understand the differences between the urinary system of birds and mammals (see Section 12.11).

Chapter 12 Urinary System

Table of Contents

12.1 Introduction
    12.1.1 Introduction to the Urinary System
    12.1.2 Introduction to the Kidneys
    12.1.3 What Are the Functions of the Kidneys?
    12.1.4 What Is Nitrogenous Waste?
    12.1.5 Introduction to the Nephrons
    12.1.6 Introduction to the Functioning of Regions of the Nephron and Adjacent Structures
        12.1.6.1 Glomerulus and Glomerular Capsule (Bowman's Capsule)
        12.1.6.2 Proximal Convoluted Tubule
        12.1.6.3 Nephron Loop (Loop of Henle)
        12.1.6.4 Distal Convoluted Tubule
        12.1.6.5 Collecting Ducts
12.2 Filtration in the Glomerular Capsule (Bowman's Capsule)
12.3 Reabsorption and Secretion in the Proximal (Convoluted) Tubule
    12.3.1 Glucose Reabsorption in the Proximal Convoluted Tubule
    12.3.2 Amino Acid Reabsorption in the Proximal Convoluted Tubule
    12.3.3 Sodium Reabsorption in the Proximal Convoluted Tubule
    12.3.4 Bicarbonate Reabsorption in the Proximal Convoluted Tubule
    12.3.5 Chloride Reabsorption in the Proximal Convoluted Tubule
    12.3.6 Reabsorption of Peptides and Small Proteins
    12.3.7 Secretion into the Proximal Tubule
    12.3.8 Water Reabsorption in the Proximal Convoluted Tubule
12.4 Reabsorption in the Descending and Ascending Nephron Loop
    12.4.1 Descending Nephron Loop
    12.4.2 Ascending Nephron Loop
    12.4.3 Aquaporins
12.5 Reabsorption in the Distal Convoluted Tubule
    12.5.1 Introduction to Reabsorption in the Distal Convoluted Tubule
    12.5.2 Effect of Hormones on Reabsorption in the Distal Convoluted Tubule
12.6 Reabsorption in the Collecting Ducts
    12.6.1 Introduction to Reabsorption in the Collecting Ducts
    12.6.2 Effect of Hormones on Reabsorption in the Collecting Ducts
12.7 Hormones Produced by the Kidneys
    12.7.1 Introduction to Hormones Produced by the Kidneys
    12.7.2 Renin-Angiotensin System
    12.7.3 Erythropoietin
    12.7.4 1,25-Dihydroxyvitamin D_3 (1,25 $(OH)_2$ Vitamin D_3)
12.8 Ureter
12.9 Bladder
12.10 Micturition (Urination)
12.11 Differences Between Mammals (Livestock and Companion Animals) and Birds (Poultry)
12.12 Abnormalities and Diseases of the Urinary System

12.1 Introduction

12.1.1 Introduction to the Urinary System

Textbox 12.1 provides a glossary of terms used relative to the urinary system.

The urinary system is composed on the following:

- Kidneys are the main urinary organs, and they produce urine and some hormones.
- Ureters are ducts from the kidneys to the bladder (mammals or livestock and companion animals) or the cloaca (birds such as poultry). The urine passes through the ureters.
- The bladder stores urine in mammals (livestock and companion animals) but not in birds (e.g., poultry) as they have no bladder.
- Urethra is a duct from the bladder to the exterior of livestock and companion animals together with other mammals. Urine passes from the bladder to the outside of animals' bodies through the urethra. The urethra is not present in birds.

Textbox 12.1   Glossary of Terms Used Relative to the Urinary System

- **Afferent arteriole**: brings blood to the glomerulus.
- **Ammonia**: highly soluble nitrogenous waste derived from the metabolism of amino acids.
- **Anions**: negatively charged ions such as chloride (Cl^-) and bicarbonate (HCO_3^-).
- **Aquaporin**: channels that allow water to enter a cell.
- **Bladder**: organ in mammals that stores urine.
- **Carbonic anhydrase**: enzyme catalyzing the formation of carbonic acid.
- **Cations**: positively charged ions such as sodium (Na^+), potassium (Na^+), and calcium (Ca^{2+}).
- **Collecting duct**: fluid passes from the nephron into the collecting ducts, where there is absorption of water and ions.
- **Distal convoluted tubule**: portion of the nephron between the ascending loop of Henle and the collecting ducts.
- **Diuretic**: agent that induces diuresis; increase loss of water from the animal's body.
- **Efferent arteriole**: takes blood from the glomerulus.
- **Endocytosis**: uptake of proteins, etc., by cells.
- **Excretion**: ridding the body of waste.
- **Filtrate**: the fluid resulting from filtration of the blood.
- **Glomerular filtration**: blood plasma is filtered from the glomerulus into glomerular capsule (Bowman's capsule). Large proteins remain in the plasma.
- **Glomerular capsule (a.k.a. Bowman's capsule)**: filtrate from glomerulus passes into this, the first part of the nephron.
- **Glomerulus**: blood vessels adjacent to Bowman's capsule.
- **Glucose transporters**: transporters that allow the movement of glucose across cell membranes.
- **Kidneys**: primary organs of liquid excretion.
- **Micturition**: process of urination.
- **Nephron**: functional unit of the kidney.
- **Nephron loop** (a.k.a. **Loop of Henle**): part of nephron establishing an osmotic gradient.
- **Nephritis**: inflammation of the kidneys.
- **Nitrogenous waste**: waste from the metabolism of amino acids.
- **Proximal convoluted tubule**: portion of the nephron between the glomerular capsule (Bowman's capsule) and the descending nephron loop (loop of Henle).
- **Reabsorption**: movement of water, ions, nutrients, etc., from the filtrate.
- **Renal cortex**: outer layer of the kidney.
- **Renal medulla**: center of the kidney.
- **Secretion**: movement of chemicals out of a cell.
- **Sodium pump** (sodium potassium pump or sodium potassium ATPase): mechanism moving sodium ions in exchange for potassium ions across the cell membrane in a 3:2 ratio.
- **Urea**: main nitrogenous waste in mammals.
- **Ureter**: ducts from each kidney to the bladder in mammals or cloaca in birds.
- **Urethra**: single duct from the bladder to the exterior.
- **Uric acid**: main nitrogenous waste in birds.
- **Urine**: production of excretion produced by the kidney, passing to the bladder in mammals or into cloaca in birds

12.1.2 Introduction to the Kidneys

Figure 12.1 illustrates the structure of the kidney in sheep. There is a similar structure of the kidneys in other mammals but the orientations may differ. The kidneys are composed of large numbers of nephrons (Figure 12.2). The number of nephrons per kidney varies in different species (see Textbox 12.2). Urine passes from the collecting ducts to the **major and minor calyx**, to the renal pelvis and to the ureter.

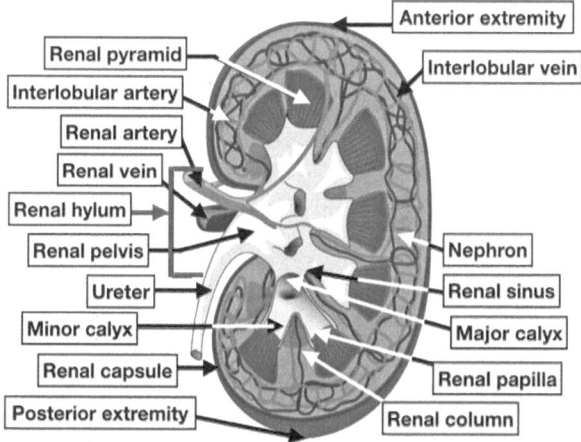

Figure 12.1 Section through the kidney of sheep showing structures.

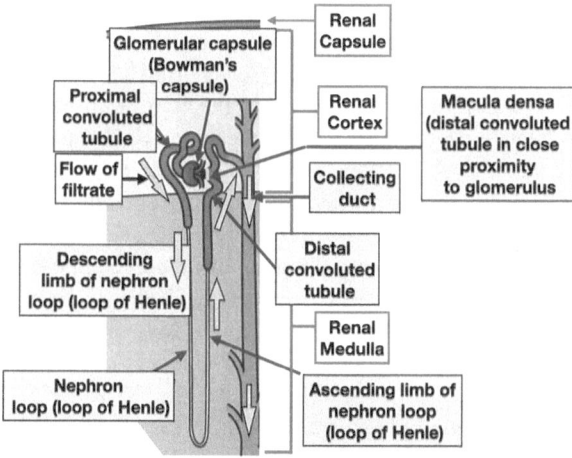

Figure 12.2 Structure of the nephron. Arrows indicate flow of filtrate.
(Adapted from Sunshineconnoly, CC BY-SA 3.0)

> **Textbox 12.2 Interesting Factoids About the Anatomy of Kidneys in Selected Domestic Animals**
>
> There are two kidneys in livestock, companion animals, and poultry. Saying that a kidney is "kidney-shaped" or "bean-shaped" (see Figure 12.1) is an anthropocentric viewpoint. It is found in some domestic animals, such as dogs, and the left kidney in horses. In contrast, the shape of a kidney differs in other domestic animals:
>
> - Cats: oval.
> - Horse: right kidney is heart-shaped.
> - Cattle: ellipsoidal.
> - Chickens: three lobes.
>
> The kidney is long and thin with, for instance, the following dimensions:
>
> - Dog: 5.9 cm long, 3.0 cm wide, 2.5 cm thick.
> - Horse: 17 cm long, 14 cm wide, 5 cm thick.
> - Cattle: 9.4 cm long, 5.1 cm wide, 2 cm thick.
>
> Table 12.1 provides information on the number of nephrons per kidney in different species of livestock, companion animals, and poultry.
>
> **Table 12.1** Weights and number of nephrons of a single kidney in selected livestock, companion animals, and poultry.
>
> | Species | Weight (g) | Number of nephrons |
> |---|---|---|
> | Dog | 11 | 400,000 |
> | Cat | 39 | 200,000 |
> | Horse | 990 | 1,000,000 |
> | Cattle | 500 | 4,000,000 |
> | Ducks | 12.5 | 2,400,000 |

12.1.3 What Are the Functions of the Kidneys?

The functions of the kidneys are the following:

- Ridding animals' bodies of nitrogenous waste.
- Removal of excess water.

- Maintaining blood pressure.
- Maintaining blood concentrations of glucose.
- Maintaining blood plasma pH.
- Maintaining blood ionic composition.
- Producing hormones.

Blood is filtered into the **renal corpuscle**. Much of the filtrate (99%) is reabsorbed in the nephron and collecting ducts (see Textbox 12.3). Some compounds, including waste materials, are secreted into the tubules from the blood.

The blood vessels supplying and draining the nephron are shown in Figure 12.3. Plasma is filtered into the glomerular capsule (a.k.a. Bowman's capsule). The filtrate is reabsorbed with the fluid, ions, glucose, and amino acids restored into the blood.

12.1.4 What Is Nitrogenous Waste?

Animals produce ammonia (NH_3) as a waste material from protein breakdown (see Figure 12.5). Proteins in cells and the extracellular matrix are turned over. Moreover, proteins in the feed are digested into amino acids. Excess amino acids are deamidated and used to produce glucose (in the process of gluconeogenesis discussed in detail in Chapter 7). This is particularly the case in cats and dogs, companion animals that are domesticated carnivores. So, what happens to the ammonia, bearing in mind that ammonia is toxic?

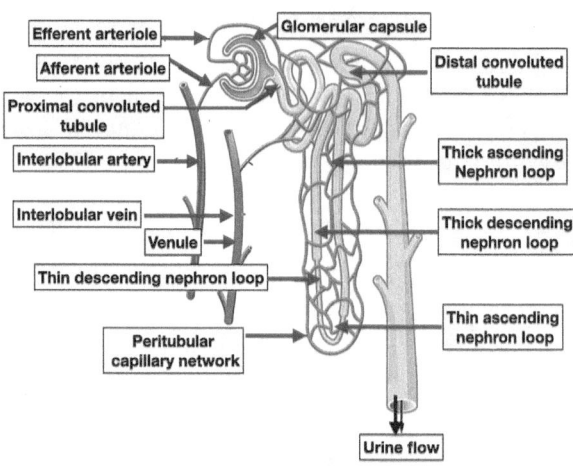

Figure 12.4 Blood vessels supplying and draining the nephron together with the regions of the nephron loop.
(Adapted from OpenStax, CC BY 4.0)

Mammals convert ammonia to urea in the liver (see Figure 12.5). This has low toxicity (LD_{50} 15 g kg^{-1}) and high solubility in water (>100 g L^{-1}). Birds convert ammonia to uric acid, together with some urea in the liver (see Figure 12.5). Uric acid has low toxicity but also low solubility in water (~60 mg L^{-1}).

12.1.5 Introduction to the Nephrons

Nephrons in livestock and companion animals are composed of the following (as in Figure 12.2):

- Renal corpuscle made up of a glomerular capsule (a.k.a. Bowman's capsule) and associated glomerulus in the cortex of the kidney (see Figure 12.3).
- Proximal convoluted tubule in the cortex of the kidney.
- Descending nephron loop (a.k.a. loop of Henle) in the medulla of the kidney.
- Thick region of descending nephron loop composed of **cuboidal epithelium** (see Figure 12.4)
- Thin region of descending nephron loop composed of **squamous epithelium** (see Figure 12.4)
- Ascending nephron loop in the medulla of the kidney.
- Thin region of ascending nephron loop composed of squamous epithelium (see Figure 12.4)
- Thick region of ascending nephron loop composed of cuboidal epithelium (see Figure 12.4)
- Distal convoluted tubule in the cortex of the kidney.

Poultry (birds) have both mammalian and reptilian nephrons (discussed in Section 12.11).

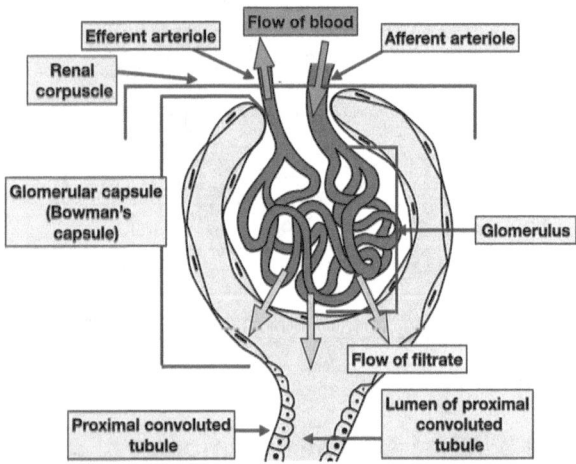

Figure 12.3 Structure of the renal corpuscle comprising the glomerular capsule (Bowman's capsule) and the glomerulus. Arrows indicate the flow of filtrate.
(Adapted from Sunshineconnoly, CC BY 3.0)

Chapter 12 Urinary System

12.1.6.2 Proximal Convoluted Tubule

The proximal convoluted tubule is important for the removal of water, glucose, and amino acids from the filtrate. This is reabsorption. There is also secretion of urea, ammonia, protons, drugs, and toxins, together with exocytosis of peptides.

12.1.6.3 Nephron Loop (Loop of Henle)

The nephron loop establishes and maintains an **osmotic gradient**.

12.1.6.4 Distal Convoluted Tubule

The distal convoluted tubule reabsorbs ions such as sodium and calcium.

12.1.6.5 Collecting Ducts

The collecting ducts not only collect urine but also removes waters from the urine.

12.2 Filtration in the Glomerular Capsule (Bowman's Capsule)

Blood is filtered in the renal corpuscle. This is due to the following:

- Blood pressure pushing fluid from the glomerulus into the glomerular capsule via pores.
- Less the hydrostatic pressure of the filtrate in the lumen of glomerular capsule (see Figure 12.3).

The pores that allow plasma to pass into the glomerular capsule do not allow the passage of the large components of blood, such as cells and large proteins. The filtrate is similar to blood plasma except that there are very few proteins (the composition of blood plasma is discussed in Chapter 2). The magnitude of the amount of filtrate in one species, the dog, is shown in Textbox 12.3. The **glomerular filtration rate (GFR)** is the amount of blood filtered by the kidneys per hour or per day. The filtrate contains the following:

- Water.
- Ions—cations: mainly sodium (Na^+) and potassium (K^+); anions: mainly chloride (Cl^-) and bicarbonate (HCO_3^-).
- Nitrogenous waste (ammonium ions, urea, and uric acid).
- Nutrients such as glucose and amino acids.
- Creatinine (this is used to determine GFR as the creatinine clearance).

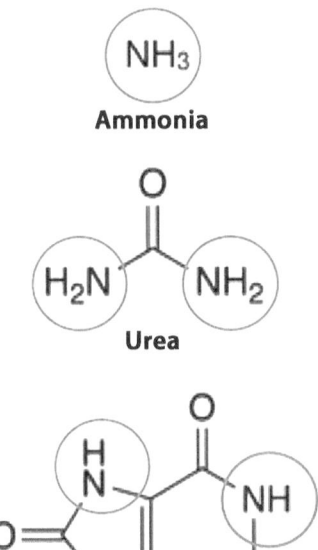

Figure 12.5 Chemical structure of nitrogenous waste.

The nephrons drain into the collecting ducts. The nephrons have an extensive blood supply that is essential to their functioning (Figure 12.4).

12.1.6 Introduction to the Functioning of Regions of the Nephron and Adjacent Structures

The functioning of the nephron may seem strange. Firstly, it filters plasma such that most of what is in plasma, except large proteins, goes into the nephron. The nephron (plus collecting duct) then reabsorbs 99% of the filtrate. There is also the following:

- Endocytosis that brings compounds such as peptides into nephron cells.
- Secretion of drugs and toxins into the nephron.

The functions of regions of the nephron and adjacent structures follows.

12.1.6.1 Glomerulus and Glomerular Capsule (Bowman's Capsule)

Glomerulus and glomerular capsule function to filter blood plasma due to hydrostatic pressure.

Obviously, if the filtrate was the same as the urine, there would be a great loss of so much water, nutrients, and sodium and this would be a major problem (see Textbox 12.3). This does not happen as *almost 99% of the filtrate is reabsorbed.*

> **Textbox 12.3   Importance of Glomerular Filtration and Reabsorption**
>
> **Glomerular Filtrate**
> Glomerular filtrate in a dog (12 kg):
>
> - Water: 64.8 L per day^{-1}.
> - Glucose: 73.5 g per day^{-1} (based on plasma concentration of 6.3 mMoles L^{-1}).
> - Sodium: 216 g per day^{-1} (based on plasma concentration of 144 mMoles L^{-1}).
>
> This is equivalent to 712 g of salt per day^{-1}.
>
> **Consequences**
> If water, glucose, and sodium were not reabsorbed in a dog, they would be urinating 65 L of urine each day. To continue living, they would need to drink 65 L (17.2 US gallons) of water each day to stay alive. This is totally unrealistic! They would also have to consume 712 g (1.6 pounds) of salt per day. Again, this is totally unrealistic! However, over 95% of the filtered water, sodium, and glucose is reabsorbed in the proximal convoluted tubule, the loop of Henle, the distal convoluted tubule, and collecting ducts.

12.3 Reabsorption and Secretion in the Proximal (Convoluted) Tubule

Proximal convoluted tubules have brush borders on their luminal (apical) membrane. This greatly increases the surface area for reabsorption of water, ions, glucose, and amino acids.

12.3.1 Glucose Reabsorption in the Proximal Convoluted Tubule

All the glucose in the filtrate (100%) is reabsorbed into the proximal convoluted tubule cells from the lumen by **sodium dependent glucose transporters** (SGLT) (see Figures 12.6 and 12.7). The glucose passes out of the proximal convoluted tubule cells into the interstitial space due to glucose transporters (GLUT). The glucose then passes into the peritubular capillaries (see Figure 12.5). Sodium is removed from the proximal convoluted tubule cells by the **sodium pump (Na/K ATPase)** in the basolateral membrane (see Figures 12.7. and 12.8). This prevents accumulation of sodium ions in the proximal convoluted tubule cells. Reabsorption of glucose is accompanied by reabsorption of sodium.

In the first part of the proximal convoluted tubule, SGLT is SGLT2—a low affinity, high capacity, glucose transport—and the glucose transporter (GLUT) is GLUT 2. In the last portion of the proximal convoluted tubule, SGLT is SGLT1—a high affinity, low-capacity glucose transport—and the GLUT is GLUT 1.

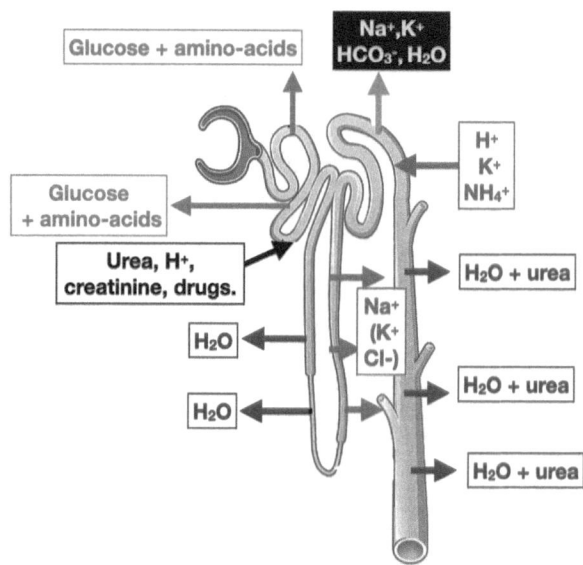

Figure 12.6   Secretion into and reabsorption in the nephron and collecting duct.
(Adapted from Sunshineconnoly, CC BY 3.0)

> **Textbox 12.4   Definitions Related the Nephron**
>
> - **Apical cell membrane**: the cell membrane of nephron epithelial cells adjacent to the lumen.
> - **Basolateral cell membrane**: the cell membrane of nephron epithelial cells adjacent to the capillaries.
> - **Capillary space**: the lumen of capillaries.
> - **Luminal space**: the space within the lumen.

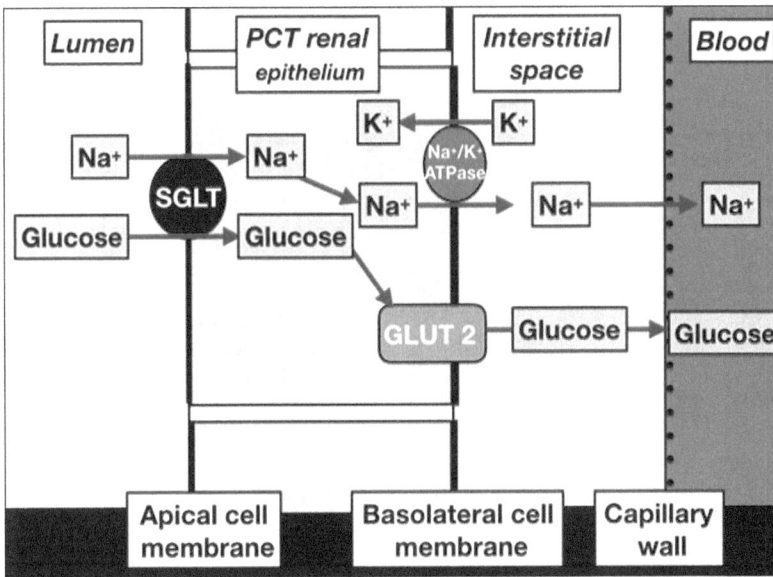

Figure 12.7 The mechanism for reabsorption of glucose. Note: The sodium pump does not have a one-to-one relationship between sodium ions and potassium ions. Instead, for every three sodium ions transiting, there are two potassium ions. See Textbox 12.4 for definitions.
Key:
Na^+ = the sodium ion
SGLT = the sodium dependent glucose cotransporter
K^+ = the potassium ion
Na^+/K^+ ATPase = the sodium pump
GLUT = the glucose transporter

12.3.2 Amino Acid Reabsorption in the Proximal Convoluted Tubule

Virtually all the amino acids in the filtrate (>95%) are reabsorbed into the proximal convoluted tubule cells from the lumen. This is achieved by a series of sodium amino acid cotransporters in the apical cell membrane. Amino acids in the proximal convoluted tubule cells pass to the interstitial space by amino acid transporters in the **basolateral cell membrane** and then on to the capillary, flowing down a concentration gradient. Reabsorption of amino acids is accompanied by sodium reabsorption.

12.3.3 Sodium Reabsorption in the Proximal Convoluted Tubule

About 65% of the sodium in the filtrate is reabsorbed in the proximal convoluted tubule. This is due to sodium accompanying glucose and amino acids in the **cotransporter-mediated reabsorption** across the apical cell membrane (Figure 12.7), together with the sodium pump in the basolateral membrane.

12.3.4 Bicarbonate Reabsorption in the Proximal Convoluted Tubule

About 80% of bicarbonate in the filtrate is reabsorbed in the proximal convoluted tubule. Figure 12.9 summarizes bicarbonate reabsorption in the proximal convoluted tubule. In the presence of hydrogen ions, bicarbonate in the lumen is converted to carbonic acid. This in turn breaks down to generate carbon dioxide (CO_2). The CO_2 is readily soluble in the cell membrane and passes into the proximal convoluted tubule cell. Its conversion to carbonic acid is catalyzed by carbonic anhydrase. The carbonic acid dissociates to generate bicarbonate ions plus a hydrogen ion. The bicarbonate passes through the basolateral membrane due to **bicarbonate chloride exchange**. The hydrogen ions pass into the lumen of the proximal convoluted tubule due to the **sodium–hydrogen ion antiporter**. The sodium ions are pumped out of the proximal convoluted tubule cell by the sodium pump.

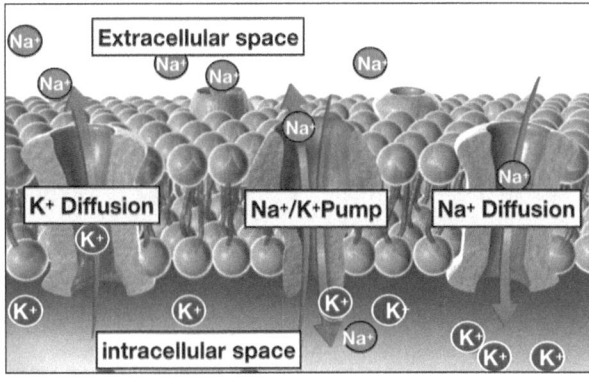

Figure 12.8 The sodium pump (Na/K ATPase), together with sodium and potassium channels. Note: The sodium pump does not have a one-to-one relationship between sodium ions and potassium ions. Instead for every three sodium ions transiting, there are two potassium ions. (Adapted from BruceBlaus, CC BY 3.0)

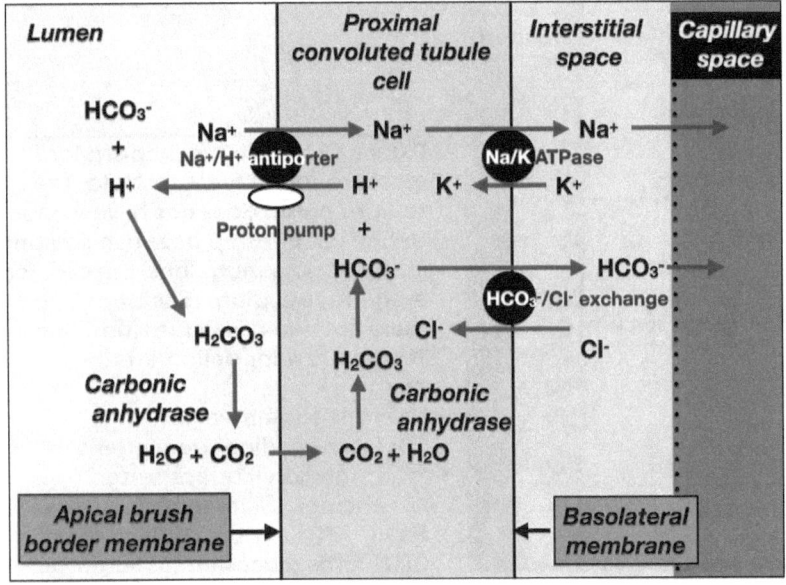

Figure 12.9 The mechanism for reabsorption of bicarbonate in the proximal convoluted tubule. Note: The sodium pump does not have a one-to-one relationship between sodium ions and potassium ions. Instead for every three sodium ions transiting, there are two potassium ions.
Key:
CO_2 = carbon dioxide
HCO_3^- = bicarbonate
Na^+/H^+ antiporter = the sodium-hydrogen ion exchanger
Na/K ATPase = the sodium pump
H^+ = hydrogen ions
HCO_3^-/Cl^- = exchange
Cl^- = chloride ions
H_2CO_3 = carbonic acid
H_2O = water

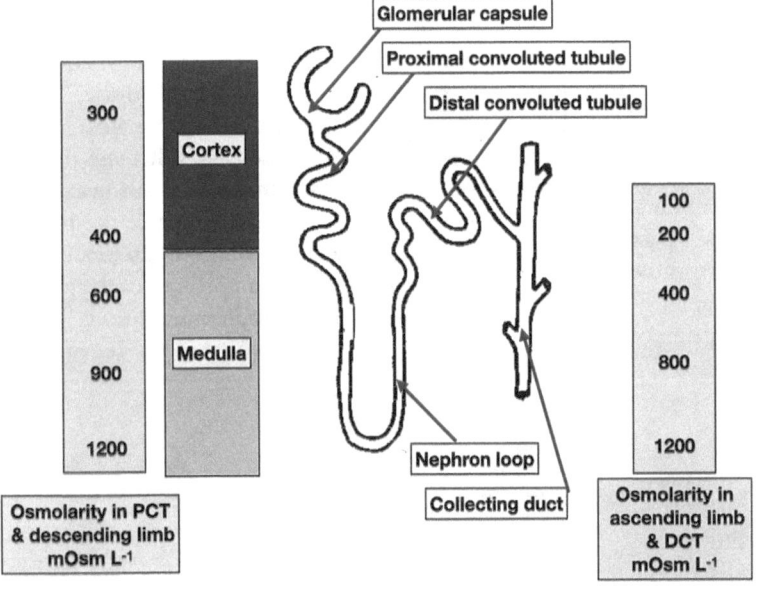

Figure 12.10 Changes in osmolarity in the proximal convoluted tubule, descending limb of the nephron loop (loop of Henle), the ascending limb of the nephron loop (loop of Henle), and distal convoluted tubule.
Note: The sodium pump does not have a one-to-one relationship between sodium ions and potassium ions. Instead, for every three sodium ions transiting, there are two potassium ions.
(Adapted from Sunshineconnoly, CC BY 3.0)

12.3.5 Chloride Reabsorption in the Proximal Convoluted Tubule

Chloride (65%) is reabsorbed in the proximate convoluted tubules. Reabsorption across the apical membrane is achieved by a series of transporters, including chloride-base exchange, chloride-formate exchange, and chloride-oxalate exchange. The functioning of these is facilitated by sodium-hydrogen ion exchange, oxalate-sulfate ion exchange and sodium-sulfate ion exchange. There is also some passive reabsorption of chloride.

12.3.6 Reabsorption of Peptides and Small Proteins

There are proteins and peptides smaller than albumin in the filtrate. These are taken up by the proximal convoluted tubule cells by endocytosis. In addition, peptides such as insulin are taken up by sodium-coupled active transport. In the proximal convoluted tubule cells, the peptides/protein are degraded by proteases, such as cathepsin B and L in the lysosomes. The amino acids are either used by proximal convoluted tubule cells or exported into the blood capillaries.

12.3.7 Secretion into the Proximal Tubule

Is there secretion into the filtrate? There is movement of multiple chemicals into the lumen of the proximal convoluted tubule. Examples include urea, antibiotics (e.g., tetracycline), other drugs (e.g., captopril), and organic acids (e.g., uric acid). This is mediated by a series of organic anion transporters.

12.3.8 Water Reabsorption in the Proximal Convoluted Tubule

Water (65% of the filtrate) is reabsorbed in the proximate convoluted tubules. This is accomplished by the water moving due to the following:

1. Moving through the "leaky" tight junctions between cells in the proximal convoluted tubules.
2. Water moving through water channels, specifically, aquaporin (AQP) 1. AQP1 is found on both the apical and basolateral membranes.
3. An osmotic effect due to the movement of glucose, amino acids, and sodium into the proximal convoluted tubule cells and into the interstitial space. The osmolarity of the lumen of the proximal convoluted tubules declines when that of the cytoplasm of the proximal convoluted tubules and the interstitial space increases. Water flows from low to high osmolarity.

12.4 Reabsorption in the Descending and Ascending Nephron Loop

12.4.1 Descending Nephron Loop

Water is reabsorbed in the descending nephron loop due to the presence of AQP (water channels), specifically AQP1, in both the apical and basolateral membranes. As water transits out of the lumen of the descending nephron loop, the osmolarity increases (Figure 12.8). Water passes to the high osmolarity environment in the interstitial fluid.

12.4.2 Ascending Nephron Loop

The ascending nephron loop is another important site of sodium reabsorption. There is also absorption of magnesium ions (~65% of that in the filtrate), calcium ions, potassium ions, and chloride ions in the ascending nephron loop due to specific channels. This is accomplished by a series of transporters such as the **sodium-chloride-potassium ion cotransporters** in both the **apical** and **basolateral membranes**. There is not absorption of water because AQP is not present. As the remaining filtrate passes up the ascending nephron loop, electrolytes are lost.

> **Textbox 12.5 Interesting Factoid: Furosemide—A Diuretic Used in Horses**
>
> A diuretic causes **diuresis**, producing very high amounts of urine. Furosemide is a diuretic used to treat acute renal failure and congestive heart failure horses. In addition, furosemide can be used (illegally) to prevent pulmonary hemorrhage following exercise and improve performance in horses. The evidence that the latter effects are effective is equivocal. Furosemide acts by binding to the sodium-potassium-chloride cotransporter.

12.4.3 Aquaporins

Aquaporins (AQP) are proteins that are found in cell membranes that allow water to cross the cell membrane. These proteins act as channels or pores that allow water into or out of the cell. There are multiple isoforms of aquaporin.

Figure 12.11 shows a schematic of a nephron. A nephron is composed of a glomerulus, a proximal convoluted tubule, the nephron loop (descending loop and ascending loop), and distal convoluted tubule. These connect to the collecting duct by the connecting duct. Glucose is absorbed from the filtrate in the proximal convoluted tubule via glucose transporters. Water passes from the filtrate in the proximal convoluted duct and descending loop of Henle. This occurs because of the presence of aquaporins (AQP1 and AQP7) and water passing down an osmotic gradient. Sodium (Na+), together with potassium and chloride ions, are removed from the filtrate in the ascending Nephron loop. This requires energy (ATP) and the presence of sodium pumps. The sodium and chloride ions pumped from filtrate form an osmotic gradient with increasing osmotic pressure around the nephron from cortex to the inner region of the medulla. In the presence of the hormone vasopressin, additional aquaporins are expressed in the collecting duct. These include AQP2, AQP3, and AQP4. More water is removed from the filtrate to form the urine.

12.5 Reabsorption in the Distal Convoluted Tubule

12.5.1 Introduction to Reabsorption in the Distal Convoluted Tubule

The distal convoluted tubule is the shortest portion of the nephron. There are three regions of the distal convoluted tubule:

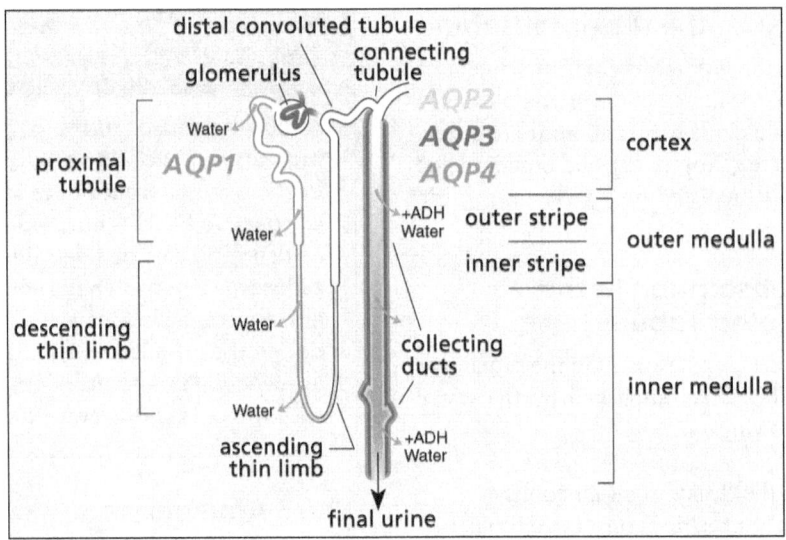

Figure 12.11 Schematic of nephron. (From Søren Nielsen, Tae-Hwan Kwon, Jørgen Frøkiær, and Mark A. Knepper (2000), "Key Roles of Renal Aquaporins in Water Balance and Water-Balance Disorders," *Physiology* 15(3): 136–143. Used by permission.)

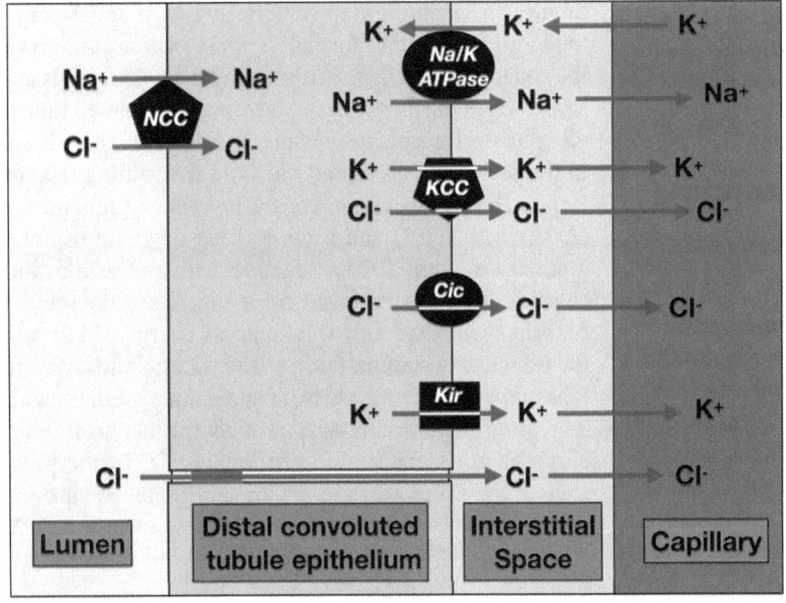

Figure 12.12 Reabsorption in the distal convoluted tubule due to transporters in the apical and basolateral membranes.
Key:
Ions:
 Cl^- = chloride
 K^+ = potassium
 Na^+ = sodium
Transporters:
 NCC = sodium chloride co-transporter
 Na/K ATPase = sodium pump
 KCC = potassium chloride co-transporter
 Cic = chloride channel
 Kir = potassium channels allowing leakage

- **Distal convoluted tubule (DCT)1** with a pronounced bush border on the apical membrane.
- **DCT 2**.
- **Collecting tubule.**

About 7% of the sodium in the filtrate is absorbed in the distal convoluted tubule. Figure 12.12 summarizes reabsorption of sodium and chloride ions. There is an additional sodium ion transporter in the DCT 2, namely thiazide sodium chloride cotransporter. There is also reabsorption of calcium with about 7% of the filtrate reabsorbed in the distal convoluted tubule. There is little water reabsorption in the distal convoluted tubules as AQPs are absent.

12.5.2 Effect of Hormones on Reabsorption in the Distal Convoluted Tubule

There are effects of hormones on absorption in the distal convoluted tubule such as the following:

- **Aldosterone** increases sodium reabsorption in the DCT 2.
- **Parathyroid hormone** (PTH) increases calcium reabsorption due to additional calcium channels.

12.6 Reabsorption in the Collecting Ducts

12.6.1 Introduction to Reabsorption in the Collecting Ducts

The collecting ducts make relatively small adjustments to the total reabsorption of urine, calibrating the quantity and composition of the urine to optimize homeostasis. The collecting ducts reabsorb the following:

- Water via AQP (see Figures 12.11 and 12.13).
- Sodium ions.
- Chloride ions.
- Urea.

12.6.2 Effect of Hormones on Reabsorption in the Collecting Ducts

Antidiuretic hormone (ADH) is produced by the posterior pituitary gland and has marked effects on water reabsorption in the collecting ducts. In the absence of ADH, large amounts of dilute urine are produced. In the presence of high concentration of ADH, small amounts of concentrated urine are produced. ADH has structural differences in different species:

- **Arginine vasopressin** in horses, cattle, sheep, cats, dogs.
- **Lysine vasopressin** in pigs.
- **Arginine vasotocin** in poultry.

ADH acts by binding to **vasopressin receptors (V2)**; in turn **adenylyl cyclase** is activated and intracellular concentrations of cAMP increase. Water channels (AQP 2) are moved from intracellular vesicles into the apical membrane increasing water transit from the lumen into the collecting duct cells (see Figure 12.13). There are AQP 3 and AQP 4 water channels in the basolateral membrane cells (see Figures 12.11 and 12.13).

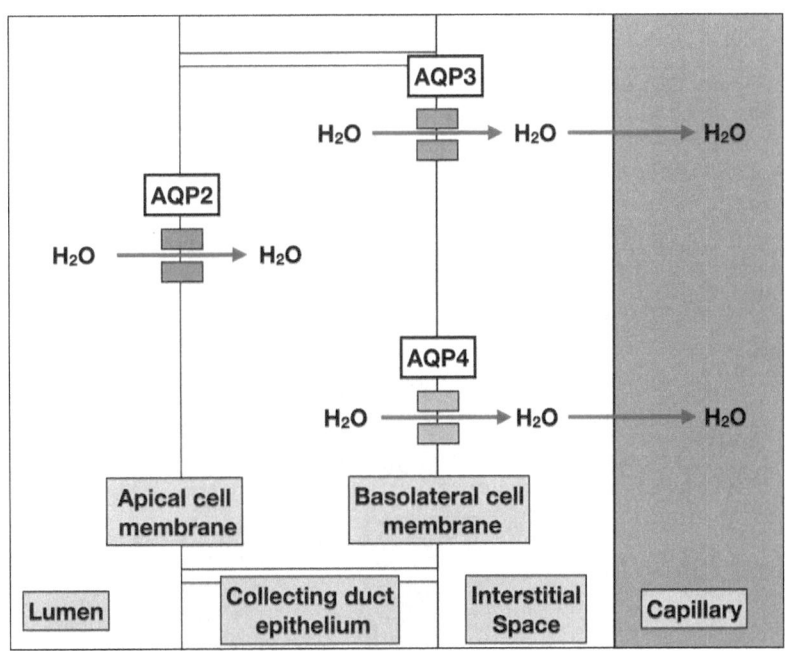

Figure 12.13 Reabsorption of water in the collecting ducts due to water channels (AQP2 in the apical, together with AQP3 and AQP4 in the basolateral membranes).

12.7 Hormones Produced by the Kidneys

12.7.1 Introduction to Hormones Produced by the Kidneys

The kidneys are not only the major urinary organs but also produce hormones and hormone activators. Major examples are the following:

- The renin-angiotensin system.
- Erythropoietin.
- The active form of vitamin D, 1,25-dihydroxyvitamin D.

12.7.2 Renin-Angiotensin System

Prorenin is activated by the juxtaglomerular apparatus to generate renin. The juxtaglomerular apparatus consists of the following (see Figure 12.14):

- **Juxtaglomerular cells** that activate and release **renin**.
- **Macula densa cells** (about 20 cells per nephron) that control the juxtaglomerular cells (see Figure 12.14).

The renin angiotensin system is summarized in Figure 12.15. The juxtaglomerular apparatus of the kidneys activates prorenin to the enzyme renin and releases it into

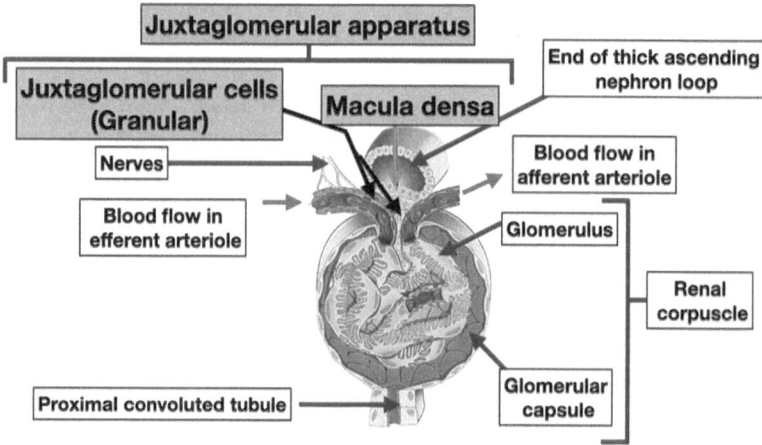

Figure 12.14 View of juxtaglomerular apparatus consisting of juxtaglomerular cells (adjacent to the glomerulus in the renal corpuscle) and the macula densa (at the end of the thick portion of the ascending loop of Henle).

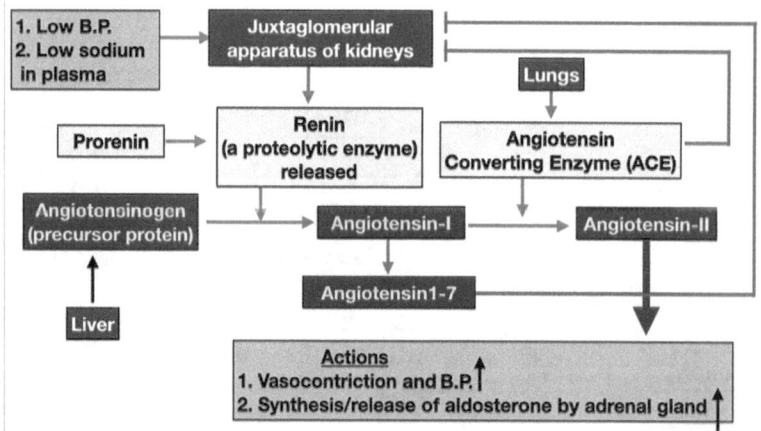

Figure 12.15 The renin-angiotensin system.
[Key: B.P. = blood pressure.]

the bloodstream (Figure 12.15). This proteolytic enzyme cleaves **angiotensinogen** in the blood to generate a small peptide, **angiotensin-I** (Figure 12.15). In turn, angiotensin-I is converted by **angiotensin-converting enzyme** (ACE) to **angiotensin-II** as it passes through the lungs (Figure 12.15). In addition, angiotensin-I is converted to angiotensin 1-7. Both angiotensin-II and angiotensin 1-7 act in a negative feedback manner, decreasing the activation of pro-renin by the juxtaglomerular.

The juxtaglomerular apparatus of the kidneys release renin in response to low blood pressure, low concentrations of sodium in plasma, and sympathetic nervous stimulation (Figure 12.15). The angiotensin-II produced by the renin-angiotensin system acts to counteract the problem. For instance, angiotensin-II induces **vasoconstriction** and consequently increases blood pressure (Figure 12.15). Moreover, angiotensin-II acts on the cortical cells of the adrenal glands to produce aldosterone (Figure 12.15). Aldosterone increases sodium reabsorption in the kidneys, correcting the low plasma concentrations of sodium and, also, increasing blood pressure.

Textbox 12.6 A Deeper Dive into the Renin-Angiotensin System

The juxtaglomerular cells activate prorenin to renin. This is controlled by the following:

- Low blood pressure as detected by baroreceptors.
- Sympathetic nerve terminals adjacent to the juxtaglomerular cells releasing norepinephrine, and this binding to β-adrenergic receptors on the juxtaglomerular cells.
- Hormones such as atrial natriuretic peptide and angiotensin-II.
- Positive effect: Low sodium concentrations act on macula densa cells that are located at the end of the ascending nephron loop and these communicate with the juxtaglomerular cells via gap junctions or nitric oxide or prostaglandin E to stimulate renin formation.
- Negative effect: High sodium concentrations act on macula densa cells and these communicate with the juxtaglomerular cells locally to produce adenosine.

12.7.3 Erythropoietin

Erythropoietin (EPO) is produced by the interstitial cells in the cortex of the kidneys in response to low blood oxygen concentrations (see Figure 12.17). EPO acts to increase the production of red blood cells by the bone marrow. Synthetic EPO has been used illegally as a performance enhancing drug in horses.

12.7.4 1,25-Dihydroxyvitamin D_3 (1,25 $(OH)_2$ Vitamin D_3)

1-alpha hydroxylase in kidney cells converts 25-dihydroxyvitamin D_3 to the active form 1,25-dihydroxyvitamin D_3 (1,25 $(OH)_2$ vitamin D_3) or calcitriol. Production of 1-alpha hydroxylase and, hence, 1,25 $(OH)_2$ vitamin D_3 occurs when plasma concentrations of calcium fall. The 1,25 $(OH)_2$ vitamin D_3 increases the following:

- Calcium absorption in the small intestine.
- Calcium reabsorption in the kidneys.
- Calcium loss from the bones.

1,25 $(OH)_2$ vitamin D_3 acts by binding to the vitamin D receptor, which is a nuclear (or steroid like) receptor.

12.8 Ureter

The **ureters** are ducts from the kidneys to the bladder in livestock and companion animals. Their walls contain smooth muscles (see Figure 12.15). The contraction of these smooth muscles pushes the urine into the bladder. In poultry and other birds, the ureters terminate in the cloaca.

12.9 Bladder

The bladder is a distensible organ (see Figure 12.18) that stores urine in mammals but is not present in birds. It has been suggested that the existence of the bladder gives the animal a selective advantage to avoid being eaten as they don't produce a trail from continuous urination.

The wall of the bladder contains smooth muscle (detrusor) (see Figure 12.18B). The capacity of the bladder varies by species:

- Horse: 4.0–4.5 liters.
- Cattle: 1.5 liters.
- Dogs: (10 kg) ~0.1 liters.
- Cat: 0.035 liters.

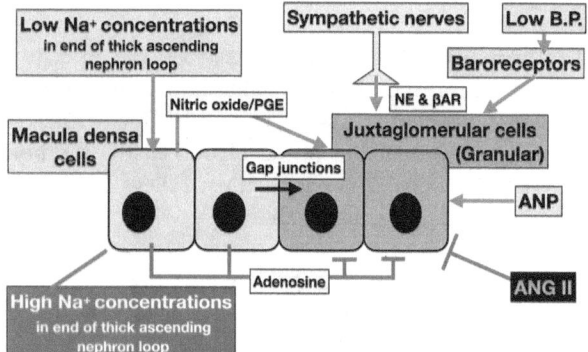

Figure 12.16 Control of juxtaglomerular cells.
Key:
B.P. = blood pressure
PGE = prostaglandin E
NE = norepinephrine
βAR = beta-adrenergic receptor
ANP = atrial natriuretic peptide
ANG II = angiotensin II
↑ is positive, ┬ is negative

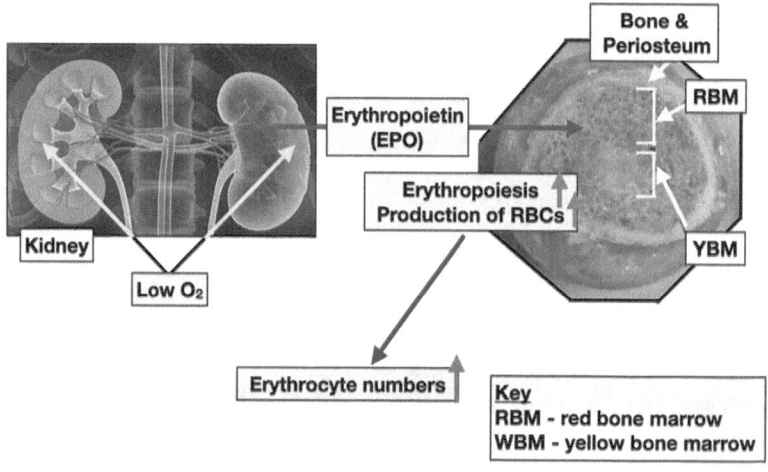

Figure 12.17 The kidneys produce the hormone erythropoietin (EPO), which in turn stimulates the production of red blood cells (RBC).

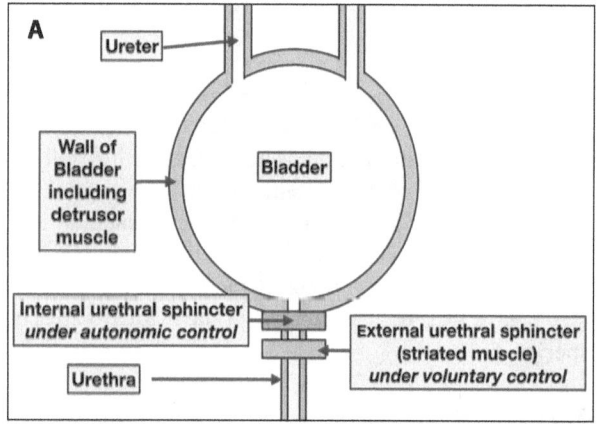

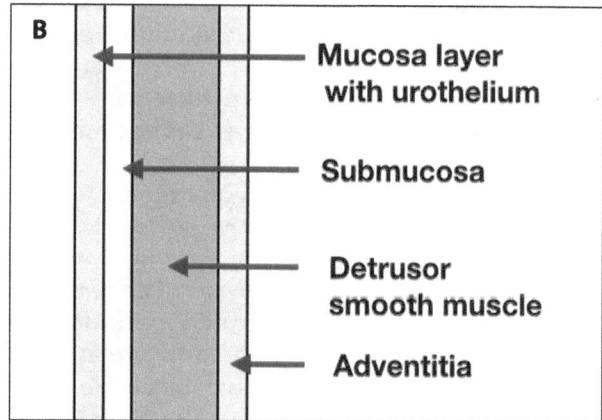

(A) Overall structure of the bladder with urine flowing into the bladder by two ureters and a single duct, the urethra, through which the urine flows to the exterior of the animal.

(B) Structure of the bladder wall.

Figure 12.18 Schematic diagrams illustrating the structure of the bladder.

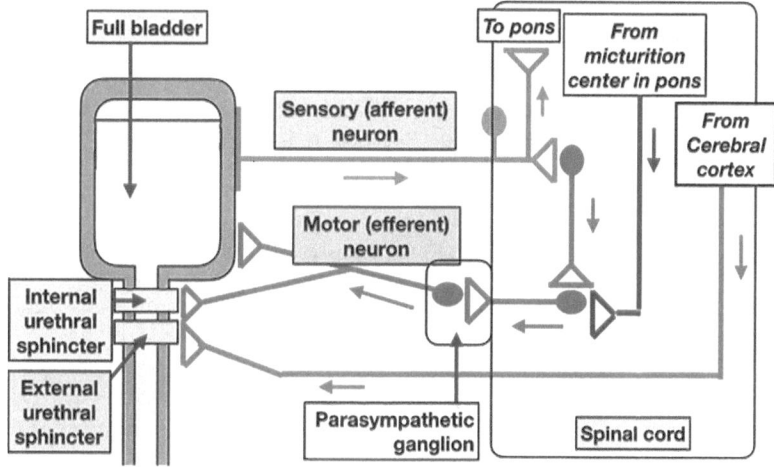

Figure 12.19 Schematic view of the neurons passing from and to the bladder and urethra.

Historically, pigs' bladders were inflated in medieval ceremonies and in footballs.

A simplified version of the afferent and efferent neurons from and to the bladder and urethra are shown in Figure 12.19.

12.10 Micturition (Urination)

Micturition, or urination, is the process by which the bladder is emptied in mammals. It is controlled by an autonomic reflex, the micturition reflex, that is present in the fetal mammal, the neonate, and the mature animal. In addition, there is a voluntary control that can be developed during the early juvenile stage of mammals.

12.11 Differences Between Mammals (Livestock and Companion Animals) and Birds (Poultry)

Table 12.2 summarizes differences in the urinary system between mammals (livestock and companion animals) with those in birds (poultry). Examples of differences are the nephrons and implications of the absence of a bladder.

Poultry have both juxtamedullary (mammalian) and cortical (reptilian) nephrons. What are they like? As the name implies, the mammalian-type nephrons in birds are like those in mammals, with a nephron capsule with a glomerulus, a proximal convoluted tubule and the distal convoluted tubule (in the cortex of the kidney), together

---

Textbox 12.7 A Deeper Dive into the Micturition

It is reasonable to assume that when the bladder is close to being full it will be emptied. In the very young, it is simple. When the bladder is close to full, stretch receptors in the bladder wall are stimulated. Action potentials pass along the sensory sympathetic neuron into the spinal cord (Figure 12.19). In turn, the motor neurons are stimulated and nerve impulses pass along motor (parasympathetic) neurons. The motor neurons act on the detrusor smooth muscles in the bladder wall via release of acetylcholine and bind to muscarinic (cholinergic) receptors. This induces contraction (note:

neurotransmitters are covered in Chapter 5). The muscles of the internal urethral sphincter are stimulated to relax, opening the sphincter. Urine then passes down the urethra to the exterior. An additional component to the system is that the terminals of the parasympathetic neurons adjacent to the muscles of the bladder release ATP to stimulate the muscles via purinergic receptors. Moreover, terminals adjacent to the urethral smooth muscle release nitric oxide, inducing relaxation.

As animals mature, cerebral controls come into play, allowing voluntary control of

(continued)

> **Textbox 12.7** (*continued*)
>
> micturition. This is a learned behavior and occurs, for instance, in puppies and kittens. To facilitate the bladder filling, there are sympathetic (efferent, post-ganglionic) neurons to the detrusor muscles in the bladder wall. These release norepinephrine that binds to inhibitory β-adrenergic receptors, causing the muscle relaxation of the detrusor muscles. Sympathetic neurons also innervate the smooth muscles of the urethra inducing their contraction. There is release of norepinephrine from neuronal terminals and this acts via α_1 adrenergic receptors
>
> There are other controls, with urothelial cells (see Figure 12.19) releasing neurotransmitters such as the following:
>
> - Neurokinin A.
> - Acetylcholine.
> - ATP.
>
> Urothelial cells have multiple receptors, such as the following:
>
> - Neurokinin receptors.
> - Nicotinic cholinergic receptors.
> - Muscarinic cholinergic receptors.

Table 12.2 Examples of differences in urinary system between livestock and companion animals with those in poultry.

	Livestock, horses, cats, & dogs	Poultry & other birds
Type(s) of nephron	Mammalian	Reptilian & Mammalian
Major nitrogenous waste	Urea	Uric acid
Ureters	Pass urine from the kidneys to the bladder	Pass urine from the kidneys to the cloaca
Bladder—storing urine	Present	Absent
Urethra—passage of urine from the bladder to the exterior	Present	Absent
Retrograde flow of urine in the colon	No	Yes
Absorption of water from urine in colon	No	Yes
Liquid urine	Yes	No
Urine mixes	No	Yes, in colon with feces, with additional water reabsorbed

with the nephron loop in the medulla of the kidneys. Reptilian nephrons are very similar to mammalian nephrons but do not have nephron loops. As such, reptilian nephrons are present in the cortex of the kidney. Both mammalian and reptilian nephrons terminate in the collecting ducts.

12.12 Abnormalities and Diseases of the Urinary System

Textbox 12.8 summarizes abnormalities and diseases of the urinary system in domestic animals.

Textbox 12.8 Abnormalities and Diseases of the Urinary System

- **Bladder stones** (uroliths or cystic calculi) are composed of struvite (magnesium ammonium phosphate—$NH_4MgPO_4 \cdot 6H_2O$) or calcium oxalate. These can be painful when voiding the bladder (cats, dogs, and horses).
- **Chronic kidney disease**: with progressive loss of kidney function (cats).
- **Cystitis**: inflammation of the bladder (pigs, dogs).
- **Diabetes** is associated with excessive thirst, production of large amounts of urine, and the presence of glucose in the urine (dogs).
- **Dysuria**: painful urination (dogs).
- **Hematuria**: blood cells in the urine (dogs with bladder stones, pigs).
- **Hemoglobinuria**: hemoglobin in the urine (pigs).
- **Kidney stones** are composed of calcium oxalate. These can be painful when voiding the bladder (cats and dogs).
- **Nephritis**: inflammation of the kidneys (pigs).
- **Polycystic kidneys**: a genetic disease with the formation of multiple cysts in the kidneys (cats).
- **Proteinuria**: protein in the urine (pigs).
- **Pyelonephritis**: inflammation of the ureters (pigs).
- **Urethritis**: inflammation of the urethra (pigs).

Veterinary Approaches to the Urinary System

- **Urethral catheterization** and **Cystocentesis** are two different techniques used by veterinarians to get a sample of urine from the bladder or to relieve pressure in the bladder of dogs and cats.

# Theme Four

Reproduction (Propagation of the Species)

Male Reproduction 13

(Including Development of Gonads, Associated Ducts, and Accessory Organs)

Learning Objectives

1. Understand the characteristics of males (see Section 13.1.1).
2. Understand why castration is performed (see Section 13.1.2).
3. Know the relevant terminology of male and castrated livestock, companion animals, and poultry (see Textbox 13.1).
4. Understand the structures and functions of the organelles of the spermatozoa (see Section 13.2).
5. Understand the roles of the epididymis, vas deferens, prostate, ampulla, seminal vesicles, and Cowper's glands (see Section 13.3).
6. Understand the genetic basis for sex determination (see Section 13.4).
7. Understand the development of the gonads and reproductive ducts and the differences between domestic mammals and poultry (see Section 13.5).
8. Understand the roles of the constituents of semen (see Section 13.3).
9. Understand the mechanisms for erection and ejaculation (see Sections 13.9. and 13.10).
10. Understand how the temperature set point of the testes is maintained (see Section 13.8).
11. Understand differences in male reproduction between livestock/companion animals and that in poultry (see Section 13.12).
12. Understand example diseases of the male reproduction in domestic animals (see Section 13.13)

Table of Contents

- 13.1 Introduction
  - 13.1.1 What Are the Major Features of Male Reproduction?
  - 13.1.2 What Is Castration?
  - 13.1.3 What Does Castration Do?
- 13.2 Spermatozoa
  - 13.2.1 What Are Characteristics That Spermatozoa Must Have?
  - 13.2.2 Structure of Spermatozoa
- 13.3 Structures of the Male Reproductive Tract and Their Functions
  - 13.3.1 Testes
  - 13.3.2 Ducts and Accessory Glands
  - 13.3.3 Penis
  - 13.3.4 Reproductive System of Poultry
- 13.4 Genetics of Sex Determination
  - 13.4.1 Are the Genetics Determining Whether an Animal Is Male or Female Across Species?
- 13.5 Development of the Gonads and Associated Ducts
  - 13.5.1 Development of the Gonads
  - 13.5.2 Development of the Gonads in Domestic Mammals
  - 13.5.3 Development of the Male and Female Reproductive Tracts
  - 13.5.4 Development of the Gonads in Poultry
  - 13.5.5 Development of the External Genitalia
- 13.6 Hormonal Control of Male Reproduction
  - 13.6.1 How Is Reproduction Controlled?
  - 13.6.2 Testosterone Mechanism of Action
  - 13.6.3 Puberty
- 13.7 Spermatogenesis
  - 13.7.1 Introduction to Spermatogenesis
  - 13.7.2 What Are the Important Characteristics of Spermatogenesis?
  - 13.7.3 Spermiogenesis
  - 13.7.4 Hormones and Testicular Functioning
    - 13.7.4.1 FSH and Sustentacular Cells
    - 13.7.4.2 Hormones and Testicular Endocrine Functioning
- 13.8 Temperature Set Point for Spermatogenesis
  - 13.8.1 Introduction to Maintaining Optimal Testicular Temperature for Spermatogenesis
  - 13.8.2 How Is the Temperature of the Testes Maintained at This Lower Set-Point in Livestock?
  - 13.8.3 How Can Blood Bathing the Testes Be at a Lower Temperature?
- 13.9 Erection
- 13.10 Ejaculation
- 13.11 Semen—Its Composition and the Roles of the Components
- 13.12 Differences Between Reproduction in Domestic Mammals and Poultry
- 13.13 Diseases of Male Reproduction in Domestic Animals

13.1 Introduction

13.1.1 What Are the Major Features of Male Reproduction?

Textbox 13.1 provides a glossary of terms used in male reproduction.

The features of male reproduction are the following:

- **Gametogenesis**: the testes producing very large numbers of the male gametes, spermatozoa. These are produced continuously.
- **Copulation and fertilization**: The presence of an organ, the penis, to deposit spermatozoa into the

---

Textbox 13.1 Glossary of Terms in Male Reproduction

- **Acrosome**: part of spermatozoa containing proteases to aid fertilization.
- **Anti-Müllerian hormone** (AHM): a hormone produced from sustentacular cells (a.k.a. Sertoli cells) of the testes. It causes the regression of the Müllerian ducts.
- **Castration**: removal of the testes.
- **Chimera**: an individual animal whose cells have at least two genotypes coming from different individuals. This can occur from the fusion of two different zygotes or can be accomplished experimentally. It is therefore possible for an animal to have cells that are XX and cells that are XY.
- **Counter current exchange**: flow of blood or other fluids in opposite directions, facilitating exchange of heat, gases, etc.
- **Coitus**: the act of mating.
- **Cryptorchidism**: presence of one or both of the testes within the body cavity.
- **Diploid**: having a set of paired chromosomes.
- **Erection**: is when the penis is in the state to allow intromission during coitus.
- **Ejaculation**: the transport of spermatozoa and secretions of accessory glands to the site of insemination.
- **Genotype**: the genetic make-up of an animal.
- **Haploid**: having a single set of unpaired chromosomes.
- **Interstitial endocrine cells** (a.k.a. **Leydig cells**): produce the male hormone testosterone.
- **Meiosis**: multiplication (reduction division) of the chromosomes generating haploid cells.
- **Mitosis**: multiplication of the chromosomes as part of cell division generating diploid cells.
- **Mosaic**: an individual animal whose cells have at least two genotypes coming from one individual but with genetically changed cells derived from a mutant cell early in development.
- **Pampiniform plexus**: region where spermatic arteries and veins are in close connection and where counter current transfer of heat occurs.
- **Phenotype**: the structures and functioning of an animal. These characteristics result from the interplay of the genotype and the environment.
- **Spermatic artery**: is the artery to the testes.
- **Spermatic vein**: is the vein from the testes.
- **Spermatogenesis**: production of the male gametes, the spermatozoa.
- **Spermatozoa** (plural; spermatozoan, singular): male gametes.
- **Spermiation**: the release of the spermatozoa from the supporting sustentacular cells and from the residual body. The excess cytoplasm that is left from the maturation of spermatids to spermatozoa is phagocytosed by the sustentacular cells and in the epididymis.
- **Spermiogenesis**: the transformation of spermatids to spermatozoa with development of the tail, midpiece, and acrosome, together with the loss of cytoplasm.
- **SRY gene**: sex-determining region on the Y chromosome.
- **SRY protein**: protein translated from the SRY gene transcript.
- **Sustentacular cells** (a.k.a. Sertoli cells): cells in the seminiferous tubules that function as nurse cells for the developing spermatozoa.
- **Testes (plural)**: male gonads producing spermatozoa and testosterone.
- **Testicle**: diminutive of testis.
- **Testis** (singular): male gonad producing spermatozoa and testosterone.
- **Testosterone**: male sex steroid hormone.
- **Vasectomy**: cutting the vas deferens so that the male can mate but not fertilize.

female reproductive tract at a site reasonably close to the site of fertilization.
- **Production of the male hormone testosterone**. This hormone, in turn, induces the following:
  - **Ducts** to allow transit of the spermatozoa from the testes to the exterior or into female tract.
  - **Growth of organs** (including the epididymis, prostate, seminal vesicles, ampulla, and bulbourethral gland) that add important constituents to the semen to allow spermatozoa motility (rudimentary in poultry).
  - **Libido**: the desire to mate.
  - Other aspects of male behavior including aggression.
- Presence of secondary sexual characteristics such as a comb in roosters and pheromones in boars.

Each of these features are discussed in more detail below.

13.1.2 What Is Castration?

Castration is removal of the testes. In animal production, livestock are frequently castrated without anesthesia. Castration is conducted with at least local anesthesia in horses, dogs, and cats. An alternative to castration is immunocastration. This entails vaccinating the animal against gonadotropin-releasing hormone (see Chapter 4). This will suppress release of luteinizing hormone and, downstream, of testosterone.

13.1.3 What Does Castration Do?

The male hormone testosterone plays a critically important role in all facets of male reproduction and the source of testosterone is removed by castration. Castration of livestock has been practiced for millennia. Castrated cattle were found to be much easier to handle as draft animals, for instance, pulling a plow or cart. In livestock and companion animals, castration is frequently performed. It results in less aggressive animals because of the loss of testosterone. It also results in dogs having a lower incidence of the undesirable behavior of mounting (objects and people). In addition, castration improves the quality of meat from male pigs as the "boar taint" is no lower present. The terminology for intact and castrated male domestic animals is given in Textbox 13.2.

Textbox 13.2 Interesting Factoids: Terminology of Intact and Castrated Animals

Table 13.1 Terminology for intact and castrated male livestock, companion animals, and poultry.

Species	Intact (uncastrated male)	Castrated
Cat	Tom or tom cat	Gib or neutered male
Cattle	Bull	Steer[1,2]
Dog	Dog	Neutered male
Horse	Stallion	Gelding
Pig	Boar	Barrow
Sheep	Ram	Wether
Chicken	Rooster[3]	Capon[4]
Turkey	Tom	Capon[4]

[1] An ox (plural oxen) is a draft animal commonly castrated.
[2] or bullock in UK English
[3] or cock in UK English
[4] Historical

13.2 Spermatozoa

13.2.1 What Are Characteristics That Spermatozoa Must Have?

Spermatozoa need the following:

- Motility.
  - The tail with its microtubules and coarse fibers propels the spermatozoan.
  - The mitochondria in the midpiece generates energy (ATP) for movement.
- Reduced weight: greatly reduced cytoplasm and no capability for protein synthesis.
- Haploid chromosomes: provide genetic material.
- Enzymes: released from the acrosome facilitate entry of the spermatozoan to the oocyte through the cumulus oophorus and zona pellucida.
- A mechanism to identify and fuse with the ovum.

13.2.2 Structure of Spermatozoa

The structure of the spermatozoan is shown in Figure 13.1.

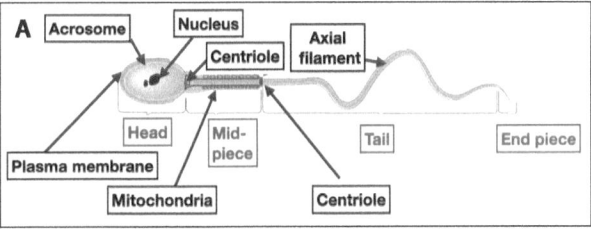

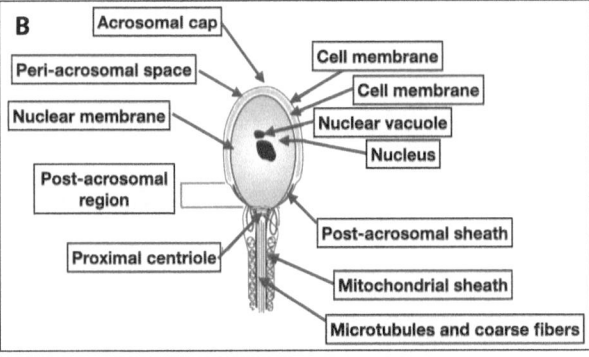

Figure 13.1 (A) Structure of a spermatozoan and (B) a spermatozoan head and midpiece. (Adapted from Bahnsen, CC BY-SA 4.0)

13.3 Structures of the Male Reproductive Tract and Their Functions

13.3.1 Testes

The testes (two) consist of the following, irrespective of whether in livestock, companion animals, or poultry:

- **Seminiferous tubules** contain developing spermatozoa, together with sustentacular cells (a.k.a. Sertoli cells) (Figures 13.2, 13.3, and 13.4). The seminiferous tubules produce large quantities of spermatozoa and the hormone **inhibin**.
- Interstitial endocrine cells (a.k.a. Leydig cells) in the interstitium between the tubules (Figure 13.3). These cells produce the male hormone testosterone (hormone controlling male reproduction is covered below in a separate section).

The testes are held in the **tunica vaginalis** (protecting them) and the **scrotum** (a sac of skin containing muscles enclosing the testes and the epididymis) in livestock and companion animals. The testes in most mammals, but not poultry, descend into the scrotum. This migration is induced by a hormone, insulin-like 3, produced by the testes. The position of the testes in the scrotum is adjusted by muscles such as the external cremaster muscle. They are brought close to the body at times when the environment is cold.

13.3.2 Ducts and Accessory Glands

Epididymis (two) is a long duct in mammals between the testes (the rete testes) and the vas deferens (see Figures 13.2 and 13.3). If uncoiled, the length of the epididymis is estimated as the following:

- Horse: 75 meters.
- Cattle: 47 meters.
- Pigs: 55 meters.

The functions of the epididymis are the following:

- To absorb fluid coming with the spermatozoa from the seminiferous tubules via the rete testis (see Figure 13.3).
- To mature spermatozoa, including removal of the remnants of the cytoplasm in the residual body.
- To store spermatozoa until needed during ejaculation.

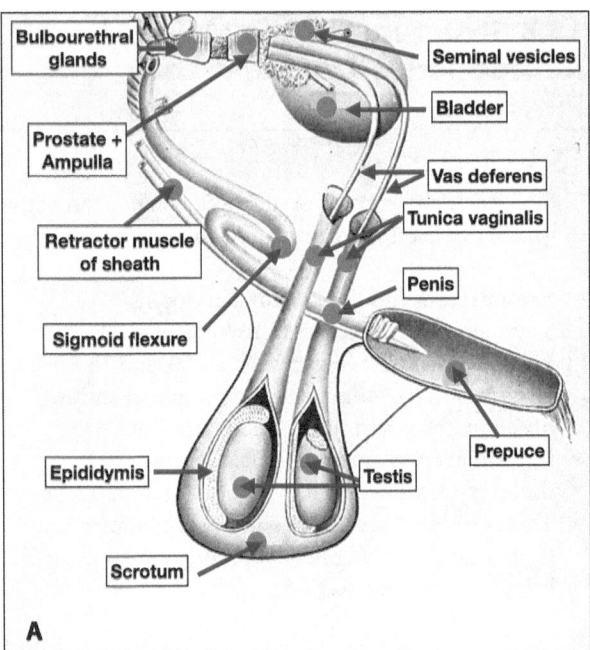

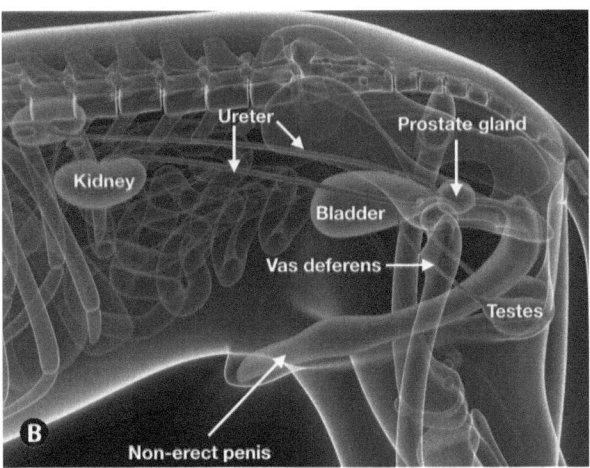

Figure 13.2 Reproductive system of male domestic animals. (A) The bull—note the sigmoid flexure of the penis; (B) The dog. (A is adapted from George Ransom White; B is adapted from Shutterstock/decade3d - anatomy online)

- To produce nutrients for spermatozoa, such as glyceryl phosphoryl choline and citric acid.
- To produce proteins to protect the spermatozoa.

In poultry, the epididymis is small.

The **vas deferens** (or ductus deferens) are two ducts that conduct spermatozoa from the epididymis to the single urethra in livestock and companion animals (Figures 13.2 and 13.3). Structurally, the vas deferens includes both circular and longitudinal smooth muscles. In some species, there is an expanded region of the vas deferens, the ampulla, adjacent to the urethra. In contrast, in poultry the vas deferens is coiled and adds fluid to the semen. It connects the epididymis to the single copulatory organ in the cloaca.

Prostate gland (one) contributes fluid to semen during ejaculation (Figure 13.2). The prostate adds fluid including **bicarbonate** and **selenium** to the semen. These aid spermatozoa motility and the pH and osmotic balance of the semen. The prostate gland is large in horses but small in bulls and not even present in poultry (see Textbox 13.2).

Seminal vesicles, or vesicular glands, (two) contribute the majority of both the fluid, and fructose, an important energy source for the spermatozoa, to semen during ejaculation (Figure 13.2). Additional energy sources in the secretions of the seminal vesicles include citric acid, glyceryl phosphoryl choline, and sorbitol. The secretions also include ions including bicarbonate, calcium, magnesium, phosphate, selenium, sodium, and zinc. These maintain the optimum pH and osmotic balance in the semen. In addition, many of the ions aid motility of spermatozoa. The seminal vesicles are not present in poultry.

Bulbourethral glands (equivalent to Cowper's glands in humans) contribute fluid; ions such as sodium, potassium, and chloride to maintain the optimum pH; and proteins to semen during pre-ejaculation (Figure 13.2). The ions such as sodium, potassium, magnesium, and chloride maintain the optimum pH and osmotic balance in the semen. In addition, the ions aid motility of spermatozoa. The Cowper's glands are not present in poultry.

13.3.3 Penis

The penis is an organ for copulation in mammals (Figure 13.2). The penis consists of three components:

- Glans.
- Shaft.
- Root.

It contains cylinders of tissue, the **corpus spongiosum** (1) and the **corpus cavernosa** (2). The urethra passes through the middle of the corpus.

There are two types of penises found in different livestock and companion animal species:

- **Musculovascular penis** (or musculocavernous).
- **Fibroelastic penis.**

Musculovascular penis is found in stallions, cats, and dogs. The corpus spongiosum and the corpus cavernosa

Textbox 13.3 Weights of Components of the Male Reproductive System

Table 13.2 provides data on the weights of male reproductive organs in horses, livestock, and chickens.

The testes in chickens are about 1% of body weight. In comparison, the testes in livestock represent a much lower percentage of body weights (0.02% of body weight in horses, 0.15% of body weight in pigs, and 0.036% of body weight in cattle). The combined weight of the accessory glands is markedly greater than the weight of the testes.

Table 13.2 Weights of male reproductive organs in horses, livestock, and horses.

Organ	Weight of organ in grams			
	Horse	Pig	Cattle	Chicken
Testes	225	312	272	20
Epididymis	190	136	23	0–not present
Seminal vesicle	320	249 (empty 144)	35	0–not present
Prostate	90	18	< 5	0–not present
Bulbourethral glands	120	137 (empty 89)	5	0–not present

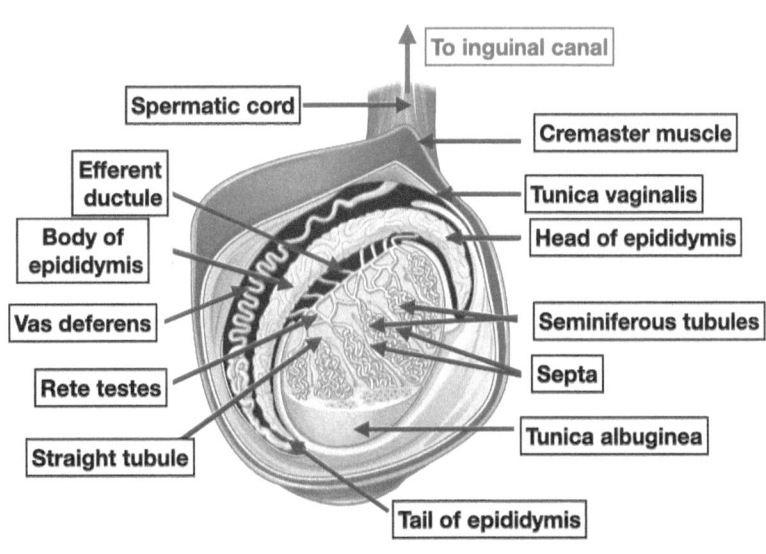

Figure 13.3 Structure of mammalian testes and associated tissues. (Adapted from OpenStax College, CC BY 4.0)

become engorged with blood during erection (erection is covered in Sections 13.9). An **os penis** (ossified penile bone) or baculum is found in dogs and cats. This enables erection. It is ossification (turned to bone) of the center of the corpora cavernosa. The os penis is not present in livestock, horses, or poultry.

Fibroelastic penis is found in bulls, boars, and rams. The corpus spongiosum and the corpus cavernosa contain collagen and fibrous tissue and has limited erectile tissue. In contrast, there is a sigmoid flexure of the penis. This is held in place by the retractor muscles in the flaccid state. During erection, the muscles relax and the

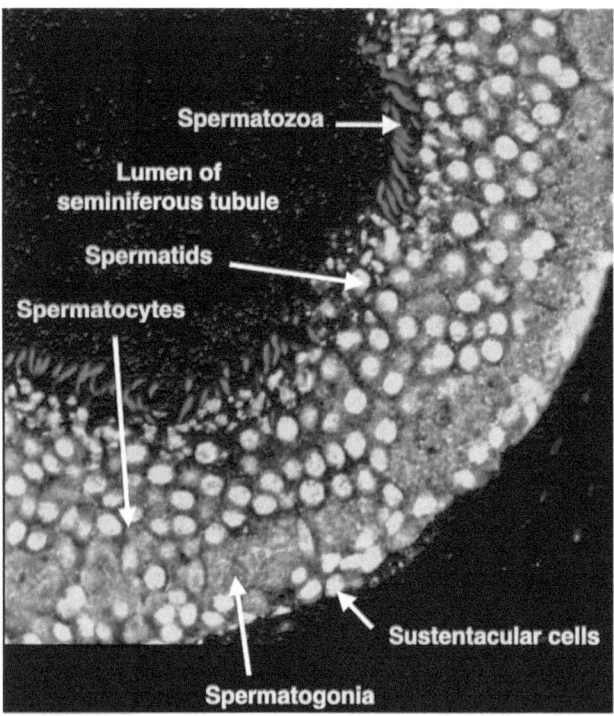 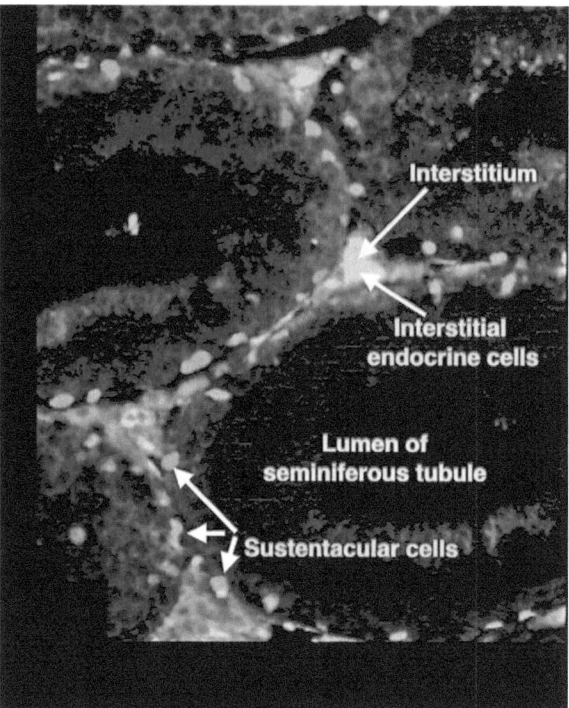

Figure 13.4 Section through the seminiferous tubule colorized by immunocytochemistry. (Adapted from Sharvari.deshpande996 and H. Li and colleagues, CC BY-SA 4.0)

penis protrudes from the body (erection is covered in Sections 13.9).

13.3.4 Reproductive System of Poultry

Male poultry do not have either a penis as an organ from intromission or the same accessory organs as in mammals. Instead, within the cloaca of a rooster or tom turkey, there are the following:

- **Rudimentary copulatory organ** (misnamed as it is not used for copulation).
- The **ejaculatory groove region** acts like an accessory organ, adding a lymph-like fluid to aid in the production of semen.
- The **papillae** serving as the mating organs.

13.4 Genetics of Sex Determination

13.4.1 Are the Genetics Determining Whether an Animal Is Male or Female Across Species?

You might expect that something as fundamental as developing a male or female reproductive system would be the same across disparate animals. This is not the case.

The system is different between livestock and companion animals (therian mammals) compared to birds, reptiles, and monotreme or egg laying mammals. The sex chromosomes determining gender in livestock and companion animals (therian mammals) are broadly the following:

- XX (females ♀).
- XY (males ♂).

The sex chromosomes determining gender in poultry are broadly the following:

- ZW (females ♀).
- ZZ (males ♂).

The male phenotype develops due to the presence of the SRY gene and its protein product in livestock and companion animals (Tables 13.3 and 13.4). Females develop because of the absence of the SRY gene. In contrast, in poultry and other birds, there is no SRY gene. So how do males or females develop? It is thought that there is a gene-dosing effect of a gene on the Z chromosome. A possible gene in birds is either the **doublesex** or **mab-3-related transcription factor-1** (DMRT1) gene.

Chapter 13 Male Reproduction

Table 13.3 Genetics of gender in livestock, companion animals, and poultry.

Phenotypes	Genotypes	
	Livestock & companion animals	Poultry
Simple model		
Male phenotype	XY	ZZ
Female phenotype	XX	ZW
More complete model		
Male phenotype	XY with functional SRY gene[1]	ZZ - no SRY gene
Female phenotype	XX without functional SRY gene[2]	ZW - no SRY gene

[1] Usually, on Y chromosome but possibly on X chromosome.
[2] The SRY gene may be nonfunctional on the Y chromosome. Under these circumstances, the female phenotype develops.

Table 13.4 Genotypes and phenotypic consequences.

Genotypes	Phenotype
Livestock and companion animals (Therian mammals)	
XY	Male if functional SRY gene
XX	Female with non-functional SRY gene
XO	Female infertile
XXY or XY/XXY mosaic[1]	Male
Poultry (Birds)	
ZZ	Male – there is no SRY gene
ZW	Female – there is no SRY gene
ZO	Embryonic lethal
WO	Embryonic lethal
ZZZ[2]	Testes but no functional spermatozoa
ZZW[2]	One testis and one ovotestes prior to sexual maturation but two testes following maturation

[1] A mosaic animal has cells of different genotypes (see Textbox 13.1).
[2] Triploidy (three chromosomes).

It is dosage-sensitive, on the Z chromosome, and its expression influences sex determination.

A male mammal having XY chromosomes with the SRY gene on the Y chromosome is broadly true. While female mammals having XX chromosomes is broadly true, there are also animals that have different genotypes such as XO (a single X chromosome), XXX (three X chromosomes), XXY (two X plus one Y chromosomes), XYY (two Y plus one X chromosomes), in livestock/companion animals, and ZZZ and ZZW in poultry (Table 13.4).

13.5 Development of the Gonads and Associated Ducts

13.5.1 Development of the Gonads

Primordial germ cells have the potential to develop into gametes. The **urogenital ridges** are the progenitors of the gonads. They are located on the surface of the mesonephros (the early kidneys). Primordial germ cells migrate from the endoderm of the yolk sac along the mesentery of the hindgut to the urogenital ridges (see Figures 13.5 and

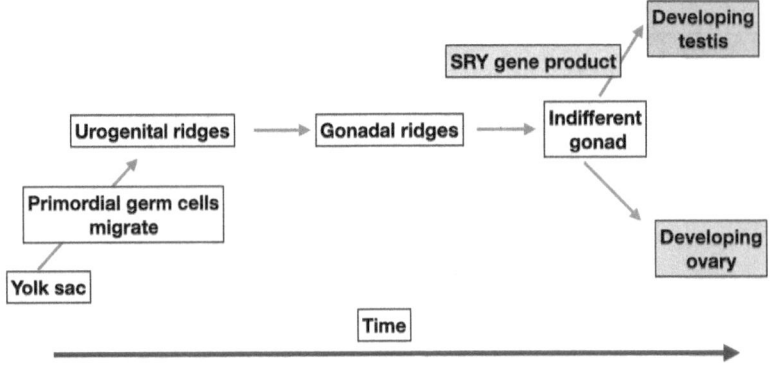

Figure 13.5 Schematic of the development of gonads in livestock and mammalian companion animals.

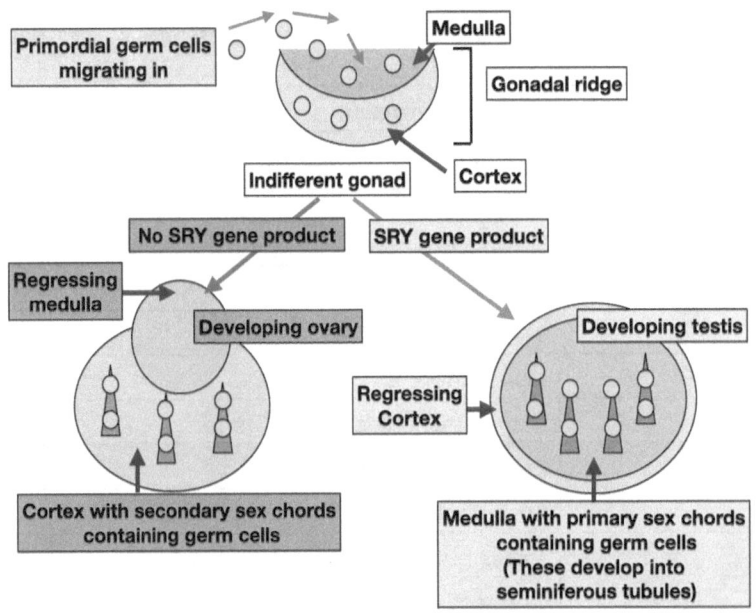

Figure 13.6 Development of gonads in males and females in mammals with primordial germ cells migrating from the yolk sac to the genital ridge forming the gonadal ridge. This develops into the indifferent gonad. The medulla of the indifferent gonads develops into the testes due to the expression of the SRY gene. In the absence of the SRY gene, the cortex develops into the ovary.

13.6). When the primordial germ cells are present in the urogenital ridges, they are then called the **gonadal ridges** (see Figures 13.5 and 13.6). These gonadal ridges develop into the indifferent gonad which has a **cortex** and a **medulla** (Figure 13.5). The primordial germ cells develop into either spermatogonia (male) or oogonia (female) and ultimately into gametes.

13.5.2 Development of the Gonads in Domestic Mammals

In male domestic mammals, the medulla develops into the testis if the SRY gene is present and its product is produced (Figure 13.5). The primary sex cords develop to the seminiferous tubules with sustentacular cells

(Figures 13.5 and 13.6). In females, the cortex develops into the ovary and the medulla regresses (Figures 13.5 and 13.7).

13.5.3 Development of the Male and Female Reproductive Tracts

In both males and females, there is two ducts from the embryonic kidneys (mesonephric kidneys) to the exterior (see Figure 13.7). These are the mesonephric ducts. As the gonads develop adjacent to the embryonic kidney, they are also drained by these ducts and they are called the **Wolffian ducts**. The **Müllerian ducts** develop alongside the Wolffian ducts (see Figure 13.7).

In males (mammals with functional SRY genes), the gonads produce both testosterone and anti-Müllerian hormone (AMH). Under the influence of testosterone, the Wolffian ducts develop into the male ducts and accessory organs. Under the influence of AMH, the Müllerian ducts regress (see Figure 13.7; for more details see Textbox 13.4).

In females (mammals lacking functional SRY genes), the gonads do not produce either testosterone or AMH. The Müllerian ducts develop into the female tract consisting of the oviducts, uteri, cervices, and posterior vagina. The Wolffian duct regresses in the absence of testosterone.

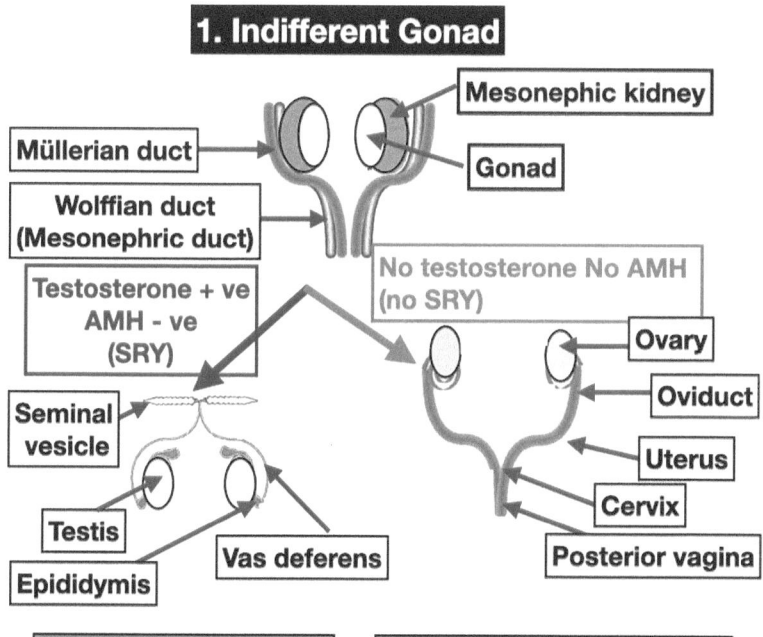

Figure 13.7 Development of male and female reproductive ducts in mammals.
Key:
AMH = anti-Müllerian hormone
SRY gene = sex-determining region of the Y chromosome

Textbox 13.4   A Deeper Dive Into the Development of the Reproductive Ducts

The Wolffian ducts develop into male reproductive ducts if there is a functioning SRY gene (mammals), and in females the Müllerian ducts develop into the oviduct, uterus, cervix, and posterior vagina in mammals, or single oviduct in poultry.

The mesonephric ducts are the drainage ducts of the embryonic/early fetal stage kidney, the mesonephros. These become repurposed as the Wolffian ducts from the gonads to the exterior. Another but separate duct develops alongside the Wolffian duct. This is called the Müllerian duct. This is the situation with an indifferent gonad and duct system (see Figure 13.8).

In males of both domestic mammals and poultry, the sustentacular cells in the developing testes produce anti-Müllerian hormone

(continued)

Textbox 13.4 (continued)

(AMH) (Figure 13.8). This causes the Müllerian duct to regress (Figure 13.8.). The interstitial endocrine cells (a.k.a. Leydig cells) of the developing testes produce testosterone and this stimulates the Wolffian duct to mature (see Figure 13.8).

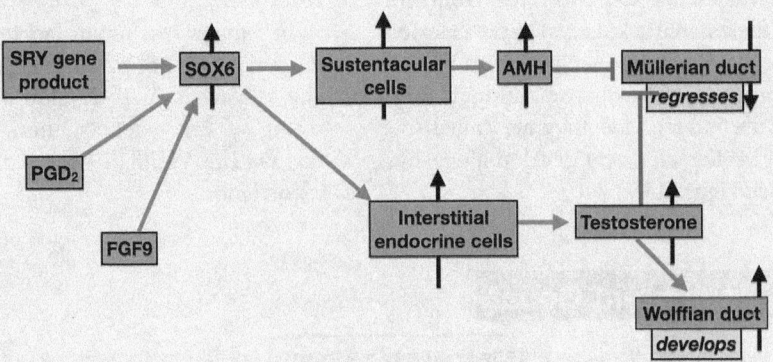

Figure 13.8 Development of male reproductive system in mammals. Key:
SYR = sex-determining region of the Y chromosome
SOX6 = sex-determining region-box 9
PGD$_2$ = prostaglandin D$_2$
FGF9 = fibroblast growth factor 9
AMH = anti-Müllerian hormone

13.5.4 Development of the Gonads in Poultry

The development of gonads in poultry is covered in Textbox 13.5.

Textbox 13.5 A Deeper Dive Into the Development of the Gonads and Ducts In Poultry

In poultry, there are also two indifferent gonads as domestic mammals and in Figure 13.5. The medulla of the indifferent gonads develops into the testes in the presence of sufficient quantities of the product of the DMRT1 gene. As in mammals, the ovary develops from the cortex of the indifferent gonad. However, there is a big difference. Only the left indifferent gonad develops into an ovary, while the right indifferent gonad regresses. In poultry, if the left ovary is removed or is destroyed by disease, the right gonad remnant will develop into a functional testis. This represents female to male sex reversal.

Genes that are critically important to the differentiation of the gonads in mammals (Figure 13.7) play a role in the development of the gonads in birds. For instance, sex-determining region-box 9 (SOX9) plays a role in the development of the testes while Wnt4 and FOXL2 (Forkhead box L2) play a role in the development of the single ovary. In addition, ovarian development is related to the presence of aromatase, an enzyme that is essential to the synthesis of estrogens.

In male chickens, the ducts develop in a manner similar to that in mammals but there is a wrinkle in the female. In female poultry, the left ovary produces anti-Müllerian hormone (AMH) together with estrogen. The estrogen protects the left Müllerian duct but the right Müllerian duct regresses under the influence of AMH.

13.5.5 Development of the External Genitalia

In domestic mammals, there are obvious differences in the external genitalia between males and females. This is not the case in poultry. In domestic mammals, there is an analogous indifferent stage for the external genitalia as with the gonads and ducts (Figure 13.9). This is also called the **ambisexual stage**. At this stage, the tissues that will become the external genitalia consists of the following (Figure 13.9):

- Genital tubercle.
- Genital swelling.
- Genital fold.
- Genital membrane.

The wonderfully named **sonic hedgehog gene** (Shh) plays an important role in the development of the genital tubercle. The external genitalia develop to the male arrangement in the presence of the product of the SRY gene (sex-determining region of the Y chromosome) and testosterone from the embryonic interstitial endocrine cells (Figure 13.10). In the

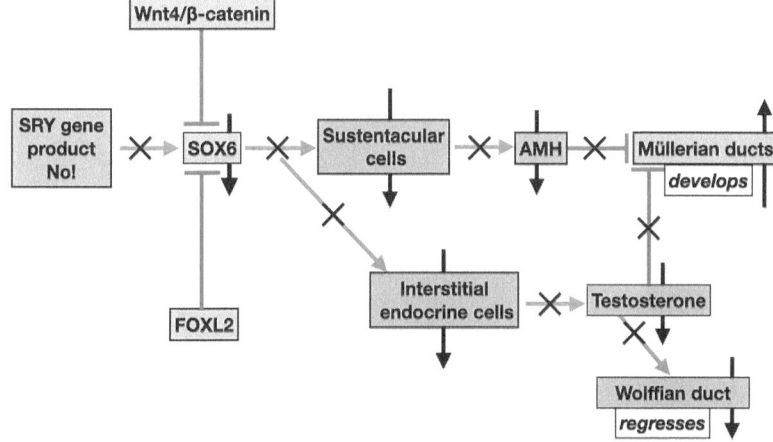

Figure 13.9 Development of female reproductive system in mammals.
Key:
Wnt4 = acts by binding to the frizzled receptor, increasing β-catenin
SYR = sex-determining region of the Y chromosome
SOX6 = sex-determining region-box 9
FOXL2 = forkhead box L2
AMH = anti-Müllerian hormone

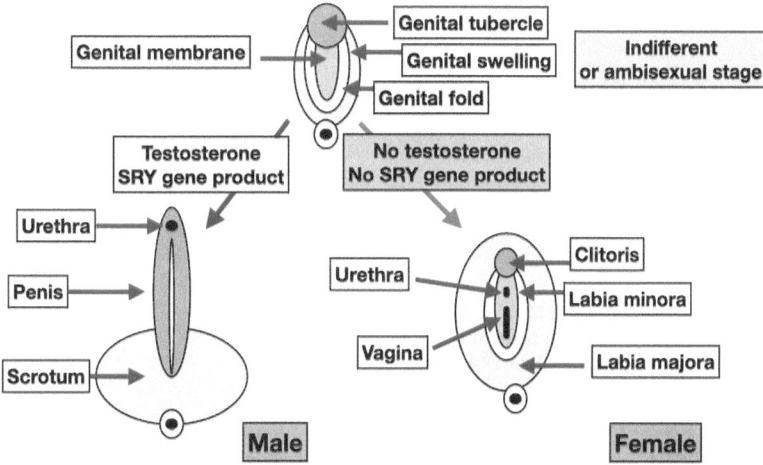

Figure 13.10 Development of the external genitalia in mammals.

absence of the SRY gene product and testosterone, female genitalia develop.

13.6 Hormonal Control of Male Reproduction

13.6.1 How Is Reproduction Controlled?

Reproduction in males is under hormonal. The control of reproduction is summarized in Figure 13.11.

The hypothalamus in the brain controls the release of the gonadotropins, luteinizing hormone (LH) and follicle-stimulating hormone (FSH) from the anterior pituitary gland. Specifically, gonadotropin-releasing hormone (GnRH) stimulates the release of both LH and FSH. In turn, FSH acts on the sustentacular cells and, hence, facilitates spermatogenesis (discussed in Section 13.7). The sustentacular cells haver the following actions:

- Producing inhibin. This inhibits release of FSH.
- Producing **androgen-binding protein** (ABP), binding testosterone allowing for high concentrations of testosterone in the lumen of the seminiferous tubule and in the fluids arriving at the head of epididymis.
- Producing AMH.

LH stimulates the interstitial endocrine cells to produce the male hormone testosterone (Figure 13.11). The testosterone has multiple effects, including the following:

- Stimulating the accessory organs, such as prostate and seminal vesicles.
- Inducing male behaviors such as libido and aggression.
- Causing the development of the secondary sexual characteristics.
- Inhibiting the release of GnRH and LH in a negative feedback loop.

13.6.2 Testosterone Mechanism of Action

Testosterone acts by binding to steroid hormone receptors that act by increasing transcription of specific genes. Testosterone either binds to androgen receptors after 5 α reductase to dihydrotestosterone or to estrogen receptors after **aromatization** (Figure 13.12).

13.6.3 Puberty

During puberty or at the onset of the breeding season, the male reproductive system becomes fully functional.

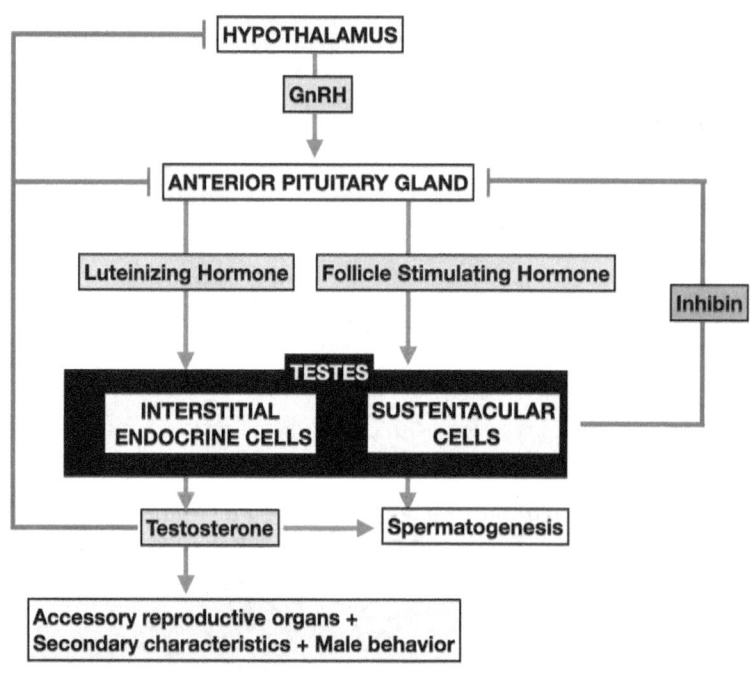

Figure 13.11 Hormonal control of reproduction in the male. Key: GnRH = gonadotropin releasing hormone ↓ is positive, T is negative

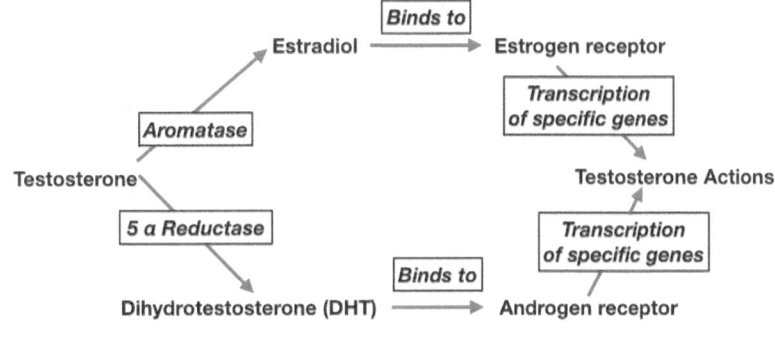

Figure 13.12 Mechanism of action of testosterone.

This is controlled by the hypothalamus releasing GnRH (see Figure 13.11). In turn, the anterior pituitary gland produces the gonadotropins, LH and FSH, that stimulate the testes to complete their maturation.

13.7 Spermatogenesis

13.7.1 Introduction to Spermatogenesis

Figures 13.13 and 13.14 summarize spermatogenesis.

13.7.2 What Are the Important Characteristics of Spermatogenesis?

There need to be very high numbers of spermatozoa that are produced continuously during the breeding season in seasonal breeding species (unlike females, where one ovum or a small number of ova are produced per estrous cycle).

Spermatogenesis is synchronized in sections of the seminiferous tubule with stages spaced along the seminiferous tubules. This allows a discontinuous process producing spermatozoa continuously. Definitions of the various aspects of spermatogenesis are given in Textboxes 13.1 and 13.7.

There are multiple mitotic divisions of **undifferentiated spermatogonia** and their **daughter cells** producing large numbers of B spermatogonia and even more primary spermatocytes (Figure 13.12). There is a progression of spermatogonia from A_1, A_2, A_3, A_4, and in (intermediate) spermatogonia. These **primary spermatocytes** undergo meiotic division producing first **secondary spermatocytes** and then **spermatids**. Retinoic acid is one of the factors required for spermatogenesis. Spermatogonia and primary spermatocytes are diploid. Secondary spermatocytes, spermatids, and spermatozoa are haploid.

13.7.3 Spermiogenesis

Spermiogenesis is a maturation/differentiation event. There is conversation of round spermatids to sleek spermatozoa. Figure 13.14 summarizes spermiogenesis. Spermiogenesis consists of the following:

- Development of the acrosome (filled with **proteolytic enzymes**) from the Golgi apparatus.
- Multiplication of mitochondria and their migration ultimately to the midpiece of the spermatozoa.
- Development of the tail or flagella from the **centriole**, which is essential to the spermatozoa being motile.

Testosterone together with retinoic acid stimulates spermiogenesis (Figure 13.13). The testosterone originates from the interstitial endocrine cells; these being located outside the seminiferous tubule and the basement membrane (also discussed below).

Textbox 13.6 The Words *Testis* and *Testicle*

Testis (plural testes) is the scientific term for the male gonads. The etymology of the word testis is as in "witness," as in "testify." Testes are witnesses of manhood or maleness! The word "testicle" is the diminutive of testis like Bob is the diminutive of Robert and Beth is a diminutive of Elizabeth.

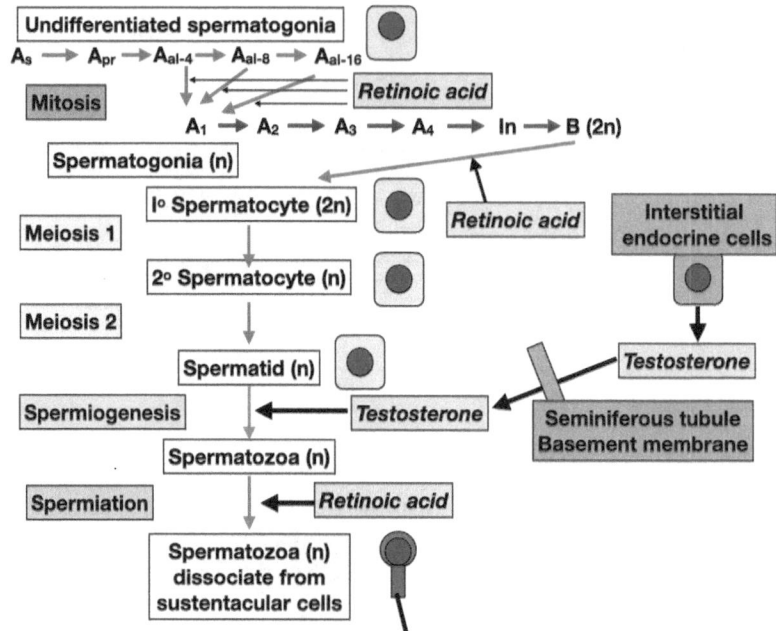

Figure 13.13 Schematic diagram of the production of spermatozoa.
Key:
A_s = spermatogenic stem cell
A_1 to B = progression of spermatogonia
(n) = haploid
(2n) = diploid

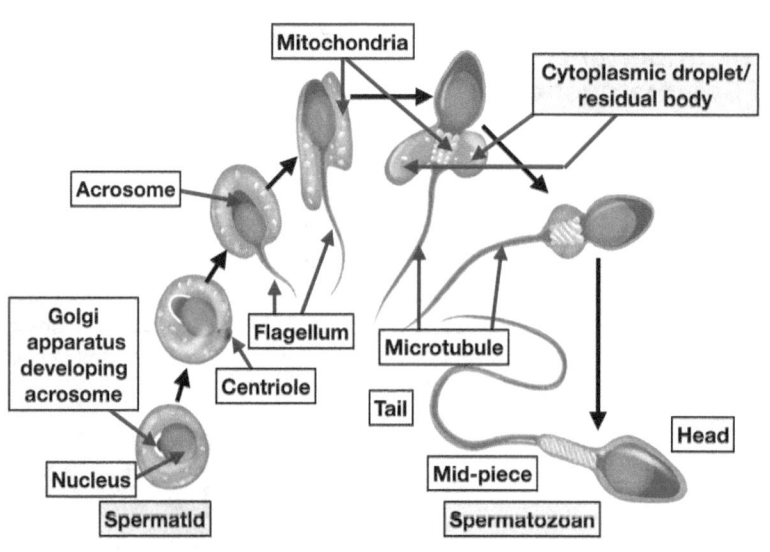

Figure 13.14 Schematic diagram of spermiogenesis. (Shutterstock/pr_camera)

13.7.4 Hormones and Testicular Functioning

13.7.4.1 FSH and Sustentacular Cells

Sustentacular cells extend from the basement membrane to the lumen of the **seminiferous tubule** (Figure 13.15). They play a critically important role in the development of the cellular precursors of the spermatozoa and the removal of the excess cytoplasm in spermiogenesis. FSH increases both the number of sustentacular cells and their functioning, including synthesis of androgen-binding protein (ABP). Binding of testosterone to ABP allows for high concentrations of testosterone in the testes.

13.7.4.2 Hormones and Testicular Endocrine Functioning

Luteinizing hormone (LH) stimulates interstitial endocrine cells to proliferate and produce testosterone by binding to LH receptors (see Figure 13.11). There are follicle-stimulating hormone (FSH) receptors on the sustentacular cells. FSH stimulates proliferation of sustentacular cells. In addition, testosterone binds to androgen receptors in sustentacular cells, stimulating their proliferation. The number of sustentacular cells has a strong effect on the number of spermatozoa processed. Testosterone and FSH additively positively impact the entry of germ cells into meiosis.

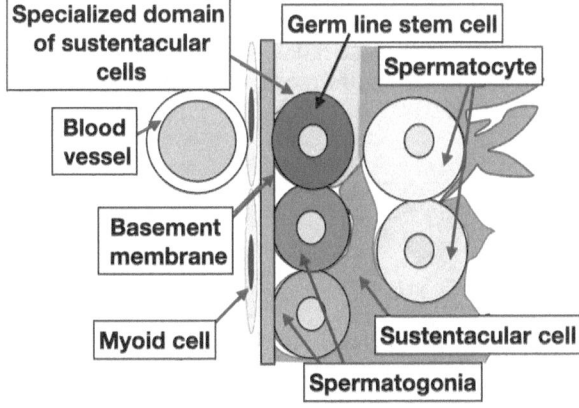

Figure 13.15 The role of sustentacular cells in spermatogenesis. The spermatogonia, spermatocytes, and spermatids are in intimate contact with sustentacular cells. (Adapted from Ting Xie, Creative Commons Attribution license)

13.8 Temperature Set Point for Spermatogenesis

13.8.1 Introduction to Maintaining Optimal Testicular Temperature for Spermatogenesis

The temperature in the testes of livestock is lower than their core body temperature. This is because there is a lower set-point temperature in the scrotum for livestock than in the rest of the body. But why? Spermatogenesis is more efficient at a lower temperature.

13.8.2 How Is the Temperature of the Testes Maintained at This Lower Set-Point in Livestock?

There are several mechanisms for maintaining the temperature of the testes, including the following:

- The blood bathing the testes is at a lower temperature than the core body temperature.
- Movement of the testes from close to the body at times of low environmental temperature to away from the body when the environmental temperature is high. The testes can be moved close to the body by the cremaster muscles and tunica dartos.

These allow the following:

- Evaporative heat lost.
- Radiative heat loss or gain.
- Conductive heat loss or gain.

13.8.3 How Can Blood Bathing the Testes Be at a Lower Temperature?

Blood exits the body cavity in the spermatic artery at the core body temperature. When it reaches the testes, it is about 4° C lower. How can that be? The spermatic artery and the spermatic vein are in close proximity in the pampiniform plexus. Heat passes from the blood in the spermatic artery from the spermatic vein in a counter-current manner (flowing in opposite directions) (Figure 13.15). This also means that the blood passing through the inguinal canal in the spermatic vein (back to the heart) is close to the core body temperature. *Note:* With internal testes in poultry, the temperature in the testes is the same as the core body temperature.

292 THEME 4 Reproduction (Propagation of the Species)

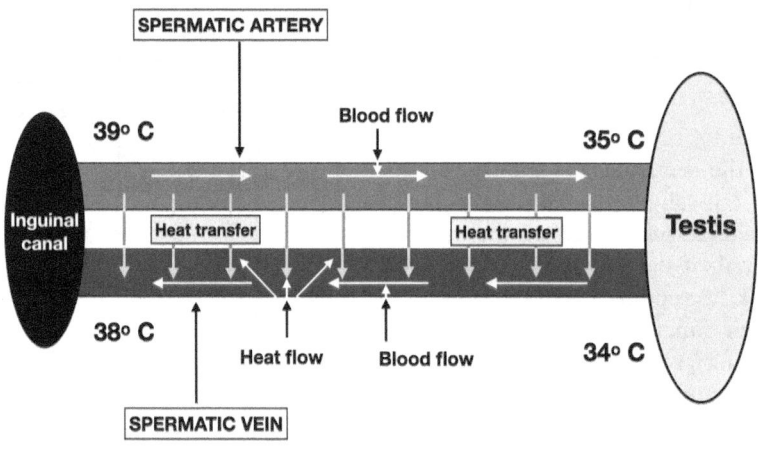

Figure 13.16 Blood bathing the testes is cooled by counter-current transfer of heat in the pampiniform plexus.

13.9 Erection

In domestic mammals during arousal by, for instance, the presence of an estrous female, the penis becomes erect (see Figures 13.16 and 13.17). There is little information on the changes in the ejaculatory groove and papillae in the sexually aroused rooster or tom turkey.

In the sexually stimulated male, there is release of nitric oxide (NO) from **efferent parasympathetic nerve terminals** (from the pelvic nerve) and from **endothelial cells** in response to acetylcholine (also in response to parasympathetic nerve terminals) (see Figure 13.18). The NO activates guanylate cyclase (GC) (see Figure 13.18). In turn, guanylate cyclase catalyzes the formation of **cyclic guanosine monophosphate** (cGMP) from guanosine-triphosphate (GTP) (see Figure 13.18). The cGMP activates **protein kinase G** (see Figure 13.18). This decreases calcium ions and causes relaxation of smooth muscles (see Figure 13.18). This relaxation allows increases blood flow and consequently causes erection in males with a vascular penis. In males with a fibroelastic penis, the decrease in calcium ions causes relaxation of muscle restraining the **sigmoid flexure** of the penis. Sympathetic nerves, if activated and releasing norepinephrine, inhibit erection.

13.10 Ejaculation

Ejaculation is the transport of spermatozoa and secretions of accessory glands to the site of insemination. Ejaculation in livestock and domestic animals is viewed as having two phases:

Figure 13.17 Stallion with erect penis attempting to mount a mare. (Peter Rohrbeck, CC BY-SA 4.0)

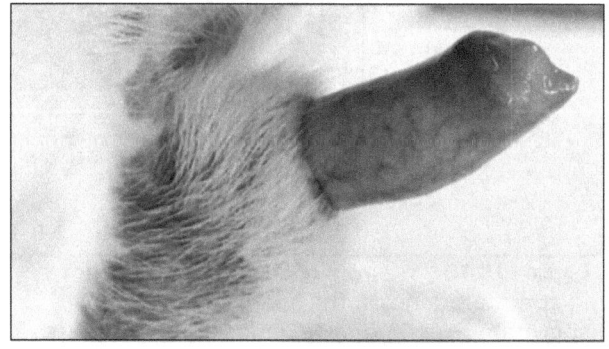

Figure 13.18 Erect penis of dog. (Shutterstock/charnsitr)

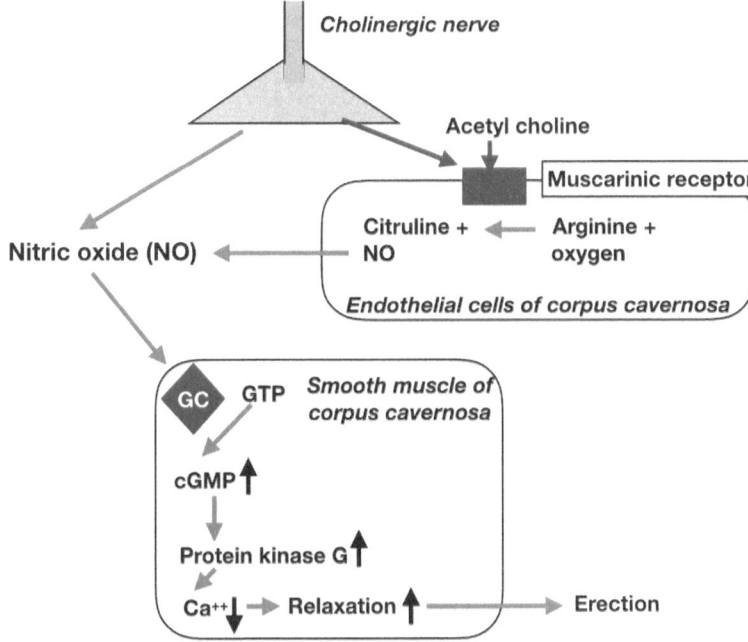

Figure 13.19 Parasympathetic control of erection mediated by nitric oxide.
Key:
NO = nitric oxide
GC = guanylate cyclase
GTP = guanosine-triphosphate
cGMP = cyclic guanosine monophosphate

Textbox 13.7   A Deeper Dive Into Ejaculation

Ejaculation is controlled by the brain. It is induced by two afferent inputs to the cerebral cortex:

- Stimulation of the penis during coitus, together with other sexual cues. There is then activation of centers in the thalamus and hypothalamus; in the case of the latter: medial **preoptic area (MPOA)**.

- The paraventricular nucleus (PVN). Finally, there are nerves passing to the genital area inducing ejaculation. Ejaculation is stimulated by the neurotransmitter dopamine but inhibited by serotonin; these acting within the brain. There is also release of oxytocin during ejaculation. Ejaculation can also be induced by electrical stimulation of pelvic nerves.

- **Emission** with spermatozoa and epididymal secretions being moved from the epididymis to the urethra and secretions of the accessory glands added. In addition, there are muscle contractions at the neck of the bladder preventing flow of semen into the bladder or urine into the semen.
- **Expulsion** with the semen expelled from the penis due to spasms of pelvic muscle contractions. The semen travels through the urethra.

There is relatively little information on the ejaculation in poultry.

Textbox 13.8   Interesting Factoid About Coitis in Dogs

During coitus of dogs, there is a "lock" of the penis and vagina for a period of about 15 minutes. This is accompanied by ejaculatory contractions.

13.11 Semen—Its Composition and the Roles of the Components

Table 13.5 summarizes semen volume in different domestic animals. Ejaculate consists of the following:

- **Pre-sperm fraction** from bulbourethral glands.
- **Sperm rich fraction** from the epididymis, seminal vesicles, and prostate.
- **Post sperm fraction** (a gel fraction) is emitted in horses and pigs.

Semen contains large numbers of spermatozoa, the male gametes. There are literally billions of spermatozoa in the ejaculate of livestock, horses, and dogs (see Table 13.5). The number of spermatozoa per ejaculation is lower in poultry and cats.

Semen contains the following:

- Ions (sodium, potassium, calcium, chloride, and zinc) that affect spermatozoa motility.
- Proteins that cause the semen to gel and proteolytic enzymes (e.g., from the prostate) that release then the spermatozoa. In addition, there are proteins called spermadhesins. These form a protective coat on the head of the spermatozoa preventing a premature acrosome reaction.
- Energy sources for the spermatozoa. These include citrate from the prostate gland, fructose from the seminal vesicles, and glycerophosphorylcholine from the epididymis (see Figure 13.2 and Table 13.6). In the case of glycerophosphorylcholine, it is split in the female reproductive track to form glyceryl phosphate. This is readily used by the spermatozoa (see Figure 13.13).
- Antioxidants including vitamin E and glutathione to prevent damage from reactive oxygen species to the membranes of the spermatozoa.
- Small RNA molecules.

There are effects of **seminal plasma** (semen without spermatozoa) on the female reproductive tract. These effects are mediated by the following:

- Ions that assure the correct osmotic balance and buffering the pH.

Table 13.5 Characteristics of semen in livestock, companion animals, and poultry.

	Volume of ejaculate (mL)	Spermatozoa concentration (million per mL)
Cattle	2–10	300–2000
Horses#	20–300	30–800
Pigs#	150–500	25–350
Sheep	0.5–2	2000–5000
Cats	0.055	41
Dogs	1–30	300–2000
Chickens	0.5	4.65

#Contains gel.

Table 13.6 Concentrations of energy sources in semen for spermatozoa motility.

	Citrate mg dL^{-1}	Glyceryl phosphoryl choline mg dL^{-1}	Fructose mg dL^{-1}
Cattle	357–1000	110–500	120–540
Horses	8–53	40–110	< 1
Pigs	36–325	119–240	20–40
Sheep	137	1600–2000	150–600

- **Prostaglandins** causing contractions of smooth muscles.
- Both pro-inflammatory and anti-inflammatory cytokines, together with growth factors.

> **Textbox 13.9 Definitions Part 2**
>
> - **Capacitation**: spermatozoa undergo capacitation in the female reproductive tract following ejaculation. This can be viewed as a maturation step so that the spermatozoa are both more active and capable of fertilizing the ovum. During capacitation, spermadhesin molecules bound to the membrane of spermatozoa are lost.
> - **Acrosome Reaction**: when the spermatozoa bind to specific proteins on the zona pellucida, the acrosome reaction occurs. The first step after binding is calcium ions flowing into the spermatozoan head. The acrosome of the spermatozoa then releases its enzymes, proteases and hyaluronidase. The action of these enzymes allows the spermatozoa to penetrate zona pellucida (a protective layer around the ovum) and, thereby, facilitate fertilization.

Of the large amount of semen in boars, most comes from the accessory glands:

- Testes and epididymis 5%.
- Prostate 40%.
- Seminal vesicle 30%.
- Bulbourethral glands 25%.

Similarly, in bulls, the semen consists of the following:

- Volume: ~50% from the seminal vesicles.
- Potassium: 75% from the seminal vesicles.
- Protein and calcium: 85% from the seminal vesicles.

The ampulla, prostate, and bulbourethral glands contribute some additional constituents to semen, such as about half the chloride and sodium. In dogs, there are no seminal vesicles but the prostate contributes about a half of the semen by volume.

13.12 Differences Between Reproduction in Domestic Mammals and Poultry

Table 13.7 summarizes differences in male reproduction between that in livestock/companion animals and that in poultry.

Table 13.7 Examples of similarities and differences in male reproduction between livestock/companion animals and poultry.

	Livestock, cats, & dogs	Poultry & other birds
Similarities		
Testes	Present	Present
Vas deferens	Present	Present
Differences		
Major genotype	XY	ZZ
Testes	Testes in scrotal sac (external)	Testes internal by kidneys
Testes	Temperature below body set-point	Temperature at body set-point
Spermatozoa number per ejaculation	> 100 million	<100 million
Prostate	Present	Absent
Seminal vesicles	Present except in dogs	Absent
Ampulla	Present	Absent
Cowper's gland	Present	Absent
Penis	Present	Absent
Semen contains nutrients for spermatozoa	Yes	No

13.13 Diseases of Male Reproduction in Domestic Animals

Textbox 13.10 summarizes reproductive diseases of male domestic animals.

> **Textbox 13.10 Pathophysiology of Male Reproduction**
>
> - **Blockage** of the ampulla is seen in stallions.
> - **Cryptorchidism** is a condition where one or both testicles have not descended. This is seen, for instance, in dogs.
> - **Heat stress** (bulls) with reduced libido and low numbers of motile spermatozoa per ejaculation.
> - **Inflammation** of the epididymis (epididymitis) or seminal vesicles (seminal vesiculitis) in bulls.
> - **Physical damage** to the penis or scrotum (for example in stallions and bulls). An example is physical damage is penile hematoma in bulls. There is rupture of the tunica albuginea of the penis with hemorrhage and hematoma at the sigmoid flexure.
> - **Poor fertility** due to low sperm count (oligospermia) or poor motility (for example in bulls and stallions). Subfertile bulls can be identified by a breeding soundness exam (BSE).
> - Male **pseudohermaphrodites** are genetically male and have testes but have female type external genitalia. This is seen in cattle.
> - **Sexually transmitted** protozoan, bacterial, and viral diseases are common in stallions, bulls, and rams.
> - **Testicular degeneration** is seen in stallions.
> - **Urospermia** is the mixing of urine in the ejaculate (stallions).

14

Female Reproduction

Learning Objectives

1. Know the terminology of female domestic animals (see Textbox 14.2).
2. Know the components of the female reproductive system and what they do (see Section 14.2).
3. Understand the processes of oogenesis, follicular development, and ovulation (see Sections 14.3 and 14.4).
4. Understand what happens during and after insemination (see Section 14.5).
5. Understand estrous cycles (see Section 14.7).
6. Understand the hormones controlling reproduction (see Sections 14.6 and 14.9).
7. Understand how and why animals breed at certain times of the year (see Section 14.7).
8. Understand the mechanisms by which the mother recognizes she is pregnant and maintains that pregnancy (see Section 14.8.3).
9. Understand the mechanisms of parturition (see Section 14.10).
10. Understand the differences in female reproduction between livestock/companion animals, and that of poultry (see Section 14.8.11).
11. Understand pathophysiology of female reproductive disorders (see Section 14.8.12).

Theme 4 Reproduction (Propagation of the Species)

Table of Contents

14.1 Introduction
14.2 Female Reproductive Organs
    14.2.1 Structures in Female Mammals
    14.2.2 What Are the Roles of the Different Parts of the Female Reproductive Tract?
14.3 Oogenesis
14.4 Follicular Development and Ovulation
    14.4.1 Overview of Follicular Development and Ovulation
    14.4.2 Two-Cell Theory of Estradiol Production
14.5 Hormones and Reproduction
    14.5.1 What Are the Hormones Controlling Reproduction?
    14.5.2 What Do the Hormones Do to Control Reproduction?
14.6 Estrous Cycles
    14.6.1 Estrus or Heat
    14.6.2 The Estrous Cycle
        14.6.2.1 Follicular Growth
        14.6.2.2 Estrus
        14.6.2.3 Ovulation
        14.6.2.4 Formation of the Corpus Luteum
        14.6.2.5 Corpus Luteum
        14.6.2.6 Luteolysis
14.7 From Insemination to Fertilization
    14.7.1 Site of Natural Insemination
    14.7.2 Following Insemination to Fertilization
14.8 Seasonal Breeding
    14.8.1 Introduction to Seasonal Breeding
    14.8.2 So How Do Animals Know What Time of Year It Is?
    14.8.3 Why Do Animals Need to Be Able to Predict the Time for Offspring to the Born (or for Chickens to Be Hatched)?
    14.8.4 Long-Day Breeders
    14.8.5 Short-Day Breeders
14.9 Pregnancy
    14.9.1 Length of Pregnancy
    14.9.2 Development of the Conceptus
    14.9.3 Maternal Recognition of Pregnancy
    14.9.4 Placenta and Its Role
        14.9.4.1 Overview of Placentation.
        14.9.4.2 What Passes Across the Placenta?
        14.9.4.3 What Does Not Pass Across the Placenta?
        14.9.4.4 Are There Other Roles for the Placenta?
14.10 Hormones and Pregnancy
    14.10.1 Progesterone
    14.10.2 Estrogens
    14.10.3 Relaxin
    14.10.4 Placental Lactogen—A Unique Ruminant Hormone
    14.10.5 Equine Chorionic Gonadotropin—A Unique Equid Hormone
14.11 Parturition
    14.11.1 Stages of Parturition
    14.11.2 Position of Fetus
    14.11.3 Hormones and Parturition
14.12 Comparison of Female Reproduction Between Livestock/Companion Animals and Poultry
14.13 Pathophysiology of Female Reproduction

14.1 Introduction

The major facets of reproduction in the female are the following, with the major organ(s) in parenthesis:

- Producing the **female gametes**, ova, plural, or ovum, singular (ovaries). These are released only at specific times.
- The presence of organ(s) to allow for **intromission** (vagina and, in some species, the cervix).
- The presence of organs to allow for **pregnancy** (uterus and oviduct).
- The presence of organs to allow for **parturition** (uterus, cervix, and vagina).
- The presence of organ(s) to allow for **lactation** (mammary glands).
- **Libido**: the desire to mate.
- Other aspects of female behavior including **estrous** and **maternal behavior**.

Textbox 14.1 provides a glossary for terms in female reproduction. The terminology for young and adult female domestic animals is given in Textbox 14.2.

14.2 Female Reproductive Organs

14.2.1 Structures in Female Mammals

The structures of the reproductive tract for two mammalian examples (dogs and horses) are shown in Figures 14.1 and 14.2. There is marked similarity between the system in livestock and companion animals. There are, however, major differences in the size of the uterine body and uterine horns; reflecting litter and single-bearing species.

Textbox 14.1 Glossary of Terms in Female Reproduction

- **Breech birth**: the rump is presented first during parturition.
- **Conceptus**: the product of fertilization.
- **Corpus luteum** (plural corpora lutea): a part of the ovary that produces progesterone to maintain pregnancy.
- **Dystocia**: difficult birth.
- **Embryo**: the stage of development when the major organs are developed. The embryonic stage starts with implantation and ends with the development to true bone.
- **Estradiol**: female sex steroid hormone.
- **Estrous** (adjective): pertaining to the entire cycle or to a specific behavior or physiology.
- **Estrus** (noun): heat or the time when a female is responsive to the male (will allow mating).
- **Fertilization**: the fusion of the male and female gametes.
- **Folliculogenesis**: the development of the ovum and cells (granulosa and theca) that support its development.
- **Gestation**: time of pregnancy from conception to parturition.
- **Haploid**: possessing half the normal complement of chromosomes.
- **Libido**: the desire to mate.
- **Luteinizing hormone**: a hormone that stimulates the corpus luteum.
- **Maternal recognition**: a chemical signal from the conceptus to the mother to prevent the breakdown of the corpus luteum.
- **Müllerian duct**: an embryonic duct that will differentiate into the following in mammals: oviduct (Fallopian tubule), uterus, cervix, and posterior vagina.
- **Oogenesis**: production of eggs.
- **Ovary**: female gonad.
- **Ovum** (singular) **ova** (plural): female gametes.
- **Oxytocin**: hormone that stimulates contractions of the uterus and myoepithelial cells in the alveoli (in mammary gland).
- **Parturition**: the process of giving birth to the offspring.
- **Placenta**: unit where gaseous exchange between embryo/fetus and mother and the transfer of nutrients from mother to fetus. Note: the bloodstreams of the embryo/fetus and mother are totally separate.

Textbox 14.2 Terminology in Female Reproduction

Table 14.1 Terminology for young and intact female livestock, companion animals, and poultry together with terms for pregnancy in different species.

Species	Young female prior to the birth of first progeny	Adult female	Term for parturition
Cat	Young female cat	Dame or queen	Kindling or queening
Cattle	Heifer or heifer calf	Cow	Calving
Dog	Bitch or young female dog	Bitch or female dog	Whelping
Horse	Filly	Mare	Foaling
Pig	Gilt	Sow	Farrowing
Sheep	Ewe lamb	Ewe	Lambing
	Young female prior to the first egg produced	Female after 1st egg laid	Process of laying an egg
Chicken	Pullet	Hen	Oviposition
Turkey	Poult	Hen	Oviposition
Duck	Duckling	Hen	Oviposition

14.2.2 What Are the Roles of the Different Parts of the Female Reproductive Tract?

In mammals, the reproductive system consists of the following:

- **Vulva**: external genitalia in female mammals. This can change at the time of estrus indicating fertility of the female and the time for coitus (equivalent to sexual intercourse in people).
- **Vagina**: site for intromission during coitus, possibly site for deposition of semen and during parturition becomes part of birth canal.
- **Cervix**: protects uterus and its contents from bacterial invasion, is the site for deposition of semen in some species, and during parturition becomes part of birth canal.
- **Uterus** (one uterine body and two uterine horns): site for development of the fetus or fetuses together with providing muscular power during parturition and also becomes part of birth canal.
- **Fallopian tubules** (oviducts) (2): transit for ovum and site for fertilization.
- **Ovaries** (two): producing ova (eggs) and releasing them.
- **Broad ligament**: supporting the uterus. This is particularly important during pregnancy as the fetus (or fetuses) grows and develops.

The uterus consists of the following tissues, each with a distinct function(s):

- **Epithelium**: conceptus passes through this as it implants.
- **Endometrium**: producing secretions as nutrition for the early conceptus, producing hormones, and being the site of implantation.
- **Myometrium**: the smooth muscles of myometrium contract, pushing the fetus out during parturition.
- **Perimetrium**: connective tissue protecting the integrity of the uterus.

Figure 14.3 shows a schematic section through the wall of the uterus of mammals.

CHAPTER 14 Female Reproduction   301

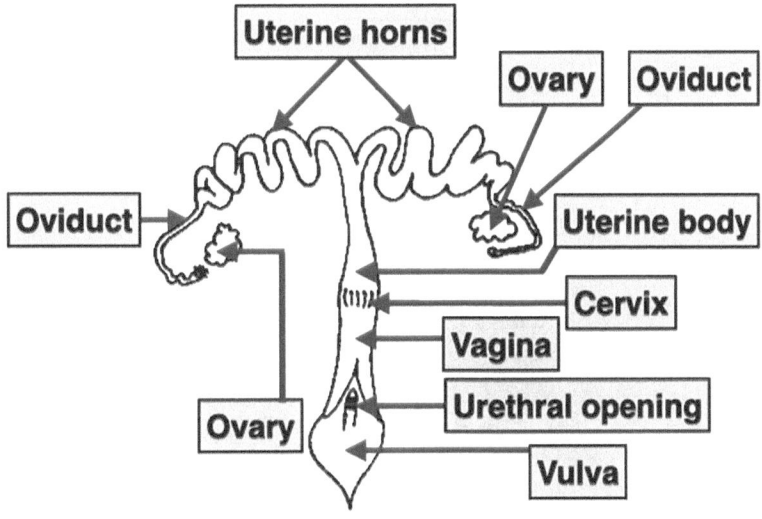

Figure 14.1 Reproductive tract of the female dog. (Adapted from J. Ruth Lawson, Otago Polytechnic)

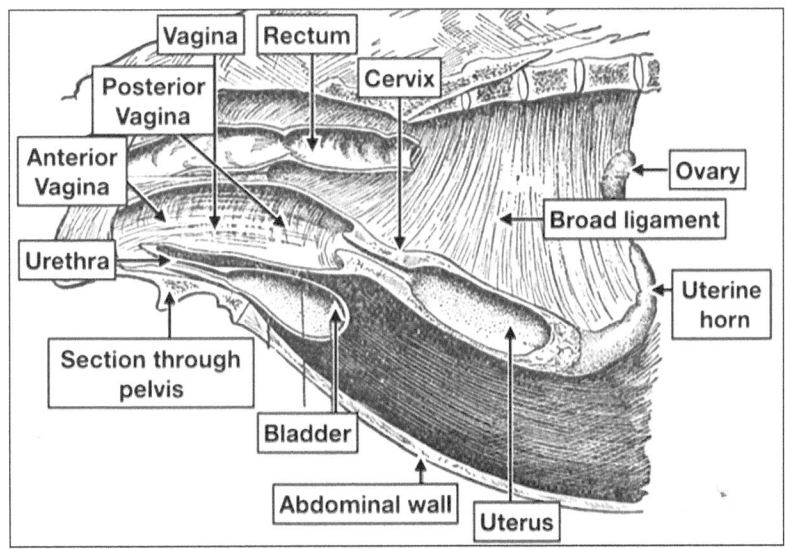

Figure 14.2 Reproductive tract of the female horse. (Adapted from Cadiot, Bitting, and Liautard, public domain)

There are marked differences between the mammalian female tract and that in chickens (Figure 14.4) as seen from the following:

- **Ovaries:**
  - There is only one functional ovary in chickens but two in domestic mammals.
  - The follicles in the ovary become progressively larger as they fill with yolk in chickens, but follicles are very small in mammals without yolk.
- **Reproductive tract:**
  - Terminology—the oviduct is the name for the entire reproductive tract in poultry but only a small part of the tract in mammals.

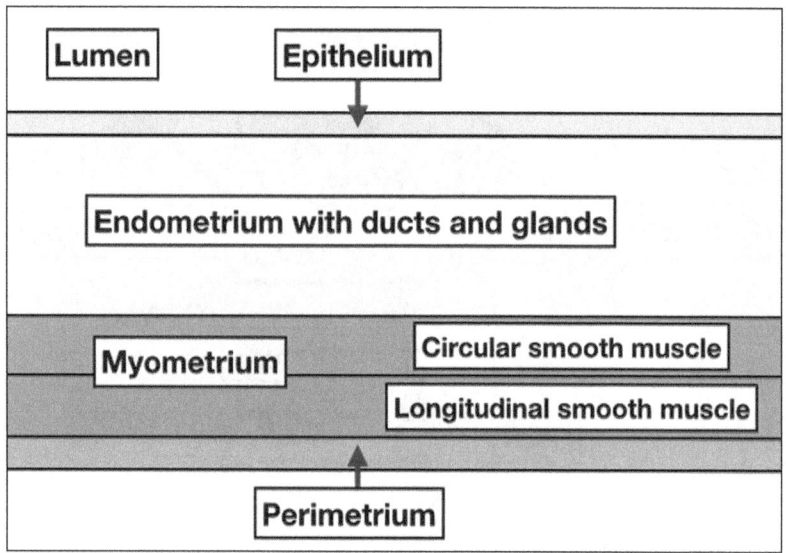

Figure 14.3 Schematic section of the structure of the uterus.

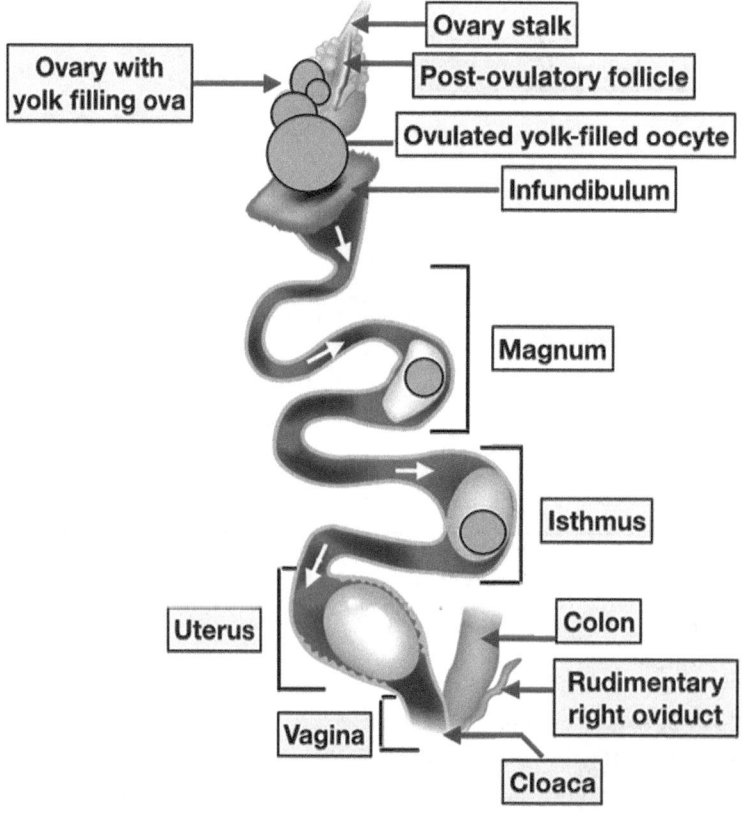

Figure 14.4 Ovary and oviduct of the chicken. Note: Multiple eggs are shown in the oviduct. In reality, there is only one egg ever in the oviduct. (Shutterstock/Sakurra)

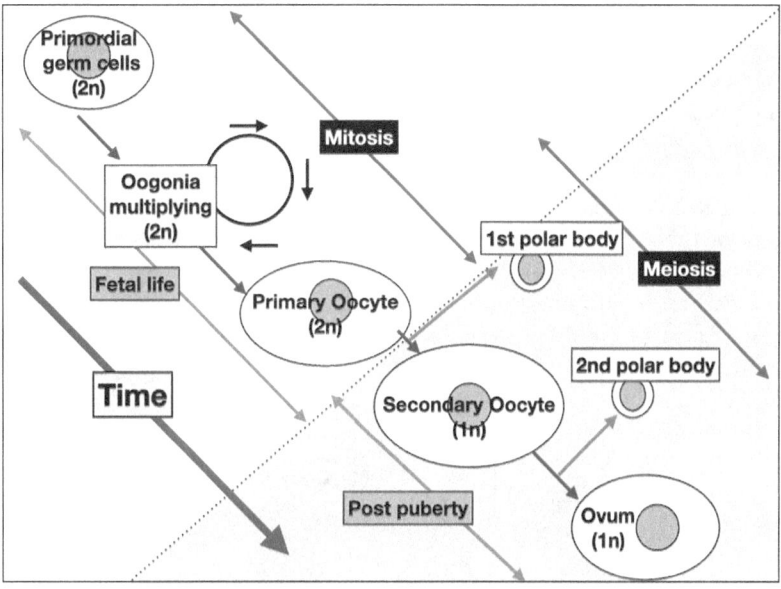

Figure 14.5 Oogenesis—meiotic divisions generating a haploid ovum and mitotic divisions of oogonia.

- o There is a single reproductive tract in female chickens. In mammals, there are two oviducts and two uterine horns.
- o There is a single cervix and single vagina.

14.3 Oogenesis

Oogenesis is the process of developing ova. The oogonia develop from the **primordial germ cells** in the **cortical sex cords** (Figure 14.5). **Oogonia** undergo tremendous proliferation (mitosis) with peak numbers in the fetus at mid-pregnancy. For example, there are the following peak numbers of oogonia in cows and sheep:

- Cows: 2,700,000.
- Sheep: 850,000.

Later in fetal development, 80–90% of the oogonia are lost due to reduced proliferation and increased cell death.

The oogonia start meiotic division during fetal life, then are arrested (see Figure 14.5). Following puberty and during each cycle, a group of oogonia undergo meiosis, developing into primary oocytes and first polar bodies (see Figure 14.5). The **first polar body** is extruded into the **perivitelline space** at the time of ovulation. **Meiosis** is completed at the time of fertilization and the **second polar body** is extruded.

14.4 Follicular Development and Ovulation

14.4.1 Overview of Follicular Development and Ovulation

Follicular development is the process of transitioning from **primordial follicles** to **primary**, then secondary to **preovulatory follicles**. This is followed by ovulation and the formation of the corpus luteum.

Initially in primary follicles, there is a single layer of **squamous epithelial cells** surrounding the oocyte (Figure 14.6). These epithelial cells divide and differentiate into cuboidal granulosa cells in the secondary follicle (Figure 14.6). The secondary follicle includes the following (Figure 14.6):

- **The zona pellucida**—noncellular layer around the oocyte.
- **Granulosa cells** in multiple layers supporting the oocyte.
- **Theca cells** made up of theca interna and theca externa.
- The **oocyte** which completes the first meiotic division releasing the first polar body.

Development and growth continue to form a large late-stage secondary follicle (Figure 14.6). The granulosa cells continue to multiply but there is the beginning of the fluid filled space (the antrum) (Figure 14.6). The theca cells also multiply (Figure 14.6).

Follicular development continues with the follicle becoming larger with a fluid filled antrum (see Figure 14.6). This is the **pre-ovulatory follicle** (or Graafian follicle). The pre-ovulatory follicle is a mature follicle (see Figure 14.7 for structure of the pre-ovulatory follicle). It produces estradiol. When the oocyte is ovulated, it is surrounded by granulosa cells of the **cumulus oophorus** and the zona pellucida (see Figure 14.7). At this stage, the **second meiotic division** starts. Ovulation occurs due to stimulation by LH (see Figure 14.7). The remaining theca interna and granulosa cells undergo luteinization and develop into the corpus luteum.

14.4.2 Two-Cell Theory of Estradiol Production

There are two cell types in the follicle that are required to produce estrogen (see Figure 14.8).

- The **theca interna cells**, which are well supplied with blood.

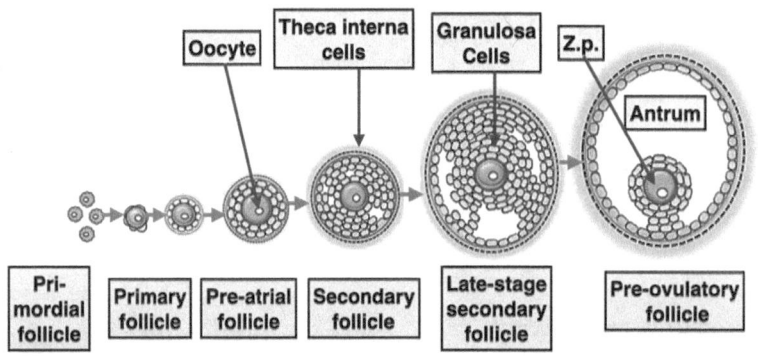

Figure 14.6 Development of the mammalian follicle. [Key: Z.p. = zona pellucida.] (Adapted from Tnhath, CC BY-SA 4.0)

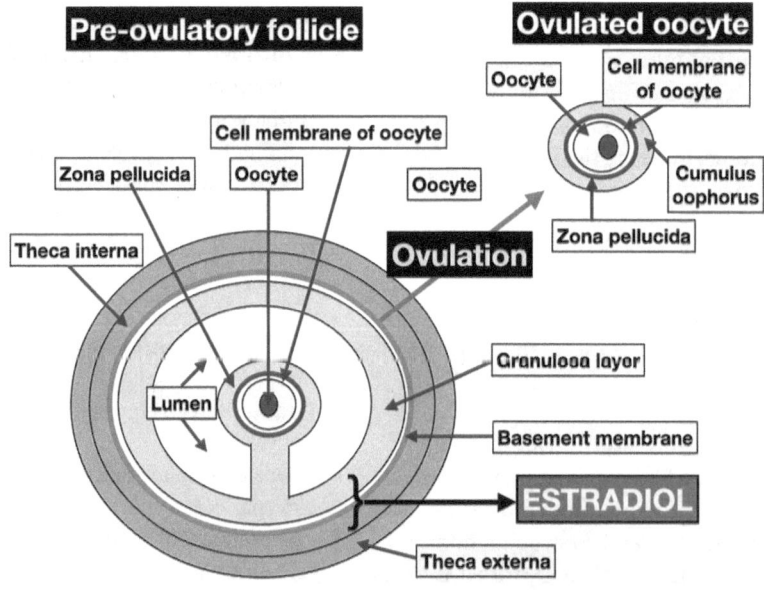

Figure 14.7 Schematic view of the structures of the pre-ovulatory (or Graafian) follicle and the ovulated oocyte.

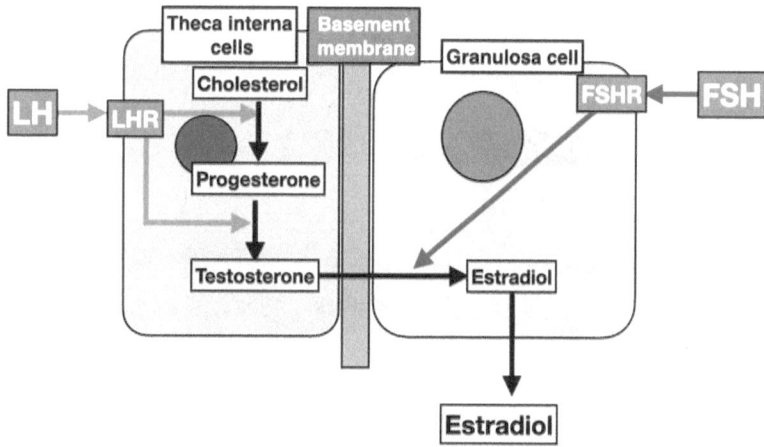

Figure 14.8 Schematic of estradiol synthesis requiring both the granulosa and theca interna cells, stimulated respectively by follicle-stimulating hormone (FSH) and luteinizing hormone (LH). [Key: LHR = luteinizing hormone receptor; FSHR = follicle-stimulating hormone receptor.]

- The granulosa cells, with no blood vessels, with oxygen and nutrients having to diffuse from the theca through the basement membrane.

The theca internal cells synthesize testosterone from progesterone. This testosterone diffuses to the granulosa cells where it is converted to estradiol (see Figure 14.8).

14.5 Hormones and Reproduction

Female reproduction, and specifically the ovary, is controlled by hormones (Figure 14.10).

14.5.1 What Are the Hormones Controlling Reproduction?

The major hormones controlling reproduction are the following:

- Gonadotropin-releasing hormone (GnRH) from the hypothalamus.
- Luteinizing hormone (LH) and follicle stimulating hormone (FSH) from the anterior pituitary gland.
- Estradiol from the mature follicle.
- Progesterone from the corpus luteum.
- Prostaglandin $F_{2\alpha}$ from the uterus causes the breakdown of the corpus luteum.

14.5.2 What Do the Hormones Do to Control Reproduction?

Luteinizing hormone (LH) has three functions (see Figure 14.9):

- Stimulating ovulation.
- Stimulating the remnants of the follicle after ovulation to be transformed into the corpus luteum.
- Stimulating the corpus luteum to synthesize progesterone.

Follicle-stimulating hormone (FSH) stimulates the follicles to produce estradiol. Gonadotropin-releasing hormone stimulates release of LH and FSH.

The estrogen estradiol has the following effects:

- In a positive feedback manner, estradiol increases release of gonadotropin-releasing hormone (GnRH) from the hypothalamus (Figure 14.9).
- Estradiol stimulates development of the uterus to facilitate pregnancy by increasing the numbers of **progesterone receptors** and, hence, increasing the effects of progesterone.

The mechanism of action of estradiol involves binding to **nuclear estrogen receptors** and increases transcription of specific genes.

Progesterone has a series of effects that promote pregnancy:

- Progesterone stimulates further development of the uterus.
- Progesterone inhibits uterine contractions.
- Progesterone allows implantation.
- Progesterone increases glycogen in the uterus to provide an energy source for the conceptus.

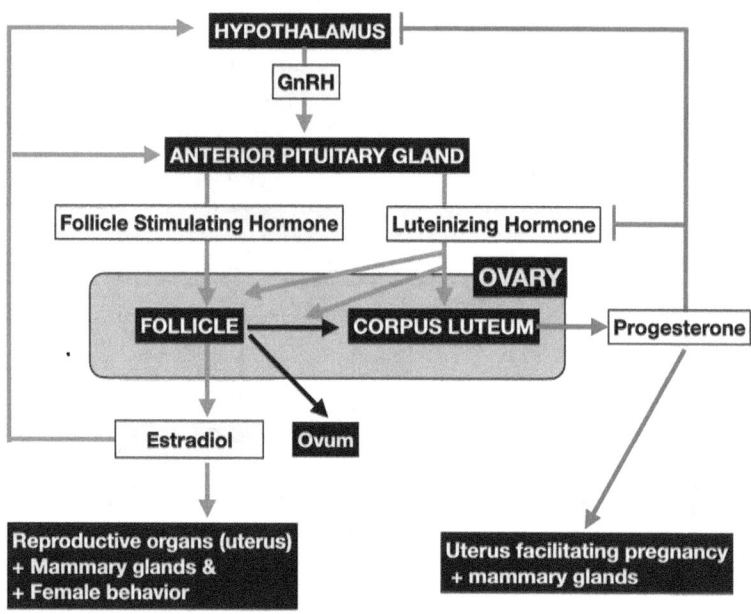

Figure 14.9 The hormonal control of reproduction in female livestock and companion animals.
[Key: GnRH = gonadotropin-releasing hormone; ↑ is positive; T is negative.]

- In a negative feedback manner, progesterone decreases release of gonadotropin-releasing hormone (GnRH) from the hypothalamus.
- Progesterone inhibits estrus (Figure 14.9).

The mechanism of action of progesterone involves binding to nuclear progesterone receptors and increases transcription of specific genes.

14.6 Estrous Cycles

14.6.1 Estrus or Heat

Estrus or heat (a word derived from the Greek for "*gadfly*" and "*frenzy*") is the time, and the only time, in an estrous cycle when the female mammal is receptive to mating. This is different from species where mating occurs at any time in their cycles and for pleasure (for instance, in women). During estrus, the females adopt the position their bodies to allow intromission. This is called **lordosis** or **standing to be mated**. When females allow mating, it is called **receptive** behavior. In addition, females may actively attract mates; exhibiting **proceptive** behaviors. For example, mares at estrus "wink" their vulvas. Cows in heat are restless and may attempt to mount other cows. Cattle exhibit estrus behavior for about 12 hours. Estrus behaviors are caused by the effects of the female sex hormone estradiol acting on the brain (see Figure 14.10).

Inspection of the vulva provides observable signs of estrus. These include swelling of the vulva due to increased blood flow to the vulva, mucus discharge from the vulva, and reduced viscosity of cervical mucus (or it becomes watery).

There is not an equivalent of estrus in poultry. Hens may or may not be responsive to a male. There is evidence that males mount females more in the afternoon than the morning.

14.6.2 The Estrous Cycle

Schematic diagrams of the estrus cycle are shown in Figures 14.11 and 14.12. The changes in estradiol and progesterone during the estrous cycle of cattle are shown in Figure 14.13. Changes in the diameter of the largest follicle and the corpus luteum during the estrous cycle of cattle are shown in Figure 14.14.

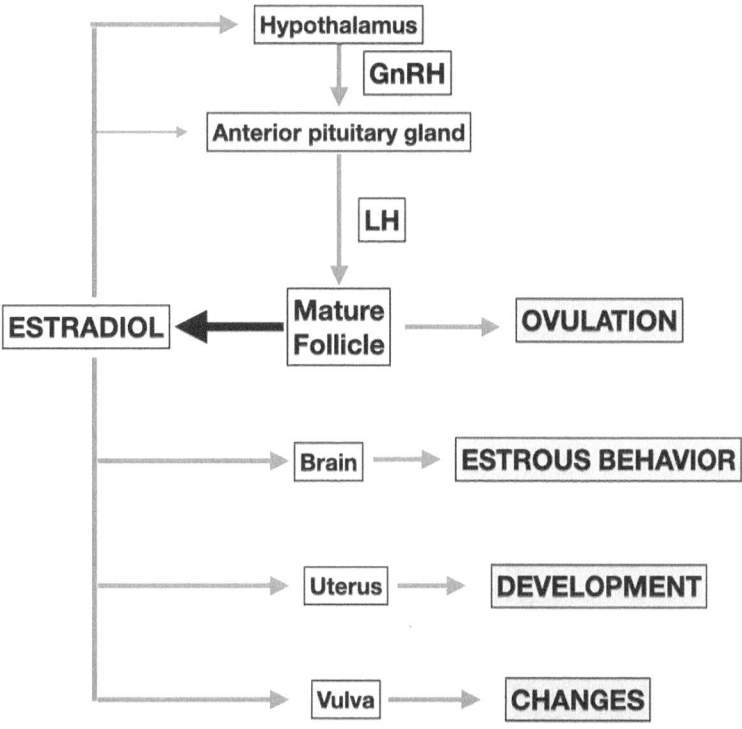

Figure 14.10 Estradiol from the mature follicle stimulates the pre-ovulatory surge in LH (luteinizing hormone) with increased release of GnRH (gonadotropin-releasing hormone), leading to the ovulation of the mature follicle. The estradiol evokes a series of estrous behaviors, including standing for mating and attracting a male. The uterus exhibits development in preparation for a possible pregnancy. Estradiol also induces changes in the vulva.

Textbox 14.3 Definitions Related to Estrus and the Corpus Luteum

- **Estrus** (noun) is heat: the time of fertility when the female allows or encourages mating (coitus).
- **Estrous** (adjective) is the type of cycle as opposed to a menstrual cycle. Both are ovarian cycles. In tandem with the ovarian cycle, there changes in the uterus. These are referred to as the uterine cycles.
- **Corpus albicans** (literally the "white body") is the remnant of the corpus luteum after luteolysis.
- **Corpus hemorrhagicum** (literally the "bloody body") forms from the ruptured blood vessels in the follicle at ovulation. The blood clots and blood vessels start to form.
- **Corpus luteum** (plural **corpora lutea**) (literally the "yellow body") produces progesterone in livestock and companion animals. This is essential for successful pregnancies.
- **Seasonal breeding** is when animals produce progeny at the optimal time of year for their survival.
  - A **long-day breeder** breeds in the Spring when the daylengths are increasing.
  - A **short-day breeder** breeds in the Fall when the daylengths are decreasing.

Textbox 14.4 A Deeper Dive Into Estrous Cycles

Length of Estrous Cycles
Length of estrous cycle varies in different livestock and companion animals:

- Cats: approximately 21 days.
- Cattle: 18–24 days (average 21 days).
- Goats: 17–24 days.
- Horses: 22 days.
- Pigs: 18–24 days.
- Sheep: 16 days.

Types of Estrous Cycles
- **Monoestrous** is the type of estrous cycle where there is a single cycle followed by a period of reproductive quiescence called anestrus. This is seen in dogs.
- **Polyestrous** is the type of estrous cycle where there is a single cycle followed by another and another if not pregnant. This is seen in horses, sheep, cattle, and pigs.
- **Reflex** or **induced ovulators**: ovulation only occurs if there is coitus. This is found in rabbits, cats, and ferrets.
- **Spontaneous ovulators** are animals that ovulate irrespective of whether or not they copulate or not (for example in livestock).

14.6.2.1 Follicular Growth

If there is no functioning corpus luteum, the follicle grows and completely matures (for discussion of follicular development, see Section 14.4.1 and Figure 14.6). This is stimulated by follicle-stimulating hormone.

14.6.2.2 Estrus

The mature follicle produces estradiol and this stimulates estrous behavior, including in cattle:

- **Standing heat** allowing coitus (receptive behavior).
- **Mounting** other cows.

At heat or estrus, estradiol also induces changes in the vulva during estrus (see Figure 14.11). Examples include the vulva winking in horses, and reddening and swelling of the vulva in cattle and dogs. There also can be discharge of mucus from the vulva at estrus.

14.6.2.3 Ovulation

Estradiol provokes release of gonadotropin-releasing hormone from the hypothalamus, then LH from the anterior pituitary gland which in turn stimulates ovulation. In addition, the LH induces both the formation of the corpus luteum and the production of progesterone.

14.6.2.4 Formation of the Corpus Luteum

LH induces ovulation. After this, the following occurs:

- Bleeding from the ruptured blood vessels with the blood clotting and the formation of the **corpus hemorrhagicum**.

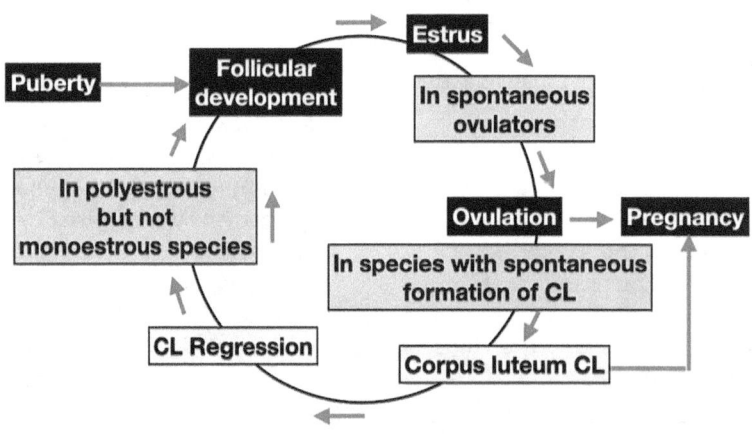

Figure 14.11 A schematic diagram of the estrous cycle.

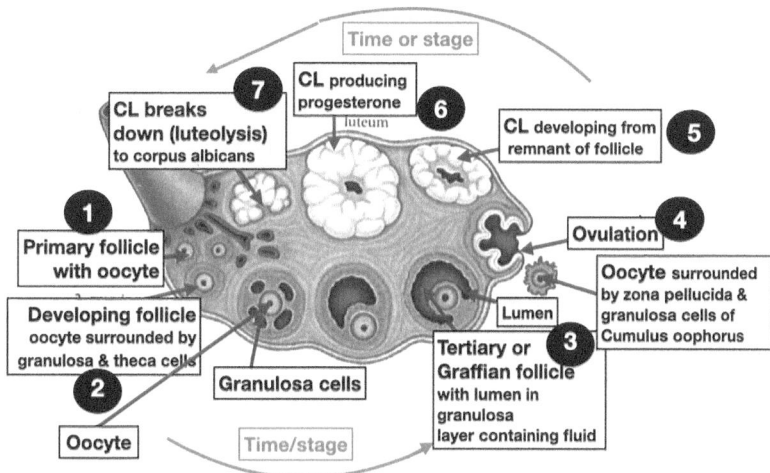

Figure 14.12 Schematic view of the development of follicles, ovulation, together with the formation and ultimate demise of the corpus luteum (CL). (Adapted from and courtesy of Kimanh Nguyen, Wikimedia Commons)

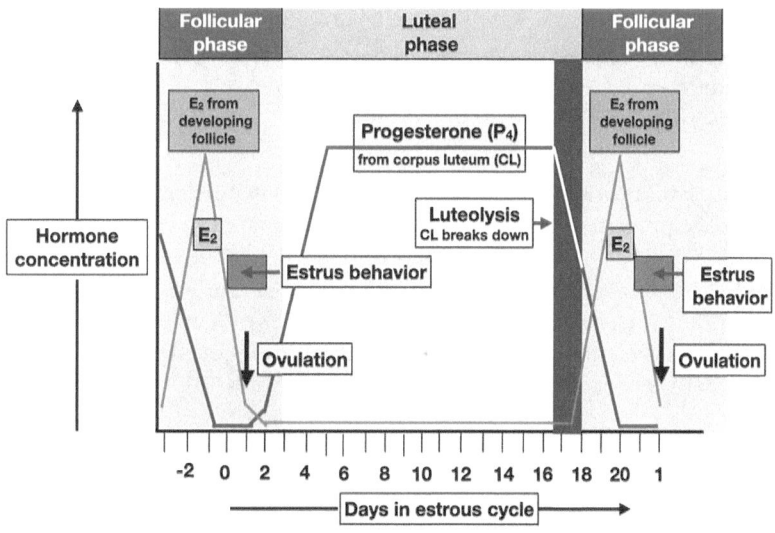

Figure 14.13 Changes in hormone concentrations during the estrous cycle of cattle.
Key:
Red line = indicates estradiol (E_2)
Blue line = progesterone (P_4)
Red square indicates time of estrus
Black down arrow = time of ovulation

- Breakage of the basement membrane.
- **Angiogenesis** (formation of new blood vessels) in the interior of the corpus hemorrhagicum. This is occurring because of the release of angiogenic factors such as vascular endothelial growth factor (VEGF) from the cells forming the corpus luteum.
- The granulosa and theca interna cells undergo luteinization forming the corpus luteum.

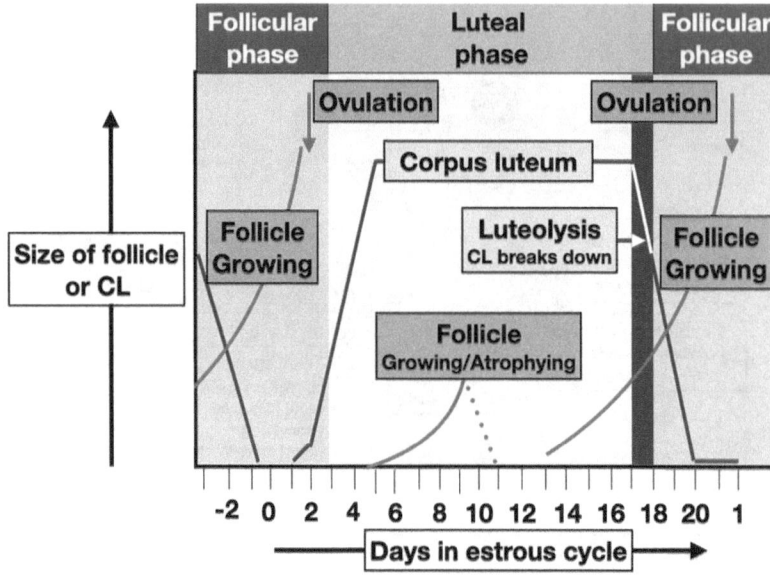

Figure 14.14 Changes in the size (diameter, volume, or weight) of the largest follicle and corpus luteum during the estrous cycle of cattle.
Key:
Red line = size of the largest follicle Blue line = size of the corpus luteum
Red down arrow = time of ovulation

- ○ The granulosa cells become **large luteal cells** and gain the ability to produce progesterone from cholesterol.
- ○ The theca interna cells become **small luteal cells**. They gain LH receptors and under LH stimulation produce progesterone from cholesterol.
- The **steroidogenic luteal cells** make contact with the capillaries.

There is now a functioning corpus luteum producing progesterone.

14.6.2.5 Corpus Luteum

Corpus luteum, or yellow body, (plural corpora lutea) forms from the remnants of the follicle after ovulation. The corpus luteum produces progesterone from cholesterol (see Figures 14.11–14.13). The corpus luteum has multiple blood vessels. There are both large and small luteal cells that produce progesterone (Figure 14.15). The small luteal cells respond to LH and increase progesterone production while the large luteal cells produce progesterone without stimulation.

Progesterone acts on the uterus, causing development of the **endometrium**. When there is pregnancy, the corpus luteum continues to produce progesterone allowing pregnancy to continue.

In the event of no pregnancy, the uterus produces prostaglandin $F_2\alpha$. This causes the corpus luteum to breakdown (luteolysis) (Figure 14.13 and 14.14). In the absence of progesterone, the next follicle develops (Figure 14.13 and 14.14).

14.6.2.6 Luteolysis

Luteolysis is the breakdown of the corpus luteum. This is induced by prostaglandin $F_2\alpha$. This comes from the uterus if the animal is not pregnant. During luteolysis, there are decreases in the vascular bed of the corpus luteum. Cytokines such as **tumor necrosis factor (TNF)-α** and **interferon (IFN)-γ** play a role inducing apoptosis of luteal cells. There are large increases in the numbers of macrophages, neutrophils, and T lymphocytes in the corpus luteum during luteolysis. These induce apoptosis of the luteal cells and also phagocytose cell debris. The remnant of the corpus luteum is called the corpus albicans.

The increase in estradiol induces release of GnRH, thereby increasing release of LH, which in turn stimulates ovulation. The LH also induces the formation of the

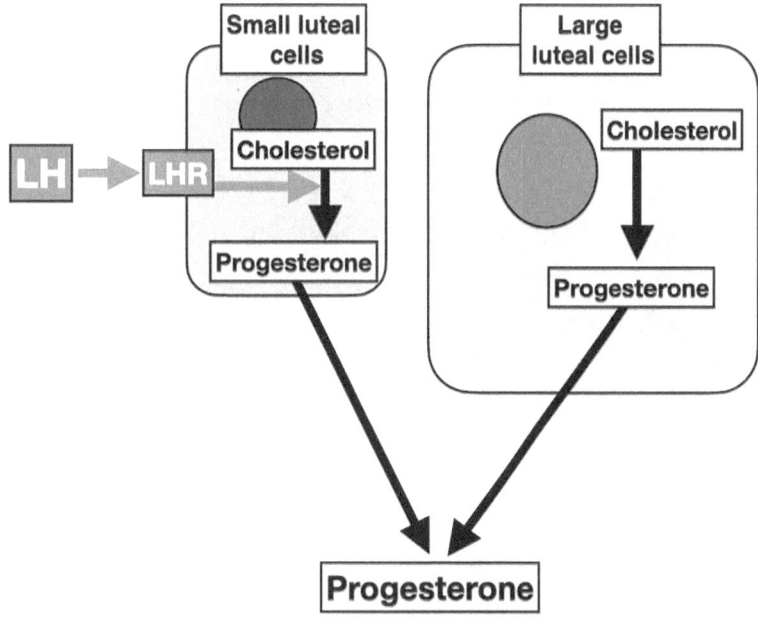

Figure 14.15 Schematic of progesterone synthesis by small and large luteal cells.
[Key: LH = luteinizing hormone; LHR = luteinizing hormone receptor]

corpus luteum and the corpus luteum to produce progesterone. The increase in estradiol also induces estrus.

The follicle grows and matures completely if there is no functioning corpus luteum producing progesterone. Alternately, if there is a functioning corpus luteum, development of the follicle is arrested, and the follicle undergoes atresia.

14.7 From Insemination to Fertilization

14.7.1 Site of Natural Insemination

In livestock, companion animals, and poultry, semen is deposited into the female reproductive tract. The location of insemination varies with species:

- Vagina (for example in female cats, cattle, dogs, and sheep).
- Cervix (pigs) with contractions of cervix stimulating ejaculation. The semen then flows into the uterus.
- Uterus (horses).
- Vagina (chickens and turkeys). The vagina is a part of the oviduct in birds where spermatozoa can be stored in sperm storage tubules.

The site of natural insemination is not necessarily the site for artificial insemination. This tends to be closer to the site of fertilization (the uterus or cervix).

14.7.2 Following Insemination to Fertilization

Spermatozoa with seminal plasma pass up along the female tract to the site of fertilization by the following mechanisms:

- Movement of the spermatozoa due to their flagella or tail (for structure of the spermatozoa see Figure 13.1)
- Cilia in the female tract.
- Spermatozoa become hyperactive in the oviducts in association with **capacitation**.

Spermatozoa are stored in a reservoir at **uterotubal** junctions in cows.

14.8 Seasonal Breeding

14.8.1 Introduction to Seasonal Breeding

While seasonal breeding applies to both male and female animals, it is covered in female reproduction. Domestic animals breed at specific times of year when held under natural light (see Textbox 14.6).

14.8.2 So How Do Animals Know What Time of Year It Is?

The most reliable indicator of time in the temperate zone is daylength (see Figure 14.18). Every year on the same date, the temperature and humidity vary considerably. What an

Textbox 14.5 A Deeper Dive: Comparison of Estrous and Menstrual Cycles

The physiological changes in menstrual cycle are shown in Figure 14.16.

There are both similarities and differences between estrous or menstrual cycle are summarized in Table 14.2.

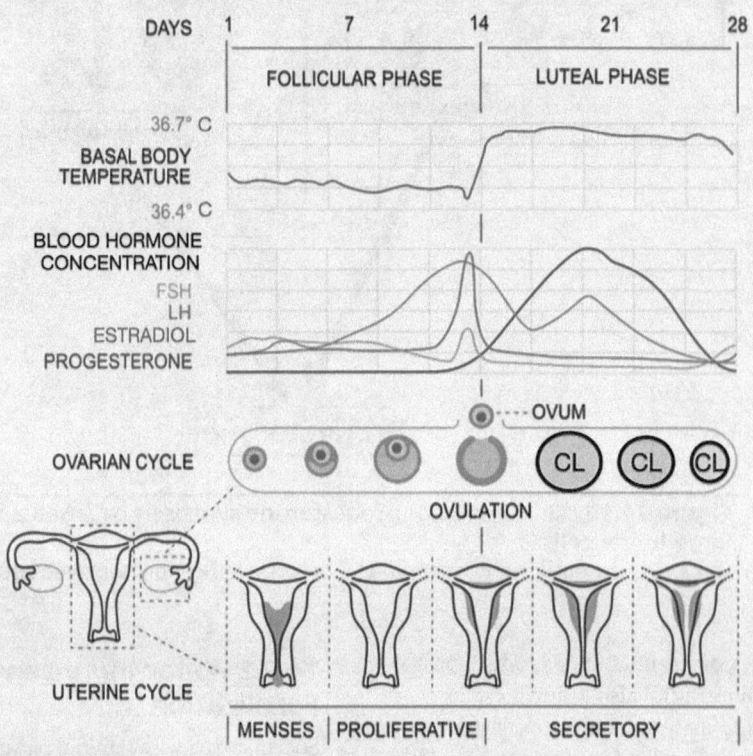

Figure 14.16 Physiological changes during the menstrual cycle.
Key:
CL = corpus luteum
FSH = follicle-stimulating hormone
LH = luteinizing hormone.
(Isometrik, CC BY-SA 3.0)

Table 14.2 Comparison of estrous and menstrual cycles.

	Estrous cycle	Menstrual cycle
Follicular phase	Present	Present
Luteal phase	Present	Present
Hormone produced by follicle	Estradiol	Estradiol
Hormone stimulating ovulation & formation of corpus luteum	Luteinizing hormone	Luteinizing hormone
Hormones produced by corpus luteum	Progesterone	Progesterone and estradiol

Chapter 14 Female Reproduction

Textbox 14.6 A Deeper Dive Into the Breeding Seasons

Table 14.3 Breeding season for domestic animals in Northern latitudes.

Species	Breeding season
Cats	February to October
Cattle	February to April
Dogs	Throughout the year
Horse	April to October
Pigs	Throughout the year
Sheep	October and November
Chicken	February to May

animal needs is a reliable indicator of what the weather and availability of food is likely to be when the offspring are born and growing. The most reliable indicator of time of year is daylength, also known as photoperiod.

14.8.3 Why Do Animals Need to Be Able to Predict the Time for Offspring to the Born (or for Chickens to Be Hatched)?

Simply put, they are seeking the optimal time for the survival and prospering of their offspring. Animals put considerable energy into breeding, and this would be wasted if they bred at the wrong time.

Textbox 14.7 Interesting Factoid: Spring Lamb and Spring Chicken

Spring used to be associated with eating new lamb and spring chicken because the neonates were first available in the spring.

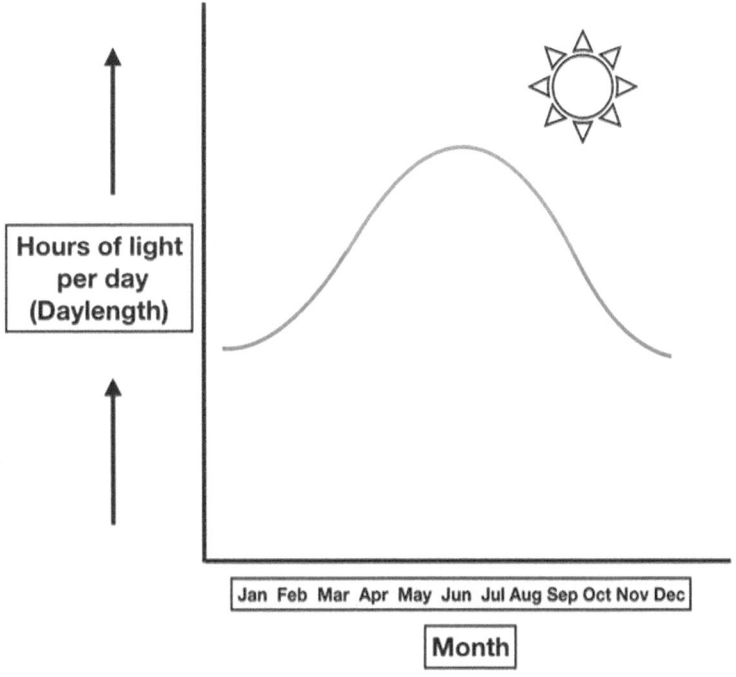

Figure 14.17 Changes in daylength, or photoperiod, during the year are a reliable indicator of season. Animals use photoperiod as a cue for breeding at the optimum time of the year for survival of the offspring.

14.8.4 Long-Day Breeders

Long-day breeders come into reproductive condition when the days are lengthening in the spring. Examples include horses and chickens.

14.8.5 Short-Day Breeders

Short-day breeders come into reproductive condition when the days are shortening in the fall. An example is sheep. Farmers and ranchers can extend the breeding season and/or induce reproduction at any time of year using artificial lighting. For example, in chickens, pullets are kept on short daylength to prevent maturation until they are fully grown. At the optimal age/size, the daylength is increased to bring all the birds into lay at the same time.

14.9 Pregnancy

14.9.1 Length of Pregnancy

There is considerable variation in the length of gestation in domesticated mammals (see Textbox 14.8).

Textbox 14.8 A Deeper Dive Into the Length of Pregnancies

Table 14.4 Gestation length in livestock and companion animals.

Species	Length of pregnancy (days)
Cat	63
Cattle	282
Dog	63
Horse	335
Pig	115*
Sheep	147

*3 months, 3 weeks, and 3 days

14.9.2 Development of the Conceptus

Following fertilization, the conceptus develops (see Figure 14.19; also see Textbox 14.9). There is cell division giving first the two-cell stage, then the four- and eight-cell stages (see Figure 14.20). Cell division continues forming a ball of cells or morula (see Figures 14.18 and 14.19). There is more cell division creating a hollow ball of cells, the blastocyst with differentiation (see Figures 14.19 and 14.20). As this is occurring within the zona pellucida, as the cells divide, they have to become smaller. The blastocyst hatches from the zona pellucida and can spread out (see Figure 14.20). There is continued cell division and the formation of the **embryonic disc** (see Figure 14.20). Development continues with the placenta developing from the trophoblast and the embryo from the embryonic disc (see Figure 14.20). At this stage implantation occurs. Development continues with the formation of organs in the embryo. Once these have been formed and true bone (calcified bone) is occurring, the conceptus is now a fetus. The fetus grows in the uterus until the end of gestation.

14.9.3 Maternal Recognition of Pregnancy

It is critically important for the mother to recognize that there is a conceptus present otherwise the conceptus will not implant and will indeed be rejected. For instance, in the absence of a conceptus in cattle and sheep, the uterus produces prostaglandin $F_{2\alpha}$, causing luteolysis (breakdown of the corpus luteum). This must be stopped or the conceptus will not implant. The conceptus produces a signal that acts on the mother's physiology to prevent it from being rejected. This signal is called the **maternal recognition of pregnancy**. In ruminants, the signal produced by the trophectoderm cells of pre-implantation blastocyst is **interferon tau** (INFT or INFτ) (see Figure 14.21). In pigs, the pregnancy recognition signal is estradiol.

LH stimulates the corpus luteum (plural corpora lutea) to produce progesterone; the hormone progesterone being essential to the continuation of pregnancy.

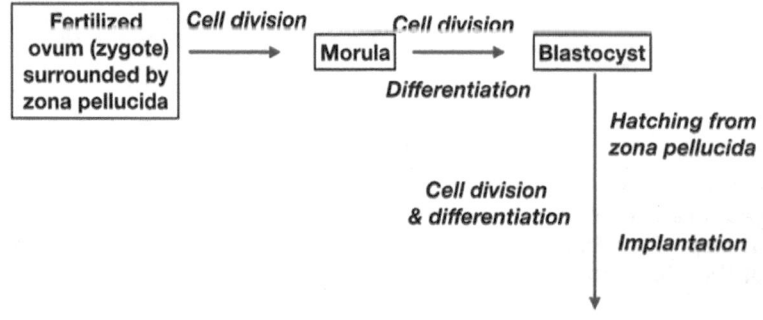

Figure 14.18 Stages in the early development of the conceptus.

Textbox 14.9 Terminology and Definitions in the Development of the Conceptus

- **Conceptus**: fertilized ovum and the subsequent stages.
- **Morula**: conceptus as a ball of cells. In sheep, the conceptus enters uterus from the oviduct as a morula on day 4.
- **Blastocyst**: stage in development when there is a hollow ball of cells (Figure 14.16 and 14.17). The cavity in the blastocyst is referred to as the "blastocoel."
- **Implantation**: the early conceptus invades the lining of the uterus.
- **Embryo**: is the stage of mammalian development from implantation to the formation of true bone. During this stage, all the major organs are formed. In birds, the embryo refers to the development from conception to hatching.
- **Fetus**: the stage of mammalian development from the formation of true bone to birth.
- **Hatching**: chick embryos leaving the egg in poultry. In addition, the trophoblast emerging from the zona pellucida is referred to as hatching.
- **Neonate**: newborn.
- **Trophoblast**: the layer around the cell mass. This layer supplies the developing embryo with nutrients and later forms part of the placenta.

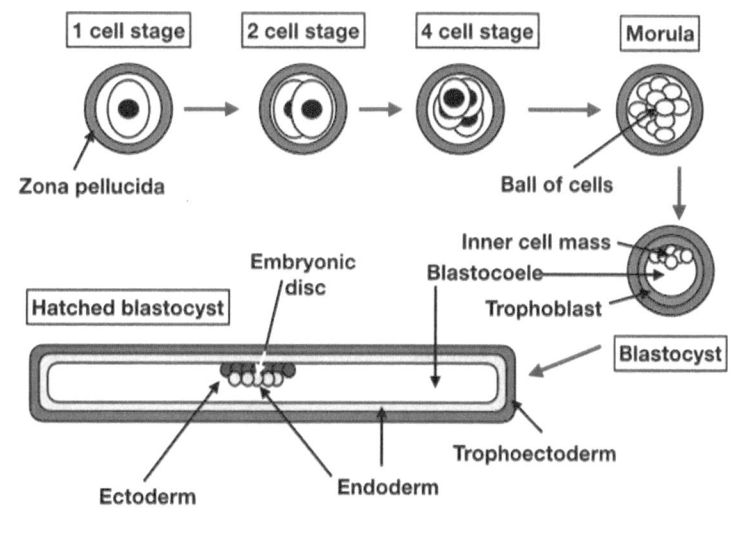

Figure 14.19 Early development of the conceptus in the pig.

If there is a conceptus in ruminants, the following occurs.

- The **trophectoderm cells** of pre-implantation blastocyst release interferon tau (IFNT or IFNτ).
- IFNτ acts on the endometrium to block the release of prostaglandin $F_2\alpha$ and, consequently, allows the corpus luteum to continue to produce progesterone.

In the nonpregnant animal, the following occurs.

- Oxytocin from the ovary stimulates the release of prostaglandin $F_2\alpha$.
- In turn, the prostaglandin $F_2\alpha$ causes the corpus luteum to regress and cease to produce progesterone.

14.9.4 Placenta and Its Role

14.9.4.1 Overview of Placentation

The presence of a placenta is essential to **viviparity**, the production of live offspring during the process of parturition or birth. The placenta is the structure facilitating the gaseous exchange between embryo/fetus and mother, as well as the transfer of nutrients from mother to fetus. Placentation involves intimate (very close) contact between maternal and embryonic/fetal tissues (also see Textbox 14.10).

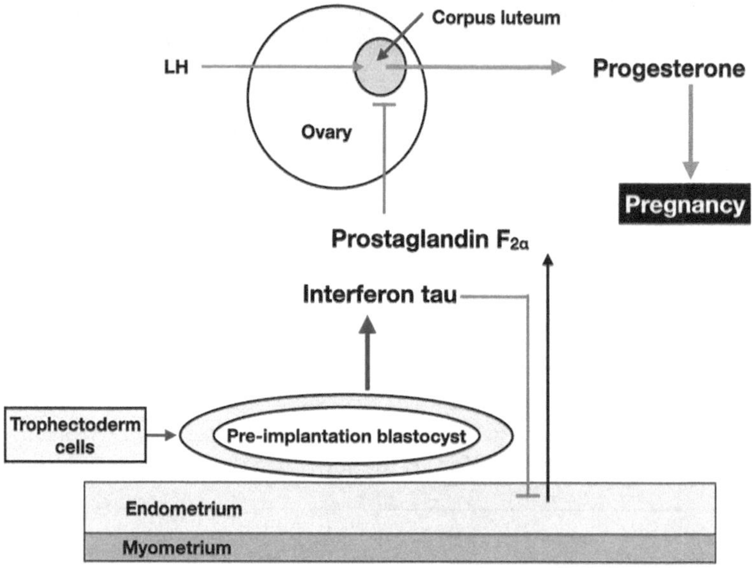

THEME 4 Reproduction (Propagation of the Species)

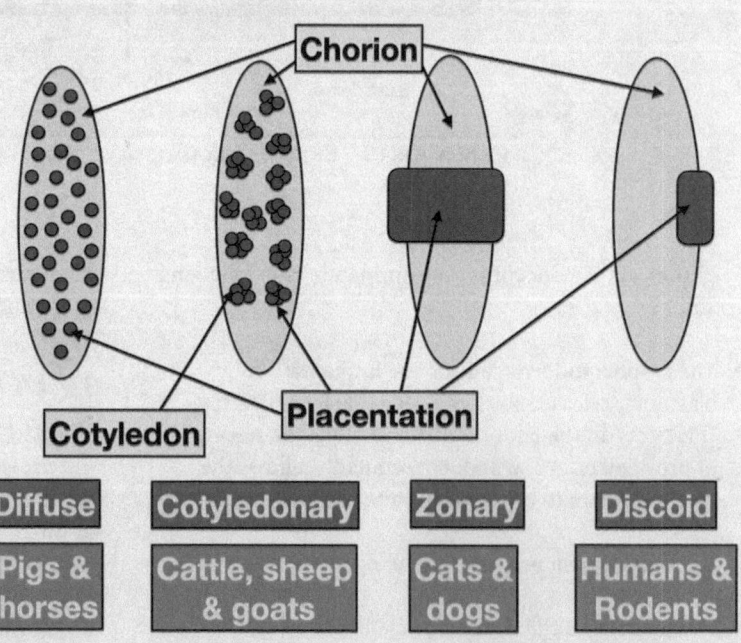

Figure 14.20 Maternal recognition of pregnancy in ruminants. [Key: ↑ is positive, T is negative.]

Textbox 14.10 A Deeper Dive Into Placental Types

Placentas are categorized under three criteria:

1. Embryonic/fetal tissues.
2. Gross appearance.
3. The number of layers.

The following are embryonic/fetal tissues contributing to the placenta:

- **Choriovitelline**: consisting of the chorion and yolk sac (providing the embryonic/fetal blood vessels). This is found in rodents and rabbits.
- **Chorioallantoic**: consisting of the chorion and allantois (providing the embryonic/fetal blood vessels). This is found in rodents (and other domesticated animals).

Gross Appearance

There are four types of placentas based on overall structure (see Figure 14.21).

Figure 14.21 Types of placentation based on gross anatomy.

Diffuse	Cotyledonary	Zonary	Discoid
Pigs & horses	Cattle, sheep & goats	Cats & dogs	Humans & Rodents

(continued)

Textbox 14.10 *(continued)*

- **Diffuse**: with placentation across the chorion. This is found in horses and pigs.
- **Cotyledonary**: with placentation in restricted circular areas (cotyledons) on the chorion. This is found in cattle, sheep, and goats.
- **Zonary**: with placentation in a restricted zone around in the chorion. This is found in dogs and cats.
- **Discoid**: with placentation restricted to a disc in the chorion. This is linked to the fetus via the umbilical cord. This type of placental is found in primates including humans, and also with rodents.

Layers

Placentas can be categorized based on the number of layers, namely three, five, or six layers from the following (Figure 14.22):

- Fetal endothelial cells surrounding allantoic blood vessels.
- Fetal connective tissue.
- Fetal epithelial cells (in chorionic villi).
- Maternal (uterine) epithelial cells.
- Maternal (uterine) connective tissue.
- Maternal (uterine) endothelial cells surrounding blood vessels.

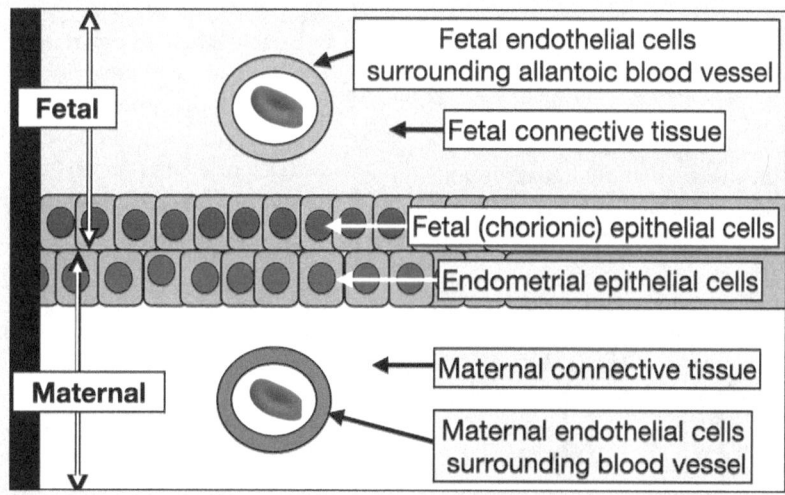

Figure 14.22 Basic layers of the placenta. (Additional layers in some categories of the placenta are not depicted.)

Table 14.5 Layers present in the placentas of domestic animals.

	Epitheliochorial	Endotheliochorial	Hemochorial
Fetal endothelial cells in allantoic blood vessels	Present	Present	Present
Fetal connective tissue	Present	Present	Present
Fetal epithelial cells	Present	Present	Present
Maternal epithelial cells	Present	Absent	Absent
Maternal connective tissue	Present	Present	Absent
Maternal endothelial cells surrounding blood vessels	Present	Present	Absent
Species found in	Cattle, sheep, pig, and horse	Cats and dogs	Humans and rodents

Although, there is not mixing of maternal and fetal blood (see Figure 14.20).

14.9.4.2 What Passes Across the Placenta?

The placenta plays a critically important role in providing the fetus with the following (see Figure 14.24):

- Oxygen passing from the maternal hemoglobin to the fetal hemoglobin.
- Water.
- Glucose and other nutrients.
- Ions such as sodium, chloride, potassium, phosphate, and calcium.

In addition, there are fluxes of carbon dioxide and nitrogenous waste (urea and ammonia) from the fetus to the mother.

14.9.4.3 What Does Not Pass Across the Placenta?

Plasma proteins generally do not pass across the placenta from mother to fetus. In livestock and companion animals, there is not transport of antibodies from mother to fetus. In contrast, this occurs in humans. Generally, hormones do not pass from mother to fetus.

14.9.4.4 Are There Other Roles for the Placenta?

The placenta produces hormones that maintain pregnancy and play a role in parturition.

14.10 Hormones and Pregnancy

14.10.1 Progesterone

Progesterone ("*pro*" and "*gestation*"), as its name implies, supports pregnancy. Progesterone is the major hormone facilitating an effective pregnancy. Progesterone exerts the following effects:

- Supporting the following:
  - Establishment of pregnancy.
  - Maintenance of pregnancy.
  - Inducing the production of nutrients for the conceptus due to glucose and amino acid transporters in the uterus.
- Inhibiting contractions of the myometrium.

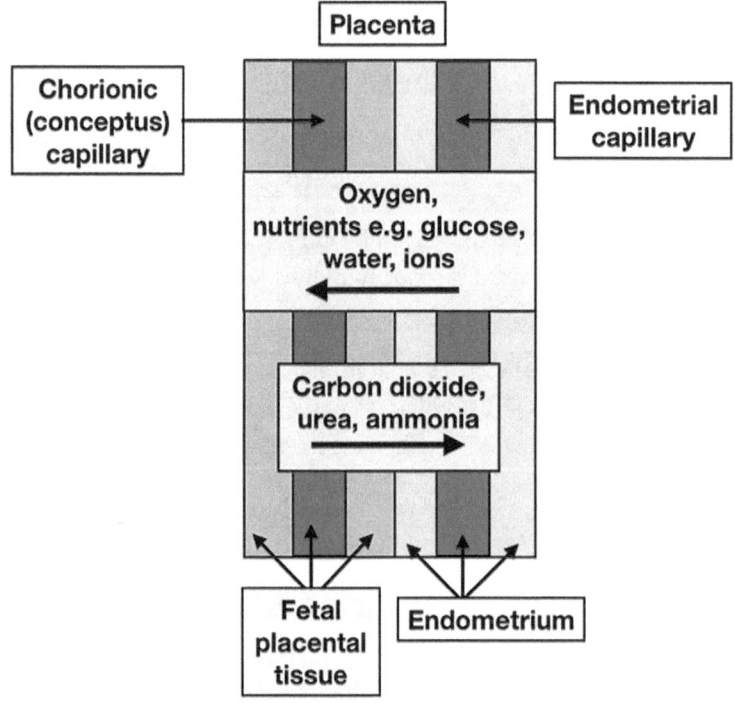

Figure 14.23 Schematic diagram of transfer across the placenta.

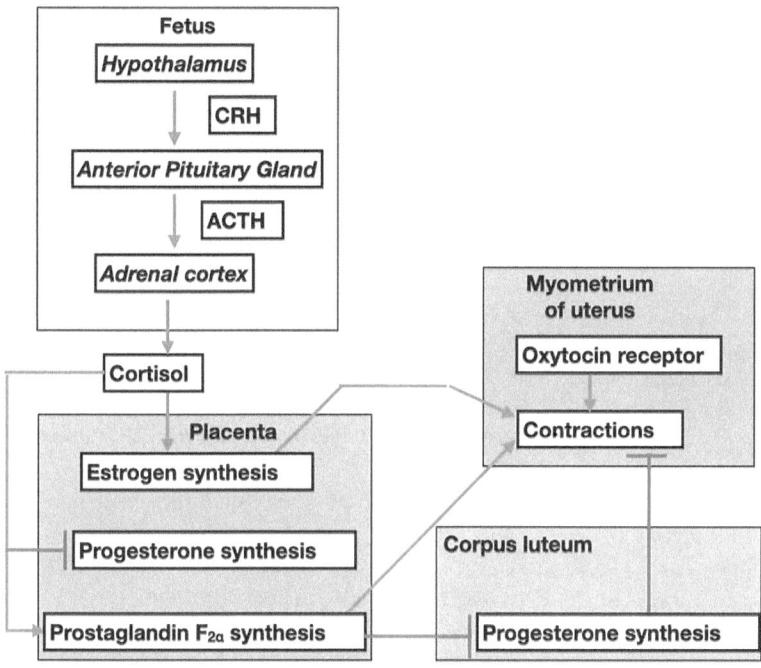

Figure 14.24 Hormones controlling parturition in sheep and cattle. [Key: CRH = corticotrophin-releasing hormone; ACTH = adrenocorticotropic hormone; ↑ is positive, ⊤ is negative.]

- Suppressing an immune response that would lead to rejection of the conceptus.

There are progesterone receptors in both the endometrial luminal epithelia and superficial glandular epithelia. These are important for implantation and establishment of pregnancy. Progesterone is produced by the corpus luteum (for instance in sheep and horses) and/or the placenta (cattle in late pregnancy).

14.10.2 Estrogens

Estrogens can play multiple roles during pregnancy. Estradiol and other estrogens stimulate contractions of the myometrium, allowing parturition. In addition, in pigs, the conceptuses produce estradiol as the pregnancy recognition signal. Estrogens are produced by the placenta.

14.10.3 Relaxin

Relaxin is important to pregnancy, stimulating uterine growth during pregnancy. Relaxin prepares the reproductive tract for the rigors of parturition causing the following:

- Breakdown of collagen in the cervix. This softening allows dilation of the cervix and is called "cervical ripening."
- Relaxation of the ligaments of the pelvis (specifically the pubic symphysis).

Relaxin also prevents premature parturition.

Despite its importance, relaxin is produced in different organs in domestic animals. For instance, the major site of relaxin production is the corpora lutea in pigs but the placental trophoblast cells in horses.

14.10.4 Placental Lactogen—A Unique Ruminant Hormone

A placental lactogen is found in ruminants. It results from gene duplication of the prolactin gene. Placental lactogen stimulates development of the mammary gland in late pregnancy.

14.10.5 Equine Chorionic Gonadotropin—A Unique Equid Hormone

Equine chorionic gonadotropin (eCG) is produced by cells of the **endometrial cups** in the placenta and induces the formation of **accessory corpora lutea** from the secondary follicles. In addition, eCG stimulates progesterone production from both the primary and accessory corpora lutea.

14.11 Parturition

Parturition can be both painful and risky to the animal, the fetus/neonate, and its owner/producer.

14.11.1 Stages of Parturition

Parturition consists of three stages:

- **Stage 1**: from beginning of labor to complete dilation of cervix.
- **Stage 2**: from complete dilation of cervix to delivery of the neonate.
- **Stage 3**: from birth to delivery of the placenta.

14.11.2 Position of Fetus

The fetus is positioned prior to parturition for optimal delivery. The legs and head are presented first; i.e., adjacent to the cervix and being pushed against the cervix with contract of the uterus. In cattle, the fetal calf should have both front legs in the birth canal, the head and neck along the front legs, and the back side up.

14.11.3 Hormones and Parturition

The fetus initiates parturition by increasing the production of cortisol (see Figure 14.24). This cortisol induces changes to the placenta including the following:

- Increased production of estrogens (which in turn enhances contraction of the myometrium).

Textbox 14.11 Hormones Used to Modify Reproduction in Female Livestock

GnRH and Synchronization of Estrus in Cattle
GnRH and prostaglandins are used to synchronize estrus in beef cattle.

Oxytocin
Oxytocin is used to aid difficult labor in livestock and companion animals.

Progestin and Synchronization of Estrus in Pigs
Progestins are used to synchronize estrus in gilts.

Prostaglandins and Reproduction
Prostaglandin $F_{2\alpha}$ or analogues are used to synchronize estrus in both dairy and beef cows. This greatly facilitates artificial insemination at the optimal time.

Textbox 14.12 Comparison of Female Reproduction Between Livestock/Companion Animals and Poultry

Table 14.6 Examples of similarities and differences in female reproduction between livestock and companion animals compared to poultry.

	Livestock, cats, & dogs	Poultry & other birds
Similarities		
Gametes	Haploid gametes—ova	
Fertilization	Internal fertilization	
Female reproduction tract	Develops from embryonic Müllerian duct(s)	
Differences		
Most common genotype	XX	ZW
Functional ovaries	2	1
Female reproductive tract(s)	2	1
Number of ova ovulated	1 in cattle, sheep, goats, & horses Multiple in pigs, cats, and dogs	1
Ovum size	Very small	Large yolk-filled ovum
Where fertilized ovum develops	Uterus	Externally

- Decreased production of progesterone as it is converted to estrogens (progesterone decreases contraction of the myometrium).
- Increased production of prostaglandin $F_{2\alpha}$. This stimulates contraction of the uterus and causes the corpus luteum to breakdown and stop producing progesterone.

As the fetus is pushed against the cervix, the mother releases oxytocin. This stimulates the uterine smooth muscles (myometrium) to contract and increase the intensity of their contractions.

14.12 Comparison of Female Reproduction Between Livestock/Companion Animals and Poultry

A comparison of female reproduction between livestock/companion animals and poultry is summarized in Textbox 14.12.

14.13 Pathophysiology of Female Reproduction

Textbox 14.13 summarizes pathophysiology of female reproduction in domestic animals.

Textbox 14.13 Pathophysiology

- **Breech birth**: neither the head nor the front legs of cattle, sheep, and goats are presented first. Frequently, the rump is presented first.
- **Dystocia** (difficult birth): fairly common in livestock and companion animals. It is associated with pain to the female and risk to both the fetus and/or mother. Dystocia is fairly common in cats and dogs. Examples of dystocia in livestock and horses include the following:
  - Cattle with (1) a 5% rate of dystocia due to abnormal positioning of the fetal calf such as posterior presentation rather than the anterior presentation; (2) presence of twins or triplets; (3) the fetus being too large and/or maternal pelvis too small; (4) a dead fetus.
  - Horses (10% dystocia) with (1) abnormal presentation, (2) a dead fetus (1-2%); (3) a failure of the chorioallantoic membrane to rupture during labor.
- Pigs if parturition is taking longer than two hours.
- **Freemartin or martin heifer**: a heifer calf with the ovaries having not develop normally and an underdeveloped female reproductive tract. They lack a cervix and the vagina is shorter. There is frequently an enlarged clitoris. This condition is due to receiving testosterone from a fraternal male twin, with the twins both sharing placental membranes. While this may seem to be a strange situation, it is estimated that about 85,000 freemartins are born each year in the United States.
- **Leptospirosis**: a cause of abortions and mastitis in repeat breeders.
- **Prolapsed uterus** (with the uterus protruding into the vagina and even out through the vulva) in the dam after delivery of the calf.
- **Stillbirth**: a dead fetus.

Lactation

15

Learning Objectives

1. Understand the composition of milk in different livestock and companion animals (see Section 15.4).
2. Understand the role of components of the mammary gland (see Section 15.2).
3. Understand the role and composition of colostrum (see Section 15.5).
4. Understand how colostrum is produced (see Section 15.5.2).
5. Understand the synthesis of lactose in milk (see Section 15.6.2).
6. Understand the synthesis of proteins in milk (see Section 15.6.3).
7. Understand the synthesis of fats in milk (see Section 15.6.4).
8. Understand the phases in the development of the mammary gland (mammogenesis) (see Section 15.3).
9. Understand the hormonal control of the initiation of milk production (lactogenesis) (see Section 15.7.2).
10. Understand the hormonal control of the maintenance of milk production (galactopoiesis) (see Section 15.7).
11. Understand the milk let-down reflex (galactokinesis) (see Section 15.8).

Table of Contents

15.1 Introduction
15.2 Functions of the Structures in the Mammary Gland
    15.2.1 Overview
15.3 Development of the Mammary Glands (Mammogenesis)
    15.3.1 Phases in the Development of Mammary Glands
    15.3.2 What Mammary Development Occurs in the Fetus?
    15.3.3 What Mammary Development Occurs Between Birth and Puberty?
15.4 Composition of Milk
    15.4.1 Overview of Composition of Milk
    15.4.2 Milk Proteins
    15.4.3 Milk Fats
    15.4.4 Milk Carbohydrate (Milk Sugar)

15.5 Colostrum
    15.5.1 Introduction to Colostrum
    15.5.2 Production of Colostrum
    15.5.3 End of Colostrum Production
15.6 Milk Synthesis
    15.6.1 Introduction to Milk Synthesis
    15.6.2 Lactose Synthesis
    15.6.3 Synthesis of Milk Proteins
    15.6.4 Synthesis of Milk Fats
15.7 Hormonal Control of Lactation
    15.7.1 Introduction to the Hormonal Control of Lactation
    15.7.2 Initiation of Milk Production (Lactogenesis)
    15.7.3 Maintenance of Milk Production (Galactopoiesis)
    15.7.4 Other Effects on Milk Production
    15.7.5 Changes in Plasma Concentrations of Hormones Around the Time of Parturition and the Onset of Lactation
15.8 Milk Let-Down (Galactokinesis)
    15.8.1 Introduction to Milk Let-Down
    15.8.2 Stress and Milk Let-Down
15.9 Involution of the Mammary Gland
15.10 Diseases Related to Lactation

15.1 Introduction

In mammals, milk is produced by mammary glands, which are modified sweat glands. Milk provides essential nutrients to the neonates (Figure 15.1). Globally, milk is an important source of high-quality proteins to the human diet and a major feature of animal agriculture. The major livestock producing milk are cows, water buffalo, sheep, and goats. In addition, milk is obtained from camels and horses. In contrast, poultry (like most birds) do not produce milk. However, pigeons and their relatives regurgitate a fatty secretion, so-called "crop milk" from the crop and this is provided to the squabs (young pigeons).

Textbox 15.1 provides a glossary of terms related to lactation.

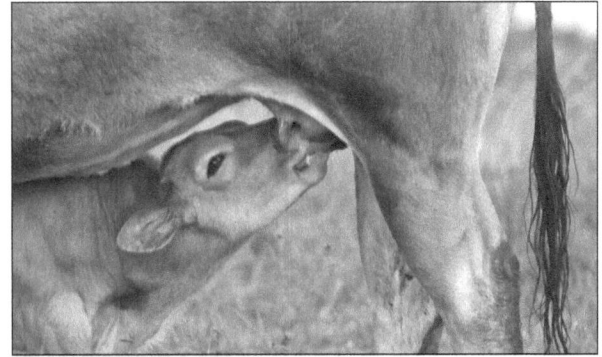

Figure 15.1 Calf suckling teat of mother. (Joaquin Cuervo, CC BY-SA 4.0)

Textbox 15.1 Glossary of Terms Related to Lactation

Structures
- **Alveolus** (plural alveoli): the structure that makes the milk and pushes it through the duct system.
- **Mammary glands** are found in all mammals. They produce milk for the newborn and young animal.
- **Myoepithelial cells** surround the alveoli. During milk let-down, these contract in response to the hormone, oxytocin.
- **Quarter**: is a single mammary gland in cattle.
- **Udder**: the name for four mammary glands in cattle.
- **Teat**: the exterior structure of the mammary gland that can contain a single or multiple openings to allow milk to pass to offspring.

Position in the Animal
- **Abdominal**: the region over the intestines.
- **Inguinal**: the groin region.
- **Thoracic**: chest region or equivalent adjacent to the ribs.

(continued)

> **Textbox 15.1** *(continued)*
>
> **Processes**
> - **Lactation** is the production of milk and lasts from several days to multiple months depending on species.
> - **Mammogenesis**: the development of the mammary gland.
> - **Lactogenesis**: the initiation of lactation.
> - **Galactopoiesis**: the maintenance of lactation.
> - **Galactokinesis or Milk let-down**: when the mammary gland is stimulated and milk is forced from the alveoli through the duct system to the teats. This happens when there is release of the hormone oxytocin, and this induces the myoepithelial cells of the alveoli to contract.
> - **Involution**: the process following a lactation cycle where the mammary gland undergoes apoptosis and the return to a non-lactating or dry state.
>
> **Milk Constituents**
> - **Casein**: the principal milk protein.
> - **Immunoglobulins**: antibodies.
> - **Lactoses**: the milk sugar. It is a disaccharide composed of glucose and galactose.
> - **Somatic cell count**: the number of cells (such as neutrophils and macrophage) in milk. High somatic count indicates an infection of the mammary gland (quarter of udder).
> - **Triglyceride**: a fat composed of one molecule of glycerol and three molecules of fatty acids.
>
> **Other Lactation-Related Factors and Issues**
> - **Colostrum**: the first "milk-like" fluid produced in a lactation. It provides antibodies to the calf or other newborn.
> - **Neonate**: a newborn.
> - **Oxytocin**: a hormone produced by the posterior pituitary gland. It induces contraction of the myoepithelial cells of the alveoli and, thereby, causes milk let-down.

Discussion of lactation will focus on the cow and somewhat on the goat. This is because there is much more research published on these species due to their economic importance. While, the same overall physiological systems are present, it is recognized that there are species differences, including with humans.

Mammary glands are paired across all mammalian species. There are marked differences in the anatomy of mammary glands in different livestock and companion animals, including the number of both **external openings** (teat canals) and teats, together with the position of the mammary glands (see Table 15.1). In cows, the four mammary glands are fused together as the four quarters of the udder (Figures 15.2), and this is located in an **inguinal region** (Figure 15.2). In goats, the two mammary glands are fused together in an inguinal udder. In contrast, there are multiple thoracic, abdominal, and inguinal mammary glands in sows (Figure 15.3) and female dogs (Figure 15.4; see Table 15.1).

Table 15.1 Comparison of position and number of mammary glands in domestic animals.

Species	Number	Position	Number of teat canals per teat
Cattle	4	Inguinal	1
Sheep	2	Inguinal	1
Goats	2	Inguinal	1
Horse	2	Inguinal	2
Pig	~10	Thoracic, abdominal, and inguinal	2
Cat	8	Thoracic and abdominal	3–7
Dog	~10	Thoracic, abdominal, and inguinal	5–6

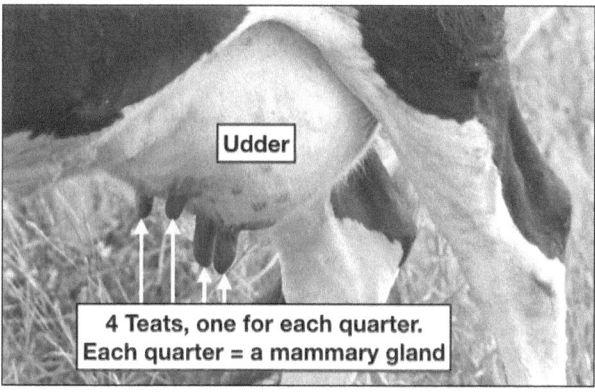

Figure 15.2 Cow's udder. (Adapted from Muhammed Mahdi Karim, GNU FDL 1.2)

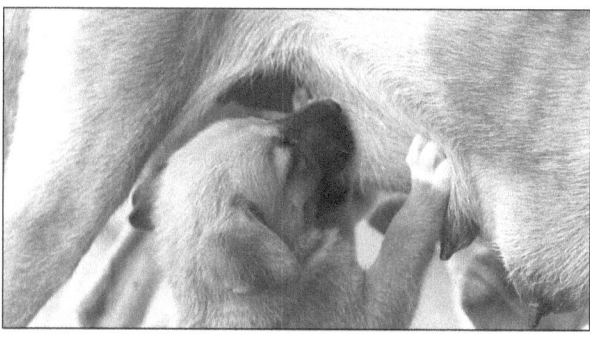

Figure 15.4 Young dog suckling teat of abdominal mammary gland of mother dog. (Shutterstock)

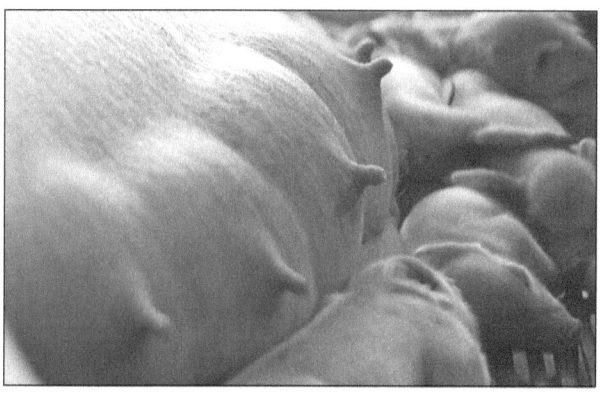

Figure 15.3 Young pigs resting by the abdominal udder of a sow. (Shutterstock/CHIRATH PHOTO)

15.2 Functions of the Structures in the Mammary Gland

15.2.1 Overview

The structure of the mammary glands in cows are shown in a series of schematic diagrams in Figures 15.2 and 15.5. The roles of the mammary gland and its structures are the following:

- Synthesis of colostrum and milk by mammary epithelial cells in the alveoli (see Figure 15.5B and 15.5D). This is discussed in more detail in Sections 15.5.2 and 15.7.2.

Textbox 15.2 Interesting and Important Information on Mastitis

Did You Know Bacteria and Other Microorganisms Can Easily Proliferate in the Mammary Gland?
From the point of view of pathogenic (disease-causing) bacteria and other microorganisms, milk is an ideal media in which to live. Just think of what happens when you leave milk out of the refrigerator.

How Does the Mammary Gland Prevent, or at Least Ameliorate, Invasion by Pathogens, and Infection (Mastitis)?
Entry of microorganisms into the teat and udder/mammary gland proper is greatly restricted by the teat or streak canal and the Furstenberg rosette. In addition, there are mechanisms to address any invading bacteria. These mechanisms include the presence of leukocytes, particularly the phagocytic neutrophils and T lymphocytes, together with macrophage (also phagocytic). There are also antibacterial peptides such as β-defensins in milk.

Is the Presence of Bacteria in Milk a Problem?
Potentially yes, if there is an infection in the mammary gland (quarter); an infection of the mammary gland being called mastitis.
  Dairy farmers use **somatic cell count (SCC)** to determine if an animal has an infection in the mammary gland. These counts are due to

(continued)

> **Textbox 15.2** *(continued)*
>
> the fact that milk contains sloughed off mammary epithelial cells together with the leukocytes (white blood cells or immune cells) that accumulate in an infected tissue. Mammary glands with an infection have markedly increased numbers of leukocytes, and thus SCC.

- Storage of milk in alveoli, ducts, teat cistern, and udder cistern.
- Milk let-down. This is achieved by contraction of the myoepithelial cells surrounding the alveoli forcing milk out of the mammary gland. This is discussed in more detail in Section 15.8.
- Maintaining its structure, e.g., in cattle by the **medial suspensory ligament**, together with the connective tissue separating the udder quarters (see Figure 15.5A).
- Preventing proliferation of bacteria and other microorganisms in the mammary gland (see Figure 15.5B).

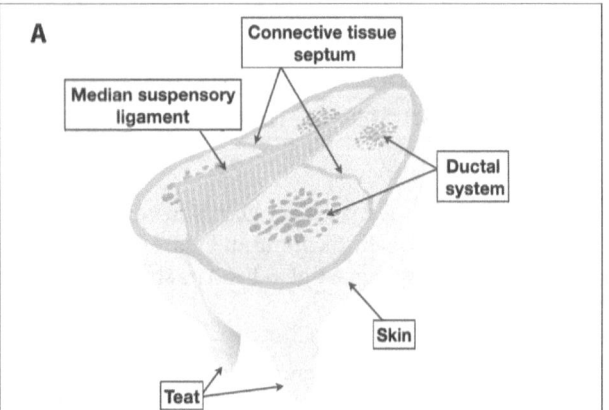
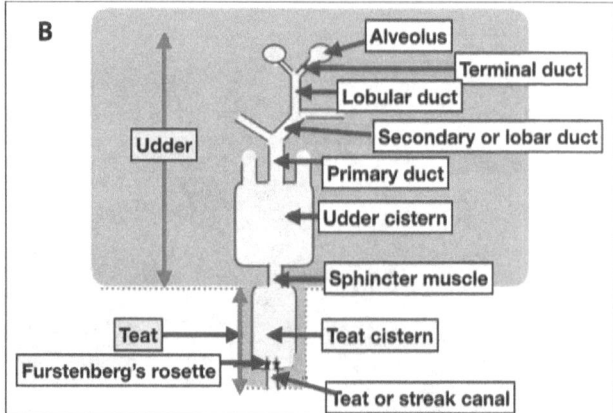
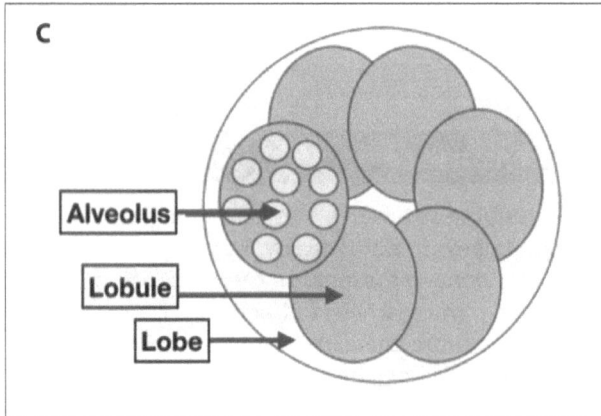
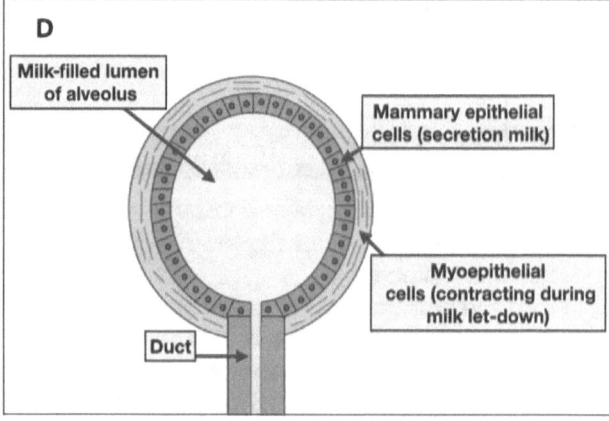

Figure 15.5 Structure of the udder of the cow.
(A) Section through the udder of a cow showing how the ligaments and connective tissue divides it into four.
(B) Schematic diagram of the overall structure of the duct system (for milk transit) and cisterns (for milk storage) within the udder (mammary glands).
(C) Schematic diagram of the lobar, lobular, and alveolar structures.
(D) Schematic diagram of the alveolus. (A is adapted from Jjost, CC BY-SA 2.0)

15.3 Development of the Mammary Glands (Mammogenesis)

15.3.1 Phases in the Development of Mammary Glands

There are five phases in the development of the mammary glands or mammogenesis:

1. **Fetal stage**, with the development of a **mammary rudiment** and teat from the mammary bud close by a fat pad. This is completed by the time of birth.
2. **Neonatal to puberty—isometric growth**. Growth of the rudimentary mammary glands is proportional to body weight and consists of adipose tissue, connective tissue, and ducts.
3. **Pubertal stage** with development of ducts.
4. **Late Pregnancy stage**, with the completion of the development of ducts and development of alveoli in preparation for lactation.
5. **End of lactation** with some involution (breakdown) of the mammary gland.

Stages 1, 3, and 4 are summarized in Figure 15.6.

15.3.2 What Mammary Development Occurs in the Fetus?

The mammary glands starts to develop early during fetal development. In cattle, development of the mammary glands starts by 30 days of fetal development (30 days of gestation).

There are two **mammary ridges** (also called mammary bands or milk lines) in the **ectoderm** (epidermis) running from anterior to posterior and displaced from the midline in the fetus. The mammary ridges condense into mammary placodes (Figure 15.7). These will ultimately be the site for the mammary glands, including the teats. Condensation of the mammary ridges into **mammary placodes** occurs in the areas where there are mammary glands in different species (Figure 15.7). For instance, the formation of mammary placodes occurs in the inguinal area in cattle, sheep, and horses. In contrast, mammary placodes are formed from the thoracic to inguinal locations in pigs and dogs. The mammary placodes go on to form **mammary buds**. These develop adjacent to the precursor to fat pads (containing adipocytes, fibroblasts, endothelial cells, and leukocytes) (Figure 15.7).

The mammary buds develop with the production of ducts and teats (see Figures 15.6 and 15.7). This production of ducts is due to the following:

- Mammary sprouting producing first the primary sprout.
- Branching producing secondary and tertiary sprouts.

These precursors of the ducts are solid structures. There is then the process of **canalization** or hollowing

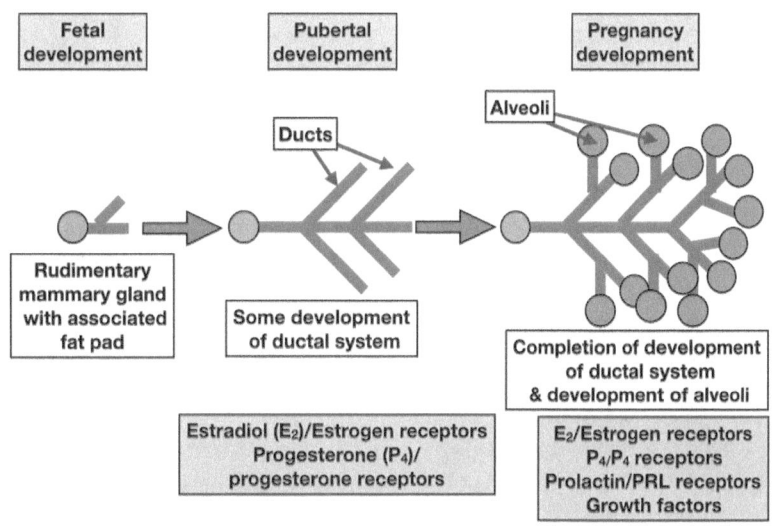

Figure 15.6 Development of the ductal alveolar system in the mammary gland.

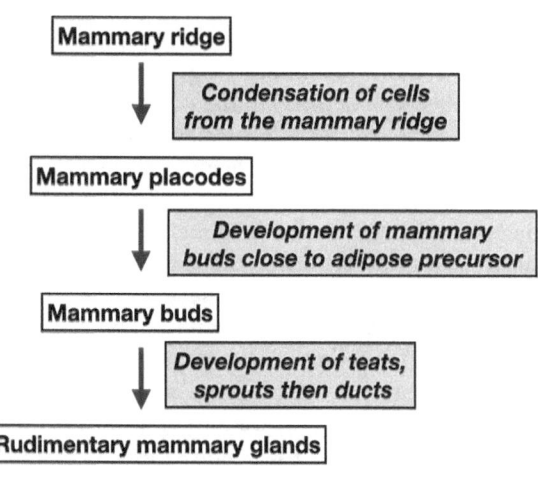

Figure 15.7 Fetal development of the mammary gland.

out of the primary, secondary, and tertiary sprouts. The process of hollowing out the sprouts occurs by approximately 100 days of fetal development in cattle. There is also formation of the supporting ligaments. These entire processes produce a rudimentary mammary gland. It is important to note that by the end of fetal development, there is still no secretory tissue (alveoli). Hormones such as growth hormone and insulin-like growth factor-1 (IGF-1) play a role in fetal mammary development. Textbox 15.3 summarizes the development of mammary glands.

15.3.3 What Mammary Development Occurs Between Birth and Puberty?

Textbox 15.3 discusses mammary development from birth through puberty.

Textbox 15.3 Development of the Mammary Glands

What Mammary Development Occurs Between Birth and Puberty?
Following parturition and prior to puberty, mammary development is limited. During this time, mammary growth is isometric growth. This is where the rate of growth of the mammary glands is similar to that of the rest of the body. During this phase, mammary growth is primarily of fat and connective tissue. The development of the mammary gland during pre-pubertal growth is regulated by growth hormone glucocorticoid, and estrogen. Again, it is important to note that during this phase of development no secretory tissue is formed.

What Mammary Development Occurs Starting at Puberty?
Following puberty, mammary growth is much greater than that of body growth (or **allometric**). There is much ductal growth and branching during this phase of mammary growth and development (Figure 15.6). This ductal development is driven by estradiol and progesterone produced by the ovaries during the estrus cycles in a cyclical manner (see Chapter 14). There are also stimulatory effects of growth hormone and IGF-1 on pubertal mammary development.

What Mammary Development Occurs During Pregnancy in Preparation for Lactation?
During pregnancy, the mammary glands are made ready for lactation. This involves the following:

- Completion of the development of the duct system due to cell division (see Figure 15.6).
- Development of the alveoli due to cell division (see Figure 15.6).
- Differentiation of the epithelial cells in the alveoli.

This results in the formation of the lobuloalveolar system that produces milk. Among the hormones critical to mammary development during pregnancy are progesterone, placental lactogen, and prolactin. There are also growth factors, such as epidermal growth factor, that act in a paracrine manner (see Chapter 6).

What Happens to the Mammary Gland at the End of Lactation?
The size of the mammary gland declines after the end of lactation, returning the mammary gland to its approximately anatomical and physiological state prior to pregnancy. There is involution (lobular alveolar collapse) of the mammary glands once suckling or milking

(continued)

> **Textbox 15.3** (*continued*)
>
> stops. There is cellular breakdown due to apoptosis (programmed cell death). Post lactation involution is due to a lack of the hormones supporting milk, such as prolactin.
>
> **When Would You Expect in the Estrous Cycle for Maximal Ductal Growth?**
> Hint: When would estradiol to be highest? Estradiol and progesterone act, respectively, by binding to the α estrogen receptors and the progestogen receptors. A key component of growth and development of the alveolar ductal system is proliferation of the **mammary stem cells** (Figure 15.8). These undergo multiple cycles of cell division (Figure 15.8). Again, it is important to note that during this phase of development no secretory tissue is formed. It is not until pregnancy that the female will start experience lobuloalveolar growth, or growth of secretory tissue which is driven by the combination of estrogens, progesterone, prolactin, growth hormone, and adrenocorticotropic hormone. Specifically, prolactin and placental lactogen play a role in the final growth and development of the ductal and alveolar system. This includes being part of the hormone complex that induces the proliferation and differentiation of the mammary stem cells towards the end of pregnancy (Figure 15.8). There are other hormones influencing both proliferation and differentiation. For instance, insulin increases both proliferation of ductal cells and differentiation.
>
>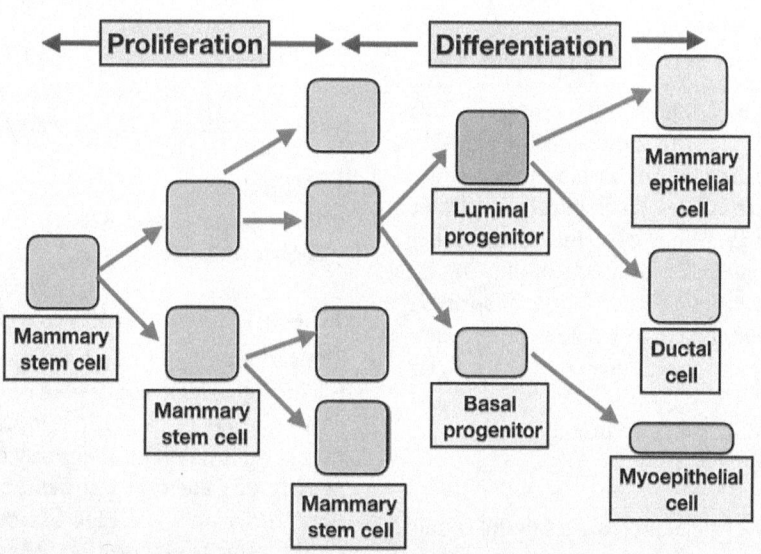
>
> **Figure 15.8** The ductal alveolar system develops due to the proliferation of the mammary stem cells and their subsequent differentiation.

15.4 Composition of Milk

15.4.1 Overview of Composition of Milk

The composition of milk in different livestock and companion animals is summarized in Table 15.2. While the specific percentages of components vary in milk from different livestock and companion animals, it is important to note that water is the greatest component. This makes it critical for producers to not only provide fresh feed, but fresh clean water to lactating females. The percentage of protein varies between a low in horses (2.7%) to a high in cats (11.1%). The percentage of lipid varies between a low in horses (1.6%) to a high in cats (10.9%) (see Table 15.2). In contrast, the concentration of lactose is consistent at about 4.6% in livestock and companion animals (see Table 15.2).

Interestingly, the concentrations of the major constituents in milk are markedly different in human (mothers) milk and that in livestock and companion animals (see Table 15.2). In particular, human milk has low concentrations of protein but high concentrations of lactose. In marine mammals, specifically dolphins, the concentrations of lipids are very high (see Table 15.2).

Theme 4 Reproduction (Propagation of the Species)

Table 15.2 Comparison in the composition of milk in different species of livestock and companion animals with the composition of human and dolphin milk included for comparison.

Species	Protein %	Lipid (fat) %	Lactose %
Cat	11.1	10.9	3.4
Cow (Holstein)	3.1	3.5	4.9
Cow (Jersey)	3.3	5.0	4.7
Dog	9.5	8.3	3.7
Goat	3.1	3.5	4.6
Horse	2.7	1.6	6.1
Pig	5.8	8.2	4.8
Sheep	5.5	5.3	4.6
Human	1.0	3.8	7.0
Dolphin	9.7	33.5	0.6

15.4.2 Milk Proteins

Milk proteins are composed of casein (80% of milk proteins) and whey proteins (20% of milk proteins). This source of nutrients provides the growing offspring with amino acids for muscle growth. The casein molecules are phosphorylated and in **micelles**, which allows for slower digestion of the nutrients. Whey proteins on the other hand are quicker to be digested and absorbed, allowing for a more readily available protein source for the offspring.

When milk is curdled (caused to coagulate by the presence of the proteolytic enzymes in rennet), there are two fractions:

- **Curds** (most of the milk solids including casein).
- Whey.

The major whey proteins (globular proteins) are the following:

- Alpha-Lactalbumin.
- Beta-lactoglobulin.

In addition, there are immunoglobulins, lactoferrin, lactoperoxidase, protease-peptones, and serum albumin in the whey fraction.

15.4.3 Milk Fats

Lipids in milk are found in fat globules and serve as an excellent source of energy. The fat globules are stabilized by a milkfat globule phospholipid membrane. The composition of lipids is the following:

- Triacylglycerol (a.k.a. triglyceride) (96%).
- Diacylglycerol (1%).
- Phospholipid (1%).
- Cholesterol (0.5%).

The fatty acids in milk lipids are the following:

- Saturated (65%).
- Monounsaturated (32%).
- Polyunsaturated (3%).

15.4.4 Milk Carbohydrate (Milk Sugar)

The carbohydrate in milk is the milk sugar **lactose**. This is a disaccharide composed of one **galactose** and one **glucose** subunits (Figure 15.9). Prolactin drives the production of lactose in the mammary gland by stimulating the activity of galactosyl transferase. It is also important to note that lactose has high osmotic properties and, thus, draws water to its location. Therefore, as lactose is produced in the mammary epithelial cells and secreted into the lumen, it will pull water along with it.

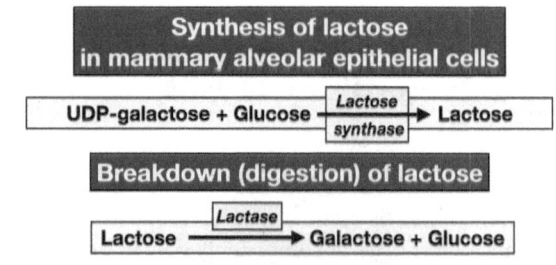

Figure 15.9 Synthesis of lactose in the mammary gland and its breakdown (digestion) in the small intestine. [Key: UDP = uridine diphosphate.]

15.5 Colostrum

15.5.1 Introduction to Colostrum

Colostrum is the first milk produced and generally produced for between 24 to 72 hours postpartum. It has a high concentration of immunoglobulins (antibodies) and other proteins particularly in the first few hours after parturition (see Tables 15.2 and 15.3). There is also markedly more calcium in colostrum than milk:

- Colostrum (0.26%).
- Milk (0.13%).

Again, there is also markedly more magnesium in colostrum than milk:

- Colostrum (0.037%).
- Milk (0.011%).

Colostrum provides the newborn with **passive immunity** to diseases the dam has been exposed to by passing immunoglobulins for these diseases to the offspring. The newborn consumes colostrum and for the first few hours after birth the immunoglobulins (antibodies) pass through the wall of the gastrointestinal tract into the bloodstream due to the high permeability of the gastrointestinal tract. After about six hours, the walls of the gastrointestinal tract start to become impermeable to the immunoglobulins. Calves should receive about 150 g of immunoglobulin in the first six hours of life and preferably in the first two hours of life. The concentrations of immunoglobulin and other proteins in the colostrum decline in the first day of after parturition (Table 15.4).

In addition to providing the newborn calf, lamb, goat kid, or piglet with antibodies, colostrum provides billions of leukocytes (neutrophils, macrophages, and lymphocytes). Colostrum contains high concentrations of insulin like growth factor-I (IGF-I) and other growth factors. These may play a role in the development of the gastrointestinal tract.

15.5.2 Production of Colostrum

Colostrum has high concentrations of protein (Tables 15.3 and 15.4). In cattle, about 90% of the protein is immunoglobulin gamma 1 (IgG_1). The concentration of IgG_1 in colostrum is five to ten times that in the plasma. In contrast, the concentration of IgG_2 in colostrum is the same as in plasma. For a deeper dive into colostrum see Textbox 15.4. Production of colostrum is controlled by hormones, specifically estradiol and progesterone.

Table 15.3 Changes in colostrum composition in dairy cattle after calving.

Time in hours after calving	Protein %	Lipid (fat) %	Lactose %	Globulin & albumen %
0	16.8	6.7	2.9	12.7
6	11.7	6.1	3.5	8.0
24	5.5	4.1	4.1	2.6
Milk	3.2	3.8	4.6	0.6

Table 15.4 Comparison in the composition of colostrum (hour zero after parturition) in different species.

Species	Protein %	Lipid (fat) %	Lactose %
Dog	13.8	13.8	2.7
Goat	8.0	9.0	2.5
Horse	19.1	0.7	4.6
Pig	18.0	7.2	2.4
Sheep	13.0	12.4	3.4

> **Textbox 15.4 A Deeper Dive Into Colostrum**
>
> **How Is This Highly Concentrated IgG1 Colostrum Produced?**
> The mammary epithelial cells of the alveoli produce colostrum prior to parturition. There is transfer of IgG_1 from the basolateral membrane across the mammary epithelial cell (in vesicles by transcytosis) and the luminal membrane into the alveolus lumen (Figure 15.10). The IgG_1 binds to the IgG_1 receptor (Fc receptor) as the first step of its movement into the alveolus (Figure 15.10). Transit of IgG_2 into the lumen does not involve the IgG_1 receptor (Figure 15.10).
>
>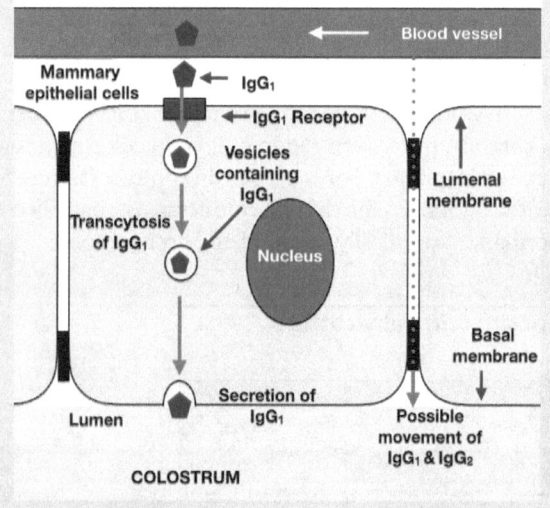
>
> **Figure 15.10** Production of colostrum with high concentrations of immunoglobulin G1 (IgG1). The IgG1 binds to IgG1 receptors in the lumen cell wall of the mammary epithelial cells. Subsequently the IgG1 is moved by transcytosis across the mammary epithelial cells in vesicles and finally secreted across the lumen membrane into the alveolus lumen.

15.5.3 End of Colostrum Production

Production of colostrum declines rapidly during the first few days after parturition (Table 15.2). This is due to the following:

- Effects of the glucocorticoid cortisol.
- Prolactin decreasing IgG_1 receptors in the mammary epithelial cells.
- A shift to the production of milk (discussed below).

15.6 Milk Synthesis

15.6.1 Introduction to Milk Synthesis

Milk is synthesized by the mammary epithelial cells in the alveolus of the mammary glands. The major solids in milk are the following:

- Lactose.
- Milk proteins.
- Milk lipid (predominantly triglyceride).

15.6.2 Lactose Synthesis

Lactose is synthesized in the **mammary epithelial cells** in the alveolus from circulating glucose (see Figure 15.11). Glucose entry into the mammary epithelial cells is mediated by a specific glucose transporter, the GLUT1 (see Figure 15.11). The glucose either passes to the Golgi apparatus, again mediated by the GLUT1, or is converted to uridine diphosphate-galactose (UDP-galactose) in a series of enzymic steps (see Figure 15.11). The UDP-galactose is moved into the Golgi apparatus by the **sodium dependent glucose/galactose transporter** (SGLT) (see Figure 15.11). The disaccharide lactose is synthesized by combining UDP-galactose with glucose; a step catalyzed by the enzyme lactose synthase (see Figure 15.11).

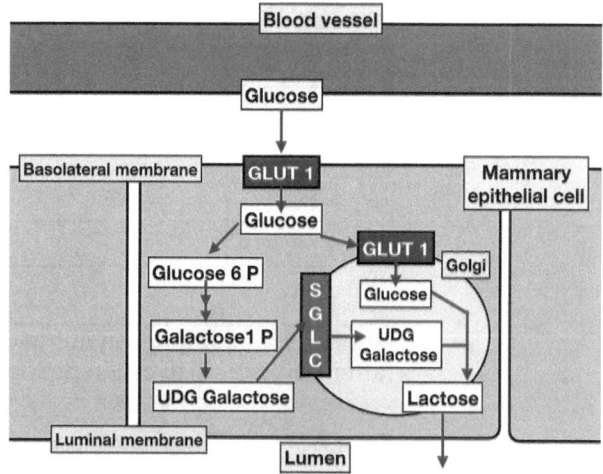

Figure 15.11 Synthesis of lactose in the alveolar epithelial cells. Within the Golgi apparatus, lactose is synthesized by the enzyme lactose synthase. [Key: GLUT1 = Glucose transporter 1; SGLT = sodium dependent glucose/galactose transporter; UDP = Uridine diphosphate. (Based on Zhang et al., 2018)

Textbox 15.5 Are There Special Problems in Lactose Synthesis in Ruminants (Cattle)?

Taking the example of a cow that is producing 6.5 gallons (or 120 lbs or 25 liters) of milk per day. The milk contains 4.9% lactose (Table 15.1).

The amount of lactose in milk per day is $25 \times 0.049 = 1.2$ kilograms (or 2.7 lbs)

This lactose is synthesized from glucose (see Figure 15.12). In turn, this has to be synthesized from the volatile fatty acid propionate, in the liver of a ruminant (cow). In contrast, in nonruminants, glucose comes from digestion of starch in the small intestine or from the storage form of glucose, glycogen, in the liver.

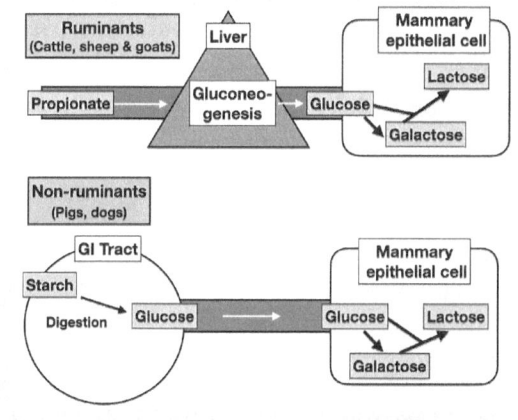

Figure 15.12 Comparison of the source of glucose for lactose in ruminants and nonruminants.

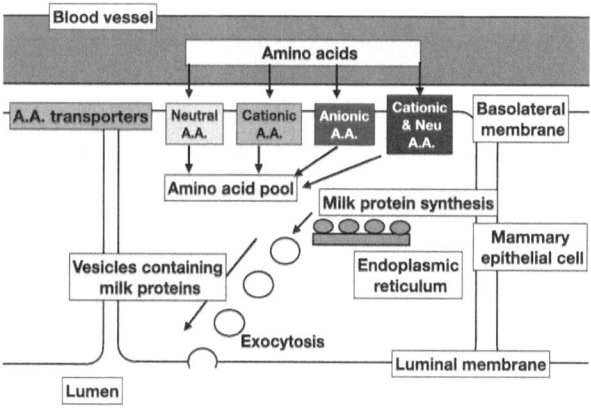

Figure 15.13 Uptake of examples of specific transporter mediated uptake of amino acids by alveolar epithelial cells and subsequent synthesis of milk proteins. [Key: A.A. = amino acid.]

15.6.3 Synthesis of Milk Proteins

The mammary epithelial cells synthesize the milk proteins. Amino acids are transferred into the mammary epithelial cells by a series of amino acid transporters (Figure 15.13). These transporters include the following:

- Anionic amino acid transporters—transporting glutamic acid and aspartic acid.
- Cationic amino acid transporters—transporting lysine and arginine.
- Neutral amino acid transporters—transporting leucine, isoleucine, methionine, phenylalanine, and valine.
- Cationic and neutral amino acid transporters—transporting histidine, threonine, and tyrosine.

The amino acids are used by the mammary epithelial cells to produce the milk proteins (Figure 15.13). The milk proteins are exported from these cells into the lumen of the alveolus.

15.6.4 Synthesis of Milk Fats

Milk fat is predominantly **triacylglyceride** (a.k.a. triglyceride). (Triacylglyceride is a compound made of glycerol and three fatty acids.) The mammary epithelial cells of the alveolus make triglyceride from fatty acids and glyceryl phosphate. The glyceryl phosphate is generated by glycolysis. The fatty acids are synthesized in the following:

- Mammary epithelial cells.
- Liver and adipose tissue.

Alternatively, fat in the animal's feed is digested to fatty acids and glycerol, these pass into the blood system and from there to the mammary epithelial cells. In ruminants, fats are fermented in the rumen, generating volatile fatty acids such as acetic acid. Fatty acids are synthesized from these volatile fatty acids. Interestingly, mutations in diacylglyceride transferase I (DGAT I) are associated with different concentrations of milk fat in dairy cattle.

15.7 Hormonal Control of Lactation

15.7.1 Introduction to the Hormonal Control of Lactation

The hormonal control of lactation is considered under two phases:

- Lactogenesis (initiation of milk production).
- Galactopoiesis (maintenance of milk production).

15.7.2 Initiation of Milk Production (Lactogenesis)

A complex of hormones is required to initiate milk production. Among the hormones required for milk production are prolactin, insulin, cortisol, and thyroid hormones. In studies with explants (small pieces of tissue incubated) from pregnant cows, prolactin alone or in combination with insulin and/or cortisol were capable of initiating the production of the following constituents of milk:

- Casein.
- α-Lactalbumin.
- Lactoferrin.
- Fatty acids.

Insulin by itself has also been demonstrated to increase casein synthesis. In addition, thyroid hormones have some role in the initiation of milk production increasing milk fat synthesis via thyroid hormone responsive protein (THRSP).

15.7.3 Maintenance of Milk Production (Galactopoiesis)

It is critically important for the offspring that there be the supply of milk that meets their needs. In addition, a goal for a dairy farmer is the production of high-quality milk. Based on studies in predominantly goats, milk production has been shown to require the presence of a series of hormones. Having a normal repertoire of metabolic hormones (cortisol, insulin, and triiodothyronine) together with sufficient prolactin (nonruminants) and somatotropin (ruminants) is essential to maintaining a full lactation.

There are a series of obvious questions. If some (level of a hormone) is good, is more better? Increasing the plasma concentrations of insulin does not affect either milk yield or composition. Stimulation of the teat also increases prolactin release from the anterior pituitary gland. However, prolactin administration has little effect on milk yield despite it being required for milk production and its critical role in lactogenesis. In contrast, bovine somatotropin (bST) increases milk yield during mid and late lactation in dairy cattle. Bovine somatotropin has been deemed as safe and effective by the US Food and Drug Administration (FDA). Biosynthetic bovine somatotropin is available commercially to dairy farmers under the brand name Posilac.

15.7.4 Other Effects on Milk Production

There are effects of management practices on milk production with increases of the following:

- Frequency of milking.
- Long photoperiods.
- Light intensity.

15.7.5 Changes in Plasma Concentrations of Hormones Around the Time of Parturition and the Onset of Lactation

There are increases in the plasma concentrations of the following at the end of pregnancy and during in the first week of lactation:

- Estradiol.
- Progesterone.
- Cortisol.
- Prolactin.

In contrast, there are decreases in the plasma concentrations of glucose and insulin in the first week of lactation. The decrease in plasma concentrations of glucose reflects transit into the mammary gland, conversion to lactose, and secretion in colostrum then the milk. As would be expected with low plasma concentrations of glucose, the concentrations of insulin decline.

15.8 Milk Let-Down (Galactokinesis)

15.8.1 Introduction to Milk Let-Down

How is milk moved from the alveoli and that stored in the ducts and cisterns? There is a well-established **neuroendocrine reflex** (see Figure 15.14). Stimulation of the teat sets off this reflex. This stimulation can be a calf or other neonate suckling or a farmer rubbing and cleaning the teats (see Figure 15.15). Nerve impulses pass along **afferent neurons** to the neurons in the spinal cord and then through the **infundibulum** to the posterior pituitary

Chapter 15 Lactation

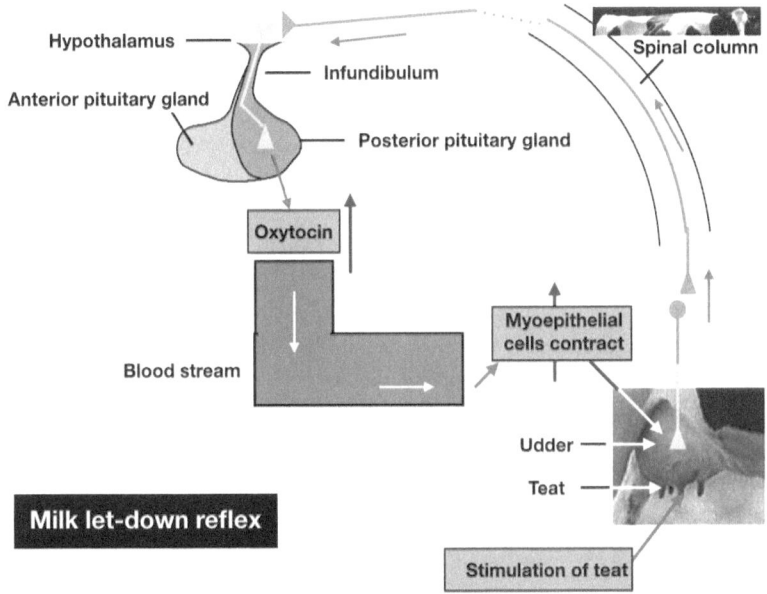

Figure 15.14 Milk let-down in cattle is stimulated in a neuroendocrine reflex. Rubbing the teat (or suckling) stimulates tactile sensory neurons. This leads to action potentials passing up neurons in the spinal cord (afferent pathway). This stimulates neuroendocrine cells with terminals in the posterior pituitary gland to release oxytocin. The oxytocin passes via the blood stream to the mammary gland where it evokes contractions of the myoepithelial cells. In turn, these contractions push milk out of the alveoli and through the ducts.

gland. **Neuroendocrine cell terminals** in the posterior pituitary gland then release oxytocin (see Figure 15.14). The oxytocin passes through bloodstream and induces contraction of the myoepithelial cells surrounding the alveoli. This forces the milk through the duct system and then to the exterior.

15.8.2 Stress and Milk Let-Down

Stress reduces milk let-down. This can be an inexperienced or rough milker or a disturbance in the area where milking or suckling is occurring.

How does stress reduce milk let-down?

In response to a stressor, there is reduced contraction of the myoepithelial cells surrounding the alveoli. This is due to the stressor inducing the chromaffin cells in the adrenal gland to release epinephrine and norepinephrine. These hormones reduce the effects of oxytocin on the myoepithelial cells. Figure 15.15 summarizes the hormones controlling mammary development, milk production, and milk let-down.

15.9 Involution of the Mammary Gland

The mammary gland undergoes involution at the end of a lactation. This brings the mammary gland back to the state prior to pregnancy. There is considerable programmed death (apoptosis) of epithelial cells and their

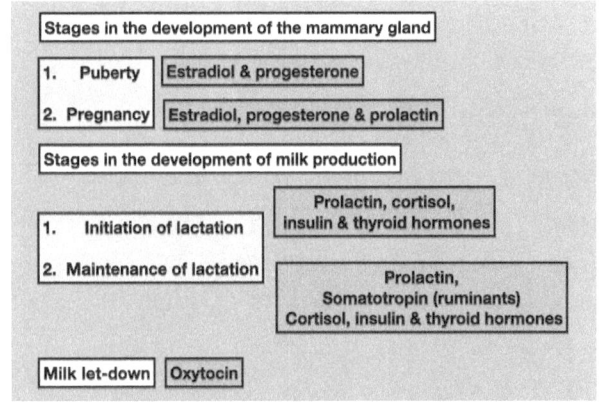

Figure 15.15 Hormones and mammary development, lactation and milk let-down. (Note: somatotropin is used by dairy scientists and biochemists. It is also called growth hormone.)

replacement with fat cells. Involution of the mammary gland occurs because of lack of stimulation of the mammary gland and loss of the hormones stimulating milk production, such as prolactin.

15.10 Diseases Related to Lactation

Textbox 15.6 summarizes diseases related to lactation in cattle.

> **Textbox 15.6 Diseases Related to Lactation**
>
> - **Mastitis** is inflammation of the mammary tissue (see Textbox 15.2). In cows, this is due to either damage to the udder (physical trauma) or microbial infection in the udder. Infections of the udder can be due to viruses, bacteria, and fungi. Examples of bacteria responsible for mastitis are Staphylococcus aureus, Pasteurella multocida, and Escherichia coli. Aspergillus fumigatus is an example of fungi responsible for mastitis. Mastitis is an infection of the udder. It is a significant loss to the farmer. The cow is treated with antibiotics but the milk is not saleable.
>
> - **Milk fever** (a.k.a. post-parturient hypocalcemia or parturient paresis downer cow): This is when the plasma concentrations of calcium and/or magnesium are abnormally low within the first three days after calving. This reflects transfer of calcium and/or magnesium from the plasma to the colostrum (first milk). High potassium feeds can result in high blood concentrations of potassium and consequently reduced calcium mobilization from bone. Altering the diet prior to parturition can reduce the risk of milk fever. Milk fever can be treated by intravenous administration of calcium gluconate followed by oral calcium supplements.

Index

Note: Page numbers followed by "*b*" and "*f*" indicate boxes and figures respectively.

A

Abdominal, 323
Abomasum, ruminants in, 249
ABP. *See* Androgen-binding protein (ABP)
Absolute refractory period, 103
Absolute temperatures, 19
Accessory corpora lutea, 319
Accessory organs, 242
ACE. *See* Angiotensin-converting enzyme (ACE)
Acetyl-coenzyme A, 145
    Acetylcholine, 31, 83, 112, 136
Acid-base homeostasis, 59
Acrosome reaction, 295
ACTH. *See* Adrenocorticotropic hormone (ACTH)
Actin, 127*b*
Actin molecules, 139*b*
Action potential, 28*b*, 97*b*
Acute inflammation, 204
Adaptive (or specific) immunity, 182, 198
Adaptive immune response, 186
Addison's disease, 91*b*
Adenosine triphosphate (ATP), 50, 127*b*
Adenylate cycle, 85
ADH. *See* Antidiuretic hormone (ADH)
Adherens junctions, 131*b*
Adipocytes, 158–159, 159*f*, 211*b*
Adipogenesis, 151*b*, 159, 160*b*
Adipose tissue, 151*b*
   adipocytes, 158–159, 159*f*
   adipogenesis, 159
   brown adipose tissue (bat), 159–160
Adrenal gland, 81–84
Adrenergic receptors, 108, 108*t*
Adrenocorticotropic hormone (ACTH), 75–76, 75*t*
Adult blow flies, 224, 224*f*
Aerobic, 151*b*
Afferent neurons, 334

Afterload, 32
Aiding movement, 59
Air
   composition of, 53
   to flow in birds, 57
Aldosterone, 76, 83–84
All-trans-retinal isomer, 116
Allergies, 205–206, 206*f*
Allometric, 328
Alopecia, 224
Alpha (α) cells, 84
α-Melanocyte-stimulating hormone (α-MSH), 211*b*, 218*b*
Alternative complement pathways, 195–196, 195–196*b*
Alveoli, 54
Alveolus, 54, 54*f*, 60*b*, 323
Ambisexual stage, 287
Amino acids, 13
   hormones, 70, 79
      synthesis of, 71
   reabsorption, 261
Ampulla, blockage of, 296
Amylases, 239
Anaerobic, 151*b*
Anaerobic fermentation, 248
Anagen, 214, 215*f*
Anaphylatoxins, 195
Androgen-binding protein (ABP), 76, 288
Androgens, 71
Aneurism, 46*b*
Angiogenesis, 309
Angiotensin-converting enzyme (ACE), 59, 84, 267
Angular leg deformities, 175
Anhidrosis, 224
Anterior pituitary gland
   adrenocorticotropic hormone, 75–76
   follicle-stimulating hormone, 76
   growth hormone, 77

Anterior pituitary gland (*continued*)
  luteinizing hormone, 76–77
  overview of, 73–75, 75*t*
  prolactin, 77
  somatotropin, 77
  thyroid-stimulating hormone, 77
Anti-inflammatory cytokines, 188
Antibody, 198–200
Antidiuretic hormone (ADH), 78, 265
Antigen-binding site, 199
Antigen-presenting cells (APC), 186
Antimicrobial peptides, 187, 221
Aortic semilunar valve, 28
APC. *See* Antigen-presenting cells (APC)
Apical cell membrane, 260*b*
Apocrine sweat glands, 216
Appendicular skeleton, 163
Aquaporins (AQP), 109, 263
Aqueous humor, 114
Arachnoid mater, 108
Arachnoid villi, 109
Arginine, 13
Arginine vasopressin (AVP), 76, 78
Aromatization, 288
Arrector pili muscles, 211*b*
Arrhythmia, 46*b*
Arteries, 26
Articular cartilage, 169
Ascites, 46*b*
Atmospheric pressure, 60*b*
ATP. *See* Adenosine triphosphate (ATP)
Atrial depolarization, 30
Atrioventricular node (AV node), 28*b*
Atrioventricular valves (AV valves), 28–29
Atrium (plural atria), 28*b*
Autocrine, 188
Autocrine factor, 68
Autonomic nervous system, 97*b*, 113
Autorhythmic cells, 28*b*
Avian long bone, 173, 173*f*
  air sacs in, 174
AVP. *See* Arginine vasopressin (AVP)
Axial skeleton, 163
Axolemma, 100, 100*f*
Axon, 97*b*
Axon hillock, 100
Axon terminals, 100

B

B lymphocytes, and antibody production, 200–201, 201*f*
Bacterial osteomyelitis, 174
Ball-and-socket joints, 169
Baroreceptors, 112, 114
Basal keratinocytes, 212
Basolateral cell membrane, 260*b*, 261

BAT. *See* Brown adipose tissue (BAT)
Beta (β) cells, 84
β-endorphins, 123
Beta-defensins, 221
Bicarbonate, 60*b*, 280
  chloride exchange, 261
  reabsorption, 261
Bichromatic vision, 115
Bilirubin, 41
Biliverdin, 41
Bladder, 266–269, 268*f*, 269*f*
  stones, 271*b*
Blastocyst, 315
Blood, 34–35, 36*f*
  cells/formed elements, 36
  clotting
    hemostasis, 45–46, 45*f*
    vitamin K and, 46
  concentrations of glucose, 152, 152*b*
    animals attempt to maintain, 153
Blood flow
  to skin, 222
Blood supply, 173
Blood–brain barrier, 97*b*, 109, 110*f*
Bloodworms, 46*b*, 47*b*
BOAS. *See* Brachycephalic Obstructive Airway Syndrome (BOAS)
Bohr effect, 61
Bone
  composition, 166
  cortical or compact, 164*f*, 166
  growth of, 170–172, 170*f*, 171*f*, 172*f*
  remodeling or turnover, 173
  spongy, 164
Bone breakage, 174
Bone cells, 166
  osteoblasts, 167, 167*f*
  osteoclasts, 167
  osteocytes, 167
Bovine spongiform encephalopathy (BSE), 124*b*
Bowman's capsule. *See* Glomerular capsule
Boyle's law, 60*b*
BP. *See* Systolic blood pressure (BP)
Brachycephalic Obstructive Airway Syndrome (BOAS), 64*b*
Brachycephaly, 174, 174*f*
Brain, 97*b*, 109
  blood–brain barrier, 109, 110*f*
  cerebellum, 112
  cerebrospinal fluid, 109, 110*f*
  cerebrum and cerebral cortex, 110
  functions of, 110
  hypothalamus, 110–111
  introduction to, 109
  medulla oblongata, 112
  mesencephalon-derived brain structures, 112

olfactory system, 111
photoreceptors, 117
pineal gland, 111–112
pons, 112
thalamus, 112
Brain tumors, 124*b*
Branched polysaccharide, 155
Breathing, central control of, 58–59, 58*f*
Breech birth, 299, 321
Broad ligament, 300
Bronchiole, 53–54, 60*b*
Bronchus, 53, 60*b*
Broodiness, 77
Brown adipose tissue (BAT), 159–160
BSE. *See* Bovine spongiform encephalopathy (BSE)
Bulbourethral glands, 280
Bundle of His, 28*b*
Bursa of Fabricius, 184

C

C-type lectin receptors (CLRs), 188
Calcitonin, 79, 80–81
Calcitonin Gene Related Peptide (CGRP), 81
Calcium
  absorption, 240–241*b*
  homeostasis, 80
Calcium phosphate, 166
Calsequestrin, 140, 143*b*
cAMP. *See* Cyclic adenosine monophosphate (cAMP)
Canalization, 327
Canine hypothyroidism, 92*b*
Canine teeth, 233, 233*f*
Capacitation, 295, 311
Capillaries, 26, 33–34
  blood flow through, 33
  introduction, 33, 33*f*
  types of, 33–34
Capillary space, 260*b*
Carbaminohemoglobin, 62
Carbohydrate
  digestion, 245–246, 245*f*
  digestion in cats and dogs, 251
  metabolism, 151–152, 151–152*b*
Carbon dioxide, 60*b*
  binding to hemoglobin, 62
  transport of, 61–62
Carbonic anhydrase, 40, 61, 237
Cardiac contractile cells, 28*b*, 29
Cardiac muscle, 129, 130*t*
  creatine to, 145
  functional anatomy of, 132–134
  mechanism of, 138–140
  stimulation of contractions of, 136–138, 138*f*
Cardiac output (CO), 28*b*, 31–32
Cardiac oxygen demand, 33
Cardiac region, 235

Cardiac troponins, 46*b*
Carnivores, 230*b*
Carotenoid pigments, 224
Cartilage
  definition, 167–168, 168*f*
  types of, 168–169
Casein, 324
Cathepsin, 167
Cats
  carbohydrate digestion in, 251
  diabetes in, 160
  digestion in, 251
  obesity in, 161
  osteoarthritis in, 175
  protein digestion in, 251
Caveolae, 142*b*
CBG. *See* Corticosteroid binding globulin (CBG)
CCAAT/enhancer-binding proteins (C/EBPs), 159
Cecum, 241
  ruminants in, 250
Cell body, 100
Cell lysis, 195
Cell-mediated immunity, 198
Cell membrane, 12
Cells of nervous system, 100
    glial cells, 100–102
    neurons, 100
Cellular innate immunity. *See* Phagocytosis
Cellular respiration, 50
Central chemoreceptors, 63
Central Dogma Theory, 16–18
Central nervous system (CNS), 97*b*, 108, 109*f*
Centriole, 289
Cephalic phase, 247
Cerebellum, 112
Cerebral cortex, 110
Cerebrospinal fluid (CSF), 97*b*, 109, 110*f*
Cerebrum, 110
Cervix, 300
CGRP. *See* Calcitonin Gene Related Peptide (CGRP)
Charles's law, 60*b*
Chemokine ligands, 191*b*
Chemokines, 189, 190*f*
Chemoreception, 97*b*, 119
  molecular olfaction receptors, 119–120
  olfaction (smell), 119, 119–120*f*
  other chemoreception, 122
  pheromones, 122
  taste, 120–122, 121*f*, 121*t*
Chemoreceptors, 113
Chemotaxis of leukocytes, 195
Chickens
  glands in, 223
  keratocytes, 223
  pigmentation, 224
  skin of, 223

Chickens (*continued*)
    uropygial gland, 223
    water loss across, 223
Chief cells, 236
Chloride reabsorption, 262
Chloride shift, 61, 62*f*
Cholesterol, 216
Chondrocytes, 180
Chords of tendineae, 29, 29*f*
Chorioallantoic, 316
Chorion plexus, 109
Choriovitelline, 316
Chromaffin cells, 81
Chronic inflammation, 204
Chronic kidney disease, 271*b*
Chronic wasting disease, 124*b*
Chylomicrons, 34
Cilia, 53
Ciliated (pseudostratified) columnar epithelium, simple, 9
Circadian rhythm, 111, 111*f*
Circulatory system
    blood, 34–35, 36*f*
    blood cells/formed elements, 36
    blood clotting, 45–46
    capillaries
        blood flow through, 33
        introduction, 33, 33*f*
        types of, 33–34
    cardiac output, 31–32
    components of, 26–27
        arteries, 26
        capillaries, 26
        coronary blood vessels, 27
        double circulation, 26–27
        hepatic portal veins, 27
        lymph vessels, 26
        veins, 26
    in domestic animals, 46, 46*b*, 47*b*
    heart, functioning of
        blood flowing in single direction, 28–29, 29*t*, 29*f*
        causes of, 29–30
        electrocardiogram, 30–31
        intrinsic conduction system, 30
        introduction to, 27–28, 27*f*
        pressure changes in, 30
    lymph and lymph vessels, 34
    oxygen to heart functioning, importance of, 32–33
    plasma, 44–45
    platelets (thrombocytes)
        lifespan of, 44
        in poultry, 44
        production of, 44, 44*f*
        role of, 44
        structure of, 43
    red blood cells (erythrocytes)
        breakdown of, 41, 41*f*
        control of, 40, 40*f*
        functioning of, 37, 40
        introduction, 37, 38–39*t*
        lifespan of, 40
        production of, 40
    role of, 25–26
    white blood cells (leukocytes)
        function of, 42
        introduction to, 41, 42*f*
        lifespan on, 42
        number and proportion of, 39*t*, 42, 43*t*
        production of, 42, 43*f*
Citric acid cycle, 144
Classical complement pathways, 195, 195–196*b*
Claws, 217–221, 219*f*, 220*t*, 220*f*
Clotting, 45
CNS. *See* Central nervous system (CNS)
CO. *See* Cardiac output (CO)
Cochlea, 118
Colitis, 253*b*
Collagen fibers, 166
Collecting ducts, 84
    reabsorption in, 265, 265*f*
Colloidal osmotic pressure, 33
Colon, 241–242
    ruminants in, 250
Colostrum, 324, 331–332, 331*t*, 332*b*
Columnar epithelium, simple, 9
Compact bone, 164*f*, 166
Companion animals
    domestication of, 5, 6–8*t*, 6*f*
    evolutionary history of, 5
Complement, 195–196, 196–197*b*
Concentration gradient, 156
Concentration of carbon dioxide (pCO2), 56, 56*t*
Concentration of oxygen (pO2), 56, 56*t*
Conceptus, 299, 315
Conducting zone, 53
Conduction, 222
Cones, 115
Connective tissues, 10
Continuous capillaries, 33
Contractile cardiac cells, 29
Contractile proteins, 133
Contractility, 32
Convection, 222
Coprophagy, 251
Copulation, 277
Coronary blood vessels, 27
Corpora lutea, 307
Corpus albicans, 307
Corpus cavernosa, 280
Corpus hemorrhagicum, 307, 308
Corpus luteum, 77, 299, 307, 310
Corpus spongiosum, 280

Cortex, 284
Cortical bone, 164f, 166
Cortical cells, 81
Cortical sex cords, 303
Corticosteroid binding globulin (CBG), 45, 83
Corticotropin. *See* Adrenocorticotropic hormone (ACTH)
Corticotropin-releasing hormone (CRH), 76
Cortisol, 83, 111
Cotransporter-mediated reabsorption, 261
Cotyledonary, 317
Creatine
   in domestic animal, 145
   skeletal muscle functioning, 145, 145f
Creatine phosphate. *See* Phosphocreatine
CRH. *See* Corticotropin-releasing hormone (CRH)
Crop, 235
Cryptorchidism, 296
CSF. *See* Cerebrospinal fluid (CSF)
Cuboidal epithelium, 258
   simple, 9
Cumulus oophorus, 304
Curds, 330
Cushing's disease, 91–92b, 224
Cyclic adenosine monophosphate (cAMP), 71, 292
Cysteine, 13
Cystitis, 271b
Cystocentesis, 271b
Cytokines, 188
   chemokines, 189, 190f
   interferons, 189, 189b
   interleukins, 188
Cytoplasm, 71
Cytosolic DNA, 188
Cytotoxic lymphocytes, 186

D

Dalton's law, 60b
Daughter cells, 289
Defense against pathogens, 59
Degenerative joint disease, in horses, 174
Deglutition or swallowing, 232
Dementia, 124b
Dendrite, 97b
Dendritic cells, 184, 186, 186f
Dendritic processes, 167
Dentine, 233, 233f
Deoxyribonucleic acid (DNA), 15b
   structure of, 15–16, 15f, 16f
Dephosphorylation, 143
Depolarization, 103
Depth of breathing, 63
Dermal papillae, 211b
Dermis, 211b
Desmosomes, 131b
Diabetes, 92b, 271b
   in cats and dogs, 160

Diaphragm, 60b
Diastolic blood pressure, 32b
Diencephalon, 99
Differentiated B lymphocytes, 198
Differentiation, 135b
Diffuse, 317
Digestion and gastrointestinal tract
   accessory organs, 242
   carbohydrate digestion, 245–246
   in cats and dogs, 251
   cecum, 241
   colon, 241–242
   control of, 247
   digestion of proteins, 244, 245f
   in domestic animals, 253b
   esophagus, 234–235, 234f
   in horses (pseudoruminants), 250–251, 250f, 251f
   introduction
      animals need to digest food, 230
      definition, 229
      functions of, 231–232
      ingesta moved through, 231, 231f
      nutrition, types of, 230
      overall structure of, 230–231, 230f, 231b
      stages of digestion in livestock, companion animals, and poultry, 231
      types of, 229–230
   lipid (fat) digestion, 246, 246f
   liver
      bile contain, 242
      bile important to digestion, 242
      enzymes in bile, 242
      gallbladder, function of, 243
      introduction to, 242
   in livestock, companion animals, and poultry, 251, 252t
   mouth, 232–233, 233f, 234f
   pancreas
      enzymes in pancreatic fluid, 243–244, 244f
      introduction to, 243
      secrete bicarbonate, 243
      secretion, 243
   in rabbits, 251
   rectum, 242
   in ruminants, 247–250, 247b
   small intestine
      cell types in villi, 238–239
      functioning of, 239
      microvilli, functioning of, 239–240
      overall introduction to, 238, 239f
      transporters in microvilli, 240, 240–241b
   stomach
      enteroendocrine cells, 238
      epithelial cells, lifespan of, 238
      functional anatomy of, 235–236, 237b
      parietal cells, 236–237, 237f, 238f
Dihydropyridine receptor, 138

Dilution of coloration, 216, 218*b*, 218*f*
Disaccharidases, 239
Discoid, 317
Distal convoluted tubule, reabsorption in, 263–265, 264*f*
Distensible organ, 234
Disulfide bridges, 198
Diuresis, 263*b*
DNA. *See* Deoxyribonucleic acid (DNA)
Dogs
    carbohydrate digestion in, 251
    cataracts in, 124*b*
    diabetes in, 160
    digestion in, 251
    glaucoma in, 124*b*
Dogs (*continued*)
    obesity in, 161
    osteoarthritis in, 175
    protein digestion in, 251
Domestic animals
    cardiovascular system in, 46, 46*b*, 47*b*
    diseases in, 207
    diseases of gastrointestinal system in, 253*b*
    muscles of, 127–128, 128*f*
Domestication, 2
Dopamine, 77
Double circulation, 27*f*
    introduction to, 26
    pulmonary capillaries, 27
    systemic capillaries, 27
Double muscling, 146–147, 148*f*
Doublesex/mab-3-related transcription factor-1 (DMRT1) gene, 282
Ducts, 278
Ductus deferens, 280
Duodenum, 238
Dura mater, 108
Dwarfism, 92*b*
Dynorphins, 123
Dystocia, 299, 321
Dysuria, 271*b*

E

Eccrine, 216
Ectoderm, 212, 327
Edema, 46*b*
Efferent parasympathetic nerve terminals, 292
Ejaculation, 292–293, 293*f*, 293*b*
Ejaculatory groove region, 282
Elastic cartilage, 168, 168*f*
Elastic fibers, 55
Elastic recoil, 55
Electrocardiogram, 30–31
Electromagnetic radiation, 115
Ellipsoidal joints, 169, 169*f*
Embryo, 135*b*, 299, 315
Embryonic disc, 314

Emission, 293
EMS. *See* Equine metabolic syndrome (EMS)
ENaC. *See* Epithelial-type sodium channels (ENaC)
Enamel, 233, 233*f*
End of lactation, 327
Endochondral ossification, 170
Endocrine, 167
Endocrine cells, 69
Endocrine pancreas, 243
Endocrine system
    adrenal gland, 81–84
    calcitonin, 80–81
    in domestic animals, 91, 91–92*b*
    gastrointestinal hormones, 87–88
    hormone
        chemistry, synthesis, and mode of action of, 70–72
        definition of, 68
    introduction to, 68–69, 69*b*
    Islets of Langerhans, 84–87
    livestock/companion animals and poultry, 88, 89–91*t*
    negative and positive feedback, 72–73
    other, 88
    parathyroid gland, 80
    pituitary gland, 73
        anterior, 73–77
        posterior, 78
    thyroid gland, 79–80
Endometrial cups, 319
Endometrium, 300, 310
Endoplasmic reticulum, 12, 157
Enteric ganglionated plexuses, 113
Enteric nervous system, 97*b*, 113, 231
Enteroendocrine cells, 236, 238
Enterokinase, 244
Enzymic digestion, 230
Enzymic digestion, 244
Epidermis, 211*b*, 279
Epididymis, inflammation of, 296
Epidural space, 108
Epiglottis, 53
Epimysium, 132
Epinephrine, 31, 81
    neuronal control of, 83
    structure of, 82
Epiphyseal plate, 172
Epistasis, 219*b*
Epithelial barrier, 239
Epithelial cells, lifespan of, 238
Epithelial glands, 118
Epithelial tissue, 9
Epithelial-type sodium channels (ENaC), 121
Epithelium, 300
EPSP. *See* Excitatory postsynaptic potential (EPSP)
Equine asthma, 64*b*, 207*b*
Equine bone fragility syndrome, 175
Equine metabolic syndrome (EMS), 160

Erection, 292, 292f
Erythrocytes, 36. *See also* Red blood cells
Erythropoiesis, 38f, 40
Erythropoietin (EPO), 88, 267
Esophageal sphincter (cardiac sphincter), 235
Esophagus, 234–235, 234f
Estradiol, 299
    production, two-cell theory of, 304–305, 305f
Estrogens, 71
Estrous, 299, 307
    cycles, 306–311, 307–311f, 307b, 308b
Estrus, 299, 307
Euglycemia, 151b
Eumelanin, 211b, 216
Eustachian tube, 118
Excess sweating. *See* Hyperhidrosis
Excitation, 138
Excitatory postsynaptic potential (EPSP), 105
Exocrine glands, 216
Exocrine pancreas, 243
Exocytosis, 71
Expiration, 54, 55b, 60b
Expiratory reserve volume, 57b
Expulsion, 293
External ears, 118
External intercostal muscles, 55
External openings, 324
External respiration, 50
Extracellular matrix, 168b
Extracellular receptors, 71
Eyelids, 117

F
Facultative pathogens, 181b
Fainting goats, 149
Fallopian tubules, 300
Fascicles, 132
Fast glycolytic fibers, 128
Fatty acid, 151b
    synthesis, 160
Feathers, 213
    color of, 216
    overview of, 216, 217f
    pecking, 225
Feedback, 69
Female gametes, 299
Female reproduction, 297–321
    estrous cycles, 306–311, 307–311f, 307b, 308b
    female reproductive organs, 299–303, 301f, 302f
    fertilization, from insemination to, 311
    follicular development/ovulation, 303–305, 304f, 305f
    hormones, 305–306, 306f
    hormones, 318–319, 318f, 319f
    livestock/companion animals and poultry, comparison of, 320b, 321
    oogenesis, 303, 303f
    parturition, 319–321
    pathophysiology of, 321, 321b
    pregnancy, 314–318, 314–317b, 314–316f
    pregnancy, 318–319, 318f, 319f
    reproduction, 305–306, 306f
    seasonal breeding, 311–314, 312b, 313b, 313f
    terminology in, 300b
Female reproductive organs, 299–303, 301f, 302f
Fenestrated capillaries, 33–34
Fermentation, 230b
Ferrous iron, 33
Fertilization, 277, 299
    from insemination to, 311
Fetal, 135b
    stage, 327
Fetus, 315
Fever, 206–207, 207f
Fibrin fibers, 45
Fibrocartilage, 169
Fibroelastic penis, 280, 281
Fight or flight response, 83
Filament bundles, 141b
First polar body, 303
Fleas, 225
Flehmen, 122
Fluoxetine, 111
fMRI. *See* Functional magnetic resonance imaging (fMRI)
Follicle-stimulating hormone (FSH), 75t, 76
Follicular cells, 79
Follicular development/ovulation, 303–305, 304f, 305f
Folliculogenesis, 299
Foregut fermenters, 230b
Founder, 225
Four-chambered pump, 28
Freemartin/martin heifer, 321
FSH. *See* Follicle-stimulating hormone (FSH)
Functional magnetic resonance imaging (fMRI), 111
Functional units, 100
Fundus or fundic region, 235
Fur, 213

G
G-protein-coupled receptors, 122
Galactokinesis, 127, 324, 334–335, 335f
Galactopoiesis, 77, 324
Galactose, 330
Gametogenesis, 277
Gap junctions, 131b, 142b
Gas exchange, 59–60
Gas laws, 51
    overview of, 51
    partial pressure of, 52–53
Gaseous exchange, 57
Gastric lipase, 246
Gastric phase, 247

Gastrin, 235
Gastrointestinal hormones, 87–88
    control of, 88
    gastrin, 88
    introduction to, 87–88t
Gestation, 299
GH. *See* Growth hormone (GH)
Ghrelin, 235
Glandular columnar epithelium, simple, 9
Glial cells, 100–102, 101f
Glomerular capsule, 259–260, 260f
Glomerular filtrate, 260b
Glomerular filtration rate (GFR), 259
Glossophygeal nerve, 112
Glucagon, 84–85, 152
Glucagon receptor (GSGR), 85
Glucocorticoids, 71, 83
Glucogenic amino acids, 155–156
Gluconeogenesis, 83, 151b, 155–156, 156f
Glucose, 151b, 153–154, 154f, 330
    animals attempt to maintain blood concentrations of, 153
    and glycogen, 155, 155f
    transporters, 156–157
Glucose-6-phosphatase, 158
Glucose homeostasis, 127
Glucose transporter, 151b
Glucose transporter 2 (GLUT2), 156
Glucose transporter 4 (GLUT4), 156–157, 156f
Glucose transporter 7 (GLUT7), 157, 157f
Glucose transporters (GLUT), 156–157, 156f
GLUT. *See* Glucose transporters (GLUT)
Glutamine, 13
Glyceryl-3-phosphate, 159, 159f
Glycogen, 151b
Glycogen synthase, 155
Glycogenesis, 151b, 155, 155f
Glycogenolysis, 85, 151b, 155, 155f
Glycolysis, 144, 144f
Glycoprotein hormone, 76
GnRH. *See* Gonadotropin-releasing hormone (GnRH)
Gonadal ridges, 284
Gonadotropin-releasing hormone (GnRH), 76
Granulosa cells, 303
Growth factor 1 (IGF-1), 77
Growth hormone (GH), 75t
Gut-associated lymphoid tissue (GALT), 183
Guttural pouch, 53

H
Hair
    color of, 216
    functions of, 213
    overview of, 216, 217f
    structure and growth, 213–214, 213b, 215f
Hair loss. *See* Alopecia
Hair sheath, 214f
Haploid, 299
Hatching, 315
Hearing, 118, 119f
Heart atresia, 46b
Heart functioning, 27–31
    blood flowing in single direction, 28–29
    causes of, 29–30
    electrocardiogram, 30–31
    intrinsic conduction system, 30
    introduction to, 27–28
    oxygen to, importance of, 32–33
    pressure changes in, 30
Heart murmurs, 46b
Heart rate, 28b, 31
Heat shock proteins, 206
Heat stress, 296
Heat stroke, 64b
Heaves, 64b
Hematocrit, 34–35
Hematuria, 271b
Heme unit, 33
Hemoglobin, 40, 60b
Hemoglobinuria, 271b
Hemostasis, 45–46, 45f
Henry's law, 60b
Hepatic portal veins, 27
Herbivores, 230b
Hertz, 118
Hindgut fermenters, 230b
Hinge joints, 169
Hip dysplasia, in dogs, 175
Homeostasis, 237
Homothermic, 3
Homothermic quadrupeds, 5
Hooves, 217–221, 219f, 220t, 220f
Hormonal control of, 334
Hormonal stimuli, 69
Hormone, 69, 291, 305–306, 306f, 318–319, 318f, 319f
    -binding proteins, 45
    chemistry, synthesis, and mode of action of, 70–72
    definition of, 68
Horns, 217
Horses
    colic in, 253b
    digestion in, 250–251, 250f, 251f
    teeth problems in, 253b
    types of teeth in, 234f
Host innate immune defense, 191
Humoral (antibody) mediated immunity, 198–200
Humoral stimuli, 69
Hyaline cartilage, 169
Hyaluronan, 168b
Hydrochloric acid, in stomach, 187
Hydrostatic pressure, 33
Hydroxyapatite, 166, 167

Hyperglycemia, 151*b*
Hyperhidrosis, 225
Hyperplasia, 136*b*, 160*b*, 180
Hypertension, 46*b*
Hyperthyroidism, 92*b*
Hypertrophy, 136*b*, 160*b*, 180
Hypoadrenocorticism. *See* Addison's disease
Hypodermis, 211*b*, 212
Hypoglycemia, 152*b*
Hypophyseal portal blood, 74
Hypothalamus, 110–111
Hypothyroidism, 92*b*
Hypoxia, 64*b*

I
IBK. *See* Infectious Bovine Keratoconjunctivitis (IBK)
Ileum, 238
Immune cells
  dendritic cells, 186, 186*f*
  lymphocytes, 185
  natural killer cells, 186
Immune function
  adaptive (or specific) immunity, 198
  allergies, 205–206, 206*f*
  b lymphocytes and antibody production, 200–201, 201*f*
  cell-mediated immunity, 198
  cells
    dendritic cells, 186, 186*f*
    lymphocytes, 185
    natural killer cells, 186
  complement, 195–196, 196–197*b*
  cytokines, 188
    chemokines, 189, 190*f*
    interferons, 189, 189*b*
    interleukins, 188
  diseases in domestic animals, 207
  fever, 206–207, 207*f*
  humoral (antibody) mediated immunity, 198–200
  immune organs, 183
    bursa of Fabricius, 184
    lymph nodes, 184–185, 185*f*
    spleen, 185
    thymus, 184, 184*f*
  immune-related molecules, 183–184
  inflammation, 203–205, 203*f*, 204–205*b*
  innate (nonspecific) immunity, 186
    antimicrobial peptides, 187
    barriers to infection, 187
    iron and immunity, 187
    oxidative or respiratory burst, 187, 187*f*, 188*f*
    pattern recognition receptors (PRR), 188
  introduction and key concepts, 181–182
  opsonization, 194–195
  phagocytes, platelets/thrombocytes, and immune response, 191–192
  phagocytosis (cellular innate immunity), 193–194, 193*f*
  vaccines, 201–203
Immune-mediated hemolytic anemia, 207*b*
Immune-mediated thrombocytopenia, 207*b*
Immune organs, 183
  bursa of Fabricius, 184
  lymph nodes, 184–185, 185*f*
  spleen, 185
  thymus, 184, 184*f*
Immune therapy, 190*b*
Immunoglobulin (IgD), 198
Immunoglobulin (IgE), 198
Immunoglobulin (IgG), 198, 199*f*
Immunoglobulin (IgM), 199
Immunoglobulin (Ig)A, 198
Immunoglobulins, 324
Implantation, 315
Incisor teeth, 233, 233*f*
Infectious Bovine Keratoconjunctivitis (IBK), 124*b*
Inflammation, 203–205, 203*f*, 204–205*b*
Influx of calcium, 30
Infundibulum, 334
Inguinal, 323
Inguinal region, 324
Inherited diseases, 124*b*
Inhibin, 279
Inhibitory postsynaptic potential (IPSP), 105
Inhibitory releasing hormones, 76*t*
Innate (nonspecific) immunity, 186
  antimicrobial peptides, 187
  barriers to infection, 187
  iron and immunity, 187
  oxidative or respiratory burst, 187, 187*f*, 188*f*
  pattern recognition receptors (PRR), 188
Innate immune system, 182
Innate sentinels, 186
Innervate muscle fibers, 136
Inorganic bone, 166
Inositol trisphosphate, 142
Insalvation, 232
Inspiration, 54, 55*b*, 60*b*
Inspiratory reserve volume, 57*b*
Insulation, impact of, 222
Insulin, 84, 85–86*t*, 152
  intestinal hormones influence, 87
  release of, 87
  role of, 86–87
  structure and synthesis, 86
Intercalated discs, 129, 130*b*
Interferon (IFN)-γ, 310
Interferon tau, 314
Interferons, 189, 189*b*
Interleukins, 188
Internal intercostal muscles, 55
Internal respiration, 50
International Units (SI) units, 18

Intervertebral discs, 169
Intestinal parasites, 253b
Intracellular structure
   cytoskeleton, 12, 12f
   introduction, 10
   mitochondria, 11–12
   nucleus, 10–11
   other organelles, 12, 12f
   tight junctions, 12, 12–13b
Intrinsic conduction system, 28b
Intrinsic conduction system, 30
Intromission, 299
Involuntary muscles, 131
Involution, 324
Iodide, 79
Iodine deficiency, 92b
IPSP. See Inhibitory postsynaptic potential (IPSP)
Iron and immunity, 187
Islets of Langerhans, 84–87
Isoforms, 144b
Isometric growth, 327
Isovolumetric contraction, 28b
Isovolumetric manner, 30

J
Jejunum, 238
Joints
   cartilage in, 170, 170f
   ligaments, 170
   synovial fluid, 170
   tendons, 170
   types of, 169
Juxtaglomerular cells, 265f, 266, 267b

K
Keratin, 212, 213b
Keratinization, 212, 213f
Keratinized epithelium, stratified, 9
Keratinocytes, 212
Ketosis (cattle), 161
Kidneys, 88, 257, 257f
   functions of, 257–258, 258f
   hormones produced by, 266–267, 266f
   stones, 271b

L
Lacrimal fluid (tears), 117–118
Lactation, 299, 322–336
   colostrum, 331–332, 331t, 332b
   diseases related to, 335, 336t
   galactokinesis, 334–335, 335f
   hormonal control of, 334
   mammary gland
      development of, 327–329, 327f, 328–329b, 328f
      functions of the structures in, 325–326, 325f, 326f
      involution of, 335
   milk
      composition of, 329–330, 330t, 330f
      let-down, 334
      synthesis, 332–333, 332f, 333f, 333b
Lactation, 324
Lactogenesis, 324
Lactose, 330
Lacunae of bones, 167
Lameness, in cattle, 225
Laminitis, 225
Langerhans cells, 211b
Lanolin, 216
Large luteal cells, 310
Larynx (voice box), 53
Late Pregnancy stage, 327
Latency period, 146
Lectin pathway, 195–196b, 196
Left atrioventricular valve, 28
Leptin, 159
Leptospirosis, 321
Leukocytes, 36. See also White blood cells
   in blood vessels, 192
   recruitment, 195
LH. See Luteinizing hormone (LH)
Libido, 278, 299
Ligaments, 170
Limb buds, 164
Lingual lipase, 246
Linnaeus, Carl, 3
Lipases, 239
Lipid (fat) digestion, 246, 246f
Lipid droplets, 159
Lipid envelope, 223
Lipogenesis, 85, 152b
Lipolysis, 77, 152b, 160b
Lipopolysaccharide (LPS), 202
Liver, 88
   bile contain, 242
   bile important to digestion, 242
   enzymes in bile, 242
   gallbladder, function of, 243
   introduction to, 242
   and metabolism, 157–158, 157f
Livestock
   companion animals and poultry, comparison of, 320b, 321
   difference in hormones between, 89–90t
   domestication of, 5, 6–8t, 6f
   evolutionary history of, 5
   gastrointestinal tract and digestion in, 251, 252t
   myopathies in, 149t
   respiratory organs in, 53–54, 53f
Long-day breeder, 307
Lubricin, 168b
Lumen, 79
   of follicles, 79

Luminal space, 260b
Luteinizing hormone (LH), 75t, 76–77, 299
Luteolysis, 310–311
Lymph nodes, 184–185, 185f
Lymph vessels, 26, 34, 34f
Lymphocytes, 185
Lysosomes, 12

M

Macula densa cells, 265f, 266
Major histocompatibility protein (MHC), 186, 186f
Male reproduction, 275–296
    castration, 278
    castration do, 278, 278b
    domestic animals, diseases of male reproduction in, 296, 296b
    domestic mammals/poultry, differences between reproduction in, 295, 295t
    ejaculation, 292–293, 293f, 293b
    erection, 292, 292f
    glossary of terms in, 277b
    gonads and associated ducts, development of, 283–288, 284f, 285–286b, 285f, 287f
    hormonal control of, 288–289, 288f, 289f
    major features of, 277–278
    semen—its composition/roles of the components, 294–295, 294t, 295b
    sex determination, genetics of, 282–283, 283t, 284t
    spermatogenesis, 289–291, 289b, 290f
        temperature set point for, 291, 291f
    spermatozoa, 279
        characteristics of, 279
        structure of, 279, 279f
    tract/functions, structures of, 279–282
        ducts and accessory glands, 279–280, 280f, 281f
        penis, 280–282
        reproductive system of poultry, 282
        testes, 279
    weights of components of, 281b
Mammals, 3
Mammary buds, 327
Mammary epithelial cells, 332
Mammary gland, 233
    development of, 327–329, 327f, 328–329b, 328f
    functions of the structures in, 325–326, 325f, 326f
    involution of, 335
Mammary placodes, 327
Mammary ridges, 327
Mammary rudiment, 327
Mammary stem cells, 329
Mammogenesis, 324
Mandibular condyle, 169
Manx cat, 175
MAP. See Mean arterial pressure (MAP)
Marsupials, 3, 5
Mast cells, 204–205b

Mastication, 232
Mastitis, 336
Maternal behavior, 299
Maternal recognition, 299
    protein, 190b
Maximal contraction, 146, 147f
Maximal oxygen uptake (VO$_2$max), 57b
MCV. See Mean corpuscular volume (MCV)
Mean arterial pressure (MAP), 32b
Mean corpuscular volume (MCV), 35
Mechanical digestion, 229–230
Mechanoreceptors, 63, 97b, 113
Mechanosensory cells, 167
Medial preoptic area (MPOA), 293b
Medial suspensory ligament, 326
Median eminence, 74
Medulla, 284
Medulla oblongata, 112
Medullary bone
    formation of, 173
    important to hens, 173
Megaesophagus, 253b
Meiosis, 303
Melanin, 211b
Melanocortin 1 receptor (MC1R), 211b
Melanocytes, 211b
Melatonin, 88, 111
Merkel cells, 212
Merocrine sweat glands, 216
Mesencephalon (midbrain), 99
    -derived brain structures, 112
Messenger RNA (mRNA), 16
Met-enkephalin, 123
Metabolic diseases
    diabetes (cats and dogs), 160
    equine metabolic syndrome, 160
    ketosis (cattle), 161
    obesity (cats, dogs, and horses), 161
Metabolism
    adipose tissue
        adipocytes, 158–159, 159f
        adipogenesis, 159
        brown adipose tissue (bat), 159–160
    blood concentrations of glucose, 152, 152b
        animals attempt to maintain, 153
    carbohydrate metabolism, 151–152, 151–152b
    gluconeogenesis, 155–156, 156f
    glucose, 153–154, 154f
        and glycogen, 155, 155f
        transporters, 156–157
    liver and, 157–158, 157f
    metabolic diseases
        diabetes (cats and dogs), 160
        equine metabolic syndrome, 160
        ketosis (cattle), 161
        obesity (cats, dogs, and horses), 161

Metabolism (*continued*)
  oxidative stress and reactive oxygen species, 160
  in poultry compared to livestock/companion animals, 160
  skeletal muscle and, 158, 158*f*
Metal-binding proteins, 45
Metencephalon, 99
Methane, 249*b*
Metric units, 18–19
MHC-I receptor, 186
Micelles, 246, 330
MicroRNA (miRNA), 16
Microvilli
  functioning of, 239–240
  transporters in, 240, 240–241*b*
Micturition (urination), 269
Migration off leukocytes, 195
Milk
  composition of, 329–330, 330*t*, 330*f*
  fever, 336
  let-down, 324, 334
  proteins, 330
  synthesis, 332–333, 332*f*, 333*f*, 333*b*
Mineralocorticoids, 83
Minute ventilation, 57*b*
Mitochondria, 127*b*
Molar teeth, 233, 233*f*
Molecular olfaction receptors, 119–120
Monoestrous, 308
Monotremes, 3, 5
Morula, 315
Motor impulses, 112
Motor nervous system, 97*b*
Motor unit, 127*b*
Mouth, 232–233, 233*f*, 234*f*
  ruminants in, 248
MPOA. *See* Medial preoptic area (MPOA)
Mucosal membranes, 187
Mucous neck cells, 236
Müllerian duct, 285, 299
Multiple smooth muscles, 132
Multiple tubular myofibrils, 132
Muscle
  cardiac muscle
    creatine to, 145
    functional anatomy of, 132–134
    mechanism of, 138–140
    stimulation of contractions of, 136–138, 138*f*
  characteristics of, 126–127, 127*b*
  contractions, 146
  domestic animals, important for, 127–128, 128*f*
  double muscling, 146–147, 148*f*
  energy production for
    creatine, muscle, and ATP, 145
    glycolysis, 144
    overall, 143–144
    oxidative respiration, 144–145
  essential for life of animals, 127
  as meat, 146
  myogenesis, 134–136
  myopathy, 149, 149*t*
  sarcoplasmic reticulum, 134
  skeletal
    functional anatomy of, 132–134
    mechanism of, 138–140
    stimulation of contractions of, 136–138, 138*f*
  smooth muscle
    functional anatomy of, 140–141
    mechanism of, 142–143
  types of
    cardiac muscle, 129, 130*t*
    classification of skeletal muscles, 128–129, 129*t*
    skeletal muscle, 128, 129*f*
    smooth muscle, 129*t*, 131–132
Muscle contraction, 138
  calcium, and sarcoplasmic reticulum, 140, 140*f*
  steps in, 139–140
Muscle relaxation, 138, 139*t*
Muscle tissue, 10
Muscular fermentation vat, 248
Musculovascular penis, 280
Myelencephalon, 99
Myelin sheaths, 100
Myoepithelial cells, 323
Myogenesis, 134–136, 336
Myoglobin, role of, 33
Myometrium, 300
Myopathy, 149
Myosin, 127*b*, 143
Myostatin gene, 147

N

Nagana. *See* Trypanosomiasis
Narrow synaptic cleft, 106
Natural killer cells, 186
NE. *See* Norepinephrine (NE)
Negative feedback, 69, 72–73
Neonate, 315, 324
Nephritis, 271*b*
Nephron loop, reabsorption in, 263
Nephrons, 258–259, 258*f*
  and adjacent structures, 259
Nerve, 97*b*
Nerve conduction/action potential
  overview of, 102–103, 102*f*
  recovery after, 103
  refractory periods, 103–105
Nervous and sensory systems
  autonomic nervous system, 113
  brain, 109
    blood–brain barrier, 109
    cerebellum, 112

cerebrospinal fluid, 109
cerebrum and cerebral cortex, 110
functions of, 110
hypothalamus, 110–111
introduction to, 109
medulla oblongata, 112
mesencephalon-derived brain structures, 112
pineal gland, 111–112
pons, 112
thalamus, 112
cells of, 100
glial cells, 100–102
neurons, 100
central nervous system, 108, 109f
chemoreception, 119
molecular olfaction receptors, 119–120
olfaction (smell), 119, 119–120f
other chemoreception, 122
pheromones, 122
taste, 120–122, 121f, 121t
development of, 98–99
eyelids, 117
hearing, 118, 119f
introduction to, 98, 98f
nerve conduction/action potential
overview of, 102–103
recovery after, 103
refractory periods, 103–105
nociceptors or pain receptors, 122–123, 123f
overall introduction, 97, 97–98b
senses in livestock/mammalians companion animals, comparison of, 123, 123–124t
senses, introduction to, 113–114
spinal cord, 112–113
synapse
functioning of, 106–108
introduction to, 105
neurotransmitters and their functioning, 105–106
tears, 117–118
vision/sight, 114
photoreception, 115–117
protection of, 114–115
retina, 114
structural integrity and physiological functioning, 114
Nervous tissue, 10
Neural crest, 98
Neural progenitor cells, 98
Neural stimuli, 69
Neural tube, 97b
Neuroendocrine cells, 111
terminals, 335
Neuroendocrine reflex, 78, 334
Neuromodulators, 106
Neuronal loop, 112
Neurons, 97b, 100, 100f
Neurophysins, 78

Neurosecretory cell, 74
Neurosecretory nerve, 78
Neurotransmitter, 97b
Nicotinic cholinergic receptors, 136
Nitrogenous waste, 26, 258, 258f, 259f
Nociception, 97b
Nociceptors, 114, 122–123, 123f
Nodes of Ranvier, 100
Non-esterified fatty acids, 160b
Non-keratinized squamous epithelium, stratified, 9
Non-peptidergic, 105
Nonapeptide, 78
Norepinephrine (NE), 81, 106–107b, 114
neuronal control of, 83
structure of, 82
Nostrils, 53
Nuclear estrogen receptors, 305
Nuclear receptor, 79
Nutrients, 26

O
Obesity (cats, dogs, and horses), 161
Obligate pathogens, 181b
Oil gland. *See* Uropygial gland
Olfaction (smell), 119, 119–120f
Olfaction, 59
Omasum, ruminants in, 249
Omnivores, 230b
Oocyte, 303
Oogenesis, 299, 303, 303f
Oogonia, 303
1-cis-retinal isomer, 116
1,25-Dihydroxy vitamin D_3, 71, 267
Oppositional mechanisms, 69
Opsonins, 194
Opsonization, 194–195
Organs
basic structure of, 9–10
growth of, 278
Organum vasculosum, 206
Os penis, 281
Osselets, 175
Ossification, 170, 171f
Osteoarthritis, in dogs and cats, 175
Osteoblasts, 167, 167f
Osteoclasts, 80, 167
Osteocytes, 167, 168b
Osteoid, 166
Ova, 299
Ovaries, 299, 300, 301
Ovum, 299
Oxidative burst, 187, 187f, 188f
Oxidative phosphorylation, 160
Oxidative respiration, 144–145, 144f
Oxidative stress, 160
Oxygen, transport of, 60–61, 60f

Oxyntic cells, 236
Oxytocin, 78, 299, 324

P

Packed Cell Volume (PCV), 34–35, 35f, 36t
Pain receptors, 122–123, 123f
Pale, soft, and exudative (PSE), 31, 149
Pancreas, 244f
    enzymes in pancreatic fluid, 243–244, 244f
    introduction to, 243
    secrete bicarbonate, 243
    secretion, 243
Panting, 59
    effect of, 63
Papillae, 282
Parabronchi, 57
Paracrine, 188
Paracrine factor, 68
Parafollicular cells, 79
Paralysis, 124b
Parasympathetic nervous system, 97b, 113
Parathyroid gland, 80
Parathyroid hormone (PTH), 80
Paraventricular nucleus (PVN), 293b
Parietal cells, 236–237, 237f, 238f
Pars esophagea, 235
Pars intermedia, 78
Pars nervosa, 78
    chemistry of, 78
Partial pressure
    of gas, 50b, 52–53
    of oxygen, 52b
Parturition, 299, 319–321
Passive immunity, 331
Pattern recognition receptors (PRR), 188
PCV. See Packed Cell Volume (PCV)
Penis/scrotum, physical damage to, 296
Peptide, 69, 70
    bonds, 13
    reabsorption of, 262
    synthesis of, 70, 70f
Peptidergic, 105
Perimetrium, 300
Perimysium, 132
Peripheral chemoreceptors, 63
Peripheral nervous system, 98b
Peristalsis, 231
Permissive mechanisms, 69
Peroxisome proliferator-activated receptor γ (PPARγ), 159
Phaeomelanin, 211b, 216
Phagocytosis, 41, 193–194, 193f
    platelets/thrombocytes, and immune response, 191–192
Phagolysosome, 41
Pharynx (throat), 53
Phenylalanine, 13
Pheromone, 98b, 122

Phosphocreatine, 144
Phosphodiesterase, 116
Phosphorylation, 71
Photoreception, 98b, 115–117, 115–117f
Photoreceptors, 113
Pia mater, 108
Pineal gland, 88, 111–112
Pinealocytes, 112
Pituitary gland, 73
    anterior, 73–77
    posterior, 78
Pituitary pars intermedia dysfunction (PPID), 78
Pivot joints, 169
Placenta, 299
Placental mammals, 3, 5
Plane joints, 169
Plasma, 34, 44–45
    glucose concentrations, 160
    osmolarity, 78
Platelets
    and immune response, 191–192
    lifespan of, 44
    in poultry, 44
    production of, 44, 44f
    role of, 44
    structure of, 43
Pleural membranes, 55
Pneumatic bones, 174
Pneumonia, 64b
Pneumothorax, 64b
Polycystic kidneys, 271b
Polyestrous, 308
POMC. See Proopiomelanocortin (POMC)
Pons, 112
Pontine respiratory group (PRG), 58
Poor fertility, 296
Porosomes, 71
Portal blood vessel, 74
Positive feedback, 69, 72–73
Possible releasing hormones, 76t
Post-natal myogenesis, 136
Post sperm fraction, 294
Post-synaptic membrane, 106
Post synaptic neurotransmitter, 113
Posterior pituitary gland
    arginine vasopressin (AVP)/antidiuretic hormone (ADH), 78
    introduction to, 78
    oxytocin, 78
    pars intermedia, 78
    pars nervosa, 78
    chemistry of, 78
Posttranslational modification, 14
Poultry
    difference in hormones between, 89–90t
    domestication of, 5, 6–8t, 6f

evolutionary history of, 5
gastrointestinal tract and digestion in, 251, 252t
PPID. *See* Pituitary pars intermedia dysfunction (PPID)
Pre-ovulatory follicle, 304
Pre-sperm fraction, 294
Pre-synaptic neurotransmitter, 113
Pre-synaptic terminal, 105
Precursor zymogens, 195
Preen. *See* Uropygial gland
Pregnancy, 299, 314–318, 314–317b, 314–316f, 318–319, 318f, 319f
    maternal recognition of, 314
Prehention, 232
Premolar teeth, 233, 233f
Preovulatory follicles, 303
PRG. *See* Pontine respiratory group (PRG)
Primary follicles, 214b
Primary, primordial follicles to, 303
Primary spermatocytes, 289
Primordial germ cells, 283, 303
PRL. *See* Prolactin (PRL)
Proceptive behaviors, 306
Progesterone receptors, 305
Progestogens, 71
Proinflammatory cytokines, 188
Prolactin (PRL), 75t, 77
Prolapsed uterus, 321
Proopiomelanocortin (POMC), 75, 211b
Proprioceptors, 114
Prosencephalon (forebrain), 99
Prostaglandins, 295
Prostate gland, 280
Proteases, 239
Protein kinase A, 85
Protein kinase G, 292
Proteinases, 79
Proteins, 70
    digestion in cats and dogs, 251
    digestion of, 245f
    hormone, 69
    in plasma, 44
    structure, 13–15, 14f
    synthesis of, 70, 70f
    as transporters, 45
Proteinuria, 271b
Proteoglycans, 169
Proteolysis, 83
Proteolytic cleavage, 195
Proteolytic enzymes, 289
Protozoa, 181b
Proximal (convoluted) tubule, reabsorption in, 260–263, 262f
Proximal convoluted tubules, 84
PRR. *See* Pattern recognition receptors (PRR)
PSE. *See* Pale, soft, and exudative (PSE)
Pseudogene, 122
Pseudohermaphrodites, 296

PTH. *See* Parathyroid hormone (PTH)
Pubertal stage, 327
Pulmonary capillaries, 27
Pulmonary edema, 64b
Pulmonary semilunar valve, 28
Pulmonary ventilation, 50, 54–56
    in chickens, 57, 57f
    control of, 58–59b
Pulp, 233, 233f
Pulse pressure, 32b
Purkinje fibers, 28b
PVN. *See* Paraventricular nucleus (PVN)
Pyelonephritis, 271b
Pyloric region, 235
Pyloric sphincter, 235
Pyrogenic cytokines, 206
Pyruvate, 153

Q
Quarter, 323
Quiescent chondrocytes, 180

R
Rabbits, digestion in, 251
Radiation, 222
Rate of respiration, 63
Reabsorption and secretion
    in collecting ducts, 265, 265f
    in descending and ascending nephron loop, 263
    in distal convoluted tubule, 263–265, 264f
    in proximal (convoluted) tubule, 260–263, 262f
Reactive oxygen intermediates (ROI), 187, 187f
Reactive oxygen species (ROS), 160, 186
Receptive behavior, 306
Rectum, 242
Red blood cells
    breakdown of, 41, 41f
    control of, 40, 40f
    functioning of, 37, 40
    introduction, 37, 38–39t
    lifespan of, 40
    production of, 40
Reflex/induced ovulators, 308
Refractory period, 103–105, 104f, 105f
Relative refractory period, 103
Renal corpuscle, 258
Renin, 84, 88
Replication, 16
Repolarization, 103
Reproduction, 305–306, 306f
Reproductive tract, 301
Residue volume, 58b
Respiration
    central control of breathing, 58–59
    cycle, 58b
    diseases related to, 63, 64b

Respiration (*continued*)
  gas exchange, 59–60
  gas laws, 51
    overview of, 51
    partial pressure of, 52–53
  integrative control of, 63
  mechanism of, 54–56
  organs in livestock and companion animals, 53–54, 54*f*
  pulmonary ventilation in chickens, 57–58
  purpose of, 50–51
  rate, 58*b*
  respiratory system, roles of, 59
  transport of
    carbon dioxide, 61–62
    oxygen, 60–61
Respiratory alkalosis/acidosis, 64*b*
Respiratory burst, 187, 187*f*, 188*f*
Respiratory zone, 53
Resting length, 146
Resting membrane potential, 103, 103*f*
Reticulum, ruminants in, 249
Retina, 114
Reverse transcriptase, 18
Rhodopsin, 117
Rhombencephalon (hindbrain), 99
Ribonucleic acid (RNA), 15*b*
  structure of, 15–16, 15*f*, 16*f*
Ribosomes, 12
Right atrioventricular valve, 28
Ringworm, 226
RNA. *See* Ribonucleic acid (RNA)
RNA polymerase, 16, 16*f*
Rods, 115
ROS. *See* Reactive oxygen species (ROS)
Rotational deformities, 175
Rough endoplasmic reticulum, 17
Rough endoplasmic reticulum, 70
Rudimentary copulatory organ, 282
Rumen
  acidosis, 253*b*
  papillae, 249, 249*f*
  ruminants in, 248–249, 248*f*
Ruminal fermentation, 247*b*
Ruminants
  gastrointestinal tract and digestion in, 247–250, 247*b*
    abomasum, 249
    colon and cecum, 250
    mouth, 248
    omasum, 249
    reticulum, 249
    rumen, 248–249, 248*f*
    rumination, 249
    small intestine, 250
    stomach, 248

Rumination, ruminants in, 249
Ryanodine receptor (RyR), 143*b*
RyR. *See* Ryanodine receptor (RyR)

S

Saddle joints, 169
Sarcolemma, 127*b*
Sarcomeres, 127*b*, 132
Sarcoplasmic reticulum, 127*b*, 134, 142*b*
Sarcoptes scabiei var canis, 225–226, 226*f*
SCC. *See* Somatic cell count (SCC)
SCFA. *See* Short chain fatty acids (SCFA)
Scrotum, 279
Seasonal breeding, 307, 311–314, 312*b*, 313*b*, 313*f*
Sebaceous glands, 211*b*
  and sebum, 216
Sebum, 211*b*, 216
Second meiotic division, 304
Second polar body, 303
Secondary follicles, 214*b*
Secondary spermatocytes, 289
Secretion, of hormones, 69
Secretory granules, 12, 106
Secretory vesicles, 70
Semen, 294–295, 294*t*, 295*b*
Semicircular canals, 118
Semifluid viscous lubricant, 170
Semilunar valves, 29, 29*f*
Seminal plasma, 294
Seminal vesicles, 280
Seminiferous tubule, 279, 291
Senses, introduction to, 113–114
Sensory cells, 167
Sensory nervous impulses, 112
Sensory nervous system, 98*b*
Serum, 34
Sex steroid binding globulin, 45
Sexually transmitted, 296
SGLT. *See* Sodium-dependent glucose transporter (SGLT)
Shear stress, 167
Short chain fatty acids (SCFA), 248–249
Short-day breeder, 307
Sigmoid flexure, 292
Signal transduction system, 71
Signaling molecules, 99
Single-unit smooth muscles, 132
Sinoatrial node (SA node or pacemaker), 28*b*
Sinusoid capillaries, 34
Skeletal action potential, 137, 137–138*f*
Skeletal muscle, 128, 129*f*
  classification of, 128–129, 129*t*
  functional anatomy of, 132–134
  mechanism of, 138–140
  and metabolism, 158, 158*f*
  stimulation of contractions of, 136–138, 138*f*
Skeleton

blood supply, 173
bone
  composition, 166
  cortical or compact, 164f, 166
  growth of, 170–172, 170f, 171f, 172f
  remodeling or turnover, 173
  spongy, 164
cartilage
  definition, 167–168, 168f
  types of, 168–169
cells, 166
  osteoblasts, 167, 167f
  osteoclasts, 167
  osteocytes, 167
dead tissue, 166
diseases, abnormalities, and differences of
  bacterial osteomyelitis, 174
  bone breakage, 174
  brachycephaly, 174
  degenerative joint disease in horses, 174
  equine bone fragility syndrome, 175
  hip dysplasia in dogs, 175
  Manx cat, 175
  osselets, 175
  osteoarthritis in dogs and cats, 175
  rotational and angular leg deformities, 175
functions of bones of, 165
introduction to, 163–164, 164f, 165f
joints
  cartilage in, 170, 170f
  ligaments, 170
  synovial fluid, 170
  tendons, 170
  types of, 169
marrow, 173
of poultry and mammalian domestic animals, 173–174, 173f, 174f
Skin
  appendages as secondary sexual characteristics, 221, 221f
  and bacteria, 221
  of chickens and other poultry, 223–224
  diseases and, 224–226
  feathers, 213
    color of, 216
    overview of, 216, 217f
  functional anatomy of, 211f, 212, 212f
  functions of, 210
  fur, 213
  hair
    color of, 216
    functions of, 213
    overview of, 216, 217f
    structure and growth, 213–214, 213b, 215f
  hooves and claws, 217–221, 219f, 220t, 220f
  horns, 217
  introduction, 209–210, 210f
  keratin, 212
  keratinization, 212
  keratinocytes, 212
  in poultry with domestic mammals, 224
  sebaceous glands and sebum, 216
  sweat glands and sweating, 214t, 216
  and temperature control, 221–223, 222t
  vitamin D and, 221
  wool, 213
Sleeping sickness. *See* Trypanosomiasis
Slick gene, 216b
Sliding filament theory, 132
Slow oxidative fibers, 128
Small intestine, 231f
  cell types in villi, 238–239
  functioning of, 239
  microvilli, functioning of, 239–240
  overall introduction to, 238, 239f
  ruminants in, 250
  transporters in microvilli, 240, 240–241b
Small luteal cells, 310
Smooth muscle, 129t, 131–132
  functional anatomy of, 140–141
  functional anatomy of, 140–141
  mechanism of, 142–143
  mechanism of, 142–143
SNARE protein molecules, 71
Sodium-dependent glucose cotransporter 1 (SGLT1), 240, 246
Sodium dependent glucose transporters (SGLT), 260, 332
Sodium heparin, 36b
Sodium homeostasis, 71
Sodium pumpm, 103
Sodium reabsorption, 261, 261f
Sodium–hydrogen ion antiporter, 261
Solubility, 60
Somatic cell count (SCC), 324, 325
Somatic nervous system, 98b
Somatotropin (ST), 75t, 77
Sonic hedgehog gene, 287
Sperm rich fraction, 294
Spermatids, 289
Spermatogenesis, 289–291, 289b, 290f
  temperature set point for, 291, 291f
Spermatozoa, 279
Spleen, 185
Spongy bone, 166f
Spontaneous ovulators, 308
Squamous (pavement) epithelium, simple, 9
Squamous epithelial cells, 303
Squamous epithelium, 258
ST. *See* Somatotropin (ST)
Step-wise manner, 146
Stereocilia, 118
Steroid hormone, 69, 72
Steroidogenic luteal cells, 310

Stillbirth, 321
Stimulatory releasing hormones, 76t
Stomach
    enteroendocrine cells, 238
    epithelial cells, lifespan of, 238
    functional anatomy of, 235–236, 237b
    parietal cells, 236–237, 237f, 238f
    ruminants in, 248
Stratum corneum, 212
Stratum granulosum, 212
Stretch receptors, 63
Stroke volume, 28b
Strongylus vulgaris. *See* Bloodworms
Subarachnoid space, 108
Submucosal ganglionated plexuses, 113
Suprachiasmatic nucleus, 111
Sweat glands and sweating, 214t, 216
Sycamore poisoning, 149
Sympathetic nervous stimuli, 84
Sympathetic nervous system, 98b, 113
Synapse, 98b
    functioning of
        introduction, 105
        neurotransmitters and, 105, 105f, 106t
Synergistic mechanisms, 69
Synovial fluid, 170
Systemic capillaries, 27
Systolic blood pressure (BP), 32b

T
T-tubules, 127b
Tachycardia, 46b
Tactile nerves, 123
Target tissue, 69
Taste buds, 120–122, 121f, 121t
Taste receptors, 251
Tears, 117–118
Teat, 323
Telencephalon, 99
Telogen, 214, 215f
Tendons, 170
Terminal cisternae, 127b
Terminal cisternae, 138
Testicular degeneration, 296
Testicular endocrine functioning, 291
Tetanus (physiological), 127b
Tetrachromatic vision, 115
Texas Longhorn cattle, 219f
TGB. *See* Thyroxine-binding globulin (TGB)
Thalamus, 112
Theca cells, 303
Theca interna cells, 304
Thoracic, 323
Thrombin, 46
Thrombocytes, 36
Thymic epithelial cells, 184

Thymic involution, 184
Thymosins, 184
Thymus, 184, 184f
Thyroglobulin, 79
Thyroid gland, 79–80
Thyroid-stimulating hormone (TSH), 77
Thyrotropin-releasing hormone, 77
Thyroxine-binding globulin (TGB), 79
Thyroxine, synthesis of, 79
Ticks, 226, 226f
Tidal volume, 58b
Tissues, basic structure of, 9–10
Toll-like receptors (TLRs), 186, 189b
Total lung capacity, 58b
Trachea (wind pipe), 53
Transcription, 16
Transdermal water loss, 222
Transfer RNA (tRNA), 16
Transitional epithelium, stratified, 9
Translation, 17
Transthyretin, 79
Triacylglyceride, 34, 333
Triacylglycerol, 152b
Trichromatic vision, 115
Triglyceride, 324
    synthesis, 160b
Trophectoderm cells, 315
Trophoblast, 315
Tropomyosin, 127b
Troponin, 127b, 139b
Trypanosomiasis, 226
Tryptophan, 13
TSH. *See* Thyroid-stimulating hormone (TSH)
Tumor necrosis factor (TNF)-α, 310
Tunica vaginalis, 279
Twitch, 127b
Tympanic membrane, 118
Tyrosine, 13

U
Udder, 323
Ulcers, 253b
Ultimobranchial glands, 80–81
Umami receptor, 122
Uncoupling protein 1 (UCP1), 160
Undifferentiated spermatogonia, 289
Unfused tetanus, 146
United States Department of Transportation, 52b
Upper respiratory obstruction, 64b
Ureter, 266
Urethral catheterization, 271b
Urethritis, 271b
Uridine diphosphate glucose, 155
Urinary system
    abnormalities and diseases of, 270, 271b
    bladder, 266–269, 268f, 269f

differences between mammals and birds, 269–270, 270t
glomerular capsule (Bowman's capsule), 259–260, 260f
introduction, 256, 256b
kidneys, 257, 257f
    functions of, 257–258, 258f
    hormones produced by, 266–267, 266f
micturition (urination), 269
nephrons, 258–259, 258f
    and adjacent structures, 259
nitrogenous waste, 258, 258f, 259f
reabsorption and secretion
    in collecting ducts, 265, 265f
    in descending and ascending nephron loop, 263
    in distal convoluted tubule, 263–265, 264f
    in proximal (convoluted) tubule, 260–263, 262f
ureter, 266
Urogenital ridges, 283
Uropygial gland, 211b, 223
Urospermia, 296
Uterotubal, 311
Uterus, 300

V

Vaccines, 201–203
Vagina, 300
Variable regions, 198
Vas deferens, 280
Vascular dermal papilla, 214, 215f
Vascular endothelium, 109
Vascular permeability, 195
Vasoactive intestinal peptide, 77
Vasopressin receptors (V2), 265
Veins, 26
Ventricle, 28b
Ventricular contraction, 30
Very low-density lipoprotein (VLDL), 45
Vesicular glands selenium, 280
VFAs. See Volatile fatty acids (VFAs)
Vibrissae (whiskers), 123
Villi, cell types in, 238–239
Vision/sight, 114, 114f
    photoreception, 115–117
    protection of, 114–115
    retina, 114
    structural integrity and physiological functioning, 114
Vital capacity, 58b
Vitamin-binding proteins, 45
Vitamin D, and skin, 221
Vitamin K and blood clotting, 46
Vitellogenin, 45
Vitreous humor, 114
VLDL. See Very low-density lipoprotein (VLDL)
VNO. See Vomeronasal organ (VNO)
Vocalization, 59
Volatile fatty acids (VFAs), 145b, 152b, 247b
Voltage-gated calcium channels, 106, 131b
Vomeronasal organ (VNO), 122
Vulva, 300

W

Warfarin, 46
WAT. See White adipose tissue (WAT)
Water reabsorption, 263
Wave of depolarization, 29
Waxy esters, 216
White adipose tissue (WAT), 160b
White blood cells, 41–42, 43f
    function of, 42
    introduction to, 41, 42f
    lifespan of, 42
    number and proportion of, 39t, 42, 43t
    production of, 42, 43f
Wolff's law, 173
Wolffian ducts, 285
Wool, 213

Y

Yersinia pestis, 181b
Yolk proteins, 45

Z

Zona pellucida, 303
Zonary, 317
Zymogen, 235b
Zymogenic cells, 236